Neural Networks as Positive Linear Operators

SERIES ON CONCRETE AND APPLICABLE MATHEMATICS
ISSN: 1793-1142

Series Editor: Professor George A. Anastassiou
Department of Mathematical Sciences
University of Memphis
Memphis, TN 38152, USA

---

*Published**

Vol. 28 *Neural Networks as Positive Linear Operators*
by George A. Anastassiou

Vol. 26 *Rainbow and Properly Colored Subgraphs*
by Jie Hu & Xueliang Li

Vol. 25 *Advanced Topics on Semilinear Evolution Equations*
by Mouffak Benchohra, Gaston M. N'Guérékata & Abdelkrim Salim

Vol. 24 *Trigonometric and Hyperbolic Generated Approximation Theory*
by George A. Anastassiou

Vol. 23 *Introduction to Matrix Theory: with Applications in Economics and Engineering*
Second Edition
by Ferenc Szidarovszky, Sandor Molnar & Mark Molnar

Vol. 22 *Bloch-Type Periodic Functions: Theory and Applications to Evolution Equations*
by Yong-Kui Chang, Gaston M. N'Guérékata & Rodrigo Ponce

Vol. 21 *Functional Equations and Inequalities: Solutions and Stability Results*
by John Michael Rassias, E. Thandapani, K. Ravi & B. V. Senthil Kumar

Vol. 20 *Discrete Approximation Theory*
by George A. Anastassiou & Merve Kester

Vol. 19 *Stochastic Models with Applications to Genetics, Cancers, AIDS and Other Biomedical Systems*

*To view the complete list of the published volumes in the series, please visit:
https://www.worldscientific.com/series/scam

Series on Concrete and Applicable Mathematics – Vol. 28

Neural Networks as Positive Linear Operators

George A. Anastassiou
The University of Memphis, USA

World Scientific

W JERSEY • LONDON • SINGAPORE • BEIJING • SHANGHAI • HONG KONG • TAIPEI • CHENNAI • TOKYO

Published by

World Scientific Publishing Co. Pte. Ltd.
5 Toh Tuck Link, Singapore 596224
USA office: 27 Warren Street, Suite 401-402, Hackensack, NJ 07601
UK office: 57 Shelton Street, Covent Garden, London WC2H 9HE

British Library Cataloguing-in-Publication Data
A catalogue record for this book is available from the British Library.

Series on Concrete and Applicable Mathematics — Vol. 28
NEURAL NETWORKS AS POSITIVE LINEAR OPERATORS

ISBN 978-981-98-2618-6 (hardcover)
ISBN 978-981-98-2619-3 (ebook for institutions)
ISBN 978-981-98-2620-9 (ebook for individuals)

For any available supplementary material, please visit
https://www.worldscientific.com/worldscibooks/10.1142/14672#t=suppl

Dedicated to all those I love!

Preface

In this monograph we deal with the approximation properties of Neural Network Operators and Activated Convolution Operators to the unit operator, a quantitative approach via inequalities and the modulus of continuity. The method is unified and very general for all involved operators, as they are not treated individually as elsewhere. Instead, we apply the Positive Linear Operators (PLOs) general Approximation Theory, as our operators are indeed PLOs. This happens for the first time in the literature and the results are elegant and subtle.

The used activation functions are the symmetrized and perturbed hyperbolic tangent and generalized logistic.

The symmetrization method uses half data, thus achieving high speeds of convergence and the perturbation method covers the disturbed environments of activity with noise.

Noise in neural networks refers to random fluctuations or unwanted data that can interfere with the model's ability to learn patterns and relationships. Thus noise can negatively impact performance.

Our results extend to all possible directions like univariate and multivariate cases, fuzziness, finite and infinite domains, fractionality, Brownian motion, etc.

Consequently, this monograph is the natural and expected evolution of the author's recent original research work put in a book form for the first time. The presented approaches are original, and chapters are self-contained and can be read independently. This monograph is suitable to be used in related graduate classes and research projects.

The motivation to write this monograph came again by the various issues related to the AI, modeling and analysis of ordinary and fractional order systems that have gained an increased popularity and use in the current literature.

The entire list of discussed topics in this monograph follows:

- Neural Networks as Positive Linear Operators
- Neural Networks in Infinite Domain as Positive Linear Operators
- Generalized Logistic Neural Networks as Positive Linear Operators
- Generalized Logistic Neural Networks as Positive Linear Operators over Infinite Domain
- Fuzzy Symmetrized and Perturbed Neural Network Approximation
- Neural Networks as Positive Linear Operators, Multivariate and Abstract Case
- Multivariate Neural Networks over Infinite Domain as Positive Linear Operators and a Touch of Measure Theory
- Multivariate Generalized Logistic Neural Networks on Infinite Domain as Positive Linear Operators and a Taste of Measure Theory
- Multivariate Fuzzy-Random Symmetrized and Perturbed Neural Network Enhanced Approximation
- Generalized Logistic Neural Networks as Positive Linear Operators, Multivariate and Finite Domain Case
- Approximation by Symmetrized and Perturbed Hyperbolic Tangent Activated Convolutions as Positive Linear Operators
- Approximation by Symmetrized and Perturbed Generalized Logistic Activated Convolutions Interpreted as Positive Linear Operators
- Complete Approximation by Symmetrized and Perturbed Hyperbolic Tangent Activated Multidimensional Convolutions as Positive Linear Operators
- Complete Approximation by Symmetrized and Perturbed Generalized Logistic Activated Multidimensional Convolutions in Positive Linear Framework
- Fractional Voronovskaya Asymptotic Expansions by Quasi-Interpolation Neural Network Operators Applied to Brownian Motion
- Neural Network Approximation over Finite Dimensional Banach Spaces
- Generalized logistic Neural Network Approximation over Finite Dimension Banach Spaces

The book's results are expected to find applications in the many areas of Applied Mathematics, Computer Science, and Engineering, especially in Artificial Intelligence, Machine Learning and Deep Learning. Other possible applications can be in applied sciences like Statistics, Economics, etc. All in all what is presented here is a valuable tool for a large range of applications.

Therefore this monograph is suitable for researchers, graduate students, practitioners and seminars of the above disciplines, also to be in all Science and Engineering libraries.

The preparation of this book took place during the Spring of 2025 at the University of Memphis, USA, and the Summer of 2025 in Athens, Greece.

The author likes to thank Professor Alina Lupas of University of Oradea, Romania, for checking and reading the manuscript. Also he thanks his family for tolerating his math. habit. Last but not least, the author likes to thank Professor Seda Karateke of Istanbul Atlas University, Turkey, and Professor Dimitra Kouloumpou of Hellenic Naval Academy, Greece, for fruitful discussions during the course of writing of this monograph.

October 1, 2025

George A. Anastassiou

Department of Mathematical Sciences
University of Memphis
Memphis, TN 38152, USA

About the Author

George A. Anastassiou received his BSc degree in Mathematics from Athens University, Greece, in 1975. He received his diploma in Operations Research from Southampton University, UK, in 1976. He also received his MA in Mathematics from the University of Rochester, USA, in 1981. He was awarded his PhD in Mathematics from the University of Rochester, USA, in 1984. From 1984 to 1986, he served as a Visiting Assistant Professor at the University of Rhode Island, USA. He had been a Faculty Member at the University of Memphis, USA, since 1986. Also, he has been a Full Professor of Mathematics since 1994. Since August 2024 he has been an Emeritus Professor. His research area is Computational Analysis, in a very broad sense. He has published over 700 research articles in international mathematics journals and over 52 monographs, proceedings and textbooks in well-known publishing houses. Several awards have been conferred to George Anastassiou. In 2007, he received an Honorary Doctoral Degree from the University of Oradea, Romania. He is Associate Editor in over 80 international journals in mathematics and he has been Editor-in-Chief in 3 journals: most notably in the well-known *Journal of Computational Analysis and Applications.*

Contents

Chapter 1

Neural Networks as Positive Linear Operators

Basic neural network operators are interpreted as positive linear operators and the related general theory applies to them. These operators are induced by a symmetrized density function deriving from the parametrized and deformed hyperbolic tangent activation function. We are acting on the space of continuous functions on a compact interval of real line to the reals. We study quantitatively the rate of convergence of these neural network operators to the unit operator. Our inequalities involve the modulus of continuity of the function under approximation or its derivative. We produce uniform and L_p, $p \geq 1$, approximation results via these inequalities. Convexity of functions is also taken into consideration. It follows [31].

1.1 Introduction

The author studied extensively the quantitative approximation of positive linear operators to the unit since 1985, see for example [1]–[3], [5], which we use in this work. He originated from the quantitative weak convergence of finite positive measures to the unit Dirac measure, having as a method the geometric moment theory, see [2], and he produced best upper bounds, leading to attained (i.e. sharp Jackson type inequalities), e.g. see [1], [2]. These studies have gone to all possible directions, univariate and multivariate, though in this work we stay only on the univariate approach.

Also, this author since 1997, he started the study of the quantitative convergence of neural network operators to the unit, and he has been written since then numerous of articles and books, e.g. see [3], [4], [5], are what we use here.

The wide range of neural network operators being treated by the author all along, are indeed by nature positive linear operators.

Here, for the first time in the literature, neural network operators are treated as positive linear operators.

So, all methods of positive linear operators apply here to our summation defined neural network operators, producing new and interesting results: pointwise, uniform and L_p, $p \geq 1$, kinds. Via the Riesz representation theorem neural networks are connected to measure theory. Convexity of functions produces optimal and sharper results.

The use of perturbed hyperbolic tangent as an activation function is well established and supported in [4], Chapter 18. Hyperbolic tangent activation function is greatly behaving and one of the most commonly used activation functions on this subject.

The author's symmetrization method presented here aims that our operators converge at high speed to the unit operator using half feed of data.

This chapter establishes a bridge for the first time connecting Neural Networks Approximation (part of AI) to Positive Linear Operators (a great part of Functional Analysis).

So it is a theoretical work intended by nature mainly to related experts.

Of great inspiration always have been [7], [8]. For classical studies on neural network, we also recommend, [9]–[14].

For newer work on neural networks we also refer the reader to [15]–[24]. For recent studies in positive linear operators we refer to [25]–[30].

1.2 Basics

Here we follow [4], pp. 455–460.

Our activation function here to be used is

$$g_{q,\lambda}(x) := \frac{e^{\lambda x} - qe^{-\lambda x}}{e^{\lambda x} + qe^{-\lambda x}}, \quad \lambda, q > 0, \ x \in \mathbb{R}. \tag{1.1}$$

Above λ is the parameter and q is the deformation coefficient, typically it is $0 < \lambda, q \leq 1$.

For more read Chapter 18 of [4]: "q-Deformed and λ-Parametrized Hyperbolic Tangent based Banach space Valued Ordinary and Fractional Neural Network Approximation".

This chapter motivates our current work.

The proposed "symmetrization method" aims to use half the data to feed into our neural networks.

We will employ the following density function

$$M_{q.\lambda}(x) := \frac{1}{4}\left(g_{q,\lambda}(x+1) - g_{q,\lambda}(x-1)\right) > 0, \tag{1.2}$$

$\forall x \in \mathbb{R};\ q, \lambda > 0.$

We have that

$$M_{q,\lambda}(-x) = M_{\frac{1}{q},\lambda}(x), \ \ \forall\, x \in \mathbb{R};\ q, \lambda > 0, \tag{1.3}$$

and

$$M_{\frac{1}{q},\lambda}(-x) = M_{q,\lambda}(x), \ \ \forall\, x \in \mathbb{R};\ q, \lambda > 0. \tag{1.4}$$

Adding (1.3) and (1.4) we obtain

$$M_{q,\lambda}(-x) + M_{\frac{1}{q},\lambda}(-x) = M_{q,\lambda}(x) + M_{\frac{1}{q},\lambda}(x), \tag{1.5}$$

the key to this work.

So that

$$\Phi(x) := \frac{M_{q,\lambda}(x) + M_{\frac{1}{q},\lambda}(x)}{2} \tag{1.6}$$

is an even function, symmetric with respect to the y-axis.

By (18.18) of [4], we have

$$\begin{gathered} M_{q,\lambda}\left(\frac{\ln q}{2\lambda}\right) = \frac{\tanh(\lambda)}{2}, \\ \text{and} \\ M_{\frac{1}{q},\lambda}\left(-\frac{\ln q}{2\lambda}\right) = \frac{\tanh(\lambda)}{2}, \ \ \lambda > 0, \end{gathered} \tag{1.7}$$

sharing the same maximum at symmetric points.

By Theorem 18.1, p. 458 of [4], we have that

$$\begin{gathered} \sum_{i=-\infty}^{\infty} M_{q,\lambda}(x-i) = 1, \ \ \forall\, x \in \mathbb{R},\ \lambda, q > 0, \\ \text{and} \\ \sum_{i=-\infty}^{\infty} M_{\frac{1}{q},\lambda}(x-i) = 1, \ \ \forall\, x \in \mathbb{R},\ \lambda, q > 0. \end{gathered} \tag{1.8}$$

Consequently, we derive that

$$\sum_{i=-\infty}^{\infty} \Phi(x-i) = 1, \ \ \forall\, x \in \mathbb{R}. \tag{1.9}$$

By Theorem 18.2, p. 459 of [4], we have that

$$\begin{aligned} &\int_{-\infty}^{\infty} M_{q,\lambda}(x)\,dx = 1, \\ &\text{and} \\ &\int_{-\infty}^{\infty} M_{\frac{1}{q},\lambda}(x)\,dx = 1, \end{aligned} \tag{1.10}$$

so that

$$\int_{-\infty}^{\infty} \Phi(x)\,dx = 1, \tag{1.11}$$

therefore Φ is a density function.

By Theorem 18.3, p. 459 of [4], we have:

Let $0 < \alpha < 1$, and $n \in \mathbb{N}$ with $n^{1-\alpha} > 2$; $q, \lambda > 0$. Then

$$\sum_{\begin{cases} k = -\infty \\ : |nx - k| \geq n^{1-\alpha} \end{cases}}^{\infty} M_{q,\lambda}(nx - k) < 2\max\left\{q, \frac{1}{q}\right\} e^{4\lambda} e^{-2\lambda n^{(1-\alpha)}} = Te^{-2\lambda n^{(1-\alpha)}}, \tag{1.12}$$

where $T := 2\max\left\{q, \frac{1}{q}\right\} e^{4\lambda}$.

Similarly, we get that

$$\sum_{k=-\infty}^{\infty} M_{\frac{1}{q},\lambda}(nx - k) < Te^{-2\lambda n^{(1-\alpha)}}. \tag{1.13}$$

Consequently, we obtain that

$$\sum_{\begin{cases} k = -\infty \\ : |nx - k| \geq n^{1-\alpha} \end{cases}}^{\infty} \Phi(nx - k) < Te^{-2\lambda n^{(1-\alpha)}}, \tag{1.14}$$

where $T := 2\max\left\{q, \frac{1}{q}\right\} e^{4\lambda}$.

Here $\lceil \cdot \rceil$ denotes the ceiling of the number, and $\lfloor \cdot \rfloor$ its integral part.

We mention

Theorem 18.4 (p. 459, [4]) Let $x \in [a, b] \subset \mathbb{R}$ and $n \in \mathbb{N}$ so that $\lceil na \rceil \leq \lfloor nb \rfloor$. For $q, \lambda > 0$, we consider $\lambda_q > z_0 > 0$, such that $M_{q,\lambda}(z_0) = M_{q,\lambda}(0)$, and $\lambda_q > 1$. Then

$$\frac{1}{\sum_{k=\lceil na \rceil}^{\lfloor nb \rfloor} M_{q,\lambda}(nx - k)} < \max\left\{\frac{1}{M_{q,\lambda}(\lambda_q)}, \frac{1}{M_{\frac{1}{q},\lambda}\left(\lambda_{\frac{1}{q}}\right)}\right\} =: \Delta(q). \tag{1.15}$$

Similarly, we consider $\lambda_{\frac{1}{q}} > z_1 > 0$, such that $M_{\frac{1}{q},\lambda}(z_1) = M_{\frac{1}{q},\lambda}(0)$, and $\lambda_{\frac{1}{q}} > 1$. Thus

$$\frac{1}{\sum_{k=\lceil na \rceil}^{\lfloor nb \rfloor} M_{\frac{1}{q},\lambda}(nx-k)} < \max\left\{\frac{1}{M_{\frac{1}{q},\lambda}\left(\lambda_{\frac{1}{q}}\right)}, \frac{1}{M_{q,\lambda}(\lambda_q)}\right\} = \Delta(q). \quad (1.16)$$

Hence

$$\sum_{k=\lceil na \rceil}^{\lfloor nb \rfloor} M_{q,\lambda}(nx-k) > \frac{1}{\Delta(q)}, \quad (1.17)$$

and

$$\sum_{k=\lceil na \rceil}^{\lfloor nb \rfloor} M_{\frac{1}{q},\lambda}(nx-k) > \frac{1}{\Delta(q)}. \quad (1.18)$$

Consequently it holds

$$\sum_{k=\lceil na \rceil}^{\lfloor nb \rfloor} \frac{\left(M_{q,\lambda}(nx-k) + M_{\frac{1}{q},\lambda}(nx-k)\right)}{2} > \frac{2}{2\Delta(q)} = \frac{1}{\Delta(q)}, \quad (1.19)$$

so that

$$\frac{1}{\sum_{k=\lceil na \rceil}^{\lfloor nb \rfloor} \frac{\left(M_{q,\lambda}(nx-k)+M_{\frac{1}{q},\lambda}(nx-k)\right)}{2}} < \Delta(q), \quad (1.20)$$

that is

$$\frac{1}{\sum_{k=\lceil na \rceil}^{\lfloor nb \rfloor} \Phi(nx-k)} < \Delta(q). \quad (1.21)$$

We have proved

Theorem 1.1. *Let $x \in [a,b] \subset \mathbb{R}$ and $n \in \mathbb{N}$ so that $\lceil na \rceil \leq \lfloor nb \rfloor$. For $q, \lambda > 0$, we consider $\lambda_q > z_0 > 0$, such that $M_{q,\lambda}(z_0) = M_{q,\lambda}(0)$, and $\lambda_q > 1$. Also consider $\lambda_{\frac{1}{q}} > z_1 > 0$, such that $M_{\frac{1}{q},\lambda}(z_1) = M_{\frac{1}{q},\lambda}(0)$, and $\lambda_{\frac{1}{q}} > 1$. Then*

$$\frac{1}{\sum_{k=\lceil na \rceil}^{\lfloor nb \rfloor} \Phi(nx-k)} < \Delta(q). \quad (1.22)$$

We make

Remark 1.1. I) By Remark 18.5, p. 460 of [4], we have

$$\lim_{n\to\infty}\sum_{k=\lceil na\rceil}^{\lfloor nb\rfloor} M_{q,\lambda}\left(nx_1-k\right)\neq 1,\ \text{ for some } x_1\in[a,b], \tag{1.23}$$

and

$$\lim_{n\to\infty}\sum_{k=\lceil na\rceil}^{\lfloor nb\rfloor} M_{\frac{1}{q},\lambda}\left(nx_2-k\right)\neq 1,\ \text{ for some } x_2\in[a,b]. \tag{1.24}$$

Therefore it holds

$$\lim_{n\to\infty}\sum_{k=\lceil na\rceil}^{\lfloor nb\rfloor}\frac{\left(M_{q,\lambda}\left(nx_1-k\right)+M_{\frac{1}{q},\lambda}\left(nx_2-k\right)\right)}{2}\neq 1. \tag{1.25}$$

Hence

$$\lim_{n\to\infty}\sum_{k=\lceil na\rceil}^{\lfloor nb\rfloor}\frac{\left(M_{q,\lambda}\left(nx_1-k\right)+M_{\frac{1}{q},\lambda}\left(nx_1-k\right)\right)}{2}\neq 1, \tag{1.26}$$

even if

$$\lim_{n\to\infty}\sum_{k=\lceil na\rceil}^{\lfloor nb\rfloor} M_{\frac{1}{q},\lambda}\left(nx_1-k\right)=1, \tag{1.27}$$

because then

$$\lim_{n\to\infty}\sum_{k=\lceil na\rceil}^{\lfloor nb\rfloor}\frac{M_{q,\lambda}\left(nx_1-k\right)}{2}+\frac{1}{2}\neq 1, \tag{1.28}$$

equivalently

$$\lim_{n\to\infty}\sum_{k=\lceil na\rceil}^{\lfloor nb\rfloor}\frac{M_{q,\lambda}\left(nx_1-k\right)}{2}\neq\frac{1}{2}, \tag{1.29}$$

true by

$$\lim_{n\to\infty}\sum_{k=\lceil na\rceil}^{\lfloor nb\rfloor} M_{q,\lambda}\left(nx_1-k\right)\neq 1. \tag{1.30}$$

II) Let $[a,b]\subset\mathbb{R}$. For large n we always have $\lceil na\rceil\leq\lfloor nb\rfloor$. Also $a\leq\frac{k}{n}\leq b$, iff $\lceil na\rceil\leq k\leq\lfloor nb\rfloor$. So in general it holds

$$\sum_{k=\lceil na\rceil}^{\lfloor nb\rfloor}\Phi\left(nx-k\right)\leq 1. \tag{1.31}$$

We need

Definition 1.1. Let $f \in C([a,b])$, $x \in [a,b]$ and $n \in \mathbb{N} : \lceil na \rceil \leq \lfloor nb \rfloor$. We introduce and define the real-valued symmetrized and perturbed positive linear neural network operators

$$H_n(f,x) := \frac{\sum_{k=\lceil na \rceil}^{\lfloor nb \rfloor} f\left(\frac{k}{n}\right) \Phi(nx-k)}{\sum_{k=\lceil na \rceil}^{\lfloor nb \rfloor} \Phi(nx-k)} =$$

$$\frac{\sum_{k=\lceil na \rceil}^{\lfloor nb \rfloor} f\left(\frac{k}{n}\right) \left(M_{q,\lambda}(nx-k) + M_{\frac{1}{q},\lambda}(nx-k)\right)}{\sum_{k=\lceil na \rceil}^{\lfloor nb \rfloor} \left(M_{q,\lambda}(nx-k) + M_{\frac{1}{q},\lambda}(nx-k)\right)}. \tag{1.32}$$

In fact, it is $H_n(f) \in C([a,b])$; and $H_n(1) = 1$.

The modulus of continuity is defined by

$$\omega_1(f,\delta) := \sup_{\substack{x,y \in [a,b] \\ |x-y| \leq \delta}} |f(x) - f(y)|, \quad \delta > 0. \tag{1.33}$$

In this work $0 < \alpha < 1$, $n \in \mathbb{N} : n^{1-\alpha} > 2$; where $\|\cdot\|_\infty$ is the supremum norm.

1.3 Main Results

The following approximation results are valid.

Theorem 1.2. *Let $f \in C([a,b])$. Then*

$$\|H_n(f) - f\|_\infty \leq 2\omega_1\left(f, \sqrt{\Delta(q)}\sqrt{\frac{1}{n^{2\alpha}} + (b-a)^2 Te^{-2\lambda n^{(1-\alpha)}}}\right) \to 0,$$

as $n \to +\infty$. (1.34)

So that $\lim_{n\to\infty} H_n(f) = f$, uniformly.

Proof. We estimate

$$\left(H_n\left((t-x)^2\right)\right)(x) = \frac{\sum_{k=\lceil na \rceil}^{\lfloor nb \rfloor} \left(\frac{k}{n} - x\right)^2 \Phi(nx-k)}{\sum_{k=\lceil na \rceil}^{\lfloor nb \rfloor} \Phi(nx-k)} \overset{(1.22)}{\leq}$$

$$\Delta(q)\left(\sum_{k=\lceil na\rceil}^{\lfloor nb\rfloor}\left(\frac{k}{n}-x\right)^2\Phi(nx-k)\right)=$$

$$\Delta(q)\left[\sum_{\substack{k=\lceil na\rceil \\ :\left|\frac{k}{n}-x\right|\leq\frac{1}{n^\alpha}}}^{\lfloor nb\rfloor}\left(\frac{k}{n}-x\right)^2\Phi(nx-k)+\right. \tag{1.35}$$

$$\left.\sum_{\substack{k=\lceil na\rceil \\ :\left|\frac{k}{n}-x\right|>\frac{1}{n^\alpha}}}^{\lfloor nb\rfloor}\left(\frac{k}{n}-x\right)^2\Phi(nx-k)\right]\overset{\text{(by (1.9), (1.14))}}{\leq}$$

$$\Delta(q)\left[\frac{1}{n^{2\alpha}}+(b-a)^2Te^{-2\lambda n^{(1-\alpha)}}\right].$$

That is

$$\left(H_n\left((t-x)^2\right)\right)(x)\leq\Delta(q)\left\{\frac{1}{n^{2\alpha}}+(b-a)^2Te^{-2\lambda n^{(1-\alpha)}}\right\}. \tag{1.36}$$

Consequently, it holds that

$$\delta_n:=\left\|\left(H_n\left((t-x)^2\right)\right)(x)\right\|_\infty^{\frac{1}{2}}\leq$$

$$\sqrt{\Delta(q)}\sqrt{\frac{1}{n^{2\alpha}}+(b-a)^2Te^{-2\lambda n^{(1-\alpha)}}}. \tag{1.37}$$

By Theorem 7.1.7, p. 203, [2], Shisha-Mond inequality ([7]) for positive linear operators and $H_n(1)=1$, we have that

$$\|H_n(f)-f\|_\infty\leq 2\omega_1(f,\delta_n). \tag{1.38}$$

Thus, it holds

$$\|H_n(f)-f\|_\infty\leq 2\omega_1\left(f,\sqrt{\Delta(q)}\sqrt{\frac{1}{n^{2\alpha}}+(b-a)^2Te^{-2\lambda n^{(1-\alpha)}}}\right),$$

proving the claim. □

It holds

Theorem 1.3. *Let $f : \mathbb{R} \to \mathbb{R}$ be a continuous and 2π periodic function with modulus of continuity ω_1. Here $\|\cdot\|_\infty$ denotes the sup-norm over $[a,b] \subset \mathbb{R}$, and the operators H_n are acting on such f over $[a,b]$; $n \in \mathbb{N} : n^{1-\alpha} > 2$, $0 < \alpha < 1$. Then*

$$\|H_n(f) - f\|_\infty \leq 2\omega_1\left(f, \pi\sqrt{\Delta(q)}\sqrt{\omega_1\left(\sin^2, \frac{1}{2n^\alpha}\right) + Te^{-2\lambda n^{(1-\alpha)}}}\right). \tag{1.39}$$

Proof. We want to estimate ($x \in [a,b]$),

$$\varepsilon_n := \pi\left\|\left(H_n\left(\sin^2\left(\frac{t-x}{2}\right)\right)\right)(x)\right\|_\infty^{\frac{1}{2}} =$$

$$\pi\left\|\left(\frac{\sum_{k=\lceil na\rceil}^{\lfloor nb\rfloor}\sin^2\left(\frac{\frac{k}{n}-x}{2}\right)\Phi(nx-k)}{\sum_{k=\lceil na\rceil}^{\lfloor nb\rfloor}\Phi(nx-k)}\right)\right\|_\infty^{\frac{1}{2}} \leq$$

$$\pi\left\|\Delta(q)\sum_{k=\lceil na\rceil}^{\lfloor nb\rfloor}\sin^2\left(\frac{\frac{k}{n}-x}{2}\right)\Phi(nx-k)\right\|_\infty^{\frac{1}{2}} =$$

$$\pi\sqrt{\Delta(q)}\left\|\sum_{\substack{k=\lceil na\rceil \\ :\left|\frac{k}{n}-x\right|\leq\frac{1}{n^\alpha}}}^{\lfloor nb\rfloor}\left(\sin^2\left(\frac{\frac{k}{n}-x}{2}\right) - \sin^2(0)\right)\Phi(nx-k) + \right. \tag{1.40}$$

$$\left.\sum_{\substack{k=\lceil na\rceil \\ :\left|\frac{k}{n}-x\right|>\frac{1}{n^\alpha}}}^{\lfloor nb\rfloor}\sin^2\left(\frac{\frac{k}{n}-x}{2}\right)\Phi(nx-k)\right\|_\infty^{\frac{1}{2}} \leq$$

$$\pi\sqrt{\Delta(q)}\left\|\sum_{\substack{k=\lceil na\rceil \\ : \left|\frac{k}{n}-x\right|\leq\frac{1}{n^{\alpha}}}}^{\lfloor nb\rfloor}\omega_1\left(\sin^2,\frac{\left|\frac{k}{n}-x\right|}{2}\right)\Phi(nx-k)+\right.$$

$$\left.1\cdot\sum_{\substack{k=\lceil na\rceil \\ : \left|\frac{k}{n}-x\right|>\frac{1}{n^{\alpha}}}}^{\lfloor nb\rfloor}\Phi(nx-k)\right\|_{\infty}^{\frac{1}{2}}\leq$$

$$\pi\sqrt{\Delta(q)}\sqrt{\omega_1\left(\sin^2,\frac{1}{2n^{\alpha}}\right)+Te^{-2\lambda n^{(1-\alpha)}}}.$$

We have proved that

$$\varepsilon_n\leq\pi\sqrt{\Delta(q)}\sqrt{\omega_1\left(\sin^2,\frac{1}{2n^{\alpha}}\right)+Te^{-2\lambda n^{(1-\alpha)}}}. \tag{1.41}$$

Therefore it holds, by [8], the following inequalities,

$$\|H_n(f)-f\|_{\infty}\leq 2\omega_1(f,\varepsilon_n)\leq$$

$$2\omega_1\left(f,\pi\sqrt{\Delta(q)}\sqrt{\omega_1\left(\sin^2,\frac{1}{2n^{\alpha}}\right)+Te^{-2\lambda n^{(1-\alpha)}}}\right), \tag{1.42}$$

hence proving the claim. □

We need

Definition 1.2. Let $f\in C([a,b])$. The Peetre ([6]) K_1 functional is defined as follows:

$$K_1(f,t):=\inf\left(\|f-g\|_{\infty}+t\|g'\|_{\infty}\right),\quad g\in C^1([a,b]),\quad t\geq 0. \tag{1.43}$$

We give

Theorem 1.4. *Let $f\in C([a,b])$. Then*

$$\|H_n(f)-f\|_{\infty}\leq 2K_1\left(f,\frac{\Delta(q)}{2}\left[\frac{1}{n^{\alpha}}+(b-a)Te^{-2\lambda n^{(1-\alpha)}}\right]\right)\to 0, \tag{1.44}$$

as $n\to+\infty$.

So that $\lim_{n\to+\infty}H_n(f)=f$, uniformly.

For the proof of (1.44) we will use.

Corollary 1.1. *([1]) (Pointwise Approximation) Let $f \in C([a,b])$ and let L be a positive linear operator acting on $C([a,b])$ satisfying $L(1,x) = 1$, all $x \in [a,b]$.*

Then, we have the attainable (i.e. sharp) inequality

$$|L(f,x) - f(x)| \leq 2K_1\left(f, \frac{1}{2}L(|y-x|, x)\right), \quad \forall\, x \in [a,b]. \tag{1.45}$$

Proof of Theorem 1.4. We apply inequality (1.45). Clearly, we have that

$$|H_n(f,x) - f(x)| \leq 2K_1\left(f, \frac{1}{2}H_n(|t-x|)(x)\right), \tag{1.46}$$

which is attainable (i.w. sharp).

We estimate

$$H_n(|t-x|)(x) = \frac{\sum_{k=\lceil na \rceil}^{\lfloor nb \rfloor} \left|\frac{k}{n} - x\right| \Phi(nx-k)}{\sum_{k=\lceil na \rceil}^{\lfloor nb \rfloor} \Phi(nx-k)} \leq$$

$$\Delta(q) \sum_{k=\lceil na \rceil}^{\lfloor nb \rfloor} \left|\frac{k}{n} - x\right| \Phi(nx-k) =$$

$$\Delta(q)\left[\sum_{\substack{k = \lceil na \rceil \\ : \left|\frac{k}{n} - x\right| \leq \frac{1}{n^\alpha}}}^{\lfloor nb \rfloor} \left|\frac{k}{n} - x\right| \Phi(nx-k) + \right. \tag{1.47}$$

$$\left. \sum_{\substack{k = \lceil na \rceil \\ : \left|\frac{k}{n} - x\right| > \frac{1}{n^\alpha}}}^{\lfloor nb \rfloor} \left|\frac{k}{n} - x\right| \Phi(nx-k)\right] \leq$$

$$\Delta(q)\left[\frac{1}{n^\alpha} + (b-a)\,Te^{-2\lambda n^{(1-\alpha)}}\right].$$

That is

$$H_n(|t-x|)(x) \le \Delta(q)\left[\frac{1}{n^\alpha}+(b-a)Te^{-2\lambda n^{(1-\alpha)}}\right]. \tag{1.48}$$

We have proved

$$|H_n(f,x)-f(x)| \le 2K_1\left(f,\frac{\Delta(q)}{2}\left[\frac{1}{n^\alpha}+(b-a)Te^{-2\lambda n^{(1-\alpha)}}\right]\right), \tag{1.49}$$

$\forall\, x \in [a,b]$.

Now, the validity of (1.44) is clear. □

It follows a trigonometric result.

Theorem 1.5. *Here* $f \in C^2([a,b])$, $x_0 \in [a,b]$, $r > 0$, $\overline{D}_3(x_0) := \left(H_n\left(|\cdot - x_0|^3\right)(x_0)\right)^{\frac{1}{3}}$. *Then*

(i)

$$|H_n(f)(x_0)-f(x_0)| \le |f'(x_0)|\,|H_n(\sin(\cdot - x_0))(x_0)| +$$
$$2|f''(x_0)|\left(H_n\left(\sin^2\left(\frac{\cdot - x_0}{2}\right)\right)(x_0)\right)+$$
$$\frac{\omega_1\left(f''+f, r\overline{D}_3(x_0)\right)}{2}\overline{D}_3^2(x_0)\left[1+\frac{1}{3r}\right], \tag{1.50}$$

(ii)

$$\|H_n(f)-f\|_\infty \le \|f'\|_\infty\,\|H_n(\sin(\cdot - x_0))(x_0)\|_\infty +$$
$$2\|f''\|_\infty\left\|H_n\left(\sin^2\left(\frac{\cdot - x_0}{2}\right)\right)(x_0)\right\|_\infty +$$
$$\frac{\omega_1\left(f''+f, r\left\|\overline{D}_3\right\|_\infty\right)}{2}\left\|\overline{D}_3\right\|_\infty^2\left(1+\frac{1}{3r}\right), \tag{1.51}$$

(iii) if $f'(x_0)=f''(x_0)=0$, *we obtain*

$$|H_n(f)(x_0)-f(x_0)| \le \frac{\omega_1\left(f''+f, r\overline{D}_3(x_0)\right)}{2}\overline{D}_3^2(x_0)\left(1+\frac{1}{3r}\right), \tag{1.52}$$

and

(iv)

$$\left|H_n(f)(x_0)-f(x_0)-f'(x_0)H_n(\sin(\cdot - x_0))(x_0)-2f''(x_0)H_n\left(\sin^2\left(\frac{\cdot - x_0}{2}\right)\right)(x_0)\right|$$
$$\le \frac{\omega_1\left(f''+f, r\overline{D}_3(x_0)\right)}{2}\overline{D}_3^2(x_0)\left(1+\frac{1}{3r}\right). \tag{1.53}$$

Proof. Direct application of Theorem 12.3, p. 384, [5]. □

Next, we give the hyperbolic version of last result.

Theorem 1.6. *All as in Theorem 1.5. Then*

(i)

$$|H_n(f)(x_0) - f(x_0)| \leq |f'(x_0)| \, |H_n(\sinh(\cdot - x_0))(x_0)| + \tag{1.54}$$

$$2|f''(x_0)| \left(H_n \left(\sinh^2 \left(\frac{\cdot - x_0}{2} \right) \right) (x_0) \right) +$$

$$\frac{\cosh(b-a)\, \omega_1 \left(f'' - f, r\overline{D}_3(x_0)\right)}{2} \overline{D}_3^2(x_0) \left[1 + \frac{1}{3r}\right],$$

(ii)

$$\|H_n(f) - f\|_\infty \leq \|f'\|_\infty \, \|H_n(\sinh(\cdot - x_0))(x_0)\|_\infty +$$

$$2\|f''\|_\infty \left\| H_n \left(\sinh^2 \left(\frac{\cdot - x_0}{2} \right) \right) (x_0) \right\|_\infty +$$

$$\frac{\cosh(b-a)\, \omega_1 \left(f'' - f, r\left\|\overline{D}_3\right\|_\infty\right)}{2} \left\|\overline{D}_3\right\|_\infty^2 \left(1 + \frac{1}{3r}\right), \tag{1.55}$$

(iii) if $f'(x_0) = f''(x_0) = 0$, $x_0 \in [a, b]$, *we obtain*

$$|H_n(f)(x_0) - f(x_0)| \leq \frac{\cosh(b-a)\, \omega_1 \left(f'' - f, r\overline{D}_3(x_0)\right)}{2} \overline{D}_3^2(x_0) \left(1 + \frac{1}{3r}\right), \tag{1.56}$$

and

(iv)

$$\left| H_n(f)(x_0) - f(x_0) - f'(x_0) H_n(\sinh(\cdot - x_0))(x_0) - 2f''(x_0) H_n \left(\sinh^2 \left(\frac{\cdot - x_0}{2} \right) \right) (x_0) \right|$$

$$\leq \frac{\cosh(b-a)\, \omega_1 \left(f'' - f, r\overline{D}_3(x_0)\right)}{2} \overline{D}_3^2(x_0) \left(1 + \frac{1}{3r}\right). \tag{1.57}$$

Proof. Direct application of Theorem 12.5, p. 390, [5]. □

Remark 1.2. We similarly have that

$$\left(H_n\left(|\cdot - x_0|^3\right)(x_0)\right)^{\frac{1}{3}} \leq \sqrt[3]{\Delta(q)}\sqrt[3]{\left\{\frac{1}{n^{3\alpha}} + (b-a)^3 Te^{-2\lambda n^{(1-\alpha)}}\right\}} \to 0, \tag{1.58}$$

as $n \to +\infty$.

Hence, by each of the above Theorems 1.5, 1.6, we obtain that $H_n(f) \to f$ uniformly convergent as $n \to +\infty$, $\forall\, f \in C^2([a,b])$.

Valid, by Theorem 12.4, p. 390 and Theorem 12.6, p. 395 from [5], respectively.

We make

Remark 1.3. Let $x_0 \in (a,b)$, $f \in C([a,b])$, here H_n is a positive linear operator from $C([a,b])$ into itself.

Also $H_n(1) = 1$. By Riesz Representation theorem, there exists a probability measure μ_{x_0} on $[a,b]$, such that

$$H_n(f)(x_0) = \int_{[a,b]} f(t)\, d\mu_{x_0}(t), \quad \forall\, f \in C([a,b]). \tag{1.59}$$

Assume here that $|f(t) - f(x_0)|$ is convex in t. We consider

$$\begin{aligned} d_2(x_0) &:= \left(\int_{[a,b]} (t-x_0)^2\, d\mu_{x_0}(t)\right)^{\frac{1}{2}} \\ &= \left(\left(H_n\left((t-x_0)^2\right)\right)(x_0)\right)^{\frac{1}{2}} \quad \text{(as earlier)} \\ &\leq \sqrt{\Delta(q)}\sqrt{\frac{1}{n^{2\alpha}} + (b-a)^2 Te^{-2\lambda n^{(1-\alpha)}}} = (*). \end{aligned} \tag{1.60}$$

Clearly, here it is $d_2(x_0) > 0$ (see (1.32)).

For large enough $n \in \mathbb{N}$ we get

$$(*) \leq \min(x_0 - a, b - x_0) \leq \max(x_0 - a, b - x_0).$$

That is

$$d_2(x_0) \leq \min(x_0 - a, b - x_0). \tag{1.61}$$

By Corollary 8.1.1, p. 245, [2], we obtain:

Theorem 1.7. *It holds*

$$|H_n(f)(x_0) - f(x_0)| \leq \omega_1\left(f, \sqrt{\Delta(q)}\sqrt{\frac{1}{n^{2\alpha}} + (b-a)^2 Te^{-2\lambda n^{(1-\alpha)}}}\right), \tag{1.62}$$

for large enough $n \in \mathbb{N}$.

Remark 1.4. We have that

$$d_1(x_0) := \int_{[a,b]} |t - x_0| \, d\mu_{x_0}(t) = H_n(|t - x_0|)(x_0) \overset{(1.48)}{\leq} \tag{1.63}$$

$$\Delta(q)\left[\frac{1}{n^\alpha} + (b-a)\, T e^{-2\lambda n^{(1-\alpha)}}\right] \leq \min(x_0 - a, b - x_0),$$

for large enough $n \in \mathbb{N}$.

By (8.1.8), p. 245, [2], we get

Theorem 1.8. *Let $r \geq 1$, then*

$$|H_n(f)(x_0) - f(x_0)| \leq r\omega_1\left(f, \frac{1}{r}\Delta(q)\left\{\frac{1}{n^\alpha} + (b-a)\, T e^{-2\lambda n^{(1-\alpha)}}\right\}\right), \tag{1.64}$$

for large enough $n \in \mathbb{N}$.

By Theorem 8.1.2, p. 248, of [2] we can derive the following result.

Theorem 1.9. *Here we consider $[a,b] \subset \mathbb{R}$, and $x_0 \in (a,b)$. Let L be a positive linear operator from $C([a,b])$ into itself, such that $L(1) = 1$.*

Denote by

$$d_{m+1}(x_0) := \left(L\left(|t - x_0|^{m+1}\right)(x_0)\right)^{\frac{1}{m+1}}, \quad m \in \mathbb{N}.$$

Consider $f \in C^m([a,b])$ such that $\left|f^{(m)}(t) - f^{(m)}(x_0)\right|$ is convex in t.

Assume that

$$0 < d_{m+1}^{m+1}(x_0) \leq (\min(x_0 - a, b - x_0))\,(m+1)!. \tag{1.65}$$

Then

$$|L(f)(x_0) - f(x_0)| - \sum_{k=1}^{m} \frac{\left|f^{(k)}(x_0)\right|}{k!} \left|L\left((t - x_0)^k\right)(x_0)\right| \leq \tag{1.66}$$

$$\omega_1\left(f^{(m)}, \frac{L\left(|t - x_0|^{m+1}\right)(x_0)}{(m+1)!}\right).$$

We make

Remark 1.5. We have that

$$0 < \left(H_n\left(|t-x_0|^{(m+1)}\right)\right)(x_0) = \frac{\sum_{k=\lceil na\rceil}^{\lfloor nb\rfloor} \left|\frac{k}{n}-x_0\right|^{m+1} \Phi(nx-k)}{\sum_{k=\lceil na\rceil}^{\lfloor nb\rfloor} \Phi(nx-k)}$$

$$\leq \Delta(q) \sum_{k=\lceil na\rceil}^{\lfloor nb\rfloor} \left|\frac{k}{n}-x_0\right|^{m+1} \Phi(nx-k) \tag{1.67}$$

... (as earlier)

$$\Delta(q)\left\{\frac{1}{n^{(m+1)\alpha}} + (b-a)^{m+1} T e^{-2\lambda n^{(1-\alpha)}}\right\} \leq \min(x_0-a, b-x_0)(m+1)!, \tag{1.68}$$

for large enough $n \in \mathbb{N}$.

By Theorem 1.9, Remark 1.5 we have proved the following:

Theorem 1.10. *Here it is $[a,b] \subset \mathbb{R}$, and $x_0 \in (a,b)$. Consider $f \in C^m([a,b]) : \left|f^{(m)}(t) - f^{(m)}(x_0)\right|$ is convex in t, $m \in \mathbb{N}$.*

Then, for sufficiently large enough $n \in \mathbb{N}$, we derive that

$$|H_n(f)(x_0) - f(x_0)| - \sum_{k=1}^{m} \frac{\left|f^{(k)}(x_0)\right|}{k!} \left|H_n\left((t-x_0)^k\right)(x_0)\right| \leq \tag{1.69}$$

$$\omega_1\left(f^{(m)}, \frac{\Delta(q)}{(m+1)!}\left\{\frac{1}{n^{(m+1)\alpha}} + (b-a)^{m+1} T e^{-2\lambda n^{(1-\alpha)}}\right\}\right).$$

Next we use Theorem 18.1, p. 419, [3].

Theorem 1.11. *Denote ($N \in \mathbb{N}$)*

$$F_N := \left\|\left(H_n\left(|t-\cdot|^N\right)\right)(\cdot)\right\|_\infty^{\frac{1}{N}} < +\infty, \tag{1.70}$$

where $\|\cdot\|_\infty$ is the supremum norm.

Let $f \in C^N([a,b])$. Then

$$\|H_n f - f\|_\infty \leq \sum_{\overline{k}=1}^{N} \frac{\left\|f^{(\overline{k})}\right\|_\infty}{\overline{k}!} \left\|\left(H_n(t-\cdot)^{\overline{k}}\right)(\cdot)\right\|_\infty + \tag{1.71}$$

$$\omega_1\left(f^{(N)}, F_N\right) F_N^{N-1}\left(\frac{(b-a)}{(N+1)!} + \frac{F_N}{2N!} + \frac{F_N^2}{8(b-a)(N-1)!}\right), \quad \forall\, n \in \mathbb{N}.$$

Furthermore it holds,

$$F_N \leq \sqrt[N]{\Delta(q)} \sqrt[N]{\frac{1}{n^{N\alpha}} + (b-a)^N T e^{-2\lambda n^{(1-\alpha)}}}, \tag{1.72}$$

and for $\overline{k} = 1, ..., N$, we have, similarly, that

$$\left\| \left(H_n \left(t - \cdot \right)^{\overline{k}} \right) (\cdot) \right\|_\infty \leq \Delta(q) \left\{ \frac{1}{n^{\overline{k}\alpha}} + (b-a)^{\overline{k}} T e^{-2\lambda n^{(1-\alpha)}} \right\}. \tag{1.73}$$

By Corollary 18.1, p. 421, [3], we get

Corollary 1.2. *Here $F_1 := \| (H_n (|t - \cdot|)) (\cdot) \|_\infty < \infty$. Let $f \in C^1([a,b])$. Then*

$$\| H_n f - f \|_\infty \leq \| f' \|_\infty \| (H_n (t - \cdot)) (\cdot) \|_\infty + \tag{1.74}$$

$$\frac{1}{2} \omega_1 (f', F_1) \left((b-a) + F_1 + \frac{F_1^2}{4(b-a)} \right).$$

Here it is

$$F_1 \leq \Delta(q) \left(\frac{1}{n^\alpha} + (b-a) T e^{-2\lambda n^{(1-\alpha)}} \right). \tag{1.75}$$

By Corollary 18.2, p. 421, [3], we get

Corollary 1.3. *Here $F_2 := \left\| \left(H_n (t - \cdot)^2 \right) (\cdot) \right\|_\infty^{\frac{1}{2}} < +\infty$. Let $f \in C^2([a,b])$. Then*

$$\| H_n f - f \|_\infty \leq \| f' \|_\infty \| (H_n (t - \cdot)) (\cdot) \|_\infty + \frac{\| f'' \|_\infty}{2} \left\| \left(H_n (t - \cdot)^2 \right) (\cdot) \right\|_\infty$$

$$+ \frac{1}{2} \omega_1 (f'', F_2) F_2 \left(\frac{(b-a)}{3} + \frac{F_2}{2} + \frac{F_2^2}{4(b-a)} \right). \tag{1.76}$$

Here they are

$$F_2 \leq \sqrt{\Delta(q)} \sqrt{\frac{1}{n^{2\alpha}} + (b-a)^2 T e^{-2\lambda n^{(1-\alpha)}}}, \tag{1.77}$$

and

$$\left\| \left(H_n (t - \cdot)^2 \right) (\cdot) \right\|_\infty \leq \Delta(q) \left(\frac{1}{n^{2\alpha}} + (b-a)^2 T e^{-2\lambda n^{(1-\alpha)}} \right). \tag{1.78}$$

By Theorem 18.2, p. 422, [3], we get:

Theorem 1.12. *Let* $\varphi := \left\| \left(H_n (t - x_0)^2 \right) (x_0) \right\|_\infty^{\frac{1}{2}} < +\infty$, *and* $r > 0$. *Let also* $f \in C^1([a,b])$. *Then*

$$\|H_n f - f\|_\infty - \|f'\|_\infty \|(H_n (t - x_0))(x_0)\|_\infty \leq \tag{1.79}$$

$$\begin{cases} \frac{1}{8r} (2+r)^2 \omega_1 (f', r\varphi) \varphi, & \text{if } r \leq 2; \\ \omega_1 (f', r\varphi) \varphi, & \text{if } r > 2. \end{cases}$$

Next results are with respect to $\|\cdot\|_1$, $\|\cdot\|_2$ norms, which are with respect to Lebesgue measure.

By Theorem 18.3, p. 424, [3], we obtain

Theorem 1.13. *Let*

$$\gamma_N := \left\| \left(H_n (t - x)^N \right) (x) \right\|_2^{\frac{1}{N}} < +\infty, \tag{1.80}$$

$n, N \in \mathbb{N}$. *Let also* $f \in C^N([a,b])$. *Then*

$$\|H_n f - f\|_1 \leq \sum_{\overline{k}=1}^{N} \frac{1}{\overline{k}!} \left\| f^{(\overline{k})} \right\|_2 \left\| \left(H_n (t - \cdot)^{\overline{k}} \right) (\cdot) \right\|_2 + \frac{1}{2} \omega_1 \left(f^{(N)}, \gamma_N \right) \gamma_N^{N-1} \tag{1.81}$$

$$\left\{ \left(\frac{b-a}{(N+1)!} \right) \cdot \sqrt{\frac{7}{3} (b-a)} + \frac{\sqrt{b-a}}{N!} \gamma_N + \frac{\gamma_N^2}{4(N-1)!} \sqrt{\frac{2}{b-a}} \right\} < +\infty.$$

By Corollary 18.3, p. 426, [3], we get:

Corollary 1.4. *Set*

$$\gamma_1 := \|H_n(|t-x|)(x)\|_2 < +\infty. \tag{1.82}$$

Let $f \in C^1([a,b])$. *Then*

$$\|H_n(f) - f\|_1 \leq \|f'\|_2 \|H_n(t - \cdot)(\cdot)\|_2 + \frac{\omega_1(f', \gamma_1)}{2}$$

$$\left\{ \left(\frac{b-a}{2} \right) \sqrt{\frac{7}{3}(b-a)} + \left(\sqrt{b-a} \right) \gamma_1 + \frac{\gamma_1^2}{4} \sqrt{\frac{2}{b-a}} \right\}. \tag{1.83}$$

By Corollary 18.4, p. 427, [3], we obtain

Corollary 1.5. *Set*

$$\gamma_2 := \left\| H_n\left((t-x)^2\right)(x)\right\|_2^{\frac{1}{2}} < +\infty. \tag{1.84}$$

Let $f \in C^2([a,b])$. *Then*

$$\|H_n f - f\|_1 \le \|f'\|_2 \|H_n(t-\cdot)(\cdot)\|_2 + \frac{\|f''\|_2}{2}\gamma_2^2 + \frac{1}{4}\omega_1(f'',\gamma_2)\gamma_2 \tag{1.85}$$

$$\left\{\left(\frac{b-a}{3}\right)\cdot\sqrt{\frac{7}{3}(b-a)} + \left(\sqrt{b-a}\right)\gamma_2 + \frac{\gamma_2^2}{2}\sqrt{\frac{2}{b-a}}\right\}.$$

Setting 1. In the next Theorems 1.14, 1.15 we use:

Let $([a,b],\mathcal{B},\mu)$, $a \ne b$, be a measure space, where $\mathcal{B}$ is the Borel σ-algebra on $[a,b]$ and μ is a positive finite measure on $[a,b]$. (Note that $C([a,b]) \subset L_p([a,b],\mathcal{B},\mu)$, any $p>1$.) Here $\|\cdot\|_p$ stands for the related L_p norm with respect to μ. Let $p,q>1$ such that $\frac{1}{p}+\frac{1}{q}=1$.

Next we apply Theorem 18.4, p. 428, [3].

Theorem 1.14. *Here* $n, N \in \mathbb{N}$. *Set*

$$\delta_N := \left\|\left(H_n\left(|t-\cdot|^N\right)\right)(\cdot)\right\|_p^{\frac{1}{N}} < +\infty. \tag{1.86}$$

Let $f \in C^N([a,b])$. *Then*

$$\|H_n f - f\|_p \le \sum_{\overline{k}=1}^{N}\frac{1}{\overline{k}!}\left\|f^{(\overline{k})}\right\|_{pq}\left\|\left(H_n(t-\cdot)^{\overline{k}}\right)(\cdot)\right\|_{p^2} + \tag{1.87}$$

$$\omega_1\left(f^{(N)},\delta_N\right)\delta_N^{N-1}\left[\frac{(b-a)}{(N+1)!} + \frac{\delta_N}{2N!} + \frac{\delta_N^2}{4(N-1)!(b-a)}\right].$$

Remark 1.6. We have that

$$\left(H_n\left(|t-x_0|^N\right)\right)(x_0) \le \Delta(q)\left\{\frac{1}{n^{N\alpha}} + (b-a)^N T e^{-2\lambda n^{(1-\alpha)}}\right\} =: \tau, \tag{1.88}$$

and

$$\left\|\left(H_n\left(|t-\cdot|^N\right)\right)(\cdot)\right\|_p =$$

$$\left(\int_{[a,b]}\left(\left(H_n\left(|t-x_0|^N\right)\right)(x_0)\right)^p \mu(dx_0)\right)^{\frac{1}{p}} \le \left(\int_{[a,b]}\tau^p d\mu\right)^{\frac{1}{p}} = \mu([a,b])^{\frac{1}{p}}\cdot\tau. \tag{1.89}$$

Therefore it holds

$$\delta_N \le \mu([a,b])^{\frac{1}{Np}}\,\tau^{\frac{1}{N}}. \tag{1.90}$$

By Theorem 18.5, p. 431, [3], we get

Theorem 1.15. *Here* $n, N \in \mathbb{N}$. *Denote*

$$\delta_N := \left\| H_n \left(|t - x|^N, x \right) \right\|_p^{\frac{1}{N}} < +\infty. \tag{1.91}$$

Let $f \in C^N([a, b])$. *Then*

$$\|H_n f - f\|_p \leq \sum_{\overline{k}=1}^{N} \frac{\left\| f^{(\overline{k})} \right\|_\infty}{\overline{k}!} \left\| \left(H_n (t - x)^{\overline{k}}, x \right) \right\|_p + \tag{1.92}$$

$$\omega_1 \left(f^{(N)}, \delta_N \right) \delta_N^{N-1} \left[\frac{(b-a)}{(N+1)!} + \frac{\delta_N}{2N!} + \frac{\delta_N^2}{4(N-1)!(b-a)} \right].$$

We need

Definition 1.3. Let $f, g \in C([a, b])$ such that $|f(x)| \leq |g(x)|$, $\forall\, x \in [a, b]$. A norm $\|\cdot\|$ on $C([a, b])$ is called monotone iff $\|f\| \leq \|g\|$. We denote a monotone norm by $\|\cdot\|_m$, e.g., L_p norms in general $(1 \leq p \leq +\infty)$, Orlicz norms, etc.

Finally we apply Corollary 18.5, p. 432, [3].

Corollary 1.6. *Let* $\|\cdot\|_m$ *be a monotone norm on* $C([a, b])$. *Denote*

$$\varepsilon_N := \left\| H_n \left(|t - x|^N, x \right) \right\|_m^{\frac{1}{N}} < +\infty, \quad n, N \in \mathbb{N}. \tag{1.93}$$

Let $f \in C^N([a, b])$. *Then*

$$\|H_n f - f\|_m \leq \sum_{\overline{k}=1}^{N} \frac{\left\| f^{(\overline{k})} \right\|_\infty}{\overline{k}!} \left\| \left(H_n (t - x)^{\overline{k}}, x \right) \right\|_m +$$

$$\omega_1 \left(f^{(N)}, \varepsilon_N \right) \varepsilon_N^{N-1} \left[\frac{(b-a)}{(N+1)!} + \frac{\varepsilon_N}{2N!} + \frac{\varepsilon_N^2}{4(N-1)!(b-a)} \right]. \tag{1.94}$$

Bibliography

[1] G.A. Anastassiou, *A "K-Attainable" inequality related to the convergence of Positive Linear Operators*, J. Approximation Theory, 44, 380–383 (1985).

[2] G.A. Anastassiou, *Moments in probability and approximation theory*, Pitman Research Notes in Mathematics series / Longman Scientific & Technical, Essex, New York, UK, USA, 1993.

[3] G.A. Anastassiou, *Quantitative Approximations*, Chapmen & Hall/CRC, London, New York, 2001.

[4] G.A. Anastassiou, *Parametrized, Deformed and General Neural Networks*, Springer, Heidelberg, New York, 2023.

[5] G.A. Anastassiou, *Trigonometric and Hyperbolic Generated Approximation Theory*, World Scientific, Singapore, New York, 2025.

[6] J. Peetre, *A Theory of Interpolation of Normed Spaces*, Notes Universidade de Brasilia, 1963.

[7] O. Shisha and B. Mond, *The degree of convergence of sequences of linear positive operators*, Nat. Acad. Sci. U.S. 60 (1968), 1196–1200.

[8] O. Shisha and B. Mond, *The degree of Approximation to Periodic Functions by Linear Positive Operators*, J. Approximation Theory, 1, 335–339 (1968).

[9] Z. Chen and F. Cao, *The approximation operators with sigmoidal functions*, Computers and Mathematics with Applications, 58 (2009), 758–765.

[10] D. Costarelli, R. Spigler, *Approximation results for neural network operators activated by sigmoidal functions*, Neural Networks 44 (2013), 101–106.

[11] D. Costarelli, R. Spigler, *Multivariate neural network operators with sigmoidal activation functions*, Neural Networks 48 (2013), 72–77.

[12] S. Haykin, *Neural Networks: A Comprehensive Foundation* (2 ed.), Prentice Hall, New York, 1998.

[13] W. McCulloch and W. Pitts, *A logical calculus of the ideas immanent in nervous activity*, Bulletin of Mathematical Biophysics, 7 (1943), 115–133.

[14] T.M. Mitchell, *Machine Learning*, WCB-McGraw-Hill, New York, 1997.

[15] Yu, Dansheng; Cao, Feilong, *Construction and approximation rate for feed-forward neural network operators with sigmoidal functions*, J. Comput. Appl. Math. 453 (2025), paper No. 116150.

[16] Cen, Siyu; Jin, Bangti; Quan, Qimeng; Zhou, Zhi; *Hybrid neural-network FEM approximation of diffusion coeficient in elyptic and parabolic problems.* IMA J. Numer. Anal. 44 (2024), no. 5, 3059–3093.

[17] Coroianu Lucian; Costarelli, Danillo; Natale, Mariarosaria; Pantiş, Alexandra; *The approximation capabilities of Durrmeyer-type neural network operators.* J. Appl. Math. Comput. 70 (2024), no. 5, 4581–4599.

[18] Warin, Xavier; *The GroupMax neural network approximation of convex*

functions. IEEE Trans. Neural Netw. Learn. Syst. 35 (2024), no. 8, 11608–11612.

[19] Fabra, Arnau; Guasch, Oriol; Baiges, Joan; Codina, Ramon; *Approximation of acoustic black holes with finite element mixed formulations and artificial neural network correction terms.* Finite Elem. Anal. Des. 241 (2024), paper No. 104236.

[20] Grohs, Philipp; Voigtlaender, Felix. *Proof of the theory-to-practice gap in deep learning via sampling complexity bounds for neural network approximation spaces.* Found. Comput. Math. 24 (2024), no. 4, 1085–1143.

[21] Basteri, Andrea; Trevisan, Dario; *Quantitative Gaussian approximation of randomly initialized deep neural networks.* Mach. Learn. 113 (2024), no. 9, 6373–6393.

[22] De Ryck, T.; Mishra, S. *Error analysis for deep neural network approximations of parametric hyperbolic conservation laws.* Math. Comp. 93 (2024), no. 350, 2643–2677.

[23] Liu, Jie; Zhang, Baoji; Lai, Yuyang; Fang, Liqiau. *Hull form optimization reserach based on multi-precision back-propagation neural network approximation model.* Internal. J. Numer. Methods Fluid 96 (2024), no. 8, 1445–1460.

[24] Yoo, Jihahm; Kim, Jaywon; Gim, Minjung; Lee, Haesung. *Error estimates of physics-informed neural networks for initial value problems.* J. Korean Soc. Ind. Appl. Math. 28 (2024), no. 1, 33–58.

[25] Kaur, Jaspreet; Goyal, Meenu. *Hyers-Ulam stability of some positive linear operators.* Stud. Univ. Babeş-Bolyai Math. 70 (2025), no. 1, 105–114.

[26] Abel, Ulrich; Acu, Ana Maria; Heilmann, Margareta; Raşa, Ioan. *On some Cauchy problems and positive linear operators.* Mediterr. J. Math. 22 (2025), no. 1, Paper No. 20, 19 pp.

[27] Moradi, Hamid Reza; Furuichi, Shigeru; Sababheh, Mohammad. *Operator quadratic mean and positive linear maps.* J. Math. Inequal. 18 (2024), no. 4, 1263–1279.

[28] Bustamante, Jorge; Torres-Campos, José D. *Power series and positive linear operators in weighted spaces.* Serdica Math. J. 50 (2024), no. 3-4, 225–250.

[29] Acu, Ana-Maria; Rasa, Ioan; Sofonea, Florin. *Composition of some positive linear integral operators.* Demonstr. Math. 57 (2024), no. 1, Paper No. 20240018, 14 pp.

[30] Patel, Prashantkumar G. *On positive linear operators linking gamma,*

Mittag-Leffler and Wright functions. Int. J. Appl. Comput. Math. 10 (2024), no. 5, Paper No. 152, 12 pp.

[31] G.A. Anastassiou, *Neural Networks as Positive Linear Operators*, Mathematics, 2025, 13, 1112; https://doi.org/10.3390/math13071112.

Chapter 2

Neural Networks as Positive Linear Operators over Infinite Domain

Neural network operators in infinite domain are interpreted as positive linear operators and related general theory applies to them. These operators are induced by a symmetrized density function deriving from the parametrized and deformed hyperbolic tangent activation function. We are acting on the space of continuous and bounded functions on the real line to the reals. We study quantitatively the rate of convergence of these neural network operators to the unit operator. Our inequalities involve the modulus of continuity of the function under approximation or its derivative. It follows [59].

2.1 Introduction

The author studied extensively the quantitative approximation of positive linear operators to the unit since 1985, see for example [32]–[34], [37], which we use in this work. He originated from the quantitative weak convergence of finite positive measures to the unit Dirac measure, having as a method the geometric moment theory, see [33], and he produced best upper bounds, leading to attained (i.e. sharp Jackson type inequalities), e.g. see [32], [33]. These studies have gone to all possible directions, univariate and multivariate, though in this work we stay only on the univariate approach over infinite domain.

Also, since 1997, this author started the study of the quantitative convergence of neural network operators to the unit, and he has been written since then numerous of articles and books, e.g. see [34], [36], [37], are what we use here.

The wide range of neural network operators over $\mathbb{R}$ being treated by the author all along, are indeed by nature, positive linear operators.

Here, for the first time in the literature, neural network operators in infinite domain are treated as positive linear operators.

So, methods of positive linear operators in infinite domain apply here to our infinite summation defined neural network operators, producing new and interesting pointwise and uniform results.

Of great inspiration always have been [39], [40]. For classic studies on neural network we recommend also, [41]–[46].

For newer work on neural networks we also refer the reader to [47]–[56].

2.2 Basics

Initially we follow [36], pp. 455–460.

Our perturbed hyperbolic tangent activation function here to be used is

$$g_{q,\lambda}(x) := \frac{e^{\lambda x} - qe^{-\lambda x}}{e^{\lambda x} + qe^{-\lambda x}}, \quad \lambda, q > 0, \ x \in \mathbb{R}. \tag{2.1}$$

Here λ is the parameter and q is the deformation coefficient.

For more details read Chapter 18 of [36]: "q-Deformed and λ-Parametrized Hyperbolic Tangent based Banach space Valued Ordinary and Fractional Neural Network Approximation".

The chapters 17 and 18 of [36] motivate our current work.

The proposed "symmetrization method" aims to use half data feed to our multivariate neural networks.

We will employ the following density function

$$M_{q.\lambda}(x) := \frac{1}{4}\left(g_{q,\lambda}(x+1) - g_{q,\lambda}(x-1)\right) > 0, \tag{2.2}$$

$\forall x \in \mathbb{R};\ q, \lambda > 0$.

We have that

$$M_{q,\lambda}(-x) = M_{\frac{1}{q},\lambda}(x), \quad \forall\, x \in \mathbb{R};\ q, \lambda > 0, \tag{2.3}$$

and

$$M_{\frac{1}{q},\lambda}(-x) = M_{q,\lambda}(x), \quad \forall\, x \in \mathbb{R};\ q, \lambda > 0. \tag{2.4}$$

Adding (2.3) and (2.4) we obtain

$$M_{q,\lambda}(-x) + M_{\frac{1}{q},\lambda}(-x) = M_{q,\lambda}(x) + M_{\frac{1}{q},\lambda}(x), \tag{2.5}$$

a key to this work.

So that

$$\Phi(x) := \frac{M_{q,\lambda}(x) + M_{\frac{1}{q},\lambda}(x)}{2} > 0, \tag{2.6}$$

is an even function, symmetric with respect to the y-axis.

By (18.18) of [36], we have

$$\begin{aligned} &M_{q,\lambda}\left(\tfrac{\ln q}{2\lambda}\right) = \tfrac{\tanh(\lambda)}{2}, \\ &\text{and} \\ &M_{\frac{1}{q},\lambda}\left(-\tfrac{\ln q}{2\lambda}\right) = \tfrac{\tanh(\lambda)}{2}, \quad \lambda > 0, \end{aligned} \tag{2.7}$$

sharing the same maximum at symmetric points.

By Theorem 18.1, p. 458 of [36], we have that

$$\begin{aligned} &\sum_{i=-\infty}^{\infty} M_{q,\lambda}(x-i) = 1, \ \ \forall\, x \in \mathbb{R},\ \lambda, q > 0, \\ &\text{and} \\ &\sum_{i=-\infty}^{\infty} M_{\frac{1}{q},\lambda}(x-i) = 1, \ \ \forall\, x \in \mathbb{R},\ \lambda, q > 0. \end{aligned} \tag{2.8}$$

Consequently, we derive that

$$\sum_{i=-\infty}^{\infty} \Phi(x-i) = 1, \ \ \forall\, x \in \mathbb{R}. \tag{2.9}$$

By Theorem 18.2, p. 459 of [36], we have that

$$\begin{aligned} &\int_{-\infty}^{\infty} M_{q,\lambda}(x)\, dx = 1, \\ &\text{and} \\ &\int_{-\infty}^{\infty} M_{\frac{1}{q},\lambda}(x)\, dx = 1, \end{aligned} \tag{2.10}$$

so that

$$\int_{-\infty}^{\infty} \Phi(x)\, dx = 1, \tag{2.11}$$

therefore Φ is a density function.

By Theorem 18.3, p. 459 of [36], we have:

Let $0 < \alpha < 1$, and $n \in \mathbb{N}$ with $n^{1-\alpha} > 2$; $q, \lambda > 0$. Then

$$\sum_{\begin{cases} k = -\infty \\ : |nx-k| \geq n^{1-\alpha} \end{cases}}^{\infty} M_{q,\lambda}(nx-k) < 2\max\left\{q, \frac{1}{q}\right\} e^{4\lambda} e^{-2\lambda n^{(1-\alpha)}} = Te^{-2\lambda n^{(1-\alpha)}}, \tag{2.12}$$

where $T := 2\max\left\{q, \frac{1}{q}\right\} e^{4\lambda}$.

Similarly, we get that

$$\sum_{k=-\infty}^{\infty} M_{\frac{1}{q},\lambda}(nx-k) < Te^{-2\lambda n^{(1-\alpha)}}. \tag{2.13}$$

Consequently we obtain that

$$\sum_{\begin{cases} k=-\infty \\ : |nx-k| \geq n^{1-\alpha} \end{cases}}^{\infty} \Phi(nx-k) < Te^{-2\lambda n^{(1-\alpha)}}, \tag{2.14}$$

where $T := 2\max\left\{q, \frac{1}{q}\right\} e^{4\lambda}$.

An essential property follows:

Theorem 2.1. *([58]) Let* $0 < \alpha < 1$, $n \in \mathbb{N} : n^{1-\alpha} > 2$. *Then*

$$\int_{\{u\in\mathbb{R}:|nx-u|\geq n^{1-\alpha}\}} \Phi(nx-u)\,du < \frac{\left(q+\frac{1}{q}\right)}{e^{2\lambda(n^{1-\alpha}-1)}},\quad q,\lambda > 0. \tag{2.15}$$

In particular, by [58], we have that

$$\Phi(x) < \left(q+\frac{1}{q}\right)\lambda e^{-2\lambda(x-1)},\quad \forall\, x \geq 1. \tag{2.16}$$

We need,

Definition 2.1. In this chapter we study in an alternative way the smooth approximation properties of the following interpolation neural network operators acting on $f \in C_B(\mathbb{R})$ (continuous and bounded functions):

(i) the basic ones

$$B_N(f,x) := \sum_{k=-\infty}^{\infty} f\left(\frac{k}{N}\right)\Phi(Nx-k),\quad \forall\, x \in \mathbb{R},\ N \in \mathbb{N}, \tag{2.17}$$

(ii) the Kantorovich type operators

$$C_N(f,x) := \sum_{k=-\infty}^{\infty} \left(N\int_{\frac{k}{N}}^{\frac{k+1}{N}} f(t)\,dt\right)\Phi(Nx-k),\quad \forall\, x \in \mathbb{R},\ N \in \mathbb{N}, \tag{2.18}$$

(iii) let $\theta \in \mathbb{N}$, $w_r \geq 0$, $\sum\limits_{r=0}^{\theta} w_r = 1$, $k \in \mathbb{Z}$, and

$$\delta_{Nk}(f) := \sum_{r=0}^{\theta} w_r f\left(\frac{k}{N} + \frac{r}{N\theta}\right), \tag{2.19}$$

we consider also the quadrature type operators

$$D_N(f, x) := \sum_{k=-\infty}^{\infty} \delta_{Nk}(f)\, \Phi(Nx - k), \ \ \forall\, x \in \mathbb{R},\, N \in \mathbb{N}. \tag{2.20}$$

Notice that $B_N(1) = C_N(1) = D_N(1) = 1$. Clearly B_N, C_N, D_N, $N \in \mathbb{N}$, are positive linear operators over $C_B(\mathbb{R})$.

We will be using the first modulus of continuity:

$$\omega_1(f, \delta) := \sup_{\substack{x,y \in \mathbb{R}: \\ |x-y| \leq \delta}} |f(x) - f(y)|, \ \ \delta > 0, \tag{2.21}$$

where $f \in C(\mathbb{R})$ which is bounded and or uniformly continuous.

2.3 Auxilliary Results

We need the following important results.

Lemma 2.1. *Let here $x \in \mathbb{R}$, $j = 1, 2, n, n+1$, $n \in \mathbb{N}$; $0 < \beta < 1$, and $N \in \mathbb{N}$ with $N^{1-\beta} > 2$; $q, \lambda > 0$. Then*

$$0 < B_N\left(|\cdot - x|^j\right)(x) \leq \frac{1}{N^{\beta j}} + \frac{1}{N^j} \frac{\left(q + \frac{1}{q}\right)}{\lambda^j} 2e^{2\lambda} j! e^{-\lambda\left(N^{1-\beta}-1\right)} < +\infty. \tag{2.22}$$

Clearly, $\lim\limits_{N \to \infty} B_N\left(|\cdot - x|^j\right)(x) = 0$.

Proof. We are basing on [57].

$$B_N\left(|\cdot - x|^j\right)(x) = \sum_{k=-\infty}^{\infty} \Phi(Nx - k) \left|\frac{k}{N} - x\right|^j =$$

$$\sum_{\begin{cases} k = -\infty \\ : \left|x - \frac{k}{N}\right| < \frac{1}{N^\beta} \end{cases}}^{\infty} \Phi(Nx - k) \left|\frac{k}{N} - x\right|^j +$$

$$\sum_{\begin{cases} k=-\infty \\ :\left|x-\frac{k}{N}\right|\geq\frac{1}{N^{\beta}} \end{cases}}^{\infty} \Phi\left(Nx-k\right)\left|\frac{k}{N}-x\right|^{j} \leq$$

$$\frac{1}{N^{\beta j}}+\frac{1}{N^{j}} \sum_{\begin{cases} k=-\infty \\ :\left|Nx-k\right|\geq N^{1-\beta} \end{cases}}^{\infty} \Phi\left(\left|Nx-k\right|\right)\left|Nx-k\right|^{j} \leq \tag{2.23}$$

$$\frac{1}{N^{\beta j}}+\frac{1}{N^{j}}\frac{\left(q+\frac{1}{q}\right)}{\lambda^{j}}2e^{2\lambda}j!e^{-\lambda\left(N^{1-\beta}-1\right)}<+\infty.$$

□

Lemma 2.2. *All as in Lemma 2.1 and* $T:=2\max\left\{q,\frac{1}{q}\right\}e^{4\lambda}$. *Then*

$$0<C_{N}\left(\left|\cdot-x\right|^{j}\right)(x)\leq$$

$$\left(\frac{1}{N}+\frac{1}{N^{\beta}}\right)^{j}+\frac{2^{j-1}}{N^{j}}\left[\frac{T}{e^{2\lambda}N^{1-\beta}}+\frac{\left(q+\frac{1}{q}\right)}{\lambda^{j}}2e^{2\lambda}j!e^{-\lambda\left(N^{1-\beta}-1\right)}\right]<+\infty. \tag{2.24}$$

Clearly, $\lim_{N\to\infty}C_{N}\left(\left|\cdot-x\right|^{j}\right)(x)=0$.

Proof. We are based again on [57].

We have that

$$C_{N}\left(\left|\cdot-x\right|^{j}\right)(x)=\sum_{k=-\infty}^{\infty}\Phi\left(Nx-k\right)\left(N\int_{0}^{\frac{1}{N}}\left|t+\frac{k}{N}-x\right|^{j}dt\right)\leq$$

$$\sum_{k=-\infty}^{\infty}\Phi\left(Nx-k\right)\left(N\int_{0}^{\frac{1}{N}}\left(\left|\frac{k}{N}-x\right|+\left|t\right|\right)^{j}dt\right)\leq$$

$$\sum_{k=-\infty}^{\infty}\Phi\left(Nx-k\right)\left(\left|\frac{k}{N}-x\right|+\frac{1}{N}\right)^{j}=$$

$$\sum_{\begin{cases} k=-\infty \\ :\left|\frac{k}{N}-x\right|<\frac{1}{N^{\beta}} \end{cases}}^{\infty} \Phi\left(Nx-k\right)\left(\left|\frac{k}{N}-x\right|+\frac{1}{N}\right)^{j}+$$

$$\sum_{\left\{\begin{array}{c} k=-\infty \\ : \left|\frac{k}{N}-x\right| \geq \frac{1}{N^{\beta}} \end{array}\right.}^{\infty} \Phi\left(Nx-k\right)\left(\left|\frac{k}{N}-x\right|+\frac{1}{N}\right)^{j} \leq \tag{2.25}$$

$$\left(\frac{1}{N^{\beta}}+\frac{1}{N}\right)^{j}+\frac{1}{N^{j}} \sum_{\left\{\begin{array}{c} k=-\infty \\ : |Nx-k| \geq N^{1-\beta} \end{array}\right.}^{\infty} \Phi\left(|Nx-k|\right)\left(|Nx-k|+1\right)^{j} \leq$$

$$\left(\frac{1}{N^{\beta}}+\frac{1}{N}\right)^{j}+\frac{2^{j-1}}{N^{j}}\left(\sum_{\left\{\begin{array}{c} k=-\infty \\ : |Nx-k| \geq N^{1-\beta} \end{array}\right.}^{\infty} \Phi\left(|Nx-k|\right)\left(1+|Nx-k|^{j}\right)\right)$$

$$\leq\left(\frac{1}{N}+\frac{1}{N^{\beta}}\right)^{j}+\frac{2^{j-1}}{N^{j}}\left[\frac{T}{e^{2\lambda N^{1-\beta}}}+\frac{\left(q+\frac{1}{q}\right)}{\lambda^{j}} 2e^{2\lambda} j! e^{-\lambda\left(N^{1-\beta}-1\right)}\right]<+\infty. \tag{2.26}$$

□

Lemma 2.3. *All here are as in Lemma 2.2. Then*

$$0<D_{N}\left(|\cdot-x|^{j}\right)(x) \leq$$

$$\left(\frac{1}{N}+\frac{1}{N^{\beta}}\right)^{j}+\frac{2^{j-1}}{N^{j}}\left[\frac{T}{e^{2\lambda} N^{1-\beta}}+\frac{\left(q+\frac{1}{q}\right)}{\lambda^{j}} 2e^{2\lambda} j! e^{-\lambda\left(N^{1-\beta}-1\right)}\right]<+\infty. \tag{2.27}$$

Clearly, $\lim_{N\to\infty} D_{N}\left(|\cdot-x|^{j}\right)(x)=0.$

Proof. As similar to Lemma 2.2 is omitted, it is based on [57]. □

We need the following Hölder's type inequality for positive linear operators.

Theorem 2.2. *Let L be a positive linear operator from $C(\mathbb{R})$ into $C_B(\mathbb{R})$, and $f, g \in C(\mathbb{R})$, furthermore let $p, q>1: \frac{1}{p}+\frac{1}{q}=1$. Assume that $L\left(\left(|f(\cdot)|^{p}\right)\right)(s_{*})$, $L\left(\left(|g(\cdot)|^{q}\right)\right)(s_{*})>0$ for some $s_{*} \in \mathbb{R}$. Then*

$$L\left(|f(\cdot) g(\cdot)|\right)(s_{*}) \leq\left(L\left(\left(|f(\cdot)^{p}|\right)\right)(s_{*})\right)^{\frac{1}{p}}\left(L\left(\left(|g(\cdot)^{q}|\right)\right)(s_{*})\right)^{\frac{1}{q}}. \tag{2.28}$$

Proof. Let $a, b \geq 0$, $p.q > 1 : \frac{1}{p} + \frac{1}{q} = 1$. The Young's inequality says

$$ab \leq \frac{a^p}{p} + \frac{b^q}{q}. \tag{2.29}$$

Then

$$\frac{|f(s)|}{\left(L\left(\left(|f(\cdot)^p|\right)\right)(s_*)\right)^{\frac{1}{p}}} \cdot \frac{|g(s)|}{\left(L\left(\left(|g(\cdot)^q|\right)\right)(s_*)\right)^{\frac{1}{q}}} \leq$$

$$\frac{|f(s)|^p}{p\left(L\left(\left(|f(\cdot)^p|\right)\right)(s_*)\right)} + \frac{|g(s)|^q}{q\left(L\left(\left(|g(\cdot)^q|\right)\right)(s_*)\right)}, \quad \forall\, s \in \mathbb{R}. \tag{2.30}$$

Hence it holds

$$\frac{L\left(|f(\cdot)|\,|g(\cdot)|\right)(s_*)}{\left(L\left(\left(|f(\cdot)^p|\right)\right)(s_*)\right)^{\frac{1}{p}}\left(L\left(\left(|g(\cdot)^q|\right)\right)(s_*)\right)^{\frac{1}{q}}} \leq$$

$$\frac{L\left(\left(|f(\cdot)^p|\right)\right)(s_*)}{p\left(L\left(\left(|f(\cdot)^p|\right)\right)(s_*)\right)} + \frac{L\left(\left(|g(\cdot)^q|\right)\right)(s_*)}{q\left(L\left(\left(|g(\cdot)^q|\right)\right)(s_*)\right)} = \frac{1}{p} + \frac{1}{q} = 1, \tag{2.31}$$

for $s_* \in \mathbb{R}$, proving the claim. □

2.4 Main Results

We present the following general approximation result.

Theorem 2.3. *Let $n \in \mathbb{N}$ and $f, f^{(n)} \in C_B(\mathbb{R})$, $x \in \mathbb{R}$. Consider L_N a sequence of positive linear operators from $C_B(\mathbb{R})$ into itself, $N \in \mathbb{N}$, such that $L_N(1) = 1$. Assume $L_N\left(|\cdot - x|^{n+1}\right)(x) > 0$, and $f^{(i)}(x) = 0$, for $i = 1, ..., n$.*

Then

$$|L_N(f)(x) - f(x)| \leq \omega_1\left(f^{(n)}, \left(\left(L_N\left(|\cdot - x|^{n+1}\right)\right)(x)\right)^{\frac{1}{n+1}}\right)$$

$$\left[L_N\left(|\cdot - x|^n\right)(x) + \frac{\left(L_N\left(|\cdot - x|^{n+1}\right)(x)\right)^{\frac{n}{n+1}}}{(n+1)}\right] < +\infty, \quad \forall\, N \in \mathbb{N}. \tag{2.32}$$

Given that $\lim_{N \to +\infty} L_N\left(|\cdot - x|^{n+1}\right)(x) = 0$, then $\lim_{N \to +\infty} L_N(f)(x) = f(x)$.

If $n = 1$, we derive

$$|L_N(f)(x) - f(x)| \leq \omega_1\left(f', \left(\left(L_N\left((\cdot - x)^2\right)\right)(x)\right)^{\frac{1}{2}}\right)$$

$$\left[L_N \left(|\cdot - x| \right) (x) + \frac{\left(L_N \left((\cdot - x)^2 \right) (x) \right)^{\frac{1}{2}}}{2} \right] < +\infty, \quad \forall\, N \in \mathbb{N}. \tag{2.33}$$

Given that $\lim_{N \to +\infty} L_N \left((\cdot - x)^2 \right) (x) = 0$, *then* $\lim_{N \to +\infty} L_N (f) (x) = f (x)$.

Proof. Let $x, y \in \mathbb{R}$, by Taylor's formula we have

$$f (y) = \sum_{i=0}^{n} f^{(i)} (x) \frac{(y - x)^i}{i!} + \frac{1}{(n-1)!} \int_x^y (y - t)^{n-1} \left(f^{(n)} (t) - f^{(n)} (x) \right) dt. \tag{2.34}$$

We call the remainder in (2.34) as

$$R_n (x, y) = \frac{1}{(n-1)!} \int_x^y (y - t)^{n-1} \left(f^{(n)} (t) - f^{(n)} (x) \right) dt, \quad \forall\, x, y \in \mathbb{R}. \tag{2.35}$$

By [33], p. 217, and [35], p. 194, Chapter 7, (7.27) there, we get

$$|R_n (x, y)| \leq \frac{\omega_1 \left(f^{(n)}, \delta \right)}{n!} |x - y|^n \left(1 + \frac{|x - y|}{(n+1)\, \delta} \right), \quad \forall\, x, y \in \mathbb{R},\ \delta > 0. \tag{2.36}$$

We may write (2.36) as

$$|R_n (x, y)| \leq \frac{\omega_1 \left(f^{(n)}, \delta \right)}{n!} \left[|x - y|^n + \frac{|x - y|^{n+1}}{(n+1)\, \delta} \right], \quad \forall\, x, y \in \mathbb{R},\ \delta > 0. \tag{2.37}$$

That is

$$\left| f (y) - \sum_{i=0}^{n} f^{(i)} (x) \frac{(y - x)^i}{i!} \right| \leq \frac{\omega_1 \left(f^{(n)}, \delta \right)}{n!} \left[|x - y|^n + \frac{|x - y|^{n+1}}{(n+1)\, \delta} \right], \tag{2.38}$$

$\forall\, x, y \in \mathbb{R}$, $\delta > 0$.

Furthermore it holds

$$|f (y) - f (x)| \leq \sum_{i=0}^{n} \left| f^{(i)} (x) \right| \frac{|y - x|^i}{i!} + \frac{\omega_1 \left(f^{(n)}, \delta \right)}{n!} \left[|x - y|^n + \frac{|x - y|^{n+1}}{(n+1)\, \delta} \right], \tag{2.39}$$

$\forall\, x, y \in \mathbb{R}$, $\delta > 0$.

Here $f^{(i)} (x) = 0$, for $i = 1, ..., n$; and for a specific $x \in \mathbb{R}$, we get

$$|f (y) - f (x)| \leq \frac{\omega_1 \left(f^{(n)}, \delta \right)}{n!} \left[|x - y|^n + \frac{|x - y|^{n+1}}{(n+1)\, \delta} \right], \quad \forall y \in \mathbb{R},\ \delta > 0. \tag{2.40}$$

In case of $n = 1$, we derive

$$|f(y) - f(x)| \leq |f'(x)| |y - x| + \omega_1(f', \delta) \left[|x - y| + \frac{(x - y)^2}{2\delta} \right], \quad (2.41)$$

$\forall\, x, y \in \mathbb{R}$, $\delta > 0$.

If $n = 1$ and $f'(x) = 0$, for a specific $x \in \mathbb{R}$, we get that

$$|f(y) - f(x)| \leq \omega_1(f', \delta) \left[|x - y| + \frac{(x - y)^2}{2\delta} \right], \quad \forall y \in \mathbb{R},\ \delta > 0. \quad (2.42)$$

Let $L_N : C_B(\mathbb{R}) \to C_B(\mathbb{R})$ a sequence of positive linear operators, $N \in \mathbb{N}$.

Let $f, g \in C_B(\mathbb{R})$, we have

$$f = f - g + g \leq |f - g| + g,$$

hence

$$L_N(f) \leq L_N(|f - g|) + L_N(g), \quad (2.43)$$

and

$$L_N(f) - L_N(g) \leq L_N(|f - g|). \quad (2.44)$$

Similarly, it holds

$$L_N(g) - L_N(f) \leq L_N(|f - g|). \quad (2.45)$$

That is

$$|L_N(f) - L_N(g)| \leq L_N(|f - g|). \quad (2.46)$$

Next, we assume that $L_N(1) = 1$, $\forall\, N \in \mathbb{N}$. Then

$$|L_N(f)(x) - f(x)| = |L_N(f)(x) - f(x) L_N(1)(x)| =$$

$$|L_N(f)(x) - L_N(f(x))(x)| = |L_N(f(\cdot) - f(x))(x)| \overset{(2.46)}{\leq}$$

$$L_N(|f(\cdot) - f(x)|)(x) \overset{(2.40)}{\leq} \quad (2.47)$$

$$\frac{\omega_1(f^{(n)}, \delta)}{n!} \left[L_N(|\cdot - x|^n)(x) + \frac{L_N\left(|\cdot - x|^{n+1}\right)(x)}{(n+1)\delta} \right],$$

$\forall\, x \in \mathbb{R}$, $\delta > 0$.

Here we assume and choose

$$\delta = \left(L_N \left(|\cdot - x|^{n+1} \right) (x) \right)^{\frac{1}{n+1}} > 0. \tag{2.48}$$

Therefore we get

$$|L_N (f)(x) - f(x)| \leq \omega_1 \left(f^{(n)}, \left(L_N \left(|\cdot - x|^{n+1} \right) (x) \right)^{\frac{1}{n+1}} \right)$$

$$\left[L_N \left(|\cdot - x|^n \right) (x) + \frac{\left(L_N \left(|\cdot - x|^{n+1} \right) (x) \right)^{\frac{n}{n+1}}}{(n+1)} \right], \tag{2.49}$$

$\forall\, N \in \mathbb{N}$.

If $n = 1$, we find that

$$|L_N (f)(x) - f(x)| \leq \omega_1 \left(f', \left(L_N \left((\cdot - x)^2 \right) (x) \right)^{\frac{1}{2}} \right)$$

$$\left[L_N \left(|\cdot - x| \right) (x) + \frac{\left(L_N \left((\cdot - x)^2 \right) (x) \right)^{\frac{1}{2}}}{2} \right], \quad \forall\, N \in \mathbb{N}. \tag{2.50}$$

Using (2.48) and Theorem 2.2, for $g = 1$ and L_N such that $L_N(1) = 1$, we obtain

$$L_N \left(|\cdot - x|^n \right) (x) \leq \left(L_N \left(|\cdot - x|^{n+1} \right) (x) \right)^{\frac{n}{n+1}}. \tag{2.51}$$

In case of $n = 1$, we derive

$$L_N \left(|\cdot - x| \right) (x) \leq \sqrt{\left(L_N \left((\cdot - x)^2 \right) (x) \right)}. \tag{2.52}$$

□

An application of Theorem 2.3 follows.

Call Z_N any of the operators B_N, C_N, D_N, $N \in \mathbb{N}$.

Theorem 2.4. *Let $n \in \mathbb{N}$ and $f, f^{(n)} \in C_B(\mathbb{R})$, $x \in \mathbb{R}$. Assume $f^{(i)}(x) = 0$, $i = 1, ..., n$. Then*

$$|Z_N (f)(x) - f(x)| \leq \omega_1 \left(f^{(n)}, \left(\left(Z_N \left(|\cdot - x|^{n+1} \right) \right) (x) \right)^{\frac{1}{n+1}} \right)$$

$$\left[Z_N \left(|\cdot - x|^n \right) (x) + \frac{\left(Z_N \left(|\cdot - x|^{n+1} \right) (x) \right)^{\frac{n}{n+1}}}{(n+1)} \right] < +\infty, \quad \forall\, N \in \mathbb{N}. \tag{2.53}$$

Here $\lim_{N \to +\infty} Z_N (f)(x) = f(x)$.

If $n = 1$, we obtain

$$\left|Z_N(f)(x) - f(x)\right| \leq \omega_1\left(f', \left(\left(Z_N\left((\cdot - x)^2\right)\right)(x)\right)^{\frac{1}{2}}\right)$$

$$\left[Z_N(|\cdot - x|)(x) + \frac{\left(Z_N\left((\cdot - x)^2\right)(x)\right)^{\frac{1}{2}}}{2}\right] < +\infty, \quad \forall\, N \in \mathbb{N}. \tag{2.54}$$

Again it holds $\lim_{N \to +\infty} Z_N(f)(x) = f(x)$.

Proof. By Theorem 2.3 and Lemmas 2.1–2.3. □

We mention the following result.

Theorem 2.5. *([38]) Let $f \in C_B(\mathbb{R})$, $x \in \mathbb{R}$. Consider L_N a sequence of positive linear operators from $C_B(\mathbb{R})$ into itself, $N \in \mathbb{N}$, such that $L_N(1) = 1$.*

Assume that $L_N(|\cdot - x|)(x) > 0$. Then

$$\left|L_N(f)(x) - f(x)\right| \leq 2\omega_1\left(f, L_N(|\cdot - x|)(x)\right), \quad \forall\, N \in \mathbb{N}. \tag{2.55}$$

Given that $\lim_{N \to +\infty} L_N(|\cdot - x|)(x) = 0$, and f is also uniformly continuous, we obtain $\lim_{N \to +\infty} L_N(f)(x) = f(x)$.

An application of Theorem 2.5 comes next.

Theorem 2.6. *Let $f \in C_B(\mathbb{R})$, $x \in \mathbb{R}$. Here $Z_N = B_N, C_N, D_N$. Then*

$$\left|Z_N(f)(x) - f(x)\right| \leq 2\omega_1\left(f, Z_N(|\cdot - x|)(x)\right) < +\infty, \tag{2.56}$$

$\forall\, N \in \mathbb{N} : N^{1-\beta} > 2$, $0 < \beta < 1$.

If f is also uniformly continuous we derive $\lim_{N \to +\infty} Z_N(f)(x) = f(x)$.

Proof. Direct application of Theorem 2.5 and Lemmas 2.1–2.3. □

Corollary 2.1. *Let $f \in C_B(\mathbb{R})$. Then*

$$\left\|B_N(f) - f\right\|_\infty \leq 2\omega_1\left(f, \left(\frac{1}{N^\beta} + \frac{1}{N}\frac{\left(q + \frac{1}{q}\right)}{\lambda} 2e^{2\lambda} e^{-\lambda\left(N^{1-\beta} - 1\right)}\right)\right) < +\infty, \tag{2.57}$$

$\forall\, N \in \mathbb{N} : N^{1-\beta} > 2$, $0 < \beta < 1$; $q, \lambda > 0$.

If f is also uniformly continuous, we derive that $\lim_{N \to +\infty} B_N(f) = f$, pointwise and uniformly.

Proof. By Lemma 2.1 and Theorem 2.6. □

We finish with

Corollary 2.2. *Let* $f \in C_B(\mathbb{R})$. *Then*

$$\left\{ \begin{array}{c} \|C_N(f) - f\|_\infty \\ \|D_N(f) - f\|_\infty \end{array} \right\} \leq$$

$$2\omega_1\left(f, \left[\left(\frac{1}{N} + \frac{1}{N^\beta}\right) + \frac{1}{N}\left(\frac{T}{e^{2\lambda N^{1-\beta}}} \frac{\left(q + \frac{1}{q}\right)}{\lambda} 2e^{2\lambda} e^{-\lambda\left(N^{1-\beta}-1\right)}\right)\right]\right) < +\infty, \tag{2.58}$$

$\forall\, N \in \mathbb{N} : N^{1-\beta} > 2,\ 0 < \beta < 1;\ q, \lambda > 0,$ *and* $T = 2\max\left\{q, \frac{1}{q}\right\} e^{4\lambda}$.

If f *is also uniformly continuous, we obtain that*

$$\lim_{N \to +\infty} C_N(f) = \lim_{N \to +\infty} D_N(f) = f,$$

pointwise and uniformly.

Proof. By Lemmas 2.2, 2.3 and Theorem 2.6. □

Bibliography

[32] G.A. Anastassiou, *A "K-Attainable" inequality related to the convergence of Positive Linear Operators*, J. Approximation Theory, 44, 380–383 (1985).

[33] G.A. Anastassiou, *Moments in probability and approximation theory*, Pitman Research Notes in Mathematics series / Longman Scientific & Technical, Essex, New York, UK, USA, 1993.

[34] G.A. Anastassiou, *Quantitative Approximations*, Chapmen & Hall/CRC, London, New York, 2001.

[35] G.A. Anastassiou, *Intelligent Computations: Abstract Fractional Calculus, Inequalities, Approximations*, Springer, Heidelberg, New York, 2018.

[36] G.A. Anastassiou, *Parametrized, Deformed and General Neural Networks*, Springer, Heidelberg, New York, 2023.

[37] G.A. Anastassiou, *Trigonometric and Hyperbolic Generated Approximation Theory*, World Scientific, Singapore, New York, 2025.

[38] R.A. DeVore, *The Approximation of Continuous Functions by Positive Linear Operators*, Lecture Notes in Mathematics, Vol. 293, Springer, Heidelberg, New York, 1972.

[39] O. Shisha and B. Mond, *The degree of convergence of sequences of linear positive operators*, Nat. Acad. Sci. U.S. 60 (1968), 1196–1200.

[40] O. Shisha and B. Mond, *The degree of Approximation to Periodic Functions by Linear Positive Operators*, J. Approximation Theory, 1, 335–339 (1968).

[41] Z. Chen and F. Cao, *The approximation operators with sigmoidal functions*, Computers and Mathematics with Applications, 58 (2009), 758–765.

[42] D. Costarelli, R. Spigler, *Approximation results for neural network operators activated by sigmoidal functions*, Neural Networks 44 (2013), 101–106.

[43] D. Costarelli, R. Spigler, *Multivariate neural network operators with sigmoidal activation functions*, Neural Networks 48 (2013), 72–77.

[44] S. Haykin, *Neural Networks: A Comprehensive Foundation* (2 ed.), Prentice Hall, New York, 1998.

[45] W. McCulloch and W. Pitts, *A logical calculus of the ideas immanent in nervous activity*, Bulletin of Mathematical Biophysics, 7 (1943), 115–133.

[46] T.M. Mitchell, *Machine Learning*, WCB-McGraw-Hill, New York, 1997.

[47] Yu, Dansheng; Cao, Feilong, *Construction and approximation rate for feed-forward neural network operators with sigmoidal functions*, J. Comput. Appl. Math. 453 (2025), paper No. 116150.

[48] Cen, Siyu; Jin, Bangti; Quan, Qimeng; Zhou, Zhi; *Hybrid neural-network FEM approximation of diffusion coeficient in elyptic and parabolic problems.* IMA J. Numer. Anal. 44 (2024), no. 5, 3059–3093.

[49] Coroianu Lucian; Costarelli, Danillo; Natale, Mariarosaria; Pantiş, Alexandra; *The approximation capabilities of Durrmeyer-type neural network operators.* J. Appl. Math. Comput. 70 (2024), no. 5, 4581–4599.

[50] Warin, Xavier; *The GroupMax neural network approximation of convex functions.* IEEE Trans. Neural Netw. Learn. Syst. 35 (2024), no. 8, 11608–11612.

[51] Fabra, Arnau; Guasch, Oriol; Baiges, Joan; Codina, Ramon; *Approximation of acoustic black holes with finite element mixed formulations and artificial neural network correction terms.* Finite Elem. Anal. Des. 241 (2024), paper No. 104236.

[52] Grohs, Philipp; Voigtlaender, Felix. *Proof of the theory-to-practice gap in deep learning via sampling complexity bounds for neural network*

approximation spaces. Found. Comput. Math. 24 (2024), no. 4, 1085–1143.

[53] Basteri, Andrea; Trevisan, Dario; *Quantitative Gaussian approximation of randomly initialized deep neural networks.* Mach. Learn. 113 (2024), no. 9, 6373–6393.

[54] De Ryck, T.; Mishra, S. *Error analysis for deep neural network approximations of parametric hyperbolic conservation laws.* Math. Comp. 93 (2024), no. 350, 2643–2677.

[55] Liu, Jie; Zhang, Baoji; Lai, Yuyang; Fang, Liqiau. *Hull form optimization reserach based on multi-precision back-propagation neural network approximation model.* Internal. J. Numer. Methods Fluid 96 (2024), no. 8, 1445–1460.

[56] Yoo, Jihahm; Kim, Jaywon; Gim, Minjung; Lee, Haesung. *Error estimates of physics-informed neural networks for initial value problems.* J. Korean Soc. Ind. Appl. Math. 28 (2024), no. 1, 33–58.

[57] G.A. Anastassiou, *Smooth symmetrized and perturbed hyperbolic tangent real and complex, ordinary and fractional neural network approximations over infinite domains*, Annals of Communications in Mathematics, 7(4)(2024), 401–429.

[58] G.A. Anastassiou, *Approximation by symmetrized and perturbed hyperbolic tangent activated convolution type operators*, Mathematics, 2024, 12, 3302; `https://doi.org/10.3390/math12203302`.

[59] G.A. Anastassiou, *Neural Networks in infinite domain as Positive Linear Operators*, Annales Univ. Sci. Budapest, Sect. Comp., accepted, 2025.

Chapter 3

Generalized Logistic Neural Networks Treated as Positive Linear Operators

Essential neural network operators are interpreted as positive linear operators and the related general theory applies to them. These operators are induced by a symmetrized density function deriving from the parametrized and deformed A-generalized logistic activation function. We are acting on the space of continuous functions on a compact interval of real line to the reals. We study quantitatively the rate of convergence of these neural network operators to the unit operator. Our inequalities involve the modulus of continuity of the function under approximation or its derivative. We produce uniform and L_p, $p \geq 1$, approximation results via these inequalities. Convexity of functions is also used to derive more refined results. It follows [91].

3.1 Introduction

The author studied extensively the quantitative approximation of positive linear operators to the unit since 1985, see for example [60]–[62], [64], which we use in this work. He originated from the quantitative weak convergence of finite positive measures to the unit Dirac measure, having as a method the geometric moment theory, see [61], and he produced best upper bounds, leading to attained (i.e. sharp Jackson type inequalities), e.g. see [60], [61]. These studies have gone to all possible directions, univariate and multivariate, though in this work we stay only on the univariate approach.

Also, since 1997, this author started the study of the quantitative convergence of neural network operators to the unit, and he has been written since then numerous of articles and books, e.g. see [62], [63], [64], are what we use here.

The wide range of neural network operators being treated by the author all along, are indeed by nature positive linear operators.

Here, the author continues his study of treating neural network operators as positive linear operators. This is a totally new approach in the literature, see [90].

Different activation functions allow for different non-linearities which might work better for solving a specific problem. So the need to use neural networks with various activation functions is vivid. Thus, performing neural network approximations using different activation functions is not only necessary but fully justified. Furthermore, the brain non-symmetry has been observed in animals and humans in terms of structure, function and behavior. This lateralization is thought to reflect evolutionary, hereditary, developmental, experiential and pathological factors. Consequently, it is natural to consider for our study deformed neural network activation functions and operators. Thus, this work is a specific study under this philosophy of approaching reality as closely as possible. The author is working currently on the long project of connecting the neural network operators to positive linear operator theory. Some of the most popular activation functions derive from the generalized logistic function-which this chapter is based on, and the hyperbolic tangent which [90] is based on. Both cases have different calculations and produce different results of great interest and with important applications in my other papers.

So, all methods of positive linear operators apply here to our summation defined neural network operators, producing new and interesting results: pointwise, uniform and L_p, $p \geq 1$, kinds. Via the Riesz representation theorem neural networks are connected to measure theory. Convexity of functions produces optimal and sharper results.

The use of A-generalized logistic function as an activation function is well established and supported in [63], Chapter 16. The A-generalized logistic activation function is greatly behaving and one of the most commonly used activation functions on this subject.

The author's symmetrization method presented here aims that our operators converge at high speed to the unit operator using half feed of data.

This chapter establishes again a bridge connecting Neural Networks Approximation (part of AI) to Positive Linear Operators (a great part of Functional Analysis).

So it is a theoretical work intended by nature mainly to related experts.

Of great inspiration always have been [66], [67]. For classical studies on neural network we recommend also, [68]–[73].

For newer work on neural networks we also refer the reader to [74]–[83]. For recent studies in positive linear operators we refer to [84]–[90].

3.2 Basics

Here we follow [63], pp. 395–417.

Our activation function here to be used is the q-deformed and λ-parametrized function

$$\varphi_{q,\lambda}(x) = \frac{1}{1+qA^{-\lambda x}}, \quad x \in \mathbb{R}, \; q, \lambda > 0, \; A > 1. \tag{3.1}$$

This is the A-generalized logistic function.

For more read Chapter 16 of [63]: "Banach space valued ordinary and fractional neural network approximation based on q-deformed and λ-parametrized A-generalized logistic function".

This chapter motivates our current work.

The proposed "symmetrization technique" aims to use half data feed to our neural networks.

We will employ the following density function

$$G_{q,\lambda}(x) := \frac{1}{2}\left(\varphi_{q,\lambda}(x+1) - \varphi_{q,\lambda}(x-1)\right), \quad x \in \mathbb{R}, \; q, \lambda > 0. \tag{3.2}$$

We have that

$$G_{q,\lambda}(-x) = G_{\frac{1}{q},\lambda}(x), \tag{3.3}$$

and

$$G_{\frac{1}{q},\lambda}(-x) = G_{q,\lambda}(x), \quad \forall\, x \in \mathbb{R}. \tag{3.4}$$

Adding (3.3) and (3.4) we obtain

$$G_{q,\lambda}(-x) + G_{\frac{1}{q},\lambda}(-x) = G_{q,\lambda}(x) + G_{\frac{1}{q},\lambda}(x), \quad \forall\, x \in \mathbb{R}, \tag{3.5}$$

the key to this work.

So that

$$W(x) := \frac{G_{q,\lambda}(x) + G_{\frac{1}{q},\lambda}(x)}{2} \tag{3.6}$$

is an even function, symmetric with respect to the y-axis.

The global maximum of $G_{q,\lambda}$ is given by (16.18), p. 401 of [63] as

$$G_{q,\lambda}\left(\frac{\log_A q}{\lambda}\right)=\frac{A^\lambda-1}{2\left(A^\lambda+1\right)}. \tag{3.7}$$

And, the global max of $G_{\frac{1}{q},\lambda}$ is

$$G_{\frac{1}{q},\lambda}\left(\frac{\log_A \frac{1}{q}}{\lambda}\right)=G_{\frac{1}{q},\lambda}\left(\frac{-\log_A q}{\lambda}\right)=\frac{A^\lambda-1}{2\left(A^\lambda+1\right)}, \tag{3.8}$$

both sharing the same maximum at symmetric points.

By Theorem 16.1, p. 401 of [63], we have that

$$\sum_{i=-\infty}^{\infty} G_{q,\lambda}(x-i)=1,\ \ \forall\, x\in\mathbb{R},\ \lambda,q>0,\ A>1, \tag{3.9}$$

and

$$\sum_{i=-\infty}^{\infty} G_{\frac{1}{q},\lambda}(x-i)=1,\ \ \forall\, x\in\mathbb{R},\ \lambda,q>0,\ A>1. \tag{3.10}$$

Consequently, we derive that

$$\sum_{i=-\infty}^{\infty} W(x-i)=1,\ \ \forall\, x\in\mathbb{R}. \tag{3.11}$$

By Theorem 16.2, p. 402 of [63], we have that

$$\int_{-\infty}^{\infty} G_{q,\lambda}(x)\,dx=1,\ \ \lambda,q>0,\ A>1, \tag{3.12}$$

similarly it holds

$$\int_{-\infty}^{\infty} G_{\frac{1}{q},\lambda}(x)\,dx=1, \tag{3.13}$$

so that

$$\int_{-\infty}^{\infty} W(x)\,dx=1, \tag{3.14}$$

therefore W is a density function.

By Theorem 16.3, p. 402 of [63], we have:

Let $0<\alpha<1$, and $n\in\mathbb{N}$ with $n^{1-\alpha}>2$. Then

$$\sum_{\substack{k=-\infty \\ :\,|nx-k|\geq n^{1-\alpha}}}^{\infty} G_{q,\lambda}(nx-k)<2\max\left\{q,\frac{1}{q}\right\}\frac{1}{A^{\lambda\left(n^{1-\alpha}-2\right)}}=\gamma A^{-\lambda\left(n^{1-\alpha}-2\right)}, \tag{3.15}$$

where $\lambda,q>0,\ A>1;\ \gamma:=2\max\left\{q,\frac{1}{q}\right\}$.

Similarly, we get that

$$\sum_{\substack{k=-\infty \\ : |nx-k| \geq n^{1-\alpha}}}^{\infty} G_{\frac{1}{q},\lambda}(nx-k) < \gamma A^{-\lambda\left(n^{1-\alpha}-2\right)}. \tag{3.16}$$

Consequently we obtain that

$$\sum_{\substack{k=-\infty \\ : |nx-k| \geq n^{1-\alpha}}}^{\infty} W(nx-k) < \gamma A^{-\lambda\left(n^{1-\alpha}-2\right)}, \tag{3.17}$$

where $\gamma := 2\max\left\{q, \frac{1}{q}\right\}$.

Here $\lceil\cdot\rceil$ denotes the ceiling of the number, and $\lfloor\cdot\rfloor$ its integral part.

We mention

Theorem 16.4 (p. 402, [63]) Let $x \in [a,b] \subset \mathbb{R}$ and $n \in \mathbb{N}$ so that $\lceil na\rceil \leq \lfloor nb\rfloor$. For $q > 0$, $\lambda > 0$, $A > 1$, we consider the number $\lambda_q > z_0 > 0$ with $G_{q,\lambda}(z_0) = G_{q,\lambda}(0)$, and $\lambda_q > 1$. Then

$$\frac{1}{\sum_{k=\lceil na\rceil}^{\lfloor nb\rfloor} G_{q,\lambda}(nx-k)} < \max\left\{\frac{1}{G_{q,\lambda}(\lambda_q)}, \frac{1}{G_{\frac{1}{q},\lambda}\left(\lambda_{\frac{1}{q}}\right)}\right\} =: K(q). \tag{3.18}$$

Similarly, we consider $\lambda_{\frac{1}{q}} > z_1 > 0$, such that $G_{\frac{1}{q},\lambda}(z_1) = G_{\frac{1}{q},\lambda}(0)$, and $\lambda_{\frac{1}{q}} > 1$. Thus

$$\frac{1}{\sum_{k=\lceil na\rceil}^{\lfloor nb\rfloor} G_{\frac{1}{q},\lambda}(nx-k)} < \max\left\{\frac{1}{G_{\frac{1}{q},\lambda}\left(\lambda_{\frac{1}{q}}\right)}, \frac{1}{G_{q,\lambda}(\lambda_q)}\right\} = K(q). \tag{3.19}$$

Hence

$$\sum_{k=\lceil na\rceil}^{\lfloor nb\rfloor} G_{q,\lambda}(nx-k) > \frac{1}{K(q)}, \tag{3.20}$$

and

$$\sum_{k=\lceil na\rceil}^{\lfloor nb\rfloor} G_{\frac{1}{q},\lambda}(nx-k) > \frac{1}{K(q)}. \tag{3.21}$$

Consequently, it holds

$$\sum_{k=\lceil na \rceil}^{\lfloor nb \rfloor} \frac{\left(G_{q,\lambda}(nx-k) + G_{\frac{1}{q},\lambda}(nx-k)\right)}{2} > \frac{2}{2K(q)} = \frac{1}{K(q)}, \tag{3.22}$$

so that

$$\frac{1}{\sum_{k=\lceil na \rceil}^{\lfloor nb \rfloor} \frac{\left(G_{q,\lambda}(nx-k)+G_{\frac{1}{q},\lambda}(nx-k)\right)}{2}} < K(q), \tag{3.23}$$

that is

$$\frac{1}{\sum_{k=\lceil na \rceil}^{\lfloor nb \rfloor} W(nx-k)} < K(q). \tag{3.24}$$

We have proved

Theorem 3.1. *Let $x \in [a,b] \subset \mathbb{R}$ and $n \in \mathbb{N}$ so that $\lceil na \rceil \leq \lfloor nb \rfloor$. For $q, \lambda > 0$, $A > 1$, we consider $\lambda_q > z_0 > 0$ with $G_{q,\lambda}(z_0) = G_{q,\lambda}(0)$, and $\lambda_q > 1$. Also consider $\lambda_{\frac{1}{q}} > z_1 > 0$, such that $G_{\frac{1}{q},\lambda}(z_1) = G_{\frac{1}{q},\lambda}(0)$, and $\lambda_{\frac{1}{q}} > 1$. Then*

$$\frac{1}{\sum_{k=\lceil na \rceil}^{\lfloor nb \rfloor} W(nx-k)} < K(q). \tag{3.25}$$

We make

Remark 3.1.

I) By Remark 16.5, p. 402 of [63], we have

$$\lim_{n\to\infty} \sum_{k=\lceil na \rceil}^{\lfloor nb \rfloor} G_{q,\lambda}(nx_1-k) \neq 1, \text{ for some } x_1 \in [a,b], \tag{3.26}$$

and

$$\lim_{n\to\infty} \sum_{k=\lceil na \rceil}^{\lfloor nb \rfloor} G_{\frac{1}{q},\lambda}(nx_2-k) \neq 1, \text{ for some } x_2 \in [a,b]. \tag{3.27}$$

Therefore it holds

$$\lim_{n\to\infty} \sum_{k=\lceil na \rceil}^{\lfloor nb \rfloor} \frac{\left(G_{q,\lambda}(nx_1-k) + G_{\frac{1}{q},\lambda}(nx_2-k)\right)}{2} \neq 1. \tag{3.28}$$

Hence it is

$$\lim_{n\to\infty}\sum_{k=\lceil na\rceil}^{\lfloor nb\rfloor}\frac{\left(G_{q,\lambda}(nx_1-k)+G_{\frac{1}{q},\lambda}(nx_1-k)\right)}{2}\neq 1, \tag{3.29}$$

even if

$$\lim_{n\to\infty}\sum_{k=\lceil na\rceil}^{\lfloor nb\rfloor}G_{\frac{1}{q},\lambda}(nx_1-k)=1, \tag{3.30}$$

because then

$$\lim_{n\to\infty}\sum_{k=\lceil na\rceil}^{\lfloor nb\rfloor}\frac{G_{q,\lambda}(nx_1-k)}{2}+\frac{1}{2}\neq 1, \tag{3.31}$$

equivalently

$$\lim_{n\to\infty}\sum_{k=\lceil na\rceil}^{\lfloor nb\rfloor}\frac{G_{q,\lambda}(nx_1-k)}{2}\neq\frac{1}{2}, \tag{3.32}$$

true by

$$\lim_{n\to\infty}\sum_{k=\lceil na\rceil}^{\lfloor nb\rfloor}G_{q,\lambda}(nx_1-k)\neq 1. \tag{3.33}$$

II) Let $[a,b]\subset\mathbb{R}$. For large n we always have $\lceil na\rceil\leq\lfloor nb\rfloor$. Also $a\leq\frac{k}{n}\leq b$, iff $\lceil na\rceil\leq k\leq\lfloor nb\rfloor$. So in general it holds

$$\sum_{k=\lceil na\rceil}^{\lfloor nb\rfloor}W(nx-k)\leq 1. \tag{3.34}$$

We need:

Definition 3.1. Let $f\in C([a,b])$ and $n\in\mathbb{N}:\lceil na\rceil\leq\lfloor nb\rfloor$. We introduce and define the real-valued linear symmetrized neural network operators

$$S_n^s(f,x):=\frac{\sum_{k=\lceil na\rceil}^{\lfloor nb\rfloor}f\left(\frac{k}{n}\right)W(nx-k)}{\sum_{k=\lceil na\rceil}^{\lfloor nb\rfloor}W(nx-k)}=$$

$$\frac{\sum_{k=\lceil na\rceil}^{\lfloor nb\rfloor}f\left(\frac{k}{n}\right)\left(G_{q,\lambda}(nx-k)+G_{\frac{1}{q},\lambda}(nx-k)\right)}{\sum_{k=\lceil na\rceil}^{\lfloor nb\rfloor}\left(G_{q,\lambda}(nx-k)+G_{\frac{1}{q},\lambda}(nx-k)\right)}. \tag{3.35}$$

In fact, it is $S_n^s(f) \in C([a,b])$; and $S_n^s(1) = 1$, S_n^s are positive linear operators.

The modulus of continuity is defined by

$$\omega_1(f,\delta) := \sup_{\begin{cases} x,y \in [a,b] \\ : |x-y| \leq \delta \end{cases}} |f(x) - f(y)|, \quad \delta > 0. \tag{3.36}$$

The same is defined for $f \in C_{uB}(\mathbb{R})$ (uniformly continuous and bounded functions) and for $f \in C_B(\mathbb{R})$ (bounded and continuous functions) and for $f \in C_u(\mathbb{R})$ (uniformly continuous functions).

The fact $f \in C([a,b])$ or $f \in C_u(\mathbb{R})$, is equivalent to $\lim_{\delta\to 0}\omega_1(f,\delta) = 0$.

In this work, $0 < \alpha < 1$, $n \in \mathbb{N} : n^{1-\alpha} > 2$; where $\|\cdot\|_\infty$ is the supremum norm.

3.3 Main Results

The following approximation results are valid.

Theorem 3.2. *Let $f \in C([a,b])$. Then*

$$\|S_n^s(f) - f\|_\infty \leq 2\omega_1\left(f, \sqrt{K(q)}\sqrt{\frac{1}{n^{2\alpha}} + (b-a)^2 \gamma A^{-\lambda(n^{1-\alpha}-2)}}\right) \to 0,$$

as $n \to +\infty$. (3.37)

So that $\lim_{n\to\infty} S_n^s(f) = f$, uniformly.

Proof. We estimate

$$\left(S|_n^s\left((t-x)^2\right)\right)(x) = \frac{\sum_{k=\lceil na\rceil}^{\lfloor nb\rfloor}\left(\frac{k}{n}-x\right)^2 W(nx-k)}{\sum_{k=\lceil na\rceil}^{\lfloor nb\rfloor} W(nx-k)} \overset{(3.25)}{\leq}$$

$$K(q)\left(\sum_{k=\lceil na\rceil}^{\lfloor nb\rfloor}\left(\frac{k}{n}-x\right)^2 W(nx-k)\right) =$$

$$K(q)\left[\sum_{\begin{cases} k = \lceil na\rceil \\ : \left|\frac{k}{n}-x\right| \leq \frac{1}{n^\alpha}\end{cases}}^{\lfloor nb\rfloor}\left(\frac{k}{n}-x\right)^2 W(nx-k) + \right. \tag{3.38}$$

$$\left. \sum_{\left\{\begin{array}{c} k=\lceil na\rceil \\ :\left|\frac{k}{n}-x\right|>\frac{1}{n^{\alpha}} \end{array}\right.}^{\lfloor nb\rfloor} \left(\frac{k}{n}-x\right)^{2} W(nx-k)\right] \overset{(\text{by } (3.11),\ (3.17))}{\leq}$$

$$K(q)\left[\frac{1}{n^{2\alpha}}+(b-a)^{2}\gamma A^{-\lambda\left(n^{1-\alpha}-2\right)}\right].$$

That is

$$\left(S_{n}^{s}\left((t-x)^{2}\right)\right)(x)\leq K(q)\left\{\frac{1}{n^{2\alpha}}+(b-a)^{2}\gamma A^{-\lambda\left(n^{1-\alpha}-2\right)}\right\}. \tag{3.39}$$

Consequently, it holds that

$$\delta_{n}^{*}:=\left\|\left(S_{n}^{s}\left((t-x)^{2}\right)\right)(x)\right\|_{\infty}^{\frac{1}{2}}\leq$$

$$\sqrt{K(q)}\sqrt{\frac{1}{n^{2\alpha}}+(b-a)^{2}\gamma A^{-\lambda\left(n^{1-\alpha}-2\right)}}. \tag{3.40}$$

By Theorem 7.1.7, p. 203, [61], Shisha-Mond inequality ([66]) for positive linear operators and $S_{n}^{s}(1)=1$, we have that

$$\left\|S_{n}^{s}(f)-f\right\|_{\infty}\leq 2\omega_{1}\left(f,\delta_{n}^{*}\right). \tag{3.41}$$

Thus, it holds

$$\left\|S_{n}^{s}(f)-f\right\|_{\infty}\leq 2\omega_{1}\left(f,\sqrt{K(q)}\sqrt{\frac{1}{n^{2\alpha}}+(b-a)^{2}\gamma A^{-\lambda\left(n^{1-\alpha}-2\right)}}\right), \tag{3.42}$$

proving the claim. □

It holds

Theorem 3.3. *Let $f:\mathbb{R}\rightarrow\mathbb{R}$ be a continuous and 2π periodic function with modulus of continuity ω_1. Here $\|\cdot\|_{\infty}$ denotes the sup-norm over $[a,b]\subset\mathbb{R}$, and the operators S_{n}^{s} are acting on such f over $[a,b]$; $n\in\mathbb{N}:n^{1-\alpha}>2$, $0<\alpha<1$. Then*

$$\left\|S_{n}^{s}(f)-f\right\|_{\infty}\leq 2\omega_{1}\left(f,\pi\sqrt{K(q)}\sqrt{\omega_{1}\left(\sin^{2},\frac{1}{2n^{\alpha}}\right)+\gamma A^{-\lambda\left(n^{1-\alpha}-2\right)}}\right). \tag{3.43}$$

Proof. We want to estimate ($x \in [a,b]$),

$$\varepsilon_n^* := \pi \left\| \left(S_n^s \left(\sin^2 \left(\frac{t-x}{2} \right) \right) \right) (x) \right\|_\infty^{\frac{1}{2}} =$$

$$\pi \left\| \left(\frac{\sum_{k=\lceil na \rceil}^{\lfloor nb \rfloor} \sin^2 \left(\frac{\frac{k}{n}-x}{2} \right) W(nx-k)}{\sum_{k=\lceil na \rceil}^{\lfloor nb \rfloor} W(nx-k)} \right) \right\|_\infty^{\frac{1}{2}} \leq$$

$$\pi \left\| K(q) \sum_{k=\lceil na \rceil}^{\lfloor nb \rfloor} \sin^2 \left(\frac{\frac{k}{n}-x}{2} \right) W(nx-k) \right\|_\infty^{\frac{1}{2}} =$$

$$\pi \sqrt{K(q)} \left\| \sum_{\begin{cases} k=\lceil na \rceil \\ : \left| \frac{k}{n}-x \right| \leq \frac{1}{n^\alpha} \end{cases}}^{\lfloor nb \rfloor} \left(\sin^2 \left(\frac{\frac{k}{n}-x}{2} \right) - \sin^2(0) \right) W(nx-k) + \right.$$

$$\left. \sum_{\begin{cases} k=\lceil na \rceil \\ : \left| \frac{k}{n}-x \right| > \frac{1}{n^\alpha} \end{cases}}^{\lfloor nb \rfloor} \sin^2 \left(\frac{\frac{k}{n}-x}{2} \right) W(nx-k) \right\|_\infty^{\frac{1}{2}} \leq \tag{3.44}$$

$$\pi \sqrt{K(q)} \left\| \sum_{\begin{cases} k=\lceil na \rceil \\ : \left| \frac{k}{n}-x \right| \leq \frac{1}{n^\alpha} \end{cases}}^{\lfloor nb \rfloor} \omega_1 \left(\sin^2, \frac{\left| \frac{k}{n}-x \right|}{2} \right) W(nx-k) + \right.$$

$$\left. 1 \cdot \sum_{\begin{cases} k=\lceil na \rceil \\ : \left| \frac{k}{n}-x \right| > \frac{1}{n^\alpha} \end{cases}}^{\lfloor nb \rfloor} W(nx-k) \right\|_\infty^{\frac{1}{2}} \leq$$

$$\pi \sqrt{K(q)} \sqrt{\omega_1 \left(\sin^2, \frac{1}{2n^{\alpha}} \right) + \gamma A^{-\lambda\left(n^{1-\alpha}-2\right)}}.$$

We have proved that

$$\varepsilon_n^* \leq \pi \sqrt{K(q)} \sqrt{\omega_1 \left(\sin^2, \frac{1}{2n^{\alpha}} \right) + \gamma A^{-\lambda\left(n^{1-\alpha}-2\right)}}. \tag{3.45}$$

Therefore it holds, by [67], the following inequalities,

$$\|S_n^s(f) - f\|_{\infty} \leq 2\omega_1(f, \varepsilon_n^*) \leq$$

$$2\omega_1 \left(f, \pi \sqrt{K(q)} \sqrt{\omega_1 \left(\sin^2, \frac{1}{2n^{\alpha}} \right) + \gamma A^{-\lambda\left(n^{1-\alpha}-2\right)}} \right), \tag{3.46}$$

hence proving the claim. □

We need

Definition 3.2. Let $f \in C([a,b])$. The Peetre ([65]) K_1 functional is defined as follows:

$$K_1(f,t) := \inf\left(\|f-g\|_{\infty} + t\|g'\|_{\infty}\right), \quad g \in C^1([a,b]), \quad t \geq 0. \tag{3.47}$$

We give

Theorem 3.4. *Let $f \in C([a,b])$. Then*

$$\|S_n^s(f) - f\|_{\infty} \leq 2K_1 \left(f, \frac{K(q)}{2} \left[\frac{1}{n^{\alpha}} + (b-a)\gamma A^{-\lambda\left(n^{1-\alpha}-2\right)} \right] \right) \to 0, \tag{3.48}$$

as $n \to +\infty$.

So that $\lim_{n \to +\infty} S_n^s(f) = f$, uniformly.

For the proof of (3.48), we will use:

Corollary 3.1. *([60]) (Pointwise Approximation) Let $f \in C([a,b])$ and let L be a positive linear operator acting on $C([a,b])$ satisfying $L(1,x) = 1$, all $x \in [a,b]$.*

Then, we have the attainable (i.e. sharp) inequality

$$|L(f,x) - f(x)| \leq 2K_1 \left(f, \frac{1}{2} L(|y-x|, x) \right), \quad \forall\, x \in [a,b]. \tag{3.49}$$

Proof of Theorem 3.4. We apply inequality (3.49). Clearly, we have that

$$|S_n^s(f,x) - f(x)| \leq 2K_1\left(f, \frac{1}{2}S_n^s(|t-x|)(x)\right), \tag{3.50}$$

which is attainable (i.e. sharp).

We estimate

$$S_n^s(|t-x|)(x) = \frac{\sum_{k=\lceil na\rceil}^{\lfloor nb\rfloor} \left|\frac{k}{n}-x\right| W(nx-k)}{\sum_{k=\lceil na\rceil}^{\lfloor nb\rfloor} W(nx-k)} \leq$$

$$K(q)\sum_{k=\lceil na\rceil}^{\lfloor nb\rfloor}\left|\frac{k}{n}-x\right| W(nx-k) =$$

$$K(q)\left[\sum_{\begin{cases} k=\lceil na\rceil \\ :\left|\frac{k}{n}-x\right| \leq \frac{1}{n^\alpha}\end{cases}}^{\lfloor nb\rfloor}\left|\frac{k}{n}-x\right| W(nx-k) + \right. \tag{3.51}$$

$$\left.\sum_{\begin{cases} k=\lceil na\rceil \\ :\left|\frac{k}{n}-x\right| > \frac{1}{n^\alpha}\end{cases}}^{\lfloor nb\rfloor}\left|\frac{k}{n}-x\right| W(nx-k)\right] \leq$$

$$K(q)\left[\frac{1}{n^\alpha} + (b-a)\gamma A^{-\lambda\left(n^{1-\alpha}-2\right)}\right].$$

That is

$$S_n^s(|t-x|)(x) \leq K(q)\left[\frac{1}{n^\alpha} + (b-a)\gamma A^{-\lambda\left(n^{1-\alpha}-2\right)}\right]. \tag{3.52}$$

We have proved

$$|S_n^s(f,x) - f(x)| \leq 2K_1\left(f, \frac{K(q)}{2}\left[\frac{1}{n^\alpha} + (b-a)\gamma A^{-\lambda\left(n^{1-\alpha}-2\right)}\right]\right), \tag{3.53}$$

$\forall\, x \in [a,b]$.

Now, the validity of (3.48) is clear. □

It follows a trigonometric result.

Theorem 3.5. *Here* $f \in C^2([a,b])$, $x_0 \in [a,b]$, $r > 0$, $\widetilde{D}_3(x_0) := \left(S_n^s\left(|\cdot - x_0|^3\right)(x_0)\right)^{\frac{1}{3}}$. *Then*

(i)

$$|S_n^s(f)(x_0) - f(x_0)| \leq |f'(x_0)| \, |S_n^s(\sin(\cdot - x_0))(x_0)| +$$

$$2|f''(x_0)| \left(S_n^s\left(\sin^2\left(\frac{\cdot - x_0}{2}\right)\right)(x_0)\right) +$$

$$\frac{\omega_1\left(f'' + f, r\widetilde{D}_3(x_0)\right)}{2} \widetilde{D}_3^2(x_0)\left[1 + \frac{1}{3r}\right], \tag{3.54}$$

(ii)

$$\|S_n^s(f) - f\|_\infty \leq \|f'\|_\infty \, \|S_n^s(\sin(\cdot - x_0))(x_0)\|_\infty +$$

$$2\|f''\|_\infty \left\|S_n^s\left(\sin^2\left(\frac{\cdot - x_0}{2}\right)\right)(x_0)\right\|_\infty +$$

$$\frac{\omega_1\left(f'' + f, r\left\|\widetilde{D}_3\right\|_\infty\right)}{2} \left\|\widetilde{D}_3\right\|_\infty^2 \left(1 + \frac{1}{3r}\right), \tag{3.55}$$

(iii) if $f'(x_0) = f''(x_0) = 0$, *we obtain*

$$|S_n^s(f)(x_0) - f(x_0)| \leq \frac{\omega_1\left(f'' + f, r\widetilde{D}_3(x_0)\right)}{2} \widetilde{D}_3^2(x_0)\left(1 + \frac{1}{3r}\right), \tag{3.56}$$

and

(iv)

$$\left|S_n^s(f)(x_0) - f(x_0) - f'(x_0) S_n^s(\sin(\cdot - x_0))(x_0) - 2f''(x_0) S_n^s\left(\sin^2\left(\frac{\cdot - x_0}{2}\right)\right)(x_0)\right|$$

$$\leq \frac{\omega_1\left(f'' + f, r\widetilde{D}_3(x_0)\right)}{2} \widetilde{D}_3^2(x_0)\left(1 + \frac{1}{3r}\right). \tag{3.57}$$

Proof. Direct application of Theorem 12.3, p. 384, [64]. □

Next, we give the hyperbolic version of the last result.

Theorem 3.6. *All as in Theorem 3.5. Then*

(i)

$$\left|S_n^s(f)(x_0)-f(x_0)\right| \leq \left|f'(x_0)\right|\left|S_n^s(\sinh(\cdot-x_0))(x_0)\right|+ 2\left|f''(x_0)\right|\left(S_n^s\left(\sinh^2\left(\frac{\cdot-x_0}{2}\right)\right)(x_0)\right)+ \frac{\cosh(b-a)\,\omega_1\left(f''-f, r\widetilde{D}_3(x_0)\right)}{2}\widetilde{D}_3^2(x_0)\left[1+\frac{1}{3r}\right], \tag{3.58}$$

(ii)

$$\left\|S_n^s(f)-f\right\|_\infty \leq \left\|f'\right\|_\infty\left\|S_n^s(\sinh(\cdot-x_0))(x_0)\right\|_\infty+ 2\left\|f''\right\|_\infty\left\|S_n^s\left(\sinh^2\left(\frac{\cdot-x_0}{2}\right)\right)(x_0)\right\|_\infty+ \frac{\cosh(b-a)\,\omega_1\left(f''-f, r\left\|\widetilde{D}_3\right\|_\infty\right)}{2}\left\|\widetilde{D}_3\right\|_\infty^2\left(1+\frac{1}{3r}\right), \tag{3.59}$$

(iii) if $f'(x_0)=f''(x_0)=0$, $x_0\in[a,b]$, we obtain

$$\left|S_n^s(f)(x_0)-f(x_0)\right| \leq \frac{\cosh(b-a)\,\omega_1\left(f''-f, r\widetilde{D}_3(x_0)\right)}{2}\widetilde{D}_3^2(x_0)\left(1+\frac{1}{3r}\right), \tag{3.60}$$

and

(iv)

$$\left|S_n^s(f)(x_0)-f(x_0)-f'(x_0)S_n^s(\sinh(\cdot-x_0))(x_0)-2f''(x_0)S_n^s\left(\sinh^2\left(\frac{\cdot-x_0}{2}\right)\right)(x_0)\right|$$

$$\leq \frac{\cosh(b-a)\,\omega_1\left(f''-f, r\widetilde{D}_3(x_0)\right)}{2}\widetilde{D}_3^2(x_0)\left(1+\frac{1}{3r}\right). \tag{3.61}$$

Proof. Direct application of Theorem 12.5, p. 390, [64]. □

Remark 3.2. We similarly have that

$$\left(S_n^s\left(\left|\cdot-x_0\right|^3\right)(x_0)\right)^{\frac{1}{3}} \leq \sqrt[3]{K(q)}\sqrt[3]{\left\{\frac{1}{n^{3\alpha}}+(b-a)^3\gamma A^{-\lambda\left(n^{1-\alpha}-2\right)}\right\}} \to 0, \tag{3.62}$$

as $n\to+\infty$.

Hence, by each of the above Theorems 3.5, 3.6, we obtain that $S_n^s(f)\to f$ uniformly convergent as $n\to+\infty$, $\forall\, f\in C^2([a,b])$.

Valid, by Theorem 12.4, p. 390 and Theorem 12.6, p. 395 from [64], respectively.

We make

Remark 3.3. Let $x_0 \in (a,b)$, $f \in C([a,b])$, here S_n^s is a positive linear operator from $C([a,b])$ into itself.

Also $S_n^s(1) = 1$. By Riesz Representation theorem, there exists a probability measure μ_{x_0} on $[a,b]$, such that

$$S_n^s(f)(x_0) = \int_{[a,b]} f(t)\, d\mu_{x_0}(t), \quad \forall\, f \in C([a,b]). \tag{3.63}$$

Assume here that $|f(t) - f(x_0)|$ is convex in t. We consider

$$\widetilde{d}_2(x_0) := \left(\int_{[a,b]} (t - x_0)^2\, d\mu_{x_0}(t) \right)^{\frac{1}{2}}$$

$$= \left(\left(S_n^s\left((t - x_0)^2 \right) \right)(x_0) \right)^{\frac{1}{2}} \quad \text{(as earlier)} \tag{3.64}$$

$$\leq \sqrt{K(q)} \sqrt{\frac{1}{n^{2\alpha}} + (b-a)^2 \gamma A^{-\lambda(n^{1-\alpha}-2)}} = (*).$$

Clearly, here it is $\widetilde{d}_2(x_0) > 0$ (see (3.35)).

For large enough $n \in \mathbb{N}$ we get

$$(*) \leq \min(x_0 - a, b - x_0) \leq \max(x_0 - a, b - x_0).$$

That is

$$\widetilde{d}_2(x_0) \leq \min(x_0 - a, b - x_0). \tag{3.65}$$

By Corollary 8.1.1, p. 245, [61], we obtain:

Theorem 3.7. *It holds*

$$|S_n^s(f)(x_0) - f(x_0)| \leq \omega_1\left(f, \sqrt{K(q)} \sqrt{\frac{1}{n^{2\alpha}} + (b-a)^2 \gamma A^{-\lambda(n^{1-\alpha}-2)}} \right), \tag{3.66}$$

for large enough $n \in \mathbb{N}$.

Remark 3.4. We have that

$$\widetilde{d}_1(x_0) := \int_{[a,b]} |t - x_0|\, d\mu_{x_0}(t) = S_n^s(|t - x_0|)(x_0) \overset{(3.52)}{\leq} \tag{3.67}$$

$$K(q) \left[\frac{1}{n^{\alpha}} + (b-a) \gamma A^{-\lambda(n^{1-\alpha}-2)} \right] \leq \min(x_0 - a, b - x_0),$$

for large enough $n \in \mathbb{N}$.

By (8.1.8), p. 245, [61], we get

Theorem 3.8. *Let $r \geq 1$, then*

$$|S_n^s(f)(x_0) - f(x_0)| \leq r\omega_1\left(f, \frac{1}{r}K(q)\left\{\frac{1}{n^\alpha} + (b-a)\gamma A^{-\lambda\left(n^{1-\alpha}-2\right)}\right\}\right), \tag{3.68}$$

for large enough $n \in \mathbb{N}$.

By Theorem 8.1.2, p. 248, of [61] we can derive the following result.

Theorem 3.9. *Here we consider $[a,b] \subset \mathbb{R}$, and $x_0 \in (a,b)$. Let L be a positive linear operator from $C([a,b])$ into itself, such that $L(1) = 1$.*

Denote by

$$\widetilde{d}_{m+1}(x_0) := \left(L\left(|t-x_0|^{m+1}\right)(x_0)\right)^{\frac{1}{m+1}}, \quad m \in \mathbb{N}.$$

Consider $f \in C^m([a,b])$ such that $\left|f^{(m)}(t) - f^{(m)}(x_0)\right|$ is convex in t.

Assume that

$$0 < \widetilde{d}_{m+1}^{m+1}(x_0) \leq (\min(x_0 - a, b - x_0))(m+1)!. \tag{3.69}$$

Then

$$|L(f)(x_0) - f(x_0)| - \sum_{k=1}^{m} \frac{\left|f^{(k)}(x_0)\right|}{k!}\left|L\left((t-x_0)^k\right)(x_0)\right| \leq \tag{3.70}$$

$$\omega_1\left(f^{(m)}, \frac{L\left(|t-x_0|^{m+1}\right)(x_0)}{(m+1)!}\right).$$

We make

Remark 3.5. We have that

$$0 < \left(S_n^s\left(|t-x_0|^{(m+1)}\right)\right)(x_0) = \frac{\sum_{k=\lceil na\rceil}^{\lfloor nb\rfloor}\left|\frac{k}{n} - x_0\right|^{m+1} W(nx-k)}{\sum_{k=\lceil na\rceil}^{\lfloor nb\rfloor} W(nx-k)}$$

$$\leq K(q)\sum_{k=\lceil na\rceil}^{\lfloor nb\rfloor}\left|\frac{k}{n} - x_0\right|^{m+1} W(nx-k) \tag{3.71}$$

... (as earlier)

$$K(q)\left\{\frac{1}{n^{(m+1)\alpha}} + (b-a)^{m+1}\gamma A^{-\lambda\left(n^{1-\alpha}-2\right)}\right\} \leq \min(x_0 - a, b - x_0)(m+1)!, \tag{3.72}$$

for large enough $n \in \mathbb{N}$.

By Theorem 3.9, Remark 3.5 we have proved the following:

Theorem 3.10. *Here it is* $[a,b] \subset \mathbb{R}$, *and* $x_0 \in (a,b)$. *Consider* $f \in C^m([a,b]) : \left|f^{(m)}(t) - f^{(m)}(x_0)\right|$ *is convex in* t, $m \in \mathbb{N}$.

Then, for sufficiently large enough $n \in \mathbb{N}$, *we derive that*

$$\left|S_n^s(f)(x_0) - f(x_0)\right| - \sum_{k=1}^{m} \frac{\left|f^{(k)}(x_0)\right|}{k!} \left|S_n^s\left((t-x_0)^k\right)(x_0)\right| \leq \tag{3.73}$$

$$\omega_1\left(f^{(m)}, \frac{K(q)}{(m+1)!}\left\{\frac{1}{n^{(m+1)\alpha}} + (b-a)^{m+1}\gamma A^{-\lambda\left(n^{1-\alpha}-2\right)}\right\}\right).$$

Next we use Theorem 18.1, p. 419, [62].

Theorem 3.11. *Denote* $(N \in \mathbb{N})$

$$\widetilde{F}_N := \left\|\left(S_n^s\left(|t-\cdot|^N\right)\right)(\cdot)\right\|_\infty^{\frac{1}{N}} < +\infty, \tag{3.74}$$

where $\|\cdot\|_\infty$ *is the supremum norm.*

Let $f \in C^N([a,b])$. *Then*

$$\left\|S_n^s f - f\right\|_\infty \leq \sum_{\overline{k}=1}^{N} \frac{\left\|f^{(\overline{k})}\right\|_\infty}{\overline{k}!} \left\|\left(S_n^s(t-\cdot)^{\overline{k}}\right)(\cdot)\right\|_\infty + \tag{3.75}$$

$$\omega_1\left(f^{(N)}, \widetilde{F}_N\right)\widetilde{F}_N^{N-1}\left(\frac{(b-a)}{(N+1)!} + \frac{\widetilde{F}_N}{2N!} + \frac{\widetilde{F}_N^2}{8(b-a)(N-1)!}\right), \quad \forall\, n \in \mathbb{N}.$$

Furthermore it holds,

$$\widetilde{F}_N \leq \sqrt[N]{K(q)}\sqrt[N]{\frac{1}{n^{N\alpha}} + (b-a)^N \gamma A^{-\lambda\left(n^{1-\alpha}-2\right)}}, \tag{3.76}$$

and for $\overline{k} = 1, ..., N$, *we have, similarly, that*

$$\left\|\left(S_n^s(t-\cdot)^{\overline{k}}\right)(\cdot)\right\|_\infty \leq K(q)\left\{\frac{1}{n^{\overline{k}\alpha}} + (b-a)^{\overline{k}}\gamma A^{-\lambda\left(n^{1-\alpha}-2\right)}\right\}. \tag{3.77}$$

By Corollary 18.1, p. 421, [62], we get

Corollary 3.2. *Here* $\widetilde{F}_1 := \left\|\left(S_n^s(|t-\cdot|)\right)(\cdot)\right\|_\infty < \infty$. *Let* $f \in C^1([a,b])$. *Then*

$$\left\|S_n^s f - f\right\|_\infty \leq \left\|f'\right\|_\infty \left\|\left(S_n^s(t-\cdot)\right)(\cdot)\right\|_\infty + \tag{3.78}$$

$$\frac{1}{2}\omega_1\left(f', \widetilde{F}_1\right)\left((b-a) + \widetilde{F}_1 + \frac{\widetilde{F}_1^2}{4(b-a)}\right).$$

Here it is

$$\widetilde{F}_1 \leq K(q)\left(\frac{1}{n^\alpha} + (b-a)\gamma A^{-\lambda\left(n^{1-\alpha}-2\right)}\right). \tag{3.79}$$

By Corollary 18.2, p. 421, [62], we get

Corollary 3.3. *Here* $\widetilde{F}_2 := \left\| \left(S_n^s (t - \cdot)^2 \right) (\cdot) \right\|_\infty^{\frac{1}{2}} < +\infty$. *Let* $f \in C^2([a,b])$. *Then*

$$\|S_n^s f - f\|_\infty \leq \|f'\|_\infty \|(S_n^s (t - \cdot))(\cdot)\|_\infty + \frac{\|f''\|_\infty}{2} \left\| \left(S_n^s (t - \cdot)^2 \right)(\cdot) \right\|_\infty \tag{3.80}$$

$$+ \frac{1}{2}\omega_1\left(f'', \widetilde{F}_2\right) \widetilde{F}_2 \left(\frac{(b-a)}{3} + \frac{\widetilde{F}_2}{2} + \frac{\widetilde{F}_2^2}{4(b-a)} \right).$$

Here they are

$$\widetilde{F}_2 \leq \sqrt{K(q)} \sqrt{\frac{1}{n^{2\alpha}} + (b-a)^2 \gamma A^{-\lambda(n^{1-\alpha}-2)}}, \tag{3.81}$$

and

$$\left\| \left(S_n^s (t - \cdot)^2 \right)(\cdot) \right\|_\infty \leq K(q) \left(\frac{1}{n^{2\alpha}} + (b-a)^2 \gamma A^{-\lambda\left(n^{1-\alpha}-2\right)} \right). \tag{3.82}$$

By Theorem 18.2, p. 422, [62], we get:

Theorem 3.12. *Let* $\widetilde{\varphi} := \left\| \left(S_n^s (t - x_0)^2 \right)(x_0) \right\|_\infty^{\frac{1}{2}} < +\infty$, *and* $r > 0$. *Let also* $f \in C^1([a,b])$. *Then*

$$\|S_n^s f - f\|_\infty - \|f'\|_\infty \|(S_n^s (t - x_0))(x_0)\|_\infty \leq \tag{3.83}$$

$$\begin{cases} \frac{1}{8r}(2+r)^2 \omega_1(f', r\widetilde{\varphi})\widetilde{\varphi}, & \text{if } r \leq 2; \\ \\ \omega_1(f', r\widetilde{\varphi})\widetilde{\varphi}, & \text{if } r > 2. \end{cases}$$

Next results are with respect to $\|\cdot\|_1$, $\|\cdot\|_2$ norms, which are with respect to Lebesgue measure.

By Theorem 18.3, p. 424, [62], we obtain

Theorem 3.13. *Let*

$$\widetilde{\gamma}_N := \left\| \left(S_n^s (t - x)^N \right)(x) \right\|_2^{\frac{1}{N}} < +\infty, \tag{3.84}$$

$n, N \in \mathbb{N}$. *Let also* $f \in C^N([a,b])$. *Then*

$$\|S_n^s f - f\|_1 \leq \sum_{\overline{k}=1}^{N} \frac{1}{\overline{k}!} \left\| f^{(\overline{k})} \right\|_2 \left\| \left(S_n^s (t - \cdot)^{\overline{k}} \right)(\cdot) \right\|_2 + \frac{1}{2}\omega_1\left(f^{(N)}, \widetilde{\gamma}_N\right) \widetilde{\gamma}_N^{N-1} \tag{3.85}$$

$$\left\{ \left(\frac{b-a}{(N+1)!} \right) \cdot \sqrt{\frac{7}{3}(b-a)} + \frac{\sqrt{b-a}}{N!}\widetilde{\gamma}_N + \frac{\widetilde{\gamma}_N^2}{4(N-1)!}\sqrt{\frac{2}{b-a}} \right\} < +\infty.$$

By Corollary 18.3, p. 426, [62], we get:

Corollary 3.4. *Set*

$$\widetilde{\gamma}_1 := \left\| S_n^s \left(|t-x| \right) (x) \right\|_2 < +\infty. \tag{3.86}$$

Let $f \in C^1([a,b])$. *Then*

$$\left\| S_n^s (f) - f \right\|_1 \leq \left\| f' \right\|_2 \left\| S_n^s (t - \cdot) (\cdot) \right\|_2 + \frac{\omega_1 (f', \widetilde{\gamma}_1)}{2}$$

$$\left\{ \left(\frac{b-a}{2} \right) \sqrt{\frac{7}{3} (b-a)} + \left(\sqrt{b-a} \right) \widetilde{\gamma}_1 + \frac{\widetilde{\gamma}_1^2}{4} \sqrt{\frac{2}{b-a}} \right\}. \tag{3.87}$$

By Corollary 18.4, p. 427, [62], we obtain

Corollary 3.5. *Set*

$$\widetilde{\gamma}_2 := \left\| S_n^s \left((t-x)^2 \right) (x) \right\|_2^{\frac{1}{2}} < +\infty. \tag{3.88}$$

Let $f \in C^2([a,b])$. *Then*

$$\left\| S_n^s f - f \right\|_1 \leq \left\| f' \right\|_2 \left\| S_n^s (t - \cdot) (\cdot) \right\|_2 + \frac{\left\| f'' \right\|_2}{2} \widetilde{\gamma}_2^2 + \frac{1}{4} \omega_1 (f'', \widetilde{\gamma}_2) \widetilde{\gamma}_2 \tag{3.89}$$

$$\left\{ \left(\frac{b-a}{3} \right) \cdot \sqrt{\frac{7}{3} (b-a)} + \left(\sqrt{b-a} \right) \widetilde{\gamma}_2 + \frac{\widetilde{\gamma}_2^2}{2} \sqrt{\frac{2}{b-a}} \right\}.$$

Setting 2. In the next Theorems 3.14, 3.15 we use:

Let $([a,b], \mathcal{B}, \mu)$, $a \neq b$, be a measure space, where $\mathcal{B}$ is the Borel σ-algebra on $[a,b]$ and μ is a positive finite measure on $[a,b]$. (Note that $C([a,b]) \subset L_p([a,b], \mathcal{B}, \mu)$, any $p > 1$.) Here $\|\cdot\|_p$ stands for the related L_p norm with respect to μ. Let $p, q > 1$ such that $\frac{1}{p} + \frac{1}{q} = 1$.

Next we apply Theorem 18.4, p. 428, [62].

Theorem 3.14. *Here* $n, N \in \mathbb{N}$. *Set*

$$\widetilde{\delta}_N := \left\| \left(S_n^s \left(|t - \cdot|^N \right) \right) (\cdot) \right\|_p^{\frac{1}{N}} < +\infty. \tag{3.90}$$

Let $f \in C^N([a,b])$. *Then*

$$\left\| S_n^s f - f \right\|_p \leq \sum_{\overline{k}=1}^{N} \frac{1}{\overline{k}!} \left\| f^{(\overline{k})} \right\|_{pq} \left\| \left(S_n^s (t - \cdot)^{\overline{k}} \right) (\cdot) \right\|_{p^2} + \tag{3.91}$$

$$\omega_1 \left(f^{(N)}, \widetilde{\delta}_N \right) \widetilde{\delta}_N^{N-1} \left[\frac{(b-a)}{(N+1)!} + \frac{\widetilde{\delta}_N}{2N!} + \frac{\widetilde{\delta}_N^2}{4(N-1)!(b-a)} \right].$$

Remark 3.6. We have that

$$\left(S_n^s\left(|t-x_0|^N\right)\right)(x_0) \le K(q)\left\{\frac{1}{n^{N\alpha}} + (b-a)^N \gamma A^{-\lambda\left(n^{1-\alpha}-2\right)}\right\} =: \widetilde{\tau}, \tag{3.92}$$

and

$$\left\|\left(S_n^s\left(|t-\cdot|^N\right)\right)(\cdot)\right\|_p =$$

$$\left(\int_{[a,b]}\left(\left(S_n^s\left(|t-x_0|^N\right)\right)(x_0)\right)^p \mu(dx_0)\right)^{\frac{1}{p}} \le \left(\int_{[a,b]} \widetilde{\tau}^p d\mu\right)^{\frac{1}{p}} = \mu([a,b])^{\frac{1}{p}} \cdot \widetilde{\tau}. \tag{3.93}$$

Therefore it holds

$$\widetilde{\delta}_N \le \mu([a,b])^{\frac{1}{Np}} \, \widetilde{\tau}^{\frac{1}{N}}. \tag{3.94}$$

By Theorem 18.5, p. 431, [62], we get

Theorem 3.15. *Here $n, N \in \mathbb{N}$. Denote*

$$\widetilde{\delta}_N := \left\|S_n^s\left(|t-x|^N, x\right)\right\|_p^{\frac{1}{N}} < +\infty. \tag{3.95}$$

Let $f \in C^N([a,b])$. Then

$$\|S_n^s f - f\|_p \le \sum_{\overline{k}=1}^{N} \frac{\left\|f^{(\overline{k})}\right\|_\infty}{\overline{k}!} \left\|\left(S_n^s (t-x)^{\overline{k}}, x\right)\right\|_p + \tag{3.96}$$

$$\omega_1\left(f^{(N)}, \widetilde{\delta}_N\right) \widetilde{\delta}_N^{N-1}\left[\frac{(b-a)}{(N+1)!} + \frac{\widetilde{\delta}_N}{2N!} + \frac{\widetilde{\delta}_N^2}{4(N-1)!(b-a)}\right].$$

We need

Definition 3.3. Let $f, g \in C([a,b])$ such that $|f(x)| \le |g(x)|$, $\forall\, x \in [a,b]$. A norm $\|\cdot\|$ on $C([a,b])$ is called monotone iff $\|f\| \le \|g\|$. We denote a monotone norm by $\|\cdot\|_m$, e.g., L_p norms in general $(1 \le p \le +\infty)$, Orlicz norms, etc.

Finally we apply Corollary 18.5, p. 432, [62].

Corollary 3.6. *Let $\|\cdot\|_m$ be a monotone norm on $C([a,b])$. Denote*

$$\widetilde{\varepsilon}_N := \left\|S_n^s\left(|t-x|^N, x\right)\right\|_m^{\frac{1}{N}} < +\infty, \quad n, N \in \mathbb{N}. \tag{3.97}$$

Let $f \in C^N([a,b])$. *Then*

$$\|S_n^s f - f\|_m \leq \sum_{\overline{k}=1}^{N} \frac{\left\|f^{(\overline{k})}\right\|_\infty}{\overline{k}!} \left\|\left(S_n^s (t-x)^{\overline{k}}, x\right)\right\|_m +$$

$$\omega_1\left(f^{(N)}, \widetilde{\varepsilon}_N\right) \widetilde{\varepsilon}_N^{N-1} \left[\frac{(b-a)}{(N+1)!} + \frac{\widetilde{\varepsilon}_N}{2N!} + \frac{\widetilde{\varepsilon}_N^2}{4(N-1)!(b-a)}\right]. \quad (3.98)$$

Bibliography

[60] G.A. Anastassiou, *A "K-Attainable" inequality related to the convergence of Positive Linear Operators*, J. Approximation Theory, 44, 380–383 (1985).

[61] G.A. Anastassiou, *Moments in probability and approximation theory*, Pitman Research Notes in Mathematics series / Longman Scientific & Technical, Essex, New York, UK, USA, 1993.

[62] G.A. Anastassiou, *Quantitative Approximations*, Chapmen & Hall/CRC, London, New York, 2001.

[63] G.A. Anastassiou, *Parametrized, Deformed and General Neural Networks*, Springer, Heidelberg, New York, 2023.

[64] G.A. Anastassiou, *Trigonometric and Hyperbolic Generated Approximation Theory*, World Scientific, Singapore, New York, 2025.

[65] J. Peetre, *A Theory of Interpolation of Normed Spaces*, Notes Universidade de Brasilia, 1963.

[66] O. Shisha and B. Mond, *The degree of convergence of sequences of linear positive operators*, Nat. Acad. Sci. U.S. 60 (1968), 1196–1200.

[67] O. Shisha and B. Mond, *The degree of Approximation to Periodic Functions by Linear Positive Operators*, J. Approximation Theory, 1, 335–339 (1968).

[68] Z. Chen and F. Cao, *The approximation operators with sigmoidal functions*, Computers and Mathematics with Applications, 58 (2009), 758–765.

[69] D. Costarelli, R. Spigler, *Approximation results for neural network operators activated by sigmoidal functions*, Neural Networks 44 (2013), 101–106.

[70] D. Costarelli, R. Spigler, *Multivariate neural network operators with sigmoidal activation functions*, Neural Networks 48 (2013), 72–77.

[71] S. Haykin, *Neural Networks: A Comprehensive Foundation* (2 ed.), Prentice Hall, New York, 1998.

[72] W. McCulloch and W. Pitts, *A logical calculus of the ideas immanent in nervous activity*, Bulletin of Mathematical Biophysics, 7 (1943), 115–133.

[73] T.M. Mitchell, *Machine Learning*, WCB-McGraw-Hill, New York, 1997.

[74] Yu, Dansheng; Cao, Feilong, *Construction and approximation rate for feed-forward neural network operators with sigmoidal functions*, J. Comput. Appl. Math. 453 (2025), paper No. 116150.

[75] Cen, Siyu; Jin, Bangti; Quan, Qimeng; Zhou, Zhi; *Hybrid neural-network FEM approximation of diffusion coeficient in elyptic and parabolic problems.* IMA J. Numer. Anal. 44 (2024), no. 5, 3059–3093.

[76] Coroianu Lucian; Costarelli, Danillo; Natale, Mariarosaria; Pantiş, Alexandra; *The approximation capabilities of Durrmeyer-type neural network operators.* J. Appl. Math. Comput. 70 (2024), no. 5, 4581–4599.

[77] Warin, Xavier; *The GroupMax neural network approximation of convex functions.* IEEE Trans. Neural Netw. Learn. Syst. 35 (2024), no. 8, 11608–11612.

[78] Fabra, Arnau; Guasch, Oriol; Baiges, Joan; Codina, Ramon; *Approximation of acoustic black holes with finite element mixed formulations and artificial neural network correction terms.* Finite Elem. Anal. Des. 241 (2024), paper No. 104236.

[79] Grohs, Philipp; Voigtlaender, Felix. *Proof of the theory-to-practice gap in deep learning via sampling complexity bounds for neural network approximation spaces.* Found. Comput. Math. 24 (2024), no. 4, 1085–1143.

[80] Basteri, Andrea; Trevisan, Dario; *Quantitative Gaussian approximation of randomly initialized deep neural networks.* Mach. Learn. 113 (2024), no. 9, 6373–6393.

[81] De Ryck, T.; Mishra, S. *Error analysis for deep neural network approximations of parametric hyperbolic conservation laws.* Math. Comp. 93 (2024), no. 350, 2643–2677.

[82] Liu, Jie; Zhang, Baoji; Lai, Yuyang; Fang, Liqiau. *Hull form optimization reserach based on multi-precision back-propagation neural network approximation model.* Internal. J. Numer. Methods Fluid 96 (2024), no. 8, 1445–1430.

[83] Yoo, Jihahm; Kim, Jaywon; Gim, Minjung; Lee, Haesung. *Error estimates of physics-informed neural networks for initial value problems.* J. Korean Soc. Ind. Appl. Math. 28 (2024), no. 1, 33–58.

[84] Kaur, Jaspreet; Goyal, Meenu; *Hyers-Ulam stability of some positive linear operators.* Stud. Univ. Babeş-Bolyai Math. 70 (2025), no. 1, 105–114.

[85] Abel, Ulrich; Acu, Ana Maria; Heilmann, Margareta; Raşa, Ioan; *On some Cauchy problems and positive linear operators.* Mediterr. J. Math. 22 (2025), no. 1, Paper No. 20, 19 pp.

[86] Moradi, Hamid Reza; Furuichi, Shigeru; Sababheh, Mohammad; *Operator quadratic mean and positive linear maps.* J. Math. Inequal. 18 (2024), no. 4, 1263–1279.

[87] Bustamante, Jorge; Torres-Campos, José D. *Power series and positive linear operators in weighted spaces.* Serdica Math. J. 50 (2024), no. 3-4, 225–250.

[88] Acu, Ana-Maria; Rasa, Ioan; Sofonea, Florin; *Composition of some positive linear integral operators.* Demonstr. Math. 57 (2024), no. 1, Paper No. 20240018, 14 pp.

[89] Patel, Prashantkumar G. *On positive linear operators linking gamma, Mittag-Leffler and Wright functions.* Int. J. Appl. Comput. Math. 10 (2024), no. 5, Paper No. 152, 12 pp.

[90] G.A. Anastassiou, *Neural Networks a Positive Linear Operators*, Mathematics, 2025, 13, 1112; `https://doi.org/10.3390/math13071112`.

[91] G.A. Anastassiou, *Generalized Logistic Neural Networks as Positive Linear Operators*, Symmetry, 2025, 17, 746; `https://doi.org/10.3390/sym17050746`.

Chapter 4

Generalized Logistic Neural Networks in Infinite Domain as Positive Linear Operators

Generalized logistic neural network operators in infinite domain are interpreted as positive linear operators and related general theory applies to them. These operators are induced by a symmetrized density function deriving from the parametrized, deformed and symmetrized generalized logistic activation function. We are acting on the space of continuous and bounded functions on the real line to the reals. We study quantitatively the rate of convergence of these neural network operators to the unit operator. Our inequalities involve the modulus of continuity of the function under approximation or its derivative under initial conditions. It follows [117].

4.1 Introduction

The author studied extensively the quantitative approximation of positive linear operators to the unit since 1985, see for example [92]–[94], [97], which we use in this work. He originated from the quantitative weak convergence of finite positive measures to the unit Dirac measure, having as a method the geometric moment theory, see [93], and he produced best upper bounds, leading to attained (i.e. sharp Jackson type inequalities), e.g. see [92], [93]. These studies have gone to all possible directions, univariate and multivariate, though in this work we stay only on the univariate approach over infinite domain.

Also, since 1997, this author started the study of the quantitative convergence of neural network operators to the unit, and he has been written since then numerous of articles and books, e.g. see [94], [96], [97], are what we use here.

The wide range of neural network operators over $\mathbb{R}$ being treated by the author all along, are indeed by nature positive linear operators.

Here, for the first time in the literature, neural network operators in infinite domain are treated as positive linear operators. Now the activation function comes from the generalized logistic function.

So, methods of positive linear operators in infinite domain apply here to our infinite summation defined neural network operators, producing new and interesting pointwise and uniform results.

Of great inspiration always have been [99], [100]. For classical studies on neural network we recommend also, [101]–[106].

For newer work on neural networks we also refer the reader to [107]–[116].

4.2 Basics

Here we follow [96], pp. 395–417.

Our activation function here to be used is the q-deformed and λ-parametrized function

$$\varphi_{q,\lambda}(x) = \frac{1}{1+qA^{-\lambda x}}, \quad x \in \mathbb{R}, \ q,\lambda > 0, \ A > 1. \tag{4.1}$$

This is the A-generalized logistic function.

For more read Chapter 16 of [96]: "Banach space valued ordinary and fractional neural network approximation based on q-deformed and λ-parametrized A-generalized logistic function".

This chapter motivates our current work.

The proposed "symmetrization technique" aims to use half data feed to our neural networks.

We will employ the following density function

$$G_{q,\lambda}(x) := \frac{1}{2}\left(\varphi_{q,\lambda}(x+1) - \varphi_{q,\lambda}(x-1)\right), \quad x \in \mathbb{R}, \ q,\lambda > 0. \tag{4.2}$$

We have that

$$G_{q,\lambda}(-x) = G_{\frac{1}{q},\lambda}(x), \tag{4.3}$$

and

$$G_{\frac{1}{q},\lambda}(-x) = G_{q,\lambda}(x), \quad \forall\, x \in \mathbb{R}. \tag{4.4}$$

Adding (4.3) and (4.4) we obtain

$$G_{q,\lambda}(-x) + G_{\frac{1}{q},\lambda}(-x) = G_{q,\lambda}(x) + G_{\frac{1}{q},\lambda}(x), \quad \forall\, x \in \mathbb{R}, \tag{4.5}$$

the key to this work.

So that

$$W(x) := \frac{G_{q,\lambda}(x) + G_{\frac{1}{q},\lambda}(x)}{2} \tag{4.6}$$

is an even function, symmetric with respect to the y-axis.

The global maximum of $G_{q,\lambda}$ is given by (16.18), p. 401 of [96] as

$$G_{q,\lambda}\left(\frac{\log_A q}{\lambda}\right) = \frac{A^\lambda - 1}{2(A^\lambda + 1)}. \tag{4.7}$$

And, the global max of $G_{\frac{1}{q},\lambda}$ is

$$G_{\frac{1}{q},\lambda}\left(\frac{\log_A \frac{1}{q}}{\lambda}\right) = G_{\frac{1}{q},\lambda}\left(\frac{-\log_A q}{\lambda}\right) = \frac{A^\lambda - 1}{2(A^\lambda + 1)}, \tag{4.8}$$

both sharing the same maximum at symmetric points.

By Theorem 16.1, p. 401 of [96], we have that

$$\sum_{i=-\infty}^{\infty} G_{q,\lambda}(x-i) = 1, \ \forall\, x \in \mathbb{R}, \lambda, q > 0, \ A > 1, \tag{4.9}$$

and

$$\sum_{i=-\infty}^{\infty} G_{\frac{1}{q},\lambda}(x-i) = 1, \ \forall\, x \in \mathbb{R}, \lambda, q > 0, \ A > 1. \tag{4.10}$$

Consequently, we derive that

$$\sum_{i=-\infty}^{\infty} W(x-i) = 1, \ \forall\, x \in \mathbb{R}. \tag{4.11}$$

By Theorem 16.2, p. 402 of [95], we have that

$$\int_{-\infty}^{\infty} G_{q,\lambda}(x)\, dx = 1, \quad \lambda, q > 0, \ A > 1, \tag{4.12}$$

similarly it holds

$$\int_{-\infty}^{\infty} G_{\frac{1}{q},\lambda}(x)\, dx = 1, \tag{4.13}$$

so that

$$\int_{-\infty}^{\infty} W(x)\, dx = 1, \tag{4.14}$$

therefore $W(x)$ is a density function.

By Theorem 16.3, p. 402 of [96], we have:

Let $0<\beta<1$, and $N\in\mathbb{N}$ with $N^{1-\beta}>2$. Then

$$\sum_{\left\{\substack{k=-\infty \\ :|Nx-k|\geq N^{1-\beta}}\right.}^{\infty} G_{q,\lambda}\left(Nx-k\right)<2\max\left\{q,\frac{1}{q}\right\}\frac{1}{A^{\lambda\left(N^{1-\beta}-2\right)}}=\gamma A^{-\lambda\left(N^{1-\beta}-2\right)}, \tag{4.15}$$

where $\lambda,q>0$, $A>1$; $\gamma:=2\max\left\{q,\frac{1}{q}\right\}$.

Similarly, we get that

$$\sum_{\left\{\substack{k=-\infty \\ :|Nx-k|\geq N^{1-\beta}}\right.}^{\infty} G_{\frac{1}{q},\lambda}\left(Nx-k\right)<\gamma A^{-\lambda\left(N^{1-\beta}-2\right)}. \tag{4.16}$$

Consequently we obtain that

$$\sum_{\left\{\substack{k=-\infty \\ :|Nx-k|\geq N^{1-\beta}}\right.}^{\infty} W\left(Nx-k\right)<\gamma A^{-\lambda\left(N^{1-\beta}-2\right)}, \tag{4.17}$$

where $\gamma:=2\max\left\{q,\frac{1}{q}\right\}$.

Here $\lceil\cdot\rceil$ denotes the ceiling of the number, and $\lfloor\cdot\rfloor$ its integral part.

We make

Remark 4.1. Let $x\in[1,\infty)$, and we apply the mean value theorem. We have that

$$G_{q,\lambda}\left(x\right)=\frac{1}{2}\left(\varphi_{q,\lambda}\left(x+1\right)-\varphi_{q,\lambda}\left(q-1\right)\right)$$

$$\frac{1}{2}\varphi'_{q,\lambda}\left(\xi\right)2=\varphi'_{q,\lambda}\left(\xi\right)=\frac{q\lambda\ln A}{\left(1+qA^{-\lambda\xi}\right)^{2}A^{\lambda\xi}}, \tag{4.18}$$

where $0\leq x-1<\xi<x+1$.

We obtain that

$$G_{q,\lambda}\left(x\right)<\left(q\lambda\ln A\right)A^{-\lambda\xi}<\left(q\lambda\ln A\right)A^{-\lambda(x-1)},\quad x\geq 1. \tag{4.19}$$

Similarly it holds

$$G_{\frac{1}{q},\lambda}\left(x\right)<\left(\frac{1}{q}\lambda\ln A\right)A^{-\lambda(x-1)},\quad x\geq 1. \tag{4.20}$$

And, finally it holds

$$W\left(x\right)<\frac{1}{2}\left(q+\frac{1}{q}\right)\left(\lambda\ln A\right)A^{-\lambda(x-1)},\quad x\geq 1, \tag{4.21}$$

where $q,\lambda>0$, $A>1$.

We also make

Remark 4.2. We would like to estimate ($N, m \in \mathbb{N}$):

$$\sum_{\begin{cases} k=-\infty \\ : |Nx-k| \geq N^{1-\beta} \end{cases}}^{\infty} W(|Nx-k|)\,|Nx-k|^m \overset{(4.21)}{\leq}$$

$$\frac{1}{2}\left(q+\frac{1}{q}\right)(\lambda \ln A) \sum_{\begin{cases} k=-\infty \\ : |Nx-k| \geq N^{1-\beta} \end{cases}}^{\infty} |Nx-k|^m A^{-\lambda(|Nx-k|-1)} = \tag{4.22}$$

$$\frac{1}{2}\left(q+\frac{1}{q}\right)(\lambda \ln A) \sum_{\begin{cases} k=-\infty \\ : |Nx-k| \geq N^{1-\beta} \end{cases}}^{\infty} |Nx-k|^m e^{-\lambda(\ln A)(|Nx-k|-1)} =$$

$$\frac{1}{2}\left(q+\frac{1}{q}\right)\lambda(\ln A) A^{\lambda} \sum_{\begin{cases} k=-\infty \\ : |Nx-k| \geq N^{1-\beta} \end{cases}}^{\infty} |Nx-k|^m e^{-\lambda(\ln A)|Nx-k|} =: (*).$$

For convenience set $\mu := \lambda \ln A > 0$.

Notice that

$$e^{\frac{\mu|Nx-k|}{2}} = \sum_{\theta=0}^{\infty} \frac{\left(\frac{\mu|Nx-k|}{2}\right)^{\theta}}{\theta!} \geq \left(\frac{\mu|Nx-k|}{2}\right)^m \frac{1}{m!}. \tag{4.23}$$

Therefore we have

$$\left(\frac{\mu|Nx-k|}{2}\right)^m \leq m! e^{\frac{\mu|Nx-k|}{2}}, \quad \text{or} \tag{4.24}$$

$$(\mu|Nx-k|)^m \leq 2^m m! e^{\frac{\mu|Nx-k|}{2}}, \text{ or} \tag{4.25}$$

$$|Nx-k|^m \leq \frac{2^m}{\mu^m} m! e^{\frac{\mu|Nx-k|}{2}}. \tag{4.26}$$

Hence it holds

$$(*) \leq \frac{1}{2}\left(q+\frac{1}{q}\right)\lambda(\ln A) A^{\lambda}$$

$$\sum_{\substack{k=-\infty \\ : |Nx-k| \geq N^{1-\beta}}}^{\infty} \frac{2^m}{\lambda^m (\ln A)^m} m! e^{\frac{\lambda \ln A |Nx-k|}{2}} e^{-\lambda \ln A |Nx-k|} =$$

$$\frac{1}{2}\left(q+\frac{1}{q}\right) \lambda (\ln A) A^{\lambda} \frac{2^m}{\lambda^m (\ln A)^m} m! \sum_{\substack{k=-\infty \\ : |Nx-k| \geq N^{1-\beta}}}^{\infty} e^{-\frac{\lambda \ln A |Nx-k|}{2}} = \tag{4.27}$$

$$\frac{1}{2}\left(q+\frac{1}{q}\right) \frac{A^{\lambda} 2^m m!}{\lambda^{m-1} (\ln A)^{m-1}} \left(\sum_{\substack{k=-\infty \\ : |Nx-k| \geq N^{1-\beta}}}^{\infty} e^{-\frac{\lambda \ln A}{2} |Nx-k|} \right) \leq$$

$$\left(q+\frac{1}{q}\right) \frac{A^{\lambda} 2^m m!}{\lambda^{m-1} (\ln A)^{m-1}} \left(\int_{N^{1-\beta}-1}^{\infty} e^{-\frac{\lambda \ln A}{2} x} dx \right) =$$

$$\frac{2}{\lambda \ln A}\left(q+\frac{1}{q}\right) \frac{A^{\lambda} 2^m m!}{\lambda^{m-1} (\ln A)^{m-1}} \left(\int_{N^{1-\beta}-1}^{\infty} e^{-\frac{\lambda \ln A}{2} x} d\left(\frac{\lambda \ln A}{2}\right) x \right) = \tag{4.28}$$

$$\frac{\left(q+\frac{1}{q}\right) A^{\lambda} 2^{m+1} m!}{\lambda^m (\ln A)^m} \left(\int_{N^{1-\beta}-1}^{\infty} e^{-y} dy \right) =$$

$$\frac{\left(q+\frac{1}{q}\right) A^{\lambda} 2^{m+1} m!}{\lambda^m (\ln A)^m} \left(-e^{-y} \Big|_{N^{1-\beta}-1}^{\infty} \right) =$$

$$\frac{\left(q+\frac{1}{q}\right) A^{\lambda} 2^{m+1} m!}{\lambda^m (\ln A)^m} \left(e^{-y} \Big|_{\infty}^{N^{1-\beta}-1} \right) = \tag{4.29}$$

$$\frac{\left(q+\frac{1}{q}\right) A^{\lambda} 2^{m+1} m!}{\lambda^m (\ln A)^m} \left(e^{-\frac{\lambda \ln A}{2} x} \Big|_{\infty}^{N^{1-\beta}-1} \right) =$$

$$\frac{\left(q+\frac{1}{q}\right) A^{\lambda} 2^{m+1} m!}{\lambda^m (\ln A)^m} \left(e^{-\frac{\lambda \ln A}{2} \left(N^{1-\beta}-1\right)} \right).$$

Thus, we have proved:

$$\sum_{\begin{cases} k=-\infty \\ : |Nx-k| \geq N^{1-\beta} \end{cases}}^{\infty} W(|Nx-k|)\,|Nx-k|^m \leq$$

$$\frac{\left(q+\frac{1}{q}\right)2^{m+1}m!}{\lambda^m (\ln A)^m} A^{-\frac{\lambda}{2}\left(N^{1-\beta}-3\right)}, \tag{4.30}$$

where $m, N \in \mathbb{N}$; $q, \lambda > 0$, $A > 1$, $0 < \beta < 1$.

We also make

Remark 4.3. We also estimate ($\gamma := 2\max\left\{q, \frac{1}{q}\right\}$)

$$\sum_{\begin{cases} k=-\infty \\ : |Nx-k| \geq N^{1-\beta} \end{cases}}^{\infty} W(|Nx-k|)\,(1+|Nx-k|)^m \leq$$

$$2^{m-1} \sum_{\begin{cases} k=-\infty \\ : |Nx-k| \geq N^{1-\beta} \end{cases}}^{\infty} W(|Nx-k|)\,(1+|Nx-k|^m) = \tag{4.31}$$

$$2^{m-1}\left\{ \sum_{\begin{cases} k=-\infty \\ : |Nx-k| \geq N^{1-\beta} \end{cases}}^{\infty} W(|Nx-k|) + \right.$$

$$\left. \sum_{\begin{cases} k=-\infty \\ : |Nx-k| \geq N^{1-\beta} \end{cases}}^{\infty} W(|Nx-k|)\,|Nx-k|^m \right\} \overset{\text{(by (4.17), (4.30))}}{\leq}$$

$$2^{m-1}\left\{ \gamma A^{-\lambda\left(N^{1-\beta}-2\right)} + \left(\frac{\left(q+\frac{1}{q}\right)2^{m+1}m!}{\lambda^m (\ln A)^m}\right) A^{-\frac{\lambda}{2}\left(N^{1-\beta}-3\right)} \right\}.$$

We have proved that

$$\sum_{\left\{\begin{array}{l} k=-\infty \\ : |Nx-k| \geq N^{1-\beta} \end{array}\right.}^{\infty} W(|Nx-k|)(1+|Nx-k|)^m \leq$$

$$2^{m-1}\left\{\gamma A^{-\lambda\left(N^{1-\beta}-2\right)}+\left(\frac{\left(q+\frac{1}{q}\right)2^{m+1}m!}{\lambda^m(\ln A)^m}\right)A^{-\frac{\lambda}{2}\left(N^{1-\beta}-3\right)}\right\}, \quad (4.32)$$

where $m, N \in \mathbb{N}$; $q, \lambda > 0$, $A > 1$, $0 < \beta < 1$, $\gamma = 2\max\left\{q, \frac{1}{q}\right\}$.

We need

Definition 4.1. In this chapter we study the smooth approximation properties of the following interpolation neural network operators acting on $f \in C_B(\mathbb{R})$ (continuous and bounded functions):

(i) the basic ones

$$B_N(f,x) := \sum_{k=-\infty}^{\infty} f\left(\frac{k}{N}\right) W(Nx-k), \quad \forall\, x \in \mathbb{R},\ N \in \mathbb{N}, \quad (4.33)$$

(ii) the Kantorovich type operators

$$C_N(f,x) := \sum_{k=-\infty}^{\infty}\left(N\int_{\frac{k}{N}}^{\frac{k+1}{N}} f(t)\,dt\right) W(Nx-k), \quad \forall\, x \in \mathbb{R},\ N \in \mathbb{N}, \quad (4.34)$$

(iii) let $\theta \in \mathbb{N}$, $w_r \geq 0$, $\sum_{r=0}^{\theta} w_r = 1$, $k \in \mathbb{Z}$, and

$$\delta_{Nk}(f) := \sum_{r=0}^{\theta} w_r f\left(\frac{k}{N}+\frac{r}{N\theta}\right), \quad (4.35)$$

we consider also the quadrature type operators

$$D_N(f,x) := \sum_{k=-\infty}^{\infty} \delta_{Nk}(f)\, W(Nx-k), \quad \forall\, x \in \mathbb{R},\ N \in \mathbb{N}. \quad (4.36)$$

We will be using the first modulus of continuity:

$$\omega_1(f,\delta) := \sup_{\substack{x,y\in\mathbb{R}: \\ |x-y|\leq\delta}} |f(x)-f(y)|, \quad \delta > 0, \quad (4.37)$$

where $f \in C(\mathbb{R})$ which is bounded and or uniformly continuous.

The space $C_u(\mathbb{R})$ are the uniformly continuous functions on $\mathbb{R}$, and $C_{uB}(\mathbb{R})$ the uniformly and bounded functions on $\mathbb{R}$.

We notice that $B_N(1) = C_N(1) = D_N(1) = 1$, and B_N, C_N, D_N are positive linear operators, and they map $C_B(\mathbb{R})$ into $C_B(\mathbb{R})$; for the last see (4.6) and [96], Ch. 15, Theorem 15.15, p. 391.

4.3 Auxilliary Results

We need the following important results.

Lemma 4.1. *Let here* $x \in \mathbb{R}$, $j \in \mathbb{N}$; $0 < \beta < 1$, $A > 1$, $q, \lambda > 0$; $N \in \mathbb{N}$ *with* $N^{1-\beta} > 2$. *Then*

$$0 < B_N\left(|\cdot - x|^j\right)(x) \leq$$

$$\frac{1}{N^{\beta j}} + \frac{1}{N^j}\left(\frac{\left(q + \frac{1}{q}\right)2^{j+1} j!}{\lambda^j (\ln A)^j}\right) A^{-\frac{\lambda}{2}\left(N^{1-\beta}-3\right)} =: \psi_1(N) < +\infty, \quad (4.38)$$

and

$$0 < \left\| B_N\left(|\cdot - x|^j\right)(x)\right\|_\infty \leq \psi_1(N) \to 0, \text{ as } N \to \infty. \quad (4.39)$$

Proof. We have that

$$B_N\left(|\cdot - x|^j\right)(x) = \sum_{k=-\infty}^{\infty}\left|\frac{k}{N} - x\right|^j W(Nx - k) =$$

$$\sum_{\begin{cases} k = -\infty \\ : \left|x - \frac{k}{N}\right| < \frac{1}{N^\beta}\end{cases}}^{\infty} \left|\frac{k}{N} - x\right|^j W(Nx - k) +$$

$$\sum_{\begin{cases} k = -\infty \\ : \left|x - \frac{k}{N}\right| \geq \frac{1}{N^\beta}\end{cases}}^{\infty} \left|\frac{k}{N} - x\right|^j W(Nx - k) \overset{(4.11)}{\leq} \quad (4.40)$$

$$\frac{1}{N^{\beta j}} + \frac{1}{N^j} \sum_{\begin{cases} k = -\infty \\ : |Nx - k| \geq N^{1-\beta}\end{cases}}^{\infty} |Nx - k|^j W(|Nx - k|) \overset{(4.30)}{\leq}$$

$$\frac{1}{N^{\beta j}} + \frac{1}{N^j}\left(\frac{\left(q + \frac{1}{q}\right)2^{j+1} j!}{\lambda^j (\ln A)^j}\right) A^{-\frac{\lambda}{2}\left(N^{1-\beta}-3\right)} < +\infty.$$

□

Lemma 4.2. *All as in Lemma 4.1. Then*

$$0 < C_N\left(|\cdot - x|^j\right)(x) \leq \left(\frac{1}{N} + \frac{1}{N^\beta}\right)^j +$$

$$\frac{2^{j-1}}{N^j}\left[\gamma A^{-\lambda\left(N^{1-\beta}-2\right)} + \left(\frac{\left(q+\frac{1}{q}\right)2^{j+1}j!}{\lambda^j(\ln A)^j}\right)A^{-\frac{\lambda}{2}\left(N^{1-\beta}-3\right)}\right] =: \psi_2(N) < +\infty, \tag{4.41}$$

and

$$0 < \left\|C_N\left(|\cdot - x|^j\right)(x)\right\|_\infty \leq \psi_2(N) \to 0, \text{ as } N \to \infty. \tag{4.42}$$

Proof. We have that

$$C_N\left(|\cdot - x|^j\right)(x) = \sum_{k=-\infty}^{\infty} W(Nx-k)\left(N\int_0^{\frac{1}{N}}\left|t + \frac{k}{N} - x\right|^j dt\right) \leq$$

$$\sum_{k=-\infty}^{\infty} W(Nx-k)\left(N\int_0^{\frac{1}{N}}\left(\left|\frac{k}{N} - x\right| + |t|\right)^j dt\right) \leq$$

$$\sum_{k=-\infty}^{\infty} W(Nx-k)\left(\left|\frac{k}{N} - x\right| + \frac{1}{N}\right)^j =$$

$$\sum_{\substack{k=-\infty \\ :\left|\frac{k}{N}-x\right| < \frac{1}{N^\beta}}}^{\infty} W(Nx-k)\left(\left|\frac{k}{N} - x\right| + \frac{1}{N}\right)^j +$$

$$\sum_{\substack{k=-\infty \\ :\left|\frac{k}{N}-x\right| \geq \frac{1}{N^\beta}}}^{\infty} W(Nx-k)\left(\left|\frac{k}{N} - x\right| + \frac{1}{N}\right)^j \leq \tag{4.43}$$

$$\left(\frac{1}{N^\beta} + \frac{1}{N}\right)^j + \frac{1}{N^j}\sum_{\substack{k=-\infty \\ :|Nx-k| \geq N^{1-\beta}}}^{\infty} W(|Nx-k|)\left(|Nx-k|+1\right)^j \overset{(4.32)}{\leq}$$

$$\left(\frac{1}{N^\beta} + \frac{1}{N}\right)^j + \frac{2^{j-1}}{N^j}\left\{\gamma A^{-\lambda\left(N^{1-\beta}-2\right)} + \left(\frac{\left(q+\frac{1}{q}\right)2^{j+1}j!}{\lambda^j(\ln A)^j}\right)A^{-\frac{\lambda}{2}\left(N^{1-\beta}-3\right)}\right\} < +\infty. \tag{4.44}$$

$\square$

Lemma 4.3. *All as in Lemma 4.1. Then*

$$0 < D_N\left(|\cdot - x|^j\right)(x) \le \left(\frac{1}{N} + \frac{1}{N^\beta}\right)^j +$$

$$\frac{2^{j-1}}{N^j}\left\{\gamma A^{-\lambda\left(N^{1-\beta}-2\right)} + \left(\frac{\left(q+\frac{1}{q}\right)2^{j+1}j!}{\lambda^j (\ln A)^j}\right)A^{-\frac{\lambda}{2}\left(N^{1-\beta}-3\right)}\right\} = \psi_2(N) < +\infty, \tag{4.45}$$

and

$$0 < \left\|D_N\left(|\cdot - x|^j\right)(x)\right\|_\infty \le \psi_2(N) \to 0, \text{ as } N \to \infty. \tag{4.46}$$

Proof. As similar to Lemma 4.2 is omitted. □

We need the following Hölder's type inequality for positive linear operators.

Theorem 4.1. *Let H be a positive linear operator from $C(\mathbb{R})$ into $C_B(\mathbb{R})$, and $f, g \in C(\mathbb{R})$, furthermore let $p, q > 1 : \frac{1}{p} + \frac{1}{q} = 1$. Assume that $H((|f(\cdot)|^p))(s_*)$, $H((|g(\cdot)|^q))(s_*) > 0$ for some $s_* \in \mathbb{R}$. Then*

$$H(|f(\cdot)g(\cdot)|)(s_*) \le \left(H((|f(\cdot)^p|))(s_*)\right)^{\frac{1}{p}}\left(H((|g(\cdot)^q|))(s_*)\right)^{\frac{1}{q}}. \tag{4.47}$$

Proof. Let $a, b \ge 0$, $p.q > 1 : \frac{1}{p} + \frac{1}{q} = 1$. The Young's inequality says

$$ab \le \frac{a^p}{p} + \frac{b^q}{q}. \tag{4.48}$$

Then

$$\frac{|f(s)|}{\left(H((|f(\cdot)^p|))(s_*)\right)^{\frac{1}{p}}} \cdot \frac{|g(s)|}{\left(H((|g(\cdot)^q|))(s_*)\right)^{\frac{1}{q}}} \le$$

$$\frac{|f(s)|^p}{p\left(H((|f(\cdot)^p|))(s_*)\right)} + \frac{|g(s)|^q}{q\left(H((|g(\cdot)^q|))(s_*)\right)}, \quad \forall\, s \in \mathbb{R}. \tag{4.49}$$

Hence it holds

$$\frac{H(|f(\cdot)||g(\cdot)|)(s_*)}{\left(H((|f(\cdot)^p|))(s_*)\right)^{\frac{1}{p}}\left(H((|g(\cdot)^q|))(s_*)\right)^{\frac{1}{q}}} \le$$

$$\frac{H((|f(\cdot)^p|))(s_*)}{p\left(H((|f(\cdot)^p|))(s_*)\right)} + \frac{H((|g(\cdot)^q|))(s_*)}{q\left(H((|g(\cdot)^q|))(s_*)\right)} = \frac{1}{p} + \frac{1}{q} = 1, \tag{4.50}$$

for $s_* \in \mathbb{R}$, proving the claim. □

4.4 Main Results

We present the following general approximation result.

Theorem 4.2. *Let $n \in \mathbb{N}$ and $f, f^{(n)} \in C_B(\mathbb{R})$, $x \in \mathbb{R}$. Consider H_N a sequence of positive linear operators from $C_B(\mathbb{R})$ into itself, $N \in \mathbb{N}$, such that $H_N(1) = 1$. Assume $H_N\left(|\cdot - x|^{n+1}\right)(x) > 0$, and $f^{(i)}(x) = 0$, for $i = 1, ..., n$.*

Then

$$|H_N(f)(x) - f(x)| \leq \omega_1\left(f^{(n)}, \left(\left(H_N\left(|\cdot - x|^{n+1}\right)\right)(x)\right)^{\frac{1}{n+1}}\right)$$

$$\left[H_N\left(|\cdot - x|^n\right)(x) + \frac{\left(H_N\left(|\cdot - x|^{n+1}\right)(x)\right)^{\frac{n}{n+1}}}{(n+1)}\right] < +\infty, \quad \forall\, N \in \mathbb{N}. \tag{4.51}$$

Given that $\lim_{N\to+\infty} H_N\left(|\cdot - x|^{n+1}\right)(x) = 0$, *then* $\lim_{N\to+\infty} H_N(f)(x) = f(x)$.

If $n = 1$, we derive

$$|H_N(f)(x) - f(x)| \leq \omega_1\left(f', \left(\left(H_N\left((\cdot - x)^2\right)\right)(x)\right)^{\frac{1}{2}}\right)$$

$$\left[H_N\left(|\cdot - x|\right)(x) + \frac{\left(H_N\left((\cdot - x)^2\right)(x)\right)^{\frac{1}{2}}}{2}\right] < +\infty, \quad \forall\, N \in \mathbb{N}. \tag{4.52}$$

Given that $\lim_{N\to+\infty} H_N\left((\cdot - x)^2\right)(x) = 0$, *then* $\lim_{N\to+\infty} H_N(f)(x) = f(x)$.

Proof. Let $x, y \in \mathbb{R}$, by Taylor's formula we have

$$f(y) = \sum_{i=0}^{n} f^{(i)}(x)\frac{(y-x)^i}{i!} + \frac{1}{(n-1)!}\int_x^y (y-t)^{n-1}\left(f^{(n)}(t) - f^{(n)}(x)\right)dt. \tag{4.53}$$

We call the remainder in (4.53) as

$$R_n^*(x,y) = \frac{1}{(n-1)!}\int_x^y (y-t)^{n-1}\left(f^{(n)}(t) - f^{(n)}(x)\right)dt, \quad \forall\, x, y \in \mathbb{R}. \tag{4.54}$$

By [93], p. 217, and [95], p. 194, Chapter 7, (7.27) there, we get

$$|R_n^*(x,y)| \leq \frac{\omega_1\left(f^{(n)}, \delta\right)}{n!}|x-y|^n\left(1 + \frac{|x-y|}{(n+1)\delta}\right), \quad \forall\, x, y \in \mathbb{R},\ \delta > 0. \tag{4.55}$$

We may write (4.55) as

$$|R_n^*(x,y)| \leq \frac{\omega_1\left(f^{(n)},\delta\right)}{n!}\left[|x-y|^n + \frac{|x-y|^{n+1}}{(n+1)\delta}\right], \ \forall\, x,y \in \mathbb{R}\ ,\ \delta>0. \tag{4.56}$$

That is

$$\left|f(y) - \sum_{i=0}^{n} f^{(i)}(x)\frac{(y-x)^i}{i!}\right| \leq \frac{\omega_1\left(f^{(n)},\delta\right)}{n!}\left[|x-y|^n + \frac{|x-y|^{n+1}}{(n+1)\delta}\right], \tag{4.57}$$

$\forall\, x,y \in \mathbb{R}$, $\delta>0$.

Furthermore it holds

$$|f(y)-f(x)| \leq \sum_{i=0}^{n}\left|f^{(i)}(x)\right|\frac{|y-x|^i}{i!} + \frac{\omega_1\left(f^{(n)},\delta\right)}{n!}\left[|x-y|^n + \frac{|x-y|^{n+1}}{(n+1)\delta}\right], \tag{4.58}$$

$\forall\, x,y \in \mathbb{R}$, $\delta>0$.

Here $f^{(i)}(x)=0$, for $i=1,\ldots,n$; and for a specific $x\in\mathbb{R}$, we get

$$|f(y)-f(x)| \leq \frac{\omega_1\left(f^{(n)},\delta\right)}{n!}\left[|x-y|^n + \frac{|x-y|^{n+1}}{(n+1)\delta}\right], \ \ \forall y\in\mathbb{R},\ \delta>0. \tag{4.59}$$

In case of $n=1$, we derive

$$|f(y)-f(x)| \leq |f'(x)|\,|y-x| + \omega_1(f',\delta)\left[|x-y| + \frac{(x-y)^2}{2\delta}\right], \tag{4.60}$$

$\forall\, x,y \in \mathbb{R}$, $\delta>0$.

If $n=1$ and $f'(x)=0$, for a specific $x\in\mathbb{R}$, we get that

$$|f(y)-f(x)| \leq \omega_1(f',\delta)\left[|x-y| + \frac{(x-y)^2}{2\delta}\right], \ \ \forall y\in\mathbb{R},\ \delta>0. \tag{4.61}$$

Let $H_N : C_B(\mathbb{R}) \to C_B(\mathbb{R})$ a sequence of positive linear operators, $N\in\mathbb{N}$.

Let $f,g\in C_B(\mathbb{R})$, we have

$$f = f-g+g \leq |f-g| + g,$$

hence

$$H_N(f) \leq H_N(|f-g|) + H_N(g), \tag{4.62}$$

and

$$H_N(f) - H_N(g) \leq H_N(|f-g|). \tag{4.63}$$

Similarly, it holds

$$H_N(g) - H_N(f) \leq H_N(|f-g|). \tag{4.64}$$

That is

$$|H_N(f) - H_N(g)| \leq H_N(|f-g|). \tag{4.65}$$

Next, we assume that $H_N(1) = 1$, $\forall\ N \in \mathbb{N}$. Then

$$|H_N(f)(x) - f(x)| = |H_N(f)(x) - f(x) H_N(1)(x)| =$$

$$|H_N(f)(x) - H_N(f(x))(x)| = |H_N(f(\cdot) - f(x))(x)| \overset{(4.65)}{\leq}$$

$$H_N(|f(\cdot) - f(x)|)(x) \overset{(4.59)}{\leq} \tag{4.66}$$

$$\frac{\omega_1\left(f^{(n)}, \delta\right)}{n!}\left[H_N(|\cdot - x|^n)(x) + \frac{H_N\left(|\cdot - x|^{n+1}\right)(x)}{(n+1)\delta}\right],$$

$\forall\ x \in \mathbb{R}$, $\delta > 0$.

Here we assume and choose

$$\delta = \left(H_N\left(|\cdot - x|^{n+1}\right)(x)\right)^{\frac{1}{n+1}} > 0. \tag{4.67}$$

Therefore we get

$$|H_N(f)(x) - f(x)| \leq \omega_1\left(f^{(n)}, \left(H_N\left(|\cdot - x|^{n+1}\right)(x)\right)^{\frac{1}{n+1}}\right)$$

$$\left[H_N(|\cdot - x|^n)(x) + \frac{\left(H_N\left(|\cdot - x|^{n+1}\right)(x)\right)^{\frac{n}{n+1}}}{(n+1)}\right], \tag{4.68}$$

$\forall\ N \in \mathbb{N}$.

If $n = 1$, we find that

$$|H_N(f)(x) - f(x)| \leq \omega_1\left(f', \left(H_N\left((\cdot - x)^2\right)(x)\right)^{\frac{1}{2}}\right)$$

$$\left[H_N(|\cdot - x|)(x) + \frac{\left(H_N\left((\cdot - x)^2\right)(x)\right)^{\frac{1}{2}}}{2}\right], \quad \forall\ N \in \mathbb{N}. \tag{4.69}$$

Using (4.67) and Theorem 4.1, for $g = 1$ and H_N such that $H_N(1) = 1$, we obtain

$$H_N(|\cdot - x|^n)(x) \leq \left(H_N\left(|\cdot - x|^{n+1}\right)(x)\right)^{\frac{n}{n+1}}. \tag{4.70}$$

In case of $n = 1$, we derive

$$H_N(|\cdot - x|)(x) \leq \sqrt{\left(H_N\left((\cdot - x)^2\right)(x)\right)}. \tag{4.71}$$

□

An application of Theorem 4.2 follows.

Call θ_N any of the operators B_N, C_N, D_N, $N \in \mathbb{N}$.

Theorem 4.3. *Let $n \in \mathbb{N}$ and $f, f^{(n)} \in C_B(\mathbb{R})$, $x \in \mathbb{R}$. Assume $f^{(i)}(x) = 0$, $i = 1, ..., n$. Then*

$$|\theta_N(f)(x) - f(x)| \leq \omega_1\left(f^{(n)}, \left(\left(\theta_N\left(|\cdot - x|^{n+1}\right)\right)(x)\right)^{\frac{1}{n+1}}\right)$$

$$\left[\theta_N\left(|\cdot - x|^n\right)(x) + \frac{\left(\theta_N\left(|\cdot - x|^{n+1}\right)(x)\right)^{\frac{n}{n+1}}}{(n+1)}\right] < +\infty, \quad \forall\, N \in \mathbb{N}. \tag{4.72}$$

Here $\lim_{N\to+\infty} \theta_N(f)(x) = f(x)$.

If $n = 1$, we obtain

$$|\theta_N(f)(x) - f(x)| \leq \omega_1\left(f', \left(\left(\theta_N\left((\cdot - x)^2\right)\right)(x)\right)^{\frac{1}{2}}\right)$$

$$\left[\theta_N\left(|\cdot - x|\right)(x) + \frac{\left(\theta_N\left((\cdot - x)^2\right)(x)\right)^{\frac{1}{2}}}{2}\right] < +\infty, \quad \forall\, N \in \mathbb{N}. \tag{4.73}$$

Again it holds $\lim_{N\to+\infty} \theta_N(f)(x) = f(x)$.

Proof. By Theorem 4.2 and Lemmas 4.1–4.3. □

We mention the following result.

Theorem 4.4. *([98]) Let $f \in C_B(\mathbb{R})$, $x \in \mathbb{R}$. Consider H_N a sequence of positive linear operators from $C_B(\mathbb{R})$ into itself, $N \in \mathbb{N}$, such that $H_N(1) = 1$.*

Assume that $H_N(|\cdot - x|)(x) > 0$. Then

$$|H_N(f)(x) - f(x)| \leq 2\omega_1\left(f, H_N(|\cdot - x|)(x)\right), \quad \forall\, N \in \mathbb{N}. \tag{4.74}$$

Given that $\lim_{N\to+\infty} H_N(|\cdot - x|)(x) = 0$, and f is also uniformly continuous, we obtain $\lim_{N\to+\infty} H_N(f)(x) = f(x)$.

An application of Theorem 4.4 comes next.

Theorem 4.5. *Let $f \in C_B(\mathbb{R})$, $x \in \mathbb{R}$. Here $\theta_N = B_N, C_N, D_N$. Then*

$$|\theta_N(f)(x) - f(x)| \leq 2\omega_1\left(f, \theta_N(|\cdot - x|)(x)\right) < +\infty, \tag{4.75}$$

$\forall\, N \in \mathbb{N} : N^{1-\beta} > 2$, $0 < \beta < 1$.

If f is also uniformly continuous we derive $\lim_{N\to+\infty} \theta_N(f)(x) = f(x)$.

Proof. Direct application of Theorem 4.4 and Lemmas 4.1–4.3. □

Corollary 4.1. *Let $f \in C_B(\mathbb{R})$. Then*

$$\left\|B_N(f)-f\right\|_\infty \leq 2\omega_1\left(f,\left[\frac{1}{N^\beta}+\frac{1}{N}\left(\frac{4\left(q+\frac{1}{q}\right)}{\lambda\ln A}\right)A^{-\frac{\lambda}{2}\left(N^{1-\beta}-3\right)}\right]\right)<+\infty, \tag{4.76}$$

$\forall\, N \in \mathbb{N} : N^{1-\beta} > 2,\ 0<\beta<1;\ q,\lambda>0,\ A>1.$

If f is also uniformly continuous, we obtain $\lim_{N\to+\infty} B_N(f) = f$, pointwise and uniformly.

Proof. By Lemma 4.1 and Theorem 4.5. □

We finish with the following result.

Corollary 4.2. *Let $f \in C_B(\mathbb{R})$. Then*

$$\left\{\begin{array}{l}\left\|C_N(f)-f\right\|_\infty\\ \left\|D_N(f)-f\right\|_\infty\end{array}\right\} \leq$$

$$2\omega_1\left(f,\left[\left(\frac{1}{N^\beta}+\frac{1}{N}\right)+\frac{1}{N}\left\{\gamma A^{-\lambda\left(N^{1-\beta}-2\right)}+\left(\frac{4\left(q+\frac{1}{q}\right)}{\lambda\ln A}\right)A^{-\frac{\lambda}{2}\left(N^{1-\beta}-3\right)}\right\}\right]\right)<+\infty, \tag{4.77}$$

$\forall\, N \in \mathbb{N} : N^{1-\beta} > 2,\ 0<\beta<1;\ q,\lambda>0,\ A>1,\ \gamma := 2\max\left\{q,\frac{1}{q}\right\}.$

If f is also uniformly continuous, we derive that

$$\lim_{N\to+\infty} C_N(f) = \lim_{N\to+\infty} D_N(f) = f,$$

pointwise and uniformly.

Proof. By Lemmas 4.2, 4.3 and Theorem 4.5. □

Bibliography

[92] G.A. Anastassiou, *A "K-Attainable" inequality related to the convergence of Positive Linear Operators*, J. Approximation Theory, 44, 380–383 (1985).

[93] G.A. Anastassiou, *Moments in probability and approximation theory*, Pitman Research Notes in Mathematics series / Longman Scientific & Technical, Essex, New York, UK, USA, 1993.

[94] G.A. Anastassiou, *Quantitative Approximations*, Chapmen & Hall/CRC, London, New York, 2001.

[95] G.A. Anastassiou, *Intelligent Computations: Abstract Fractional Calculus, Inequalities, Approximations*, Springer, Heidelberg, New York, 2018.

[96] G.A. Anastassiou, *Parametrized, Deformed and General Neural Networks*, Springer, Heidelberg, New York, 2023.

[97] G.A. Anastassiou, *Trigonometric and Hyperbolic Generated Approximation Theory*, World Scientific, Singapore, New York, 2025.

[98] R.A. DeVore, *The Approximation of Continuous Functions by Positive Linear Operators*, Lecture Notes in Mathematics, Vol. 293, Springer, Heidelberg, New York, 1972.

[99] O. Shisha and B. Mond, *The degree of convergence of sequences of linear positive operators*, Nat. Acad. Sci. U.S. 60 (1968), 1196–1200.

[100] O. Shisha and B. Mond, *The degree of Approximation to Periodic Functions by Linear Positive Operators*, J. Approximation Theory, 1, 335–339 (1968).

[101] Z. Chen and F. Cao, *The approximation operators with sigmoidal functions*, Computers and Mathematics with Applications, 58 (2009), 758–765.

[102] D. Costarelli, R. Spigler, *Approximation results for neural network operators activated by sigmoidal functions*, Neural Networks 44 (2013), 101–106.

[103] D. Costarelli, R. Spigler, *Multivariate neural network operators with sigmoidal activation functions*, Neural Networks 48 (2013), 72–77.

[104] S. Haykin, *Neural Networks: A Comprehensive Foundation* (2 ed.), Prentice Hall, New York, 1998.

[105] W. McCulloch and W. Pitts, *A logical calculus of the ideas immanent in nervous activity*, Bulletin of Mathematical Biophysics, 7 (1943), 115–133.

[106] T.M. Mitchell, *Machine Learning*, WCB-McGraw-Hill, New York, 1997.

[107] Yu, Dansheng; Cao, Feilong, *Construction and approximation rate for feed-forward neural network operators with sigmoidal functions*, J. Comput. Appl. Math. 453 (2025), paper No. 116150.

[108] Cen, Siyu; Jin, Bangti; Quan, Qimeng; Zhou, Zhi; *Hybrid neural-network FEM approximation of diffusion coeficient in elyptic and parabolic problems.* IMA J. Numer. Anal. 44 (2024), no. 5, 3059–3093.

[109] Coroianu Lucian; Costarelli, Danillo; Natale, Mariarosaria; Pantiş, Alexandra; *The approximation capabilities of Durrmeyer-type neural network operators.* J. Appl. Math. Comput. 70 (2024), no. 5, 4581–4599.

[110] Warin, Xavier; *The GroupMax neural network approximation of convex functions.* IEEE Trans. Neural Netw. Learn. Syst. 35 (2024), no. 8, 11608–11612.

[111] Fabra, Arnau; Guasch, Oriol; Baiges, Joan; Codina, Ramon; *Approximation of acoustic black holes with finite element mixed formulations and artificial neural network correction terms.* Finite Elem. Anal. Des. 241 (2024), paper No. 104236.

[112] Grohs, Philipp; Voigtlaender, Felix. *Proof of the theory-to-practice gap in deep learning via sampling complexity bounds for neural network approximation spaces.* Found. Comput. Math. 24 (2024), no. 4, 1085–1143.

[113] Basteri, Andrea; Trevisan, Dario; *Quantitative Gaussian approximation of randomly initialized deep neural networks.* Mach. Learn. 113 (2024), no. 9, 6373–6393.

[114] De Ryck, T.; Mishra, S. *Error analysis for deep neural network approximations of parametric hyperbolic conservation laws.* Math. Comp. 93 (2024), no. 350, 2643–2677.

[115] Liu, Jie; Zhang, Baoji; Lai, Yuyang; Fang, Liqiau. *Hull form optimization reserach based on multi-precision back-propagation neural network approximation model.* Internal. J. Numer. Methods Fluid 96 (2024), no. 8, 1445–1460.

[116] Yoo, Jihahm; Kim, Jaywon; Gim, Minjung; Lee, Haesung. *Error estimates of physics-informed neural networks for initial value problems.* J. Korean Soc. Ind. Appl. Math. 28 (2024), no. 1, 33–58.

[117] G.A. Anastassiou, *Generalized Logistic Neural Networks as Positive Linear Operators over infinite domain*, Neural, Parallel and Scintific Computations, 33(2025), 39–54.

Chapter 5

Symmetrized and Perturbed Fuzzy Neural Network Approximation

Here we treat the univariate fuzzy quantitative approximation of fuzzy real valued functions on a compact interval by quasi-interpolation symmetrized and perturbed hyperbolic tangent and generalized logistic functions based fuzzy neural network operators. These approximations are derived by establishing fuzzy Jackson type inequalities involving the fuzzy modulus of continuity of the involved fuzzy function and the fuzzy derivative of the engaged function. The approximations are fuzzy pointwise and fuzzy uniform. The related feed-forward fuzzy neural networks are with one hidden layer. The foundation of this work is based on positive linear operators generated deterministic real neural networks approximations. It follows [139].

5.1 Introduction

The author in [118] and [119], see Chapters 2–5 was the first to establish neural network approximation to continuous functions with rates by very specifically defined neural network operators of Cardaliaguet-Euvrard and "Squashing" types ([129]), by employing the modulus of continuity of the engaged function or its high order derivative, and producing very tight Jackson type inequalities. He treats there both the univariate and multivariate cases. The defining these operators "bell-shaped" and "squashing" functions were assumed to be of compact support.

The author inspired by [130], continued his studies on neural networks approximation by introducing and using the proper quasi-interpolation operators of various sigmoidal activation functions resulted into [122]–[126], by treating both the univariate and multivariate cases.

In Section 5.2, we give the basics of symmetrized, perturbed hyperbolic tangent neural network deterministic real operators, see also [127].

In Section 5.3, we give the basics of symmetrized, perturbed generalized logistic neural network deterministic real operators, see also [128].

In Section 5.4, we present the basics of fuzzy real analysis and we introduce the symmetrized and perturbed fuzzy neural network operators corresponding to Sections 5.2, 5.3. We prove also some very important properties. See also [122].

In Section 5.5, we demonstrate our fuzzy results: fuzzy symmetrized and perturbed neural network quantitative approximations, the fuzzy analogs of the Sections 5.2, 5.3 positive linear operators generated results.

Our fuzzy feed-forward neural networks (FFNNs) are with one hidden layer. About neural networks in general study [132], [134], [135].

5.2 Background I: About Symmetrized Perturbed Hyperbolic Tangent Neural Network Operators ([127])

Here we follow [125], pp. 455–460.

Our activation function here to be used is

$$g_{q,\lambda}(x) := \frac{e^{\lambda x} - qe^{-\lambda x}}{e^{\lambda x} + qe^{-\lambda x}}, \quad \lambda, q > 0, \ x \in \mathbb{R}. \tag{5.1}$$

Above λ is the parameter and q is the deformation coefficient, typically it is $0 < \lambda, q \leq 1$.

For more read Chapter 18 of [125]: "q-Deformed and λ-Parametrized Hyperbolic Tangent based Banach space Valued Ordinary and Fractional Neural Network Approximation".

This chapter motivates our current work.

The proposed "symmetrization method" aims to use half data feed to our neural networks.

We will employ the following density function

$$M_{q,\lambda}(x) := \frac{1}{4}\left(g_{q,\lambda}(x+1) - g_{q,\lambda}(x-1)\right) > 0, \tag{5.2}$$

$\forall x \in \mathbb{R}$; $q, \lambda > 0$.

We have that

$$M_{q,\lambda}(-x) = M_{\frac{1}{q},\lambda}(x), \quad \forall\, x \in \mathbb{R};\ q, \lambda > 0, \tag{5.3}$$

and

$$M_{\frac{1}{q},\lambda}(-x) = M_{q,\lambda}(x), \quad \forall\, x \in \mathbb{R};\ q, \lambda > 0. \tag{5.4}$$

Adding (5.3) and (5.4) we obtain

$$M_{q,\lambda}(-x) + M_{\frac{1}{q},\lambda}(-x) = M_{q,\lambda}(x) + M_{\frac{1}{q},\lambda}(x), \tag{5.5}$$

the key to this work.

So that

$$\Phi(x) := \frac{M_{q,\lambda}(x) + M_{\frac{1}{q},\lambda}(x)}{2} > 0, \tag{5.6}$$

is an even function, symmetric with respect to the y-axis.

By (18.18) of [125], we have

$$\begin{aligned} &M_{q,\lambda}\left(\tfrac{\ln q}{2\lambda}\right) = \tfrac{\tanh(\lambda)}{2}, \\ &\text{and} \\ &M_{\frac{1}{q},\lambda}\left(-\tfrac{\ln q}{2\lambda}\right) = \tfrac{\tanh(\lambda)}{2}, \quad \lambda > 0. \end{aligned} \tag{5.7}$$

sharing the same maximum at symmetric points.

By Theorem 18.1, p. 458 of [125], we have that

$$\begin{aligned} &\textstyle\sum_{i=-\infty}^{\infty} M_{q,\lambda}(x - i) = 1, \quad \forall\, x \in \mathbb{R},\ \lambda, q > 0, \\ &\text{and} \\ &\textstyle\sum_{i=-\infty}^{\infty} M_{\frac{1}{q},\lambda}(x - i) = 1, \quad \forall\, x \in \mathbb{R},\ \lambda, q > 0. \end{aligned} \tag{5.8}$$

Consequently, we derive that

$$\sum_{i=-\infty}^{\infty} \Phi(x - i) = 1, \quad \forall\, x \in \mathbb{R}. \tag{5.9}$$

By Theorem 18.2, p. 459 of [125], we have that

$$\begin{aligned} &\textstyle\int_{-\infty}^{\infty} M_{q,\lambda}(x)\, dx = 1, \\ &\text{and} \\ &\textstyle\int_{-\infty}^{\infty} M_{\frac{1}{q},\lambda}(x)\, dx = 1, \end{aligned} \tag{5.10}$$

so that

$$\int_{-\infty}^{\infty} \Phi(x)\, dx = 1, \tag{5.11}$$

therefore Φ is a density function.

By Theorem 18.3, p. 459 of [125], we have:

Let $0 < \alpha < 1$, and $n \in \mathbb{N}$ with $n^{1-\alpha} > 2$; $q, \lambda > 0$. Then

$$\sum_{\substack{k=-\infty \\ : |nx-k| \geq n^{1-\alpha}}}^{\infty} M_{q,\lambda}(nx-k) < 2\max\left\{q, \frac{1}{q}\right\} e^{4\lambda} e^{-2\lambda n^{(1-\alpha)}} = Te^{-2\lambda n^{(1-\alpha)}}, \tag{5.12}$$

where $T := 2\max\left\{q, \frac{1}{q}\right\} e^{4\lambda}$.

Similarly, we get that

$$\sum_{k=-\infty}^{\infty} M_{\frac{1}{q},\lambda}(nx-k) < Te^{-2\lambda n^{(1-\alpha)}}. \tag{5.13}$$

Consequently we obtain that

$$\sum_{\substack{k=-\infty \\ : \ nx-k| \geq n^{1-\alpha}}}^{\infty} \Phi(nx-k) < Te^{-2\lambda n^{(1-\alpha)}}, \tag{5.14}$$

where $T := 2\max\left\{q, \frac{1}{q}\right\} e^{4\lambda}$.

Here $\lceil \cdot \rceil$ denotes the ceiling of the number, and $\lfloor \cdot \rfloor$ its integral part.

We mention

Theorem 18.4 (p. 459, [125]) Let $x \in [a,b] \subset \mathbb{R}$ and $n \in \mathbb{N}$ so that $\lceil na \rceil \leq \lfloor nb \rfloor$. For $q, \lambda > 0$, we consider $\lambda_q > z_0 > 0$, such that $M_{q,\lambda}(z_0) = M_{q,\lambda}(0)$, and $\lambda_q > 1$. Then

$$\frac{1}{\sum_{k=\lceil na \rceil}^{\lfloor nb \rfloor} M_{q,\lambda}(nx-k)} < \max\left\{\frac{1}{M_{q,\lambda}(\lambda_q)}, \frac{1}{M_{\frac{1}{q},\lambda}\left(\lambda_{\frac{1}{q}}\right)}\right\} =: \Delta(q). \tag{5.15}$$

Similarly, we consider $\lambda_{\frac{1}{q}} > z_1 > 0$, such that $M_{\frac{1}{q},\lambda}(z_1) = M_{\frac{1}{q},\lambda}(0)$, and $\lambda_{\frac{1}{q}} > 1$. Thus

$$\frac{1}{\sum_{k=\lceil na \rceil}^{\lfloor nb \rfloor} M_{\frac{1}{q},\lambda}(nx-k)} < \max\left\{\frac{1}{M_{\frac{1}{q},\lambda}\left(\lambda_{\frac{1}{q}}\right)}, \frac{1}{M_{q,\lambda}(\lambda_q)}\right\} = \Delta(q). \tag{5.16}$$

Hence

$$\sum_{k=\lceil na \rceil}^{\lfloor nb \rfloor} M_{q,\lambda}(nx-k) > \frac{1}{\Delta(q)}, \tag{5.17}$$

and

$$\sum_{k=\lceil na\rceil}^{\lfloor nb\rfloor} M_{\frac{1}{q},\lambda}(nx-k) > \frac{1}{\Delta(q)}. \tag{5.18}$$

Consequently it holds

$$\sum_{k=\lceil na\rceil}^{\lfloor nb\rfloor} \frac{\left(M_{q,\lambda}(nx-k) + M_{\frac{1}{q},\lambda}(nx-k)\right)}{2} > \frac{2}{2\Delta(q)} = \frac{1}{\Delta(q)}, \tag{5.19}$$

so that

$$\frac{1}{\sum_{k=\lceil na\rceil}^{\lfloor nb\rfloor} \frac{\left(M_{q,\lambda}(nx-k)+M_{\frac{1}{q},\lambda}(nx-k)\right)}{2}} < \Delta(q), \tag{5.20}$$

that is

$$\frac{1}{\sum_{k=\lceil na\rceil}^{\lfloor nb\rfloor} \Phi(nx-k)} < \Delta(q). \tag{5.21}$$

We have proved

Theorem 5.1. *Let $x \in [a,b] \subset \mathbb{R}$ and $n \in \mathbb{N}$ so that $\lceil na\rceil \leq \lfloor nb\rfloor$. For $q,\lambda > 0$, we consider $\lambda_q > z_0 > 0$, such that $M_{q,\lambda}(z_0) = M_{q,\lambda}(0)$, and $\lambda_q > 1$. Also consider $\lambda_{\frac{1}{q}} > z_1 > 0$, such that $M_{\frac{1}{q},\lambda}(z_1) = M_{\frac{1}{q},\lambda}(0)$, and $\lambda_{\frac{1}{q}} > 1$. Then*

$$\frac{1}{\sum_{k=\lceil na\rceil}^{\lfloor nb\rfloor} \Phi(nx-k)} < \Delta(q). \tag{5.22}$$

We make

Remark 5.1. I) By Remark 18.5, p. 460 of [125], we have

$$\lim_{n\to\infty} \sum_{k=\lceil na\rceil}^{\lfloor nb\rfloor} M_{q,\lambda}(nx_1-k) \neq 1, \text{ for some } x_1 \in [a,b], \tag{5.23}$$

and

$$\lim_{n\to\infty} \sum_{k=\lceil na\rceil}^{\lfloor nb\rfloor} M_{\frac{1}{q},\lambda}(nx_2-k) \neq 1, \text{ for some } x_2 \in [a,b]. \tag{5.24}$$

Therefore it holds

$$\lim_{n\to\infty}\sum_{k=\lceil na\rceil}^{\lfloor nb\rfloor}\frac{\left(M_{q,\lambda}(nx_1-k)+M_{\frac{1}{q},\lambda}(nx_2-k)\right)}{2}\neq 1. \tag{5.25}$$

Hence

$$\lim_{n\to\infty}\sum_{k=\lceil nc\rceil}^{\lfloor nb\rfloor}\frac{\left(M_{q,\lambda}(nx_1-k)+M_{\frac{1}{q},\lambda}(nx_1-k)\right)}{2}\neq 1, \tag{5.26}$$

even if

$$\lim_{n\to\infty}\sum_{k=\lceil na\rceil}^{\lfloor nb\rfloor}M_{\frac{1}{q},\lambda}(nx_1-k)=1, \tag{5.27}$$

because then

$$\lim_{n\to\infty}\sum_{k=\lceil na\rceil}^{\lfloor nb\rfloor}\frac{M_{q,\lambda}(nx_1-k)}{2}+\frac{1}{2}\neq 1, \tag{5.28}$$

equivalently

$$\lim_{n\to\infty}\sum_{k=\lceil na\rceil}^{\lfloor nb\rfloor}\frac{M_{q,\lambda}(nx_1-k)}{2}\neq\frac{1}{2}, \tag{5.29}$$

true by

$$\lim_{n\to\infty}\sum_{k=\lceil na\rceil}^{\lfloor nb\rfloor}M_{q,\lambda}(nx_1-k)\neq 1. \tag{5.30}$$

II) Let $[a,b]\subset\mathbb{R}$. For large n we always have $\lceil na\rceil\leq\lfloor nb\rfloor$. Also $a\leq\frac{k}{n}\leq b$, iff $\lceil na\rceil\leq k\leq\lfloor nb\rfloor$. So in general it holds

$$\sum_{k=\lceil na\rceil}^{\lfloor nb\rfloor}\Phi(nx-k)\leq 1. \tag{5.31}$$

We need

Definition 5.1. Let $f\in C([a,b])$, $x\in[a,b]$ and $n\in\mathbb{N}:\lceil na\rceil\leq\lfloor nb\rfloor$. We introduce and define the real valued symmetrized and perturbed hyperbolic tangent positive linear neural network operators

$$H_n(f,x):=\frac{\sum_{k=\lceil na\rceil}^{\lfloor nb\rfloor}f\left(\frac{k}{n}\right)\Phi(nx-k)}{\sum_{k=\lceil na\rceil}^{\lfloor nb\rfloor}\Phi(nx-k)}=$$

$$\frac{\sum_{k=\lceil na\rceil}^{\lfloor nb\rfloor} f\left(\frac{k}{n}\right)\left(M_{q,\lambda}(nx-k)+M_{\frac{1}{q},\lambda}(nx-k)\right)}{\sum_{k=\lceil na\rceil}^{\lfloor nb\rfloor}\left(M_{q,\lambda}(nx-k)+M_{\frac{1}{q},\lambda}(nx-k)\right)}. \tag{5.32}$$

In fact, it is $H_n(f) \in C([a,b])$; and $H_n(1) = 1$.

The modulus of continuity is defined by

$$\omega_1(f,\delta) := \sup_{\substack{x,y\in[a,b]\\ |x-y|\le\delta}} |f(x)-f(y)|, \quad \delta>0. \tag{5.33}$$

In this work $0<\alpha<1$, $n\in\mathbb{N}: n^{1-\alpha}>2$; where $\|\cdot\|_\infty$ is the supremum norm.

The following approximation results, all from [127], are valid.

Theorem 5.2. *Let $f \in C([a,b])$. Then*

$$\|H_n(f)-f\|_\infty \le 2\omega_1\left(f, \sqrt{\Delta(q)}\sqrt{\frac{1}{n^{2\alpha}}+(b-a)^2 Te^{-2\lambda n^{(1-\alpha)}}}\right) \to 0,$$

as $n\to+\infty$. (5.34)

So that $\lim_{n\to\infty} H_n(f) = f$, uniformly.

It holds

Theorem 5.3. *Let $f:\mathbb{R}\to\mathbb{R}$ be a continuous and 2π periodic function with modulus of continuity ω_1. Here $\|\cdot\|_\infty$ denotes the sup-norm over $[a,b]\subset\mathbb{R}$, and the operators H_n are acting on such f over $[a,b]$; $n\in\mathbb{N}: n^{1-\alpha}>2$, $0<\alpha<1$. Then*

$$\|H_n(f)-f\|_\infty \le 2\omega_1\left(f, \pi\sqrt{\Delta(q)}\sqrt{\omega_1\left(\sin^2, \frac{1}{2n^\alpha}\right)+Te^{-2\lambda n^{(1-\alpha)}}}\right). \tag{5.35}$$

We also mention, all from [127].

Theorem 5.4. *Denote $(N\in\mathbb{N})$*

$$F_N := \left\|\left(H_n\left(|t-\cdot|^N\right)\right)(\cdot)\right\|_\infty^{\frac{1}{N}} < +\infty, \tag{5.36}$$

where $\|\cdot\|_\infty$ is the supremum norm.

Let $f \in C^N([a,b])$. *Then*

$$\|H_n f - f\|_\infty \le \sum_{\overline{k}=1}^{N} \frac{\left\|f^{(\overline{k})}\right\|_\infty}{\overline{k}!} \left\|\left(H_n(t-\cdot)^{\overline{k}}\right)(\cdot)\right\|_\infty + \tag{5.37}$$

$$\omega_1\left(f^{(N)}, F_N\right) F_N^{N-1}\left(\frac{(b-a)}{(N+1)!} + \frac{F_N}{2N!} + \frac{F_N^2}{8(b-a)(N-1)!}\right), \quad \forall\, n \in \mathbb{N}.$$

Furthermore it holds,

$$F_N \le \sqrt[N]{\Delta(q)}\sqrt[N]{\frac{1}{n^{N\alpha}} + (b-a)^N T e^{-2\lambda n^{(1-\alpha)}}}, \tag{5.38}$$

and for $\overline{k} = 1, ..., N$, *we have, similarly, that*

$$\left\|\left(H_n(t-\cdot)^{\overline{k}}\right)(\cdot)\right\|_\infty \le \Delta(q)\left\{\frac{1}{n^{\overline{k}\alpha}} + (b-a)^{\overline{k}} T e^{-2\lambda n^{(1-\alpha)}}\right\}. \tag{5.39}$$

Corollary 5.1. *Here* $F_1 := \|(H_n(|t-\cdot|))(\cdot)\|_\infty < \infty$. *Let* $f \in C^1([a,b])$. *Then*

$$\|H_n f - f\|_\infty \le \|f'\|_\infty \|(H_n(t-\cdot))(\cdot)\|_\infty + \tag{5.40}$$

$$\frac{1}{2}\omega_1(f', F_1)\left((b-a) + F_1 + \frac{F_1^2}{4(b-a)}\right).$$

Here it is

$$F_1 \le \Delta(q)\left(\frac{1}{n^\alpha} + (b-a) T e^{-2\lambda n^{(1-\alpha)}}\right). \tag{5.41}$$

Corollary 5.2. *Here* $F_2 := \left\|\left(H_n(t-\cdot)^2\right)(\cdot)\right\|_\infty^{\frac{1}{2}} < +\infty$. *Let* $f \in C^2([a,b])$. *Then*

$$\|H_n f - f\|_\infty \le \|f'\|_\infty \|(H_n(t-\cdot))(\cdot)\|_\infty + \frac{\|f''\|_\infty}{2}\left\|\left(H_n(t-\cdot)^2\right)(\cdot)\right\|_\infty \tag{5.42}$$

$$+\frac{1}{2}\omega_1(f'', F_2) F_2\left(\frac{(b-a)}{3} + \frac{F_2}{2} + \frac{F_2^2}{4(b-a)}\right).$$

Here they are

$$F_2 \le \sqrt{\Delta(q)}\sqrt{\frac{1}{n^{2\alpha}} + (b-a)^2 T e^{-2\lambda n^{(1-\alpha)}}}, \tag{5.43}$$

and

$$\left\|\left(H_n(t-\cdot)^2\right)(\cdot)\right\|_\infty \le \Delta(q)\left(\frac{1}{n^{2\alpha}} + (b-a)^2 T e^{-2\lambda n^{(1-\alpha)}}\right). \tag{5.44}$$

Theorem 5.5. *Let $\varphi := \left\| \left(H_n \left(t - \cdot \right)^2 \right) (\cdot) \right\|_\infty^{\frac{1}{2}} < +\infty$, and $r > 0$. Let also $f \in C^1 ([a,b])$. Then*

$$\|H_n f - f\|_\infty - \|f'\|_\infty \|(H_n (t - \cdot)) (\cdot)\|_\infty \leq \tag{5.45}$$

$$\begin{cases} \frac{1}{8r} (2+r)^2 \omega_1 (f', r\varphi) \varphi, & \text{if } r \leq 2; \\ \\ \omega_1 (f', r\varphi) \varphi, & \text{if } r > 2. \end{cases}$$

We derive the following useful estimates:

Corollary 5.3. *For $\overline{k} = 1, ..., N \in \mathbb{N}$, we denote by*

$$\psi_{1,\overline{k},n} := \Delta (q) \left\{ \frac{1}{n^{\overline{k}\alpha}} + (b-a)^{\overline{k}} T e^{-2\lambda n^{(1-\alpha)}} \right\}, \tag{5.46}$$

and $\|\cdot\|_\infty$ is the supremum norm.

Let $f \in C^N ([a,b])$. Then

$$\|H_n f - f\|_\infty \leq \sum_{\overline{k}=1}^{N} \frac{\left\| f^{(\overline{k})} \right\|_\infty}{\overline{k}!} \psi_{1,\overline{k},n} + \tag{5.47}$$

$$\omega_1\left(f^{(N)}, \psi_{1,N,n}^{\frac{1}{N}}\right) \psi_{1,N,n}^{\frac{N-1}{N}} \left(\frac{(b-a)}{(N+1)!} + \frac{\psi_{1,N,n}^{\frac{1}{N}}}{2N!} + \frac{\psi_{1,N,n}^{\frac{2}{N}}}{8(b-a)(N-1)!} \right), \forall\, n \in \mathbb{N}.$$

Proof. By Theorem 5.4. □

Corollary 5.4. *Let $f \in C^1 ([a,b])$. Then*

$$\|H_n f - f\|_\infty \leq \|f'\|_\infty \psi_{1,1,n} +$$

$$\frac{1}{2} \omega_1 (f', \psi_{1,1,n}) \left((b-a) + \psi_{1,1,n} + \frac{\psi_{1,1,n}^2}{4(b-a)} \right). \tag{5.48}$$

Proof. By Corollary 5.1 and (5.46). □

Corollary 5.5. *Let $f \in C^2 ([a,b])$. Then*

$$\|H_n f - f\|_\infty \leq \|f'\|_\infty \psi_{1,1,n} + \frac{\|f''\|_\infty}{2} \psi_{1,2,n}$$

$$+ \frac{1}{2} \omega_1 \left(f'', \psi_{1,2,n}^{\frac{1}{2}} \right) \psi_{1,2,n}^{\frac{1}{2}} \left(\frac{(b-a)}{3} + \frac{\psi_{1,2,n}^{\frac{1}{2}}}{2} + \frac{\psi_{1,2,n}}{4(b-a)} \right), \ \forall\, n \in \mathbb{N}. \tag{5.49}$$

Proof. By Corollary 5.2 and (5.46). □

Corollary 5.6. *Let* $f \in C^1([a,b])$, $r > 0$. *Then*

$$\|H_n f - f\|_\infty \leq \|f'\|_\infty \psi_{1,1,n} +$$
$$\begin{cases} \frac{1}{8r}(2+r)^2 \omega_1\left(f', r\psi_{1,2,n}^{\frac{1}{2}}\right)\psi_{1,2,n}^{\frac{1}{2}}, & \text{if } r \leq 2; \\ \omega_1\left(f', r\psi_{1,2,n}^{\frac{1}{2}}\right)\psi_{1,2,n}^{\frac{1}{2}}, & \text{if } r > 2. \end{cases} \tag{5.50}$$

Proof. By Theorem 5.5 and (5.46). □

5.3 Background II: About Symmetrized, Perturbed Generalized Logistic Neural Network Operators ([128])

Here we follow [125], pp. 395–417.

Our activation function here to be used is the q-deformed and λ-parametrized function

$$\varphi_{q,\lambda}(x) = \frac{1}{1 + qA^{-\lambda x}}, \quad x \in \mathbb{R}, \ q, \lambda > 0, \ A > 1. \tag{5.51}$$

This is the A-generalized logistic function.

For more read Chapter 16 of [125]: "Banach space valued ordinary and fractional neural network approximation based on q-deformed and λ-parametrized A-generalized logistic function".

This chapter motivates our current work.

The proposed "symmetrization technique" aims to use half data feed to our neural networks.

We will employ the following density function

$$G_{q,\lambda}(x) := \frac{1}{2}\left(\varphi_{q,\lambda}(x+1) - \varphi_{q,\lambda}(x-1)\right), \quad x \in \mathbb{R}, \ q, \lambda > 0. \tag{5.52}$$

We have that

$$G_{q,\lambda}(-x) = G_{\frac{1}{q},\lambda}(x), \tag{5.53}$$

and

$$G_{\frac{1}{q},\lambda}(-x) = G_{q,\lambda}(x), \ \forall \, x \in \mathbb{R}. \tag{5.54}$$

Adding (5.3) and (5.4) we obtain

$$G_{q,\lambda}(-x) + G_{\frac{1}{q},\lambda}(-x) = G_{q,\lambda}(x) + G_{\frac{1}{q},\lambda}(x), \ \forall \, x \in \mathbb{R}, \tag{5.55}$$

the key to this work.

So that

$$W(x) := \frac{G_{q,\lambda}(x) + G_{\frac{1}{q},\lambda}(x)}{2} > 0, \tag{5.56}$$

is an even function, symmetric with respect to the y-axis.

The global maximum of $G_{q,\lambda}$ is given by (16.18), p. 401 of [125] as

$$G_{q,\lambda}\left(\frac{\log_A q}{\lambda}\right) = \frac{A^\lambda - 1}{2(A^\lambda + 1)}. \tag{5.57}$$

And, the global max of $G_{\frac{1}{q},\lambda}$ is

$$G_{\frac{1}{q},\lambda}\left(\frac{\log_A \frac{1}{q}}{\lambda}\right) = G_{\frac{1}{q},\lambda}\left(\frac{-\log_A q}{\lambda}\right) = \frac{A^\lambda - 1}{2(A^\lambda + 1)}, \tag{5.58}$$

both sharing the same maximum at symmetric points.

By Theorem 16.1, p. 401 of [125], we have that

$$\sum_{i=-\infty}^{\infty} G_{q,\lambda}(x - i) = 1, \ \ \forall\, x \in \mathbb{R},\, \lambda, q > 0, \ A > 1, \tag{5.59}$$

and

$$\sum_{i=-\infty}^{\infty} G_{\frac{1}{q},\lambda}(x - i) = 1, \ \ \forall\, x \in \mathbb{R},\, \lambda, q > 0, \ A > 1. \tag{5.60}$$

Consequently, we derive that

$$\sum_{i=-\infty}^{\infty} W(x - i) = 1, \ \ \forall\, x \in \mathbb{R}. \tag{5.61}$$

By Theorem 16.2, p. 402 of [125], we have that

$$\int_{-\infty}^{\infty} G_{q,\lambda}(x)\, dx = 1, \quad \lambda, q > 0, \ A > 1, \tag{5.62}$$

similarly it holds

$$\int_{-\infty}^{\infty} G_{\frac{1}{q},\lambda}(x)\, dx = 1, \tag{5.63}$$

so that

$$\int_{-\infty}^{\infty} W(x)\, dx = 1, \tag{5.64}$$

therefore W is a density function.

By Theorem 16.3, p. 402 of [125], we have:

Let $0 < \alpha < 1$, and $n \in \mathbb{N}$ with $n^{1-\alpha} > 2$. Then

$$\sum_{\left\{\substack{k=-\infty \\ : |nx-k| \geq n^{1-\alpha}}\right.}^{\infty} G_{q,\lambda}(nx-k) < 2\max\left\{q, \frac{1}{q}\right\} \frac{1}{A^{\lambda(n^{1-\alpha}-2)}} = \gamma A^{-\lambda(n^{1-\alpha}-2)}, \tag{5.65}$$

where $\lambda, q > 0$, $A > 1$; $\gamma := 2\max\left\{q, \frac{1}{q}\right\}$.

Similarly, we get that

$$\sum_{\left\{\substack{k=-\infty \\ : |nx-k| \geq n^{1-\alpha}}\right.}^{\infty} G_{\frac{1}{q},\lambda}(nx-k) < \gamma A^{-\lambda(n^{1-\alpha}-2)}. \tag{5.66}$$

Consequently we obtain that

$$\sum_{\left\{\substack{k=-\infty \\ : |nx-k| \geq n^{1-\alpha}}\right.}^{\infty} W(nx-k) < \gamma A^{-\lambda(n^{1-\alpha}-2)}, \tag{5.67}$$

where $\gamma := 2\max\left\{q, \frac{1}{q}\right\}$.

We mention

Theorem 16.4 (p. 402, [125]) Let $x \in [a,b] \subset \mathbb{R}$ and $n \in \mathbb{N}$ so that $\lceil na \rceil \leq \lfloor nb \rfloor$. For $q > 0$, $\lambda > 0$, $A > 1$, we consider the number $\lambda_q > z_0 > 0$ with $G_{q,\lambda}(z_0) = G_{q,\lambda}(0)$, and $\lambda_q > 1$. Then

$$\frac{1}{\sum_{k=\lceil na \rceil}^{\lfloor nb \rfloor} G_{q,\lambda}(nx-k)} < \max\left\{\frac{1}{G_{q,\lambda}(\lambda_q)}, \frac{1}{G_{\frac{1}{q},\lambda}\left(\lambda_{\frac{1}{q}}\right)}\right\} =: K(q). \tag{5.68}$$

Similarly, we consider $\lambda_{\frac{1}{q}} > z_1 > 0$, such that $G_{\frac{1}{q},\lambda}(z_1) = G_{\frac{1}{q},\lambda}(0)$, and $\lambda_{\frac{1}{q}} > 1$. Thus

$$\frac{1}{\sum_{k=\lceil na \rceil}^{\lfloor nb \rfloor} G_{\frac{1}{q},\lambda}(nx-k)} < \max\left\{\frac{1}{G_{\frac{1}{q},\lambda}\left(\lambda_{\frac{1}{q}}\right)}, \frac{1}{G_{q,\lambda}(\lambda_q)}\right\} = K(q). \tag{5.69}$$

Hence

$$\sum_{k=\lceil na \rceil}^{\lfloor nb \rfloor} G_{q,\lambda}(nx-k) > \frac{1}{K(q)}, \tag{5.70}$$

and

$$\sum_{k=\lceil na\rceil}^{\lfloor nb\rfloor} G_{\frac{1}{q},\lambda}(nx-k) > \frac{1}{K(q)}. \tag{5.71}$$

Consequently it holds

$$\sum_{k=\lceil na\rceil}^{\lfloor nb\rfloor} \frac{\left(G_{q,\lambda}(nx-k) + G_{\frac{1}{q},\lambda}(nx-k)\right)}{2} > \frac{2}{2K(q)} = \frac{1}{K(q)}, \tag{5.72}$$

so that

$$\frac{1}{\sum_{k=\lceil na\rceil}^{\lfloor nb\rfloor} \frac{\left(G_{q,\lambda}(nx-k)+G_{\frac{1}{q},\lambda}(nx-k)\right)}{2}} < K(q), \tag{5.73}$$

that is

$$\frac{1}{\sum_{k=\lceil na\rceil}^{\lfloor nb\rfloor} W(nx-k)} < K(q). \tag{5.74}$$

We have proved

Theorem 5.6. *Let $x \in [a,b] \subset \mathbb{R}$ and $n \in \mathbb{N}$ so that $\lceil na\rceil \leq \lfloor nb\rfloor$. For $q, \lambda > 0$, $A > 1$, we consider $\lambda_q > z_0 > 0$ with $G_{q,\lambda}(z_0) = G_{q,\lambda}(0)$, and $\lambda_q > 1$. Also consider $\lambda_{\frac{1}{q}} > z_1 > 0$, such that $G_{\frac{1}{q},\lambda}(z_1) = G_{\frac{1}{q},\lambda}(0)$, and $\lambda_{\frac{1}{q}} > 1$. Then*

$$\frac{1}{\sum_{k=\lceil na\rceil}^{\lfloor nb\rfloor} W(nx-k)} < K(q). \tag{5.75}$$

We make

Remark 5.2. I) By Remark 16.5, p. 402 of [125], we have

$$\lim_{n\to\infty} \sum_{k=\lceil na\rceil}^{\lfloor nb\rfloor} G_{q,\lambda}(nx_1-k) \neq 1, \text{ for some } x_1 \in [a,b], \tag{5.76}$$

and

$$\lim_{n\to\infty} \sum_{k=\lceil na\rceil}^{\lfloor nb\rfloor} G_{\frac{1}{q},\lambda}(nx_2-k) \neq 1, \text{ for some } x_2 \in [a,b]. \tag{5.77}$$

Therefore it holds

$$\lim_{n\to\infty}\sum_{k=\lceil na\rceil}^{\lfloor nb\rfloor}\frac{\left(G_{q,\lambda}\left(nx_1-k\right)+G_{\frac{1}{q},\lambda}\left(nx_2-k\right)\right)}{2}\neq 1. \tag{5.78}$$

Hence it is

$$\lim_{n\to\infty}\sum_{k=\lceil na\rceil}^{\lfloor nb\rfloor}\frac{\left(G_{q,\lambda}\left(nx_1-k\right)+G_{\frac{1}{q},\lambda}\left(nx_1-k\right)\right)}{2}\neq 1, \tag{5.79}$$

even if

$$\lim_{n\to\infty}\sum_{k=\lceil na\rceil}^{\lfloor nb\rfloor}G_{\frac{1}{q},\lambda}\left(nx_1-k\right)=1, \tag{5.80}$$

because then

$$\lim_{n\to\infty}\sum_{k=\lceil na\rceil}^{\lfloor nb\rfloor}\frac{G_{q,\lambda}\left(nx_1-k\right)}{2}+\frac{1}{2}\neq 1, \tag{5.81}$$

equivalently

$$\lim_{n\to\infty}\sum_{k=\lceil na\rceil}^{\lfloor nb\rfloor}\frac{G_{q,\lambda}\left(nx_1-k\right)}{2}\neq\frac{1}{2}, \tag{5.82}$$

true by

$$\lim_{n\to\infty}\sum_{k=\lceil na\rceil}^{\lfloor nb\rfloor}G_{q,\lambda}\left(nx_1-k\right)\neq 1. \tag{5.83}$$

II) Let $[a,b]\subset\mathbb{R}$. For large n we always have $\lceil na\rceil\leq\lfloor nb\rfloor$. Also $a\leq\frac{k}{n}\leq b$, iff $\lceil na\rceil\leq k\leq\lfloor nb\rfloor$. So in general it holds

$$\sum_{k=\lceil na\rceil}^{\lfloor nb\rfloor}W\left(nx-k\right)\leq 1. \tag{5.84}$$

We need:

Definition 5.2. Let $f\in C\left([a,b]\right)$ and $n\in\mathbb{N}:\lceil na\rceil\leq\lfloor nb\rfloor$. We introduce and define the real valued symmetrized and perturbed generalized logistic positive linear neural network operators

$$S_n^s\left(f,x\right):=\frac{\sum_{k=\lceil na\rceil}^{\lfloor nb\rfloor}f\left(\frac{k}{n}\right)W\left(nx-k\right)}{\sum_{k=\lceil na\rceil}^{\lfloor nb\rfloor}W\left(nx-k\right)}=$$

$$\frac{\sum_{k=\lceil na\rceil}^{\lfloor nb\rfloor} f\left(\frac{k}{n}\right)\left(G_{q,\lambda}(nx-k)+G_{\frac{1}{q},\lambda}(nx-k)\right)}{\sum_{k=\lceil na\rceil}^{\lfloor nb\rfloor}\left(G_{q,\lambda}(nx-k)+G_{\frac{1}{q},\lambda}(nx-k)\right)}. \tag{5.85}$$

In fact, it is $S_n^s(f) \in C([a,b])$; and $S_n^s(1) = 1$.
The modulus of continuity is again defined by (5.33).

The same ω_1 is defined for $f \in C_{uB}(\mathbb{R})$ (uniformly continuous and bounded functions) and for $f \in C_B(\mathbb{R})$ (bounded and continuous functions) and for $f \in C_u(\mathbb{R})$ (uniformly continuous functions).

The fact $f \in C([a,b])$ or $f \in C_u(\mathbb{R})$, is equivalent to $\lim_{\delta\to 0}\omega_1(f,\delta) = 0$.

The following approximation results, all from [128], are valid.

Theorem 5.7. *Let $f \in C([a,b])$. Then*

$$\|S_n^s(f) - f\|_\infty \le 2\omega_1\left(f, \sqrt{K(q)}\sqrt{\frac{1}{n^{2\alpha}} + (b-a)^2\gamma A^{-\lambda(n^{1-\alpha}-2)}}\right) \to 0,$$

as $n \to +\infty$. (5.86)

So that $\lim_{n\to\infty} S_n^s(f) = f$, uniformly.

It holds

Theorem 5.8. *Let $f : \mathbb{R} \to \mathbb{R}$ be a continuous and 2π periodic function with modulus of continuity ω_1. Here $\|\cdot\|_\infty$ denotes the sup-norm over $[a,b] \subset \mathbb{R}$, and the operators S_n^s are acting on such f over $[a,b]$; $n \in \mathbb{N} : n^{1-\alpha} > 2$, $0 < \alpha < 1$. Then*

$$\|S_n^s(f) - f\|_\infty \le 2\omega_1\left(f, \pi\sqrt{K(q)}\sqrt{\omega_1\left(\sin^2, \frac{1}{2n^\alpha}\right) + \gamma A^{-\lambda(n^{1-\alpha}-2)}}\right). \tag{5.87}$$

Also from [128] come:

Theorem 5.9. *Denote $(N \in \mathbb{N})$*

$$\widetilde{F}_N := \left\|\left(S_n^s\left(|t - \cdot|^N\right)\right)(\cdot)\right\|_\infty^{\frac{1}{N}} < +\infty, \tag{5.88}$$

where $\|\cdot\|_\infty$ is the supremum norm.

Let $f \in C^N([a,b])$. Then

$$\|S_n^s f - f\|_\infty \leq \sum_{\overline{k}=1}^{N} \frac{\left\|f^{(\overline{k})}\right\|_\infty}{\overline{k}!} \left\|\left(S_n^s (t-\cdot)^{\overline{k}}\right)(\cdot)\right\|_\infty + \tag{5.89}$$

$$\omega_1\left(f^{(N)}, \widetilde{F}_N\right) \widetilde{F}_N^{N-1} \left(\frac{(b-a)}{(N+1)!} + \frac{\widetilde{F}_N}{2N!} + \frac{\widetilde{F}_N^2}{8(b-a)(N-1)!}\right), \quad \forall\, n \in \mathbb{N}.$$

Furthermore it holds,

$$\widetilde{F}_N \leq \sqrt[N]{K(q)} \sqrt[N]{\frac{1}{n^{N\alpha}} + (b-a)^N \gamma A^{-\lambda(n^{1-\alpha}-2)}}, \tag{5.90}$$

and for $\overline{k} = 1, ..., N$, we have, similarly, that

$$\left\|\left(S_n^s (t-\cdot)^{\overline{k}}\right)(\cdot)\right\|_\infty \leq K(q) \left\{\frac{1}{n^{\overline{k}\alpha}} + (b-a)^{\overline{k}} \gamma A^{-\lambda\left(n^{1-\alpha}-2\right)}\right\}. \tag{5.91}$$

Corollary 5.7. *Here $\widetilde{F}_1 := \|(S_n^s(|t-\cdot|))(\cdot)\|_\infty < \infty$. Let $f \in C^1([a,b])$. Then*

$$\|S_n^s f - f\|_\infty \leq \|f'\|_\infty \|(S_n^s(t-\cdot))(\cdot)\|_\infty + \tag{5.92}$$

$$\frac{1}{2}\omega_1\left(f', \widetilde{F}_1\right)\left((b-a) + \widetilde{F}_1 + \frac{\widetilde{F}_1^2}{4(b-a)}\right).$$

Here it is

$$\widetilde{F}_1 \leq K(q)\left(\frac{1}{n^\alpha} + (b-a)\gamma A^{-\lambda\left(n^{1-\alpha}-2\right)}\right). \tag{5.93}$$

Corollary 5.8. *Here $\widetilde{F}_2 := \left\|\left(S_n^s(t-\cdot)^2\right)(\cdot)\right\|_\infty^{\frac{1}{2}} < +\infty$. Let $f \in C^2([a,b])$. Then*

$$\|S_n^s f - f\|_\infty \leq \|f'\|_\infty \|(S_n^s(t-\cdot))(\cdot)\|_\infty + \frac{\|f''\|_\infty}{2}\left\|\left(S_n^s(t-\cdot)^2\right)(\cdot)\right\|_\infty \tag{5.94}$$

$$+\frac{1}{2}\omega_1\left(f'', \widetilde{F}_2\right)\widetilde{F}_2\left(\frac{(b-a)}{3} + \frac{\widetilde{F}_2}{2} + \frac{\widetilde{F}_2^2}{4(b-a)}\right).$$

Here they are

$$\widetilde{F}_2 \leq \sqrt{K(q)}\sqrt{\frac{1}{n^{2\alpha}} + (b-a)^2 \gamma A^{-\lambda(n^{1-\alpha}-2)}}, \tag{5.95}$$

and

$$\left\|\left(S_n^s(t-\cdot)^2\right)(\cdot)\right\|_\infty \leq K(q)\left(\frac{1}{n^{2\alpha}} + (b-a)^2 \gamma A^{-\lambda\left(n^{1-\alpha}-2\right)}\right). \tag{5.96}$$

Theorem 5.10. *Let* $\widetilde{\varphi} := \left\| \left(S_n^s (t - \cdot)^2 \right) (\cdot) \right\|_\infty^{\frac{1}{2}} < +\infty$, *and* $r > 0$. *Let also* $f \in C^1([a,b])$. *Then*

$$\|S_n^s f - f\|_\infty - \|f'\|_\infty \|(S_n^s (t - \cdot))(\cdot)\|_\infty \leq \tag{5.97}$$

$$\begin{cases} \frac{1}{8r}(2+r)^2 \omega_1(f', r\widetilde{\varphi})\widetilde{\varphi}, & \text{if } r \leq 2; \\ \omega_1(f', r\widetilde{\varphi})\widetilde{\varphi}, & \text{if } r > 2. \end{cases}$$

We derive the following useful estimates:

Corollary 5.9. *For* $\overline{k} = 1, \ldots, N \in \mathbb{N}$, *we denote by*

$$\psi_{2,\overline{k},n} := K(q) \left\{ \frac{1}{n^{\overline{k}\alpha}} + (b-a)^{\overline{k}} \gamma A^{-\lambda(n^{1-\alpha}-2)} \right\}. \tag{5.98}$$

Let $f \in C^N([a,b])$. *Then*

$$\|S_n^s f - f\|_\infty \leq \sum_{\overline{k}=1}^{N} \frac{\left\| f^{(\overline{k})} \right\|_\infty}{\overline{k}!} \psi_{2,\overline{k},n} +$$

$$\omega_1\left(f^{(N)}, \psi_{2,N,n}^{\frac{1}{N}}\right) \psi_{2,N,n}^{\frac{N-1}{N}} \left(\frac{(b-a)}{(N+1)!} + \frac{\psi_{2,N,n}^{\frac{1}{N}}}{2N!} + \frac{\psi_{2,N,n}^{\frac{2}{N}}}{8(b-a)(N-1)!} \right), \forall\, n \in \mathbb{N}. \tag{5.99}$$

Proof. By Theorem 5.9. □

Corollary 5.10. *Let* $f \in C^1([a,b])$. *Then*

$$\|S_n^s f - f\|_\infty \leq \|f'\|_\infty \psi_{2,1,n} + \tag{5.100}$$

$$\frac{1}{2}\omega_1(f', \psi_{2,1,n}) \left((b-a) + \psi_{2,1,n} + \frac{\psi_{2,1,n}^2}{4(b-a)} \right).$$

Proof. By Corollary 5.7 and (5.98). □

Corollary 5.11. *Let* $f \in C^2([a,b])$. *Then*

$$\|S_n^s f - f\|_\infty \leq \|f'\|_\infty \psi_{2,1,n} + \frac{\|f''\|_\infty}{2} \psi_{2,2,n}$$

$$+ \frac{1}{2}\omega_1\left(f'', \psi_{2,2,n}^{\frac{1}{2}}\right) \psi_{2,2,n}^{\frac{1}{2}} \left(\frac{(b-a)}{3} + \frac{\psi_{2,2,n}^{\frac{1}{2}}}{2} + \frac{\psi_{2,2,n}}{4(b-a)} \right), \quad \forall\, n \in \mathbb{N}. \tag{5.101}$$

Proof. By Corollary 5.8 and (5.98). □

Corollary 5.12. *Let $f \in C^1([a,b])$, $r > 0$. Then*

$$\|S_n^s f - f\|_\infty \le \|f'\|_\infty \psi_{2,1,n} +$$

$$\begin{cases} \frac{1}{8r}(2+r)^2 \omega_1\left(f', r\psi_{2,2,n}^{\frac{1}{2}}\right)\psi_{2,2,n}^{\frac{1}{2}}, & \text{if } r \le 2; \\ \omega_1\left(f', r\psi_{2,2,n}^{\frac{1}{2}}\right)\psi_{2,2,n}^{\frac{1}{2}}, & \text{if } r > 2. \end{cases} \tag{5.102}$$

Proof. By Theorem 5.10 and (5.98). □

5.4 Background III: Fuzzy Mathematical Analysis

We need the following basic background

Definition 5.3. (see [137]) Let $\mu : \mathbb{R} \to [0,1]$ with the following properties:

(i) is normal, i.e., $\exists\ x_0 \in \mathbb{R}$; $\mu(x_0) = 1$.

(ii) $\mu(\lambda x + (1-\lambda) y) \ge \min\{\mu(x), \mu(y)\}$, $\forall\ x, y \in \mathbb{R}$, $\forall\ \lambda \in [0,1]$ (μ is called a convex fuzzy subset).

(iii) μ is upper semicontinuous on $\mathbb{R}$, i.e. $\forall\ x_0 \in \mathbb{R}$ and $\forall\ \varepsilon > 0$, $\exists$ neighborhood $V(x_0) : \mu(x) \le \mu(x_0) + \varepsilon$, $\forall\ x \in V(x_0)$.

(iv) The set $\overline{\text{supp}(\mu)}$ is compact in $\mathbb{R}$ (where $\text{supp}(\mu) := \{x \in \mathbb{R} : \mu(x) > 0\}$).

We call μ a fuzzy real number. Denote the set of all μ with $\mathbb{R}_{\mathcal{F}}$.

E.g. $\chi_{\{x_0\}} \in \mathbb{R}_{\mathcal{F}}$, for any $x_0 \in \mathbb{R}$, where $\chi_{\{x_0\}}$ is the characteristic function at x_0.

For $0 < r \le 1$ and $\mu \in \mathbb{R}_{\mathcal{F}}$ define

$$[\mu]^r := \{x \in \mathbb{R} : \mu(x) \ge r\}$$

and

$$[\mu]^0 := \overline{\{x \in \mathbb{R} : \mu(x) \ge 0\}}.$$

Then it is well known that for each $r \in [0,1]$, $[\mu]^r$ is a closed and bounded interval on $\mathbb{R}$ ([131]).

For $u, v \in \mathbb{R}_{\mathcal{F}}$ and $\lambda \in \mathbb{R}$, we define uniquely the sum $u \oplus v$ and the product $\lambda \odot u$ by

$$[u \oplus v]^r = [u]^r + [v]^r, \quad [\lambda \odot u]^r = \lambda [u]^r, \quad \forall\ r \in [0,1],$$

where

$[u]^r + [v]^r$ means the usual addition of two intervals (as subsets of $\mathbb{R}$) and

$\lambda [u]^r$ means the usual product between a scalar and a subset of $\mathbb{R}$ (see, e.g. [137]).

Notice $1 \odot u = u$ and it holds

$$u \oplus v = v \oplus u, \; \lambda \odot u = u \odot \lambda.$$

If $0 \leq r_1 \leq r_2 \leq 1$ then

$$[u]^{r_2} \subseteq [u]^{r_1}.$$

Actually $[u]^r = \left[u_-^{(r)}, u_+^{(r)}\right]$, where $u_-^{(r)} \leq u_+^{(r)}$, $u_-^{(r)}$, $u_+^{(r)} \in \mathbb{R}$, $\forall \; r \in [0,1]$.

For $\lambda > 0$ one has $\lambda u_\pm^{(r)} = (\lambda \odot u)_\pm^{(r)}$, respectively.

Define $D : \mathbb{R}_\mathcal{F} \times \mathbb{R}_\mathcal{F} \rightarrow \mathbb{R}_\mathcal{F}$ by

$$D(u,v) := \sup_{r \in [0,1]} \max \left\{ \left| u_-^{(r)} - v_-^{(r)} \right|, \left| u_+^{(r)} - v_+^{(r)} \right| \right\},$$

where

$$[v]^r = \left[v_-^{(r)}, v_+^{(r)}\right]; \quad u, v \in \mathbb{R}_\mathcal{F}.$$

We have that D is a metric on $\mathbb{R}_\mathcal{F}$.

Then $(\mathbb{R}_\mathcal{F}, D)$ is a complete metric space, see [137], [138].

Here $\sum^{*}$ stands for fuzzy summation and $\widetilde{o} := \chi_{\{0\}} \in \mathbb{R}_\mathcal{F}$ is the neural element with respect to $\oplus$, i.e.,

$$u \oplus \widetilde{0} = \widetilde{0} \oplus u = u, \quad \forall \; u \in \mathbb{R}_\mathcal{F}.$$

Denote

$$D^*(f,g) = \sup_{x \in X \subseteq \mathbb{R}} D(f(x), g(x)),$$

where $f, g : X \rightarrow \mathbb{R}_\mathcal{F}$.

We mention

Definition 5.4. Let $f : X \subseteq \mathbb{R} \rightarrow \mathbb{R}_\mathcal{F}$, X interval, we define the (first) fuzzy modulus of continuity of f by

$$\omega_1^{(\mathcal{F})}(f,\delta)_X = \sup_{x,y \in X, \; |x-y| \leq \delta} D(f(x), f(y)), \quad \delta > 0. \tag{5.103}$$

When $g : X \subseteq \mathbb{R} \rightarrow \mathbb{R}$, we define

$$\omega_1(g,\delta) = \omega_1(g,\delta)_X = \sup_{x,y \in X, \; |x-y| \leq \delta} |g(x) - g(y)|.$$

We define by $C_{\mathcal{F}}^{U}(\mathbb{R})$ the space of fuzzy uniformly continuous functions from $\mathbb{R} \to \mathbb{R}_{\mathcal{F}}$, also $C_{\mathcal{F}}(\mathbb{R})$ is the space of fuzzy continuous functions on $\mathbb{R}$, and $C_b(\mathbb{R}, \mathbb{R}_{\mathcal{F}})$ is the fuzzy continuous and bounded functions.

We mention

Proposition 5.1. *([121]) Let $f \in C_{\mathcal{F}}^{U}(X)$. Then $\omega_1^{(\mathcal{F})}(f, \delta)_X < \infty$, for any $\delta > 0$.*

By [122], p. 129 we have that $C_{\mathcal{F}}^{U}([a,b]) = C_{\mathcal{F}}([a,b])$, fuzzy continuous functions on $[a,b] \subset \mathbb{R}$.

Proposition 5.2. *([121]) It holds*

$$\lim_{\delta \to 0} \omega_1^{(\mathcal{F})}(f, \delta)_X = \omega_1^{(\mathcal{F})}(f, 0)_X = 0,$$

iff $f \in C_{\mathcal{F}}^{U}(X)$, where X is a compact interval.

Proposition 5.3. *([121]) Here $[f]^r = \left[f_-^{(r)}, f_+^{(r)}\right]$, $r \in [0,1]$. Let $f \in C_{\mathcal{F}}(\mathbb{R})$. Then $f_{\pm}^{(r)}$ are equicontinuous with respect to $r \in [0,1]$ over $\mathbb{R}$, respectively in $\pm$.*

Note 5.1. It is clear by Propositions 5.2, 5.3, that if $f \in C_{\mathcal{F}}^{U}(\mathbb{R})$, then $f_{\pm}^{(r)} \in C_U(\mathbb{R})$ (uniformly continuous on $\mathbb{R}$). Also if $f \in C_b(\mathbb{R}, \mathbb{R}_{\mathcal{F}})$ implies $f_{\pm}^{(r)} \in C_b(\mathbb{R})$ (continuous and bounded functions on $\mathbb{R}$).

Proposition 5.4. *Let $f : \mathbb{R} \to \mathbb{R}_{\mathcal{F}}$. Assume that $\omega_1^{\mathcal{F}}(f, \delta)_X$, $\omega_1\left(f_-^{(r)}, \delta\right)_X$, $\omega_1\left(f_+^{(r)}, \delta\right)_X$ are finite for any $\delta > 0$, $r \in [0,1]$, where X any interval of $\mathbb{R}$.*

Then

$$\omega_1^{(\mathcal{F})}(f, \delta)_X = \sup_{r \in [0,1]} \max\left\{\omega_1\left(f_-^{(r)}, \delta\right)_X, \omega_1\left(f_+^{(r)}, \delta\right)_X\right\}. \tag{5.104}$$

Proof. Similar to Proposition 14.15, p. 246 of [122]. □

Definition 5.5. Let $x, y \in \mathbb{R}_{\mathcal{F}}$. If there exists $z \in \mathbb{R}_{\mathcal{F}} : x = y \oplus z$, then we call z the H-difference on x and y, denoted $x - y$.

Definition 5.6. ([136]) Let $T := [x_0, x_0 + \beta] \subset \mathbb{R}$, with $\beta > 0$. A function $f : T \to \mathbb{R}_{\mathcal{F}}$ is H-differential at $x \in T$ if there exists an $f'(x) \in \mathbb{R}_{\mathcal{F}}$ such that the limits (with respect to D)

$$\lim_{h \to 0+} \frac{f(x+h) - f(x)}{h}, \quad \lim_{h \to 0+} \frac{f(x) - f(x-h)}{h} \tag{5.105}$$

exist and are equal to $f'(x)$.

We call f' the H-derivative or fuzzy derivative of f at x.

Above is assumed that the H-differences $f(x+h) - f(x)$, $f(x) - f(x-h)$ exists in $\mathbb{R}_{\mathcal{F}}$ in a neighborhood of x.

Higher order H-fuzzy derivatives are defined the obvious way, like in the real case.

We denote by $C_{\mathcal{F}}^{N}(\mathbb{R})$, $N \geq 1$, the space of all N-times continuously H-fuzzy differentiable functions from $\mathbb{R}$ into $\mathbb{R}_{\mathcal{F}}$, similarly is defined $C_{\mathcal{F}}^{N}([a,b])$, $[a,b] \subset \mathbb{R}$.

We mention

Theorem 5.11. *([133]) Let $f : \mathbb{R} \to \mathbb{R}_{\mathcal{F}}$ be H-fuzzy differentiable. Let $t \in \mathbb{R}$, $0 \leq r \leq 1$. Clearly*

$$[f(t)]^{r} = \left[f(t)_{-}^{(r)}, f(t)_{+}^{(r)}\right] \subseteq \mathbb{R}.$$

Then $(f(t))_{\pm}^{(r)}$ are differentiable and

$$[f'(t)]^{r} = \left[\left(f(t)_{-}^{(r)}\right)', \left(f(t)_{+}^{(r)}\right)'\right].$$

I.e.

$$(f')_{\pm}^{(r)} = \left(f_{\pm}^{(r)}\right)', \quad \forall\, r \in [0,1]. \tag{5.106}$$

Remark 5.3. ([120]) Let $f \in C_{\mathcal{F}}^{N}(\mathbb{R})$, $N \geq 1$. Then by Theorem 5.11 we obtain

$$\left[f^{(i)}(t)\right]^{r} = \left[\left(f(t)_{-}^{(r)}\right)^{(i)}, \left(f(t)_{+}^{(r)}\right)^{(i)}\right],$$

for $i = 0, 1, 2, ..., N$, and in particular we have that

$$\left(f^{(i)}\right)_{\pm}^{(r)} = \left(f_{\pm}^{(r)}\right)^{(i)}, \tag{5.107}$$

for any $r \in [0,1]$, all $i = 0, 1, 2, ..., N$.

Note 5.2. ([120]) Let $f \in C_{\mathcal{F}}^{N}(\mathbb{R})$, $N \geq 1$. Then by Theorem 5.11 we have $f_{\pm}^{(r)} \in C^{N}(\mathbb{R})$, for any $r \in [0,1]$.

Items 36–38 are valid also on $[a,b]$.

By [122], p. 131, if $f \in C_{\mathcal{F}}([a,b])$, then f is a fuzzy bounded function.

Remark 5.4. We try to determine and use ($W \subset \mathbb{R}$):

$$D^{*}(f, \widetilde{o}) = \sup_{x \in W} D(f(x), \widetilde{o}) =$$

$$\sup_{x \in W} \sup_{r \in [0,1]} \max\left\{\left|f_{-}^{(r)}(x)\right|, \left|f_{+}^{(r)}(x)\right|\right\}.$$

By the principle of iterated suprema we find that

$$D^*(f,\widetilde{o}) = \sup_{r\in[0,1]} \max\left\{\left\|f_-^{(r)}\right\|_\infty, \left\|f_+^{(r)}\right\|_\infty\right\}, \tag{5.108}$$

under the assumption $D^*(f,\widetilde{o}) < \infty$, that is f is assumed a fuzzy bounded function.

Above $\|\cdot\|_\infty$ is the supremum norm of the function over $W \subseteq \mathbb{R}$.

Another direct proof of (5.108) follows:

We easily see that

$$D^*(f,\widetilde{o}) \leq \sup_{r\in[0,1]} \max\left\{\left\|f_-^{(r)}\right\|_\infty, \left\|f_+^{(r)}\right\|_\infty\right\}.$$

On the other hand we observe that $\forall\, x \in W$: each

$$\left|f_\pm^{(r)}(x)\right| \leq \max\left\{\left|f_\pm^{(r)}(x)\right|\right\} \leq \sup_{r\in[0,1]} \max\left\{\left|f_\pm^{(r)}(x)\right|\right\} \leq$$

$$\sup_{x\in W}\sup_{r\in[0,1]} \max\left\{\left|f_\pm^{(r)}(x)\right|\right\} = D^*(f,\widetilde{o}).$$

That is, each

$$\left|f_\pm^{(r)}(x)\right| \leq D^*(f,\widetilde{o}), \quad \forall\, x \in W, \tag{5.109}$$

hence each

$$\left\|f_\pm^{(r)}\right\|_\infty \leq D^*(f,\widetilde{o}), \tag{5.110}$$

and

$$\max\left\{\left\|f_\pm^{(r)}\right\|_\infty\right\} \leq D^*(f,\widetilde{o}), \tag{5.111}$$

and

$$\sup_{r\in[0,1]} \max\left\{\left\|f_-^{(r)}\right\|_\infty, \left\|f_+^{(r)}\right\|_\infty\right\} \leq D^*(f,\widetilde{o}), \tag{5.112}$$

proving (5.108).

The assumption $D^*(f,\widetilde{o}) < \infty$ implies $\left\|f_\pm^{(r)}\right\|_\infty < \infty, \forall\, r \in [0,1]$.

Clearly, it holds

$$\left|\left(f^{(j)}\right)_\pm^{(r)}(x)\right| \leq D\left(f^{(j)}(x), \widetilde{o}\right), \quad j = 1,...,N, \tag{5.113}$$

and

$$\left\|\left(f^{(N)}\right)_\pm^{(r)}\right\|_\infty \leq D^*\left(f^{(N)}, \widetilde{o}\right). \tag{5.114}$$

We make

Definition 5.7. Let $f \in C_{\mathcal{F}}([a,b])$ (fuzzy continuous functions on $[a,b] \subset \mathbb{R}$), $n \in \mathbb{N}$. We define the following Fuzzy symmetrized and perturbed neural network operators

$$H_n^{\mathcal{F}}(f,x) := \sum_{k=\lceil na \rceil}^{\lfloor nb \rfloor *} f\left(\frac{k}{n}\right) \odot \frac{\Phi(nx-k)}{\sum_{k=\lceil na \rceil}^{\lfloor nb \rfloor} \Phi(nx-k)}, \tag{5.115}$$

and

$$S_n^{s\mathcal{F}}(f,x) := \sum_{k=\lceil na \rceil}^{\lfloor nb \rfloor *} f\left(\frac{k}{n}\right) \odot \frac{W(nx-k)}{\sum_{k=\lceil na \rceil}^{\lfloor nb \rfloor} W(nx-k)}, \tag{5.116}$$

$\forall\, x \in [a,b]$, both are finite fuzzy sums.

Call

$$\Theta_1(x) := \Phi(x), \quad \Theta_2(x) := W(x), \tag{5.117}$$

and

$$L_{1n}^{\mathcal{F}}(f,x) := H_n^{\mathcal{F}}(f,x) = \sum_{k=\lceil na \rceil}^{\lfloor nb \rfloor *} f\left(\frac{k}{n}\right) \odot \frac{\Theta_1(nx-k)}{\sum_{k=\lceil na \rceil}^{\lfloor nb \rfloor} \Theta_1(nx-k)}, \tag{5.118}$$

and

$$L_{2n}^{\mathcal{F}}(f,x) := S_n^{s\mathcal{F}}(f,x) := \sum_{k=\lceil na \rceil}^{\lfloor nb \rfloor *} f\left(\frac{k}{n}\right) \odot \frac{\Theta_2(nx-k)}{\sum_{k=\lceil na \rceil}^{\lfloor nb \rfloor} \Theta_2(nx-k)}, \tag{5.119}$$

$\forall\, x \in [a,b]$, $n \in \mathbb{N}$.

Also, we call

$$L_{1n}(f,x) := H_n(f,x) = \frac{\sum_{k=\lceil na \rceil}^{\lfloor nb \rfloor} f\left(\frac{k}{n}\right) \Theta_1(nx-k)}{\sum_{k=\lceil na \rceil}^{\lfloor nb \rfloor} \Theta_1(nx-k)}, \tag{5.120}$$

and

$$L_{2n}(f,x) := S_n^{s}(f,x) = \frac{\sum_{k=\lceil na \rceil}^{\lfloor nb \rfloor} f\left(\frac{k}{n}\right) \Theta_2(nx-k)}{\sum_{k=\lceil na \rceil}^{\lfloor nb \rfloor} \Theta_2(nx-k)}, \tag{5.121}$$

$\forall\, x \in [a,b]$, $n \in \mathbb{N}$.

We give the following important result:

Theorem 5.12. *Let $f \in C_{\mathcal{F}}([a,b])$. Then*

$$\left(L_{jn}^{\mathcal{F}}(f,x)\right)_{\pm}^{(r)} = L_{jn}\left(f_{\pm}^{(r)}, x\right), \tag{5.122}$$

respectively, $\forall\, r \in [0,1]$, $\forall\, x \in [a,b]$, $j = 1,2$, $n \in \mathbb{N}$.

Therefore we get

$$D\left(L_{jn}^{\mathcal{F}}(f,x), f(x)\right) =$$

$$\sup_{r\in[0,1]} \max\left\{\left|L_{jn}\left(f_{-}^{(r)},x\right) - f_{-}^{(r)}(x)\right|, \left|L_{jn}\left(f_{+}^{(r)},x\right) - f_{+}^{(r)}(x)\right|\right\}, \tag{5.123}$$

$\forall\, x \in [a,b]$, $j = 1,2$.

Proof. Let $r \in [0,1]$, we observe that

$$\left[L_{jn}^{\mathcal{F}}(f,x)\right]^r = \sum_{k=\lceil na\rceil}^{\lfloor nb\rfloor}\left[f\left(\frac{k}{n}\right)\right]^r \left(\frac{\Theta_j(nx-k)}{\sum_{k=\lceil na\rceil}^{\lfloor nb\rfloor}\Theta_j(nx-k)}\right) =$$

$$\sum_{k=\lceil na\rceil}^{\lfloor nb\rfloor}\left[f_{-}^{(r)}\left(\frac{k}{n}\right), f_{+}^{(r)}\left(\frac{k}{n}\right)\right]\left(\frac{\Theta_j(nx-k)}{\sum_{k=\lceil na\rceil}^{\lfloor nb\rfloor}\Theta_j(nx-k)}\right) =$$

$$\left[\sum_{k=\lceil na\rceil}^{\lfloor nb\rfloor} f_{-}^{(r)}\left(\frac{k}{n}\right)\left(\frac{\Theta_j(nx-k)}{\sum_{k=\lceil na\rceil}^{\lfloor nb\rfloor}\Theta_j(nx-k)}\right), \sum_{k=\lceil na\rceil}^{\lfloor nb\rfloor} f_{+}^{(r)}\left(\frac{k}{n}\right)\left(\frac{\Theta_j(nx-k)}{\sum_{k=\lceil na\rceil}^{\lfloor nb\rfloor}\Theta_j(nx-k)}\right)\right] \tag{5.124}$$

$$= \left[L_{jn}\left(f_{-}^{(r)},x\right), L_{jn}\left(f_{+}^{(r)},x\right)\right], \quad \forall x \in [a,b],\; n \in \mathbb{N},$$

proving the claim. □

Definition 5.8. Let $f \in C_{\mathcal{F}}(\mathbb{R})$. We call it fuzzy 2π periodic iff

$$f(x+2\pi) = f(x), \quad \forall\, x \in \mathbb{R}. \tag{5.125}$$

We have that ($r \in [0,1]$)

$$[f(x)]^r = [f(x+2\pi)]^r = \left[f_{-}^{(r)}(x+2\pi), f_{+}^{(r)}(x+2\pi)\right] = \left[f_{-}^{(r)}(x), f_{+}^{(r)}(x)\right]. \tag{5.126}$$

Hence $f_{\pm}^{(r)}$ are continuous and 2π periodic over $\mathbb{R}$, $\forall\, r \in [0,1]$.

5.5 Main Results

Here we present fuzzy symmetrized and perturbed positive linear operators generated neural network approximation results:

Theorem 5.13. *Let $f \in C_{\mathcal{F}}([a,b])$, $0<\alpha<1$, $x \in [a,b]$, $n \in \mathbb{N}$ with $n^{1-\alpha}>2$. Then*

1)

$$D\left(H_n^{\mathcal{F}}(f,x), f(x)\right) \leq 2\omega_1^{(\mathcal{F})}\left(f, \sqrt{\Delta(q)}\sqrt{\frac{1}{n^{2\alpha}}+(b-a)^2 Te^{-2\lambda n^{(1-\alpha)}}}\right)$$

$$=: \rho_{1n}, \tag{5.127}$$

$\forall\, x \in [a,b]$,

2)

$$D^*\left(H_n^{\mathcal{F}}(f), f\right) \leq \rho_{1n}. \tag{5.128}$$

Thus $\lim_{n\to\infty}\left(H_n^{\mathcal{F}}(f)\right)(x) \overset{D}{\to} f(x)$, $\lim_{n\to\infty} H_n^{\mathcal{F}}(f) \overset{D^*}{\to} f$, *pointwise and uniformly.*

Proof. We have that $f_{\pm}^{(r)} \in C([a,b])$, $\forall\, r \in [0,1]$. Hence by (5.34) we obtain

$$\left|H_n\left(f_{\pm}^{(r)}, x\right) - f_{\pm}^{(r)}(x)\right| \leq$$

$$2\omega_1\left(f_{\pm}^{(r)}, \sqrt{\Delta(q)}\sqrt{\frac{1}{n^{2\alpha}}+(b-a)^2 Te^{-2\lambda n^{(1-\alpha)}}}\right) \overset{(5.104)}{\leq}$$

$$2\omega_1^{(\mathcal{F})}\left(f, \sqrt{\Delta(q)}\sqrt{\frac{1}{n^{2\alpha}}+(b-a)^2 Te^{-2\lambda n^{(1-\alpha)}}}\right). \tag{5.129}$$

Taking into account (5.123) the proof of the claim is completed. □

The counterpart of Theorem 5.13 follows.

Theorem 5.14. *All as in Theorem 5.13. Then*

1)

$$D\left(S_n^{s\mathcal{F}}(f,x), f(x)\right) \leq 2\omega_1^{(\mathcal{F})}\left(f, \sqrt{K(q)}\sqrt{\frac{1}{n^{2\alpha}}+(b-a)^2 \gamma A^{-\lambda(n^{1-\alpha}-2)}}\right)$$

$$=: \rho_{2n}, \tag{5.130}$$

$\forall\, x \in [a,b]$,

2)

$$D^{*}\left(S_{n}^{s\mathcal{F}}\left(f\right),f\right)\leq\rho_{2n}. \tag{5.131}$$

Thus $\lim_{n\to\infty}\left(S_{n}^{s\mathcal{F}}\left(f\right)\right)\left(x\right)\overset{D}{\to}f\left(x\right)$, $\lim_{n\to\infty}S_{n}^{s\mathcal{F}}\left(f\right)\overset{D^{*}}{\to}f$, *pointwise and uniformly.*

Proof. Similar to Theorem 5.13. □

We continue with:

Theorem 5.15. *Let* $f\in C_{\mathcal{F}}\left(\mathbb{R}\right)$ *be fuzzy* 2π *periodic with a modulus of continuity* $\omega_{1}^{(\mathcal{F})}$. *Here we consider* $f|_{[a,b]}$, $[a,b]\subset\mathbb{R}$, *and* $H_{n}^{\mathcal{F}}$ *acts on it;* $n\in\mathbb{N}:n^{1-\alpha}>2$, $0<\alpha<1$. *Then*

1)

$$D\left(H_{n}^{\mathcal{F}}\left(f,x\right),f\left(x\right)\right)\leq$$

$$2\omega_{1}^{(\mathcal{F})}\left(f,\pi\sqrt{\Delta\left(q\right)}\sqrt{\omega_{1}\left(\sin^{2},\frac{1}{2n^{2\alpha}}\right)+Te^{-2\lambda n^{(1-\alpha)}}}\right)=:\sigma_{1n}, \tag{5.132}$$

$\forall\ x\in\left[a,b\right]$,

2)

$$D^{*}\left(H_{n}^{\mathcal{F}}\left(f\right),f\right)\leq\sigma_{1n}. \tag{5.133}$$

Thus $\lim_{n\to\infty}\left(H_{n}^{\mathcal{F}}\left(f\right)\right)\left(x\right)\overset{D}{\to}f\left(x\right)$, $\lim_{n\to\infty}H_{n}^{\mathcal{F}}\left(f\right)\overset{D^{*}}{\to}f$, *pointwise and uniformly.*

Proof. Similar to Theorem 5.13. □

Its counterpart follows:

Theorem 5.16. *Here all as in Theorem 5.15. Then*

1)

$$D\left(S_{n}^{s\mathcal{F}}\left(f,x\right),f\left(x\right)\right)\leq$$

$$2\omega_{1}^{(\mathcal{F})}\left(f,\pi\sqrt{K\left(q\right)}\sqrt{\omega_{1}\left(\sin^{2},\frac{1}{2n^{2\alpha}}\right)+\gamma A^{-\lambda\left(n^{1-\alpha}-2\right)}}\right)=:\sigma_{2n}, \tag{5.134}$$

$\forall\ x\in\left[a,b\right]$,

2)

$$D^{*}\left(S_{n}^{s\mathcal{F}}\left(f\right),f\right)\leq\sigma_{2n}. \tag{5.135}$$

Thus $\lim_{n\to\infty}\left(S_{n}^{s\mathcal{F}}\left(f\right)\right)\left(x\right)\overset{D}{\to}f\left(x\right)$, $\lim_{n\to\infty}S_{n}^{s\mathcal{F}}\left(f\right)\overset{D^{*}}{\to}f$, *pointwise and uniformly.*

Proof. Similar to Theorem 5.13. □

We continue with:

Theorem 5.17. *Here* $f \in C_{\mathcal{F}}^{N}([a,b])$, $N \in \mathbb{N}$. *For* $\overline{k} = 1, ..., N \in \mathbb{N}$, *we denote by*

$$\psi_{1,\overline{k},n} := \Delta(q)\left\{\frac{1}{n^{\overline{k}\alpha}} + (b-a)^{\overline{k}} T e^{-2\lambda n^{(1-\alpha)}}\right\},$$

$0 < \alpha < 1$, $n \in \mathbb{N} : n^{1-\alpha} > 2$. *Then*

$$D^*\left(H_n^{\mathcal{F}}(f), f\right) \leq \sum_{\overline{k}=1}^{N} \frac{D^*\left(f^{(\overline{k})}, \widetilde{o}\right)}{\overline{k}!} \psi_{1,\overline{k},n} +$$

$$\omega_1^{(\mathcal{F})}\left(f^{(N)}, \psi_{1,N,n}^{\frac{1}{N}}\right) \psi_{1,N,n}^{\frac{N-1}{N}} \left(\frac{(b-a)}{(N+1)!} + \frac{\psi_{1,N,n}^{\frac{1}{N}}}{2N!} + \frac{\psi_{1,N,n}^{\frac{2}{N}}}{8(b-a)(N-1)!}\right). \tag{5.136}$$

Clearly it holds $\lim_{n\to\infty} H_n^{\mathcal{F}}(f) \overset{D^*}{\to} f$, *uniformly, and* $\lim_{n\to\infty}\left(H_n^{\mathcal{F}}(f)\right)(x) \overset{D}{\to} f(x)$, *pointwise,* $\forall\, x \in [a,b]$.

Proof. Since $f \in C_{\mathcal{F}}^{N}([a,b])$, $N \in \mathbb{N}$, we have that $f_{\pm}^{(r)} \in C^{N}([a,b])$, $\forall\, r \in [0,1]$.

Using (5.47) we get

$$\left|H_n\left(f_{\pm}^{(r)}, x\right) - f_{\pm}^{(r)}(x)\right| \leq \sum_{\overline{k}=1}^{N} \frac{\left\|\left(f_{\pm}^{(r)}\right)^{(\overline{k})}\right\|_{\infty}}{\overline{k}!} \psi_{1,\overline{k},n} +$$

$$\omega_1\left(\left(f_{\pm}^{(r)}\right)^{(N)}, \psi_{1,N,n}^{\frac{1}{N}}\right) \psi_{1,N,n}^{\frac{N-1}{N}} \left(\frac{(b-a)}{(N+1)!} + \frac{\psi_{1,N,n}^{\frac{1}{N}}}{2N!} + \frac{\psi_{1,N,n}^{\frac{2}{N}}}{8(b-a)(N-1)!}\right) \overset{(5.107)}{=} \tag{5.137}$$

$$\sum_{\overline{k}=1}^{N} \frac{\left\|\left(f^{(\overline{k})}\right)_{\pm}^{(r)}\right\|_{\infty}}{\overline{k}!} \psi_{1,\overline{k},n} +$$

$$\omega_1\left(\left(f^{(N)}\right)_{\pm}^{(r)}, \psi_{1,N,n}^{\frac{1}{N}}\right) \psi_{1,N,n}^{\frac{N-1}{N}} \left(\frac{(b-a)}{(N+1)!} + \frac{\psi_{1,N,n}^{\frac{1}{N}}}{2N!} + \frac{\psi_{1,N,n}^{\frac{2}{N}}}{8(b-a)(N-1)!}\right)$$

$$\text{(by (5.114) and (5.104))} \leq$$

$$\sum_{\overline{k}=1}^{N} \frac{D^*\left(f^{(\overline{k})}, \widetilde{o}\right)}{\overline{k}!} \psi_{1,\overline{k},n} +$$

$$\omega_1^{(\mathcal{F})}\left(f^{(N)}, \psi_{1,N,n}^{\frac{1}{N}}\right) \psi_{1,N,n}^{\frac{N-1}{N}} \left(\frac{(b-a)}{(N+1)!} + \frac{\psi_{1,N,n}^{\frac{1}{N}}}{2N!} + \frac{\psi_{1,N,n}^{\frac{2}{N}}}{8(b-a)(N-1)!}\right), \tag{5.138}$$

$\forall\, x \in [a,b]$, $n \in \mathbb{N}$.

Next, we take into account (5.123).

The theorem is proved. □

Corollary 5.13. *Here $f \in C_{\mathcal{F}}^1([a,b])$. Then*

$$D^*\left(H_n^{\mathcal{F}}(f), f\right) \leq D^*\left(f^{(1)}, \widetilde{o}\right) \psi_{1,1,n} +$$

$$\frac{1}{2}\omega_1^{(\mathcal{F})}\left(f^{(1)}, \psi_{1,1,n}\right)\left((b-a) + \psi_{1,1,n} + \frac{\psi_{1,1,n}^2}{4(b-a)}\right). \tag{5.139}$$

Proof. By Theorem 5.17, $N = 1$ case. □

Corollary 5.14. *Here $f \in C_{\mathcal{F}}^2([a,b])$. Then*

$$D^*\left(H_n^{\mathcal{F}}(f), f\right) \leq D^*\left(f^{(1)}, \widetilde{o}\right) \psi_{1,1,n} + \frac{D^*\left(f^{(2)}, \widetilde{o}\right)}{2} \psi_{1,2,n} +$$

$$\frac{1}{2}\omega_1^{(\mathcal{F})}\left(f^{(2)}, \psi_{1,2,n}^{\frac{1}{2}}\right) \psi_{1,2,n}^{\frac{1}{2}} \left(\frac{(b-a)}{3} + \frac{\psi_{1,2,n}^{\frac{1}{2}}}{2} + \frac{\psi_{1,2,n}}{4(b-a)}\right), \tag{5.140}$$

$\forall\, n \in \mathbb{N}$.

Proof. By Theorem 5.17, $N = 2$ case. □

Corollary 5.15. *Let $f \in C_{\mathcal{F}}^1([a,b])$, $r > 0$. Then*

$$D^*\left(H_n^{\mathcal{F}}(f), f\right) \leq D^*\left(f^{(1)}, \widetilde{o}\right) \psi_{1,1,n} +$$

$$\begin{cases} \frac{1}{8r}(2+r)^2 \omega_1^{(\mathcal{F})}\left(f^{(1)}, r\psi_{1,2,n}^{\frac{1}{2}}\right) \psi_{1,2,n}^{\frac{1}{2}}, & \text{if } r \leq 2; \\ \omega_1^{(\mathcal{F})}\left(f^{(1)}, r\psi_{1,2,n}^{\frac{1}{2}}\right) \psi_{1,2,n}^{\frac{1}{2}}, & \text{if } r > 2. \end{cases} \tag{5.141}$$

Proof. Similar proof as in Theorem 5.17 based on Corollary 5.6. □

We continue with:

Theorem 5.18. *Here $f \in C_{\mathcal{F}}^{N}([a,b])$, $N \in \mathbb{N}$. For $\overline{k} = 1, ..., N \in \mathbb{N}$, we denote by*

$$\psi_{2,\overline{k},n} := K(q)\left\{\frac{1}{n^{\overline{k}\alpha}} + (b-a)^{\overline{k}}\gamma A^{-\lambda\left(n^{1-\alpha}-2\right)}\right\},$$

$0 < \alpha < 1$, $n \in \mathbb{N} : n^{1-\alpha} > 2$. Then

$$D^{*}\left(S_{n}^{s\mathcal{F}}(f), f\right) \leq \sum_{\overline{k}=1}^{N} \frac{D^{*}\left(f^{(\overline{k})}, \widetilde{o}\right)}{\overline{k}!}\psi_{2,\overline{k},n} +$$

$$\omega_{1}^{(\mathcal{F})}\left(f^{(N)}, \psi_{2,N,n}^{\frac{1}{N}}\right)\psi_{2,N,n}^{\frac{N-1}{N}}\left(\frac{(b-a)}{(N+1)!} + \frac{\psi_{2,N,n}^{\frac{1}{N}}}{2N!} + \frac{\psi_{2,N,n}^{\frac{2}{N}}}{8(b-a)(N-1)!}\right). \tag{5.142}$$

Clearly it holds $\lim_{n\to\infty} S_{n}^{s\mathcal{F}}(f) \overset{D^{*}}{\to} f$, uniformly, and $\lim_{n\to\infty} S_{n}^{s\mathcal{F}}(f)(x) \overset{D}{\to} f(x)$, pointwise, $\forall\ x \in [a,b]$.

Proof. Since $f \in C_{\mathcal{F}}^{N}([a,b])$, $N \in \mathbb{N}$, we have that $f_{\pm}^{(r)} \in C^{N}([a,b])$, $\forall$ $r \in [0,1]$.

Using (5.99) we get

$$\left|S_{n}^{s}\left(f_{\pm}^{(r)}, x\right) - f_{\pm}^{(r)}(x)\right| \leq \sum_{\overline{k}=1}^{N} \frac{\left\|\left(f_{\pm}^{(r)}\right)^{(\overline{k})}\right\|_{\infty}}{\overline{k}!}\psi_{2,\overline{k},n} +$$

$$\omega_{1}\left(\left(f_{\pm}^{(r)}\right)^{(N)}, \psi_{2,N,n}^{\frac{1}{N}}\right)\psi_{2,N,n}^{\frac{N-1}{N}}\left(\frac{(b-a)}{(N+1)!} + \frac{\psi_{2,N,n}^{\frac{1}{N}}}{2N!} + \frac{\psi_{2,N,n}^{\frac{2}{N}}}{8(b-a)(N-1)!}\right) \overset{(5.107)}{=} \tag{5.143}$$

$$\sum_{\overline{k}=1}^{N} \frac{\left\|\left(f^{(\overline{k})}\right)_{\pm}^{(r)}\right\|_{\infty}}{\overline{k}!}\psi_{2,\overline{k},n} +$$

$$\omega_{1}\left(\left(f^{(N)}\right)_{\pm}^{(r)}, \psi_{2,N,n}^{\frac{1}{N}}\right)\psi_{2,N,n}^{\frac{N-1}{N}}\left(\frac{(b-a)}{(N+1)!} + \frac{\psi_{2,N,n}^{\frac{1}{N}}}{2N!} + \frac{\psi_{2,N,n}^{\frac{2}{N}}}{8(b-a)(N-1)!}\right)$$

$$\text{(by (5.114) and (5.104))} \leq$$

$$\sum_{\overline{k}=1}^{N} \frac{D^*\left(f^{(\overline{k})}, \widetilde{o}\right)}{\overline{k}!} \psi_{2,\overline{k},n} +$$

$$\omega_1^{(\mathcal{F})}\left(f^{(N)}, \psi_{2,N,n}^{\frac{1}{N}}\right) \psi_{2,N,n}^{\frac{N-1}{N}} \left(\frac{(b-a)}{(N+1)!} + \frac{\psi_{2,N,n}^{\frac{1}{N}}}{2N!} + \frac{\psi_{2,N,n}^{\frac{2}{N}}}{8(b-a)(N-1)!}\right), \tag{5.144}$$

$\forall\, x \in [a,b]$, $n \in \mathbb{N}$.

Next, we take into account (5.123).

The theorem is proved. □

Corollary 5.16. *Here $f \in C^1_{\mathcal{F}}([a,b])$. Then*

$$D^*\left(S_n^{s\mathcal{F}}(f), f\right) \leq D^*\left(f^{(1)}, \widetilde{o}\right) \psi_{2,1,n} +$$

$$\frac{1}{2}\omega_1^{(\mathcal{F})}\left(f^{(1)}, \psi_{2,1,n}\right)\left((b-a) + \psi_{2,1,n} + \frac{\psi_{2,1,n}^2}{4(b-a)}\right). \tag{5.145}$$

Proof. By Theorem 5.18, $N = 1$ case. □

Corollary 5.17. *Here $f \in C^2_{\mathcal{F}}([a,b])$. Then*

$$D^*\left(S_n^{s\mathcal{F}}(f), f\right) \leq D^*\left(f^{(1)}, \widetilde{o}\right) \psi_{2,1,n} + \frac{D^*\left(f^{(2)}, \widetilde{o}\right)}{2} \psi_{2,2,n} +$$

$$\frac{1}{2}\omega_1^{(\mathcal{F})}\left(f^{(2)}, \psi_{2,2,n}^{\frac{1}{2}}\right) \psi_{2,2,n}^{\frac{1}{2}} \left(\frac{(b-a)}{3} + \frac{\psi_{2,2,n}^{\frac{1}{2}}}{2} + \frac{\psi_{2,2,n}}{4(b-a)}\right), \tag{5.146}$$

$\forall\, n \in \mathbb{N}$.

Proof. By Theorem 5.18, $N = 2$ case. □

Corollary 5.18. *Let $f \in C^1_{\mathcal{F}}([a,b])$, $r > 0$. Then*

$$D^*\left(S_n^{s\mathcal{F}}(f), f\right) \leq D^*\left(f^{(1)}, \widetilde{o}\right) \psi_{2,1,n} +$$

$$\begin{cases} \frac{1}{8r}(2+r)^2 \omega_1^{(\mathcal{F})}\left(f^{(1)}, r\psi_{2,2,n}^{\frac{1}{2}}\right) \psi_{2,2,n}^{\frac{1}{2}}, & \text{if } r \leq 2; \\[2ex] \omega_1^{(\mathcal{F})}\left(f^{(1)}, r\psi_{2,2,n}^{\frac{1}{2}}\right) \psi_{2,2,n}^{\frac{1}{2}}, & \text{if } r > 2. \end{cases} \tag{5.147}$$

Proof. Similar proof as in Theorem 5.18 based on Corollary 5.12. □

Bibliography

[118] G.A. Anastassiou, *Rate of convergence of some neural network operators to the unit-univariate case*, Journal of Mathematical Analysis and Application, Vol. 212 (1997), 237–262.

[119] G.A. Anastassiou, *Quantitative Approximation*, Chapmann and Hall/CRC, Boca Raton, New York, 2001.

[120] G.A. Anastassiou, *Higher order Fuzzy Korovkin Theory via inequalities*, Communications in Applied Analysis, 10(2006), No. 2, 359–392.

[121] G.A. Anastassiou, *Fuzzy Korovkin Theorems and Inequalities*, Journal of Fuzzy Mathematics, 15(2007), No. 1, 169–205.

[122] G.A. Anastassiou, *Fuzzy Mathematics: Approximation Theory*, Springer, Heildelberg, New York, 2010.

[123] G.A. Anastassiou, *Intelligent Systems: Approximation by Artificial Neural Networks*, Intelligent Systems Reference Library, Vol. 19, Springer, Heidelberg, 2011.

[124] G.A. Anastassiou, *Intelligent Systems II: Complete Approximation by Neural Network Operators*, Springer, Heidelberg, New York, 2016.

[125] G.A. Anastassiou, *Parametrized, Deformed and General Neural Networks*, Springer, Heidelberg, New York, 2023.

[126] G.A. Anastassiou, *Trigonometric and Hyperbolic Generated Approximation Theory*, World Scientific, Singapore, New York, 2025.

[127] G.A. Anastassiou, *Neural Networks as Positive Linear Operators*, Mathematics, 2025, 13, 1112; `https://doi.org/10.3390/math13071112`.

[128] G.A. Anastassiou, *Generalized Logistic Neural Networks in Positive Linear Framework*, Symmetry, 2025, 17, 746, `https://doi.org/10.3390/sym17050746`.

[129] P. Cardaliaguet, G. Euvrard, *Approximation of a function and its derivarive with a neural network*, Neural Networks 5 (1992), 207–220.

[130] Z. Chen and F. Cao, *The approximation operators with sigmoidal functions*, Computers and Mathematics with Applications, 58 (2009), 758–765.

[131] R. Goetschel Jr., W. Voxman, *Elementary fuzzy calculus*, Fuzzy Sets and Systems, 18(1986), 31–43.

[132] S. Haykin, *Neural Networks: A Comprehensive Foundation* (2 ed.), Prentice Hall, New York, 1998.

[133] O. Kaleva, *Fuzzy differential equations*, Fuzzy Sets and Systems, 24(1987), 301–317.

[134] W. McCulloch and W. Pitts, *A logical calculus of the ideas immanent in nervous activity*, Bulletin of Mathematical Biophysics, 7 (1943), 115–133.

[135] T.M. Mitchell, *Machine Learning*, WCB-McGraw-Hill, New York, 1997.

[136] Wu Congxin, Gong Zengtai, *On Henstock integrals of interval-valued functions and fuzzy valued functions*, Fuzzy Sets and Systems, Vol. 115, No. 3, 2000, 377–391.

[137] C. Wu, Z. Gong, *On Henstock integral of fuzzy-number-valued functions (I)*, Fuzzy Sets and Systems, 120, No. 3, (2001), 523–532.

[138] C. Wu, M. Ma, *On embedding problem of fuzzy number spaces: Part 1*, Fuzzy Sets and Systems, 44 (1991), 33–38.

[139] G.A. Anastassiou, *Fuzzy Symmetrized and Perturbed Neural Network Approximation*, Neural, Parallel and Scientific Computations, 33 (2025), 55–82.

Chapter 6

Multivariate and Abstract Neural Networks as Positive Linear Operators

Basic multivariate-abstract neural network operators are interpreted as positive linear operators and the related general theory applies to them. These operators are induced by a symmetrized density function deriving from the parametrized and deformed hyperbolic tangent activation function. We are acting on the space of continuous functions on a box in $\mathbb{R}^N$, $N \in \mathbb{N}$, with values to a general Banach space. We study quantitatively the rate of convergence of these neural network operators to the unit operator. Our inequalities involve the modulus of continuity of the function under approximation or its Fréchet derivative. Convexity of functions is also taken into consideration. It follows [155].

6.1 Introduction

The author studied extensively the quantitative approximation of positive linear operators to the unit since 1985, see for example [140]–[144], [146], which motivate this work. He originated from the quantitative weak convergence of finite positive measures to the unit Dirac measure, having as a method the geometric moment theory, see [141], and he produced best upper bounds, leading to attained (i.e. sharp Jackson type inequalities), e.g. see [140], [141]. These studies have gone to all possible directions, univariate and multivariate, though in this work we treat only the multivariate-abstract case.

Also, this author since 1997, he started the study of the quantitative convergence of neural network operators to the unit, and he has been written since then numerous of articles and books, e.g. see [142], [145], [146], which are what we use here.

The wide range of neural network operators being treated by the author all along, are indeed by nature positive linear operators.

Here, for the first time in the literature, multivariate-abstract neural network operators are treated as positive linear operators.

So, important methods of positive linear operators apply here to our finite summation defined neural network operators, producing new and interesting results: pointwise and uniform. Via the Riesz representation theorem neural networks are connected to measure theory. Convexity of functions produces optimal and sharper results.

The use of perturbed hyperbolic tangent as an activation function is well-established and supported in [145], Chapters 17, 18. Hyperbolic tangent activation function is greatly behaving and one of the most commonly used activation functions on this subject.

The author's symmetrization method presented here aims that our operators converge at high speed to the unit operator using half feed of data.

This chapter establishes a bridge for the first time connecting Neural Networks Approximation (part of AI) to Positive Linear Operators (a great part of Functional Analysis).

So it is a theoretical work intended by nature mainly to related experts.

Of great inspiration always have been [152], [153]. For classical studies on neural network we recommend also, [148]–[150], and [154]–[157].

For newer work on neural networks we also refer the reader to [158]–[167].

6.2 Basics

Initially we follow [145], pp. 455–460.

Our perturbed hyperbolic tangent activation function here to be used is

$$g_{q,\lambda}(x) := \frac{e^{\lambda x} - qe^{-\lambda x}}{e^{\lambda x} + qe^{-\lambda x}}, \quad \lambda, q > 0, \; x \in \mathbb{R}. \tag{6.1}$$

Above λ is the parameter and q is the deformation coefficient, typically it is $0 < \lambda, q \leq 1$.

For more read Chapter 18 of [145]: "q-Deformed and λ-Parametrized Hyperbolic Tangent based Banach space Valued Ordinary and Fractional Neural Network Approximation".

The chapters 17 and 18 of [145] motivate our current work.

The proposed "symmetrization method" aims to use half data to feed into our multivariate neural networks.

We will employ the following density function

$$M_{q.\lambda}(x) := \frac{1}{4}\left(g_{q,\lambda}(x+1) - g_{q,\lambda}(x-1)\right) > 0, \tag{6.2}$$

$\forall x \in \mathbb{R};\ q, \lambda > 0.$

We have that

$$M_{q,\lambda}(-x) = M_{\frac{1}{q},\lambda}(x),\ \ \forall\ x \in \mathbb{R};\ q, \lambda > 0, \tag{6.3}$$

and

$$M_{\frac{1}{q},\lambda}(-x) = M_{q,\lambda}(x),\ \ \forall\ x \in \mathbb{R};\ q, \lambda > 0. \tag{6.4}$$

Adding (6.3) and (6.4) we obtain

$$M_{q,\lambda}(-x) + M_{\frac{1}{q},\lambda}(-x) = M_{q,\lambda}(x) + M_{\frac{1}{q},\lambda}(x), \tag{6.5}$$

a key to this work.

So that

$$\Phi(x) := \frac{M_{q,\lambda}(x) + M_{\frac{1}{q},\lambda}(x)}{2} \tag{6.6}$$

is an even function, symmetric with respect to the y-axis.

By (18.18) of [145], we have

$$\begin{aligned} &M_{q,\lambda}\left(\tfrac{\ln q}{2\lambda}\right) = \tfrac{\tanh(\lambda)}{2}, \\ &\text{and} \\ &M_{\frac{1}{q},\lambda}\left(-\tfrac{\ln q}{2\lambda}\right) = \tfrac{\tanh(\lambda)}{2},\ \ \lambda > 0, \end{aligned} \tag{6.7}$$

sharing the same maximum at symmetric points.

By Theorem 18.1, p. 458 of [145], we have that

$$\begin{aligned} &\sum_{i=-\infty}^{\infty} M_{q,\lambda}(x-i) = 1,\ \ \forall\ x \in \mathbb{R},\ \lambda, q > 0, \\ &\text{and} \\ &\sum_{i=-\infty}^{\infty} M_{\frac{1}{q},\lambda}(x-i) = 1,\ \ \forall\ x \in \mathbb{R},\ \lambda, q > 0. \end{aligned} \tag{6.8}$$

Consequently, we derive that

$$\sum_{i=-\infty}^{\infty} \Phi(x-i) = 1,\ \ \forall\ x \in \mathbb{R}. \tag{6.9}$$

By Theorem 18.2, p. 459 of [145], we have that

$$\begin{aligned} &\int_{-\infty}^{\infty} M_{q,\lambda}(x)\,dx = 1, \\ &\text{and} \\ &\int_{-\infty}^{\infty} M_{\frac{1}{q},\lambda}(x)\,dx = 1, \end{aligned} \tag{6.10}$$

so that

$$\int_{-\infty}^{\infty} \Phi(x)\,dx = 1, \tag{6.11}$$

therefore Φ is a density function.

By Theorem 18.3, p. 459 of [145], we have:

Let $0 < \alpha < 1$, and $n \in \mathbb{N}$ with $n^{1-\alpha} > 2$; $q, \lambda > 0$. Then

$$\sum_{\left\{\begin{array}{c} k=-\infty \\ : |nx-k| \geq n^{1-\alpha} \end{array}\right.}^{\infty} M_{q,\lambda}(nx-k) < 2\max\left\{q, \frac{1}{q}\right\} e^{4\lambda} e^{-2\lambda n^{(1-\alpha)}} = Te^{-2\lambda n^{(1-\alpha)}}, \tag{6.12}$$

where $T := 2\max\left\{q, \frac{1}{q}\right\} e^{4\lambda}$.

Similarly, we get that

$$\sum_{k=-\infty}^{\infty} M_{\frac{1}{q},\lambda}(nx-k) < Te^{-2\lambda n^{(1-\alpha)}}. \tag{6.13}$$

Consequently we obtain that

$$\sum_{\left\{\begin{array}{c} k=-\infty \\ : |nx-k| \geq n^{1-\alpha} \end{array}\right.}^{\infty} \Phi(nx-k) < Te^{-2\lambda n^{(1-\alpha)}}, \tag{6.14}$$

where $T := 2\max\left\{q, \frac{1}{q}\right\} e^{4\lambda}$.

Here $\lceil \cdot \rceil$ denotes the ceiling of the number, and $\lfloor \cdot \rfloor$ its integral part.

We mention

Theorem 18.4 (p. 459, [145]) Let $x \in [a,b] \subset \mathbb{R}$ and $n \in \mathbb{N}$ so that $\lceil na \rceil \leq \lfloor nb \rfloor$. For $q, \lambda > 0$, we consider $\lambda_q > z_0 > 0$, such that $M_{q,\lambda}(z_0) = M_{q,\lambda}(0)$, and $\lambda_q > 1$. Then

$$\frac{1}{\sum_{k=\lceil na \rceil}^{\lfloor nb \rfloor} M_{q,\lambda}(nx-k)} < \max\left\{\frac{1}{M_{q,\lambda}(\lambda_q)}, \frac{1}{M_{\frac{1}{q},\lambda}\left(\lambda_{\frac{1}{q}}\right)}\right\} =: \Delta(q). \tag{6.15}$$

Similarly, we consider $\lambda_{\frac{1}{q}} > z_1 > 0$, such that $M_{\frac{1}{q},\lambda}(z_1) = M_{\frac{1}{q},\lambda}(0)$, and $\lambda_{\frac{1}{q}} > 1$. Thus

$$\frac{1}{\sum_{k=\lceil na \rceil}^{\lfloor nb \rfloor} M_{\frac{1}{q},\lambda}(nx-k)} < \max\left\{\frac{1}{M_{\frac{1}{q},\lambda}\left(\lambda_{\frac{1}{q}}\right)}, \frac{1}{M_{q,\lambda}(\lambda_q)}\right\} = \Delta(q). \quad (6.16)$$

Hence

$$\sum_{k=\lceil na \rceil}^{\lfloor nb \rfloor} M_{q,\lambda}(nx-k) > \frac{1}{\Delta(q)}, \quad (6.17)$$

and

$$\sum_{k=\lceil na \rceil}^{\lfloor nb \rfloor} M_{\frac{1}{q},\lambda}(nx-k) > \frac{1}{\Delta(q)}. \quad (6.18)$$

Consequently it holds

$$\sum_{k=\lceil na \rceil}^{\lfloor nb \rfloor} \frac{\left(M_{q,\lambda}(nx-k) + M_{\frac{1}{q},\lambda}(nx-k)\right)}{2} > \frac{2}{2\Delta(q)} = \frac{1}{\Delta(q)}, \quad (6.19)$$

so that

$$\frac{1}{\sum_{k=\lceil na \rceil}^{\lfloor nb \rfloor} \frac{\left(M_{q,\lambda}(nx-k)+M_{\frac{1}{q},\lambda}(nx-k)\right)}{2}} < \Delta(q), \quad (6.20)$$

that is

$$\frac{1}{\sum_{k=\lceil na \rceil}^{\lfloor nb \rfloor} \Phi(nx-k)} < \Delta(q). \quad (6.21)$$

We have proved

Theorem 6.1. *Let $x \in [a,b] \subset \mathbb{R}$ and $n \in \mathbb{N}$ so that $\lceil na \rceil \le \lfloor nb \rfloor$. For $q, \lambda > 0$, we consider $\lambda_q > z_0 > 0$, such that $M_{q,\lambda}(z_0) = M_{q,\lambda}(0)$, and $\lambda_q > 1$. Also consider $\lambda_{\frac{1}{q}} > z_1 > 0$, such that $M_{\frac{1}{q},\lambda}(z_1) = M_{\frac{1}{q},\lambda}(0)$, and $\lambda_{\frac{1}{q}} > 1$. Then*

$$\frac{1}{\sum_{k=\lceil na \rceil}^{\lfloor nb \rfloor} \Phi(nx-k)} < \Delta(q). \quad (6.22)$$

We make

Remark 6.1.

I) By Remark 18.5, p. 460 of [145], we have

$$\lim_{n\to\infty}\sum_{k=\lceil na\rceil}^{\lfloor nb\rfloor} M_{q,\lambda}\left(nx_1-k\right)\neq 1, \text{ for some } x_1\in[a,b], \tag{6.23}$$

and

$$\lim_{n\to\infty}\sum_{k=\lceil na\rceil}^{\lfloor nb\rfloor} M_{\frac{1}{q},\lambda}\left(nx_2-k\right)\neq 1, \text{ for some } x_2\in[a,b]. \tag{6.24}$$

Therefore it holds

$$\lim_{n\to\infty}\sum_{k=\lceil na\rceil}^{\lfloor nb\rfloor} \frac{\left(M_{q,\lambda}\left(nx_1-k\right)+M_{\frac{1}{q},\lambda}\left(nx_2-k\right)\right)}{2}\neq 1. \tag{6.25}$$

Hence

$$\lim_{n\to\infty}\sum_{k=\lceil na\rceil}^{\lfloor nb\rfloor} \frac{\left(M_{q,\lambda}\left(nx_1-k\right)+M_{\frac{1}{q},\lambda}\left(nx_1-k\right)\right)}{2}\neq 1, \tag{6.26}$$

even if

$$\lim_{n\to\infty}\sum_{k=\lceil na\rceil}^{\lfloor nb\rfloor} M_{\frac{1}{q},\lambda}\left(nx_1-k\right)=1, \tag{6.27}$$

because then

$$\lim_{n\to\infty}\sum_{k=\lceil na\rceil}^{\lfloor nb\rfloor} \frac{M_{q,\lambda}\left(nx_1-k\right)}{2}+\frac{1}{2}\neq 1, \tag{6.28}$$

equivalently

$$\lim_{n\to\infty}\sum_{k=\lceil na\rceil}^{\lfloor nb\rfloor} \frac{M_{q,\lambda}\left(nx_1-k\right)}{2}\neq \frac{1}{2}, \tag{6.29}$$

true by

$$\lim_{n\to\infty}\sum_{k=\lceil na\rceil}^{\lfloor nb\rfloor} M_{q,\lambda}\left(nx_1-k\right)\neq 1. \tag{6.30}$$

II) Let $[a,b]\subset\mathbb{R}$. For large n we always have $\lceil na\rceil\leq\lfloor nb\rfloor$. Also $a\leq\frac{k}{n}\leq b$, iff $\lceil na\rceil\leq k\leq\lfloor nb\rfloor$. So in general it holds

$$\sum_{k=\lceil na\rceil}^{\lfloor nb\rfloor}\Phi\left(nx-k\right)\leq 1. \tag{6.31}$$

Next, we move on to the multivariate case, see Chapter 17 of [145], pp. 419–452, as a model of action.

We make

Remark 6.2. We introduce

$$Z_1(x_1, ..., x_N) := Z_1(x) := \prod_{i=1}^{N} \Phi(x_i) = \frac{1}{2^N} \prod_{i=1}^{N} \left(M_{q,\lambda} + M_{\frac{1}{q},\lambda}\right)(x_i), \tag{6.32}$$

$x = (x_1, ..., x_N) \in \mathbb{R}^N$, $\lambda, q > 0$, $N \in \mathbb{N}$.

Properties:

(i)

$$Z_1(x) > 0, \ \forall \ x \in \mathbb{R}^N, \tag{6.33}$$

(ii)

$$\sum_{k=-\infty}^{\infty} Z_1(x-k) := \sum_{k_1=-\infty}^{\infty} \sum_{k_2=-\infty}^{\infty} ... \sum_{k_N=-\infty}^{\infty} Z_1(x_1 - k_1, ..., x_N - k_N) = 1, \tag{6.34}$$

$k := (k_1, ..., k_n) \in \mathbb{Z}^N$, $\forall \ x \in \mathbb{R}^N$,

hence

(iii)

$$\sum_{k=-\infty}^{\infty} Z_1(nx-k) = 1, \tag{6.35}$$

$\forall \ x \in \mathbb{R}^N$, $n \in \mathbb{N}$,

(iv)

$$\int_{\mathbb{R}^N} Z_1(x)\, dx = 1, \tag{6.36}$$

that is Z_1 is a multivariate density function.

Here denote $\|x\|_\infty := \max\{|x_1|, ..., |x_N|\}$, $x \in \mathbb{R}^N$, $\infty := (\infty, ..., \infty)$, $-\infty := (-\infty, ..., -\infty)$, $\lceil na \rceil := (\lceil na_1 \rceil, ..., \lceil na_N \rceil)$, $\lfloor nb \rfloor := (\lfloor nb_1 \rfloor, ..., \lfloor nb_N \rfloor)$, $a := (a_1, ..., a_N)$, $b := (b_1, ..., b_N)$.

We obviously see that

$$\sum_{k=\lceil na \rceil}^{\lfloor nb \rfloor} Z_1(nx-k) = \sum_{k=\lceil na \rceil}^{\lfloor nb \rfloor} \left(\prod_{i=1}^{N} \Phi(nx_i - k_i) \right) = \prod_{i=1}^{N} \left(\sum_{k_i=\lceil na_i \rceil}^{\lfloor nb_i \rfloor} \Phi(nx_i - k_i) \right). \tag{6.37}$$

(v) We derive that

$$\sum_{\substack{k=\lceil na\rceil \\ \left\|\frac{k}{n}-x\right\|_{\infty}>\frac{1}{n^{\beta}}}}^{\lfloor nb\rfloor} Z_1(nx-k) < Te^{-2\lambda n^{(1-\beta)}}, \text{ where } 0<\beta<1, \tag{6.38}$$

with $n \in \mathbb{N} : n^{1-\beta} > 2$, $x \in \prod_{i=1}^{N} [a_i, b_i]$; $T := 2\max\left\{q, \frac{1}{q}\right\} e^{4\lambda}$.

(vi) It holds

$$0 < \frac{1}{\sum_{k=\lceil na\rceil}^{\lfloor nb\rfloor} Z_1(nx-k)} < (\Delta(q))^N, \tag{6.39}$$

$\forall\, x \in \left(\prod_{i=1}^{N} [a_i, b_i]\right)$, $n \in \mathbb{N}$.

It is clear that

(vii)

$$\sum_{\substack{k=-\infty \\ \left\|\frac{k}{n}-x\right\|_{\infty}>\frac{1}{n^{\beta}}}}^{\infty} Z_1(nx-k) < Te^{-2\lambda n^{(1-\beta)}}, \tag{6.40}$$

where $0<\beta<1$, $n \in \mathbb{N} : n^{1-\beta} > 2$, $x \in \mathbb{R}^N$.

Furthermore it holds

$$\lim_{n\to\infty} \sum_{k=\lceil na\rceil}^{\lfloor nb\rfloor} Z_1(nx-k) \neq 1, \tag{6.41}$$

for at least some $x \in \left(\prod_{i=1}^{N} [a_i, b_i]\right)$.

Here $\left(X, \|\cdot\|_{\gamma}\right)$ is a Banach space.

Let $f \in C\left(\prod_{i=1}^{N} [a_i, b_i], X\right)$, $x = (x_1, ..., x_N) \in \prod_{i=1}^{N} [a_i, b_i]$, $n \in \mathbb{N}$: $\lceil na_i\rceil \leq \lfloor nb_i\rfloor$, $i = 1, ..., N$.

We introduce and define the following multivariate linear normalized symmetrized neural network operator, let $x := (x_1, ..., x_N) \in \prod_{i=1}^{N} [a_i, b_i]$:

$$A_n^s(f, x_1, ..., x_N) := A_n^s(f, x) := \frac{\sum_{k=\lceil na\rceil}^{\lfloor nb\rfloor} f\left(\frac{k}{n}\right) Z_1(nx-k)}{\sum_{k=\lceil na\rceil}^{\lfloor nb\rfloor} Z_1(nx-k)} =$$

$$\frac{\sum_{k_1=\lceil na_1\rceil}^{\lfloor nb_1\rfloor} \sum_{k_2=\lceil na_2\rceil}^{\lfloor nb_2\rfloor} \cdots \sum_{k_N=\lceil na_N\rceil}^{\lfloor nb_N\rfloor} f\left(\frac{k_1}{n}, ..., \frac{k_N}{n}\right)\left(\prod_{i=1}^{N}\left(M_{q,\lambda}(nx_i-k_i) + M_{\frac{1}{q},\lambda}(nx_i-k_i)\right)\right)}{\prod_{i=1}^{N}\left(\sum_{k_i=\lceil na_i\rceil}^{\lfloor nb_i\rfloor}\left(M_{q,\lambda}(nx_i-k_i) + M_{\frac{1}{q},\lambda}(nx_i-k_i)\right)\right)}. \tag{6.42}$$

For large enough $n \in \mathbb{N}$ we always obtain $\lceil na_i \rceil \leq \lfloor nb_i \rfloor$, $i = 1, ..., N$. Also $a_i \leq \frac{k_i}{n} \leq b_i$, iff $\lceil na_i \rceil \leq k_i \leq \lfloor nb_i \rfloor$, $i = 1, ..., N$.

When $h \in C\left(\prod_{i=1}^{N} [a_i, b_i]\right)$ we define the companion operator

$$\widetilde{A}_n^s(h, x) := \frac{\sum_{k=\lceil na \rceil}^{\lfloor nb \rfloor} h\left(\frac{k}{n}\right) Z_1(nx - k)}{\sum_{k=\lceil na \rceil}^{\lfloor nb \rfloor} Z_1(nx - k)}. \tag{6.43}$$

Clearly $\widetilde{A}_n^s$ is a positive linear operator. We have that

$$\widetilde{A}_n^s(1, x) = 1, \ \ \forall\, x \in \prod_{i=1}^{N} [a_i, b_i].$$

Notice that $A_n^s(f) \in C\left(\prod_{i=1}^{N} [a_i, b_i], X\right)$ and $\widetilde{A}_n^s(h) \in C\left(\prod_{i=1}^{N} [a_i, b_i]\right)$.

Furthermore it holds

$$\left\|A_n^s(f, x)\right\|_\gamma \leq \widetilde{A}_n^s\left(\|f\|_\gamma, x\right), \tag{6.44}$$

$\forall\, x \in \prod_{i=1}^{N} [a_i, b_i]$,

and

$$A_n^s(ch, x) = c\widetilde{A}_n^s(h, x), \ \ \forall\, x \in \prod_{i=1}^{N} [a_i, b_i], \tag{6.45}$$

and

$$A_n^s(c) = c, \ \ \text{any } c \in X. \tag{6.46}$$

6.3 Abstract Korovkin Type Inequalities

All here in this section come from [143] and [144], Ch. 10, pp. 261–281.

We make

Remark 6.3. Let $\left(\mathbb{R}^k, \|\cdot\|_p\right)$, $k \in \mathbb{N}$; where $\|\cdot\|_p$ is the L_p-norm, $1 \leq p \leq +\infty$. $\mathbb{R}^k$ is a Banach space, and $\left(\mathbb{R}^k\right)^j$ denotes the j-fold product space $\mathbb{R}^k \times ... \times \mathbb{R}^k$ endowed with the max-norm $\|x\|_{(\mathbb{R}^k)^j} := \max_{1 \leq \lambda \leq j} \|x_\lambda\|_p$, where $x := (x_1, ..., x_j) \in \left(\mathbb{R}^k\right)^j$.

Let $\left(X, \|\cdot\|_\gamma\right)$ be a general Banach space. Then the space $L_j := L_j\left(\left(\mathbb{R}^k\right)^j; X\right)$ of all j-multilinear continuous maps $g : \left(\mathbb{R}^k\right)^j \to X$, $j = 1, ..., n$, is a Banach space with norm

$$\|g\| := \|g\|_{L_j} := \sup_{\left(\|x\|_{(\mathbb{R}^k)^j} = 1\right)} \|g(x)\|_\beta = \sup \frac{\|g(x)\|_\beta}{\|x_1\|_p \cdots \|x_j\|_p}. \tag{6.47}$$

Let M be a non-empty convex and compact subset of $\mathbb{R}^k$ and $x_0 \in M$ is fixed.

Let O an open subset of $\mathbb{R}^k : M \subset O$. Let $f : O \to X$ be a continuous function whose Fréchet derivatives (see [151]) $f^{(j)} : O \to L_j = L_j\left(\left(\mathbb{R}^k\right)^j ; X\right)$ exist and are continuous for $1 \leq j \leq n$, $n \in \mathbb{N}$. Call $(x - x_0)^j := (x - x_0, ..., x - x_0) \in \left(\mathbb{R}^k\right)^j$, $x \in M$.

Here we deal with $f|_M$.

Then by Taylor's formula ([147]), ([151], p. 124), we get

$$f(x) = \sum_{j=0}^{n} \frac{f^{(j)}(x_0)(x - x_0)^j}{j!} + R_n(x, x_0), \quad \text{all } x \in M, \tag{6.48}$$

where the remainder is the Riemann integral

$$R_n(x, x_0) := \int_0^1 \frac{(1-u)^{n-1}}{(n-1)!} \left(f^{(n)}(x_0 + u(x - x_0)) - f^{(n)}(x_0)\right)(x - x_0)^n du, \tag{6.49}$$

here we set $f^{(0)}(x_0)(x - x_0)^0 = f(x_0)$.

Considering

$$w := \omega_1\left(f^{(n)}, h\right) := \sup_{\substack{x,y \in M: \\ \|x-y\|_p \leq h}} \left\|f^{(n)}(x) - f^{(n)}(y)\right\|, \tag{6.50}$$

we obtain

$$\left\|\left(f^{(n)}(x_0 + u(x - x_0)) - f^{(n)}(x_0)\right)(x - x_0)^n\right\|_\gamma \leq$$

$$\left\|f^{(n)}(x_0 + u(x - x_0)) - f^{(n)}(x_0)\right\| \cdot \|x - x_0\|_p^n \leq$$

$$w \cdot \|x - x_0\|_p^n \cdot \left\lceil \frac{u\|x - x_0\|_p}{h} \right\rceil, \tag{6.51}$$

by Lemma 7.1.1, [141], p. 208, $\lceil\cdot\rceil$ is the ceiling.

Therefore for all $x \in M$ (see [141], pp. 121–122):

$$\|R_n(x, x_0)\|_\gamma \leq w\|x - x_0\|_p^n \int_0^1 \left\lceil \frac{u\|x - x_0\|_p}{h} \right\rceil \frac{(1-u)^{n-1}}{(n-1)!} du$$

$$= w\Phi_n\left(\|x - x_0\|_p\right) \tag{6.52}$$

by a change of variable, where

$$\Phi_n(t) := \int_0^{|t|} \left\lceil \frac{s}{h} \right\rceil \frac{(|t| - s)^{n-1}}{(n-1)!} ds = \frac{1}{n!} \left(\sum_{j=0}^{\infty} (|t| - jh)_+^n \right), \quad \forall\, t \in \mathbb{R}, \tag{6.53}$$

is a (polynomial) spline function, see [141], pp. 210–211. Also from there we get

$$\Phi_n(t) \leq \left(\frac{|t|^{n+1}}{(n+1)!h} + \frac{|t|^n}{2n!} + \frac{h\,|t|^{n-1}}{8\,(n-1)!} \right), \quad \forall\, t \in \mathbb{R}, \tag{6.54}$$

with equality true only at $t = 0$. Therefore it holds

$$\|R_n(x, x_0)\|_\gamma \leq w \left(\frac{\|x - x_0\|_p^{n+1}}{(n+1)!h} + \frac{\|x - x_0\|_p^n}{2n!} + \frac{h\,\|x - x_0\|_p^{n-1}}{8\,(n-1)!} \right), \forall\, x \in M. \tag{6.55}$$

We have found that

$$\left\| f(x) - \sum_{j=0}^{n} \frac{f^{(j)}(x_0)(x - x_0)^j}{j!} \right\|_\gamma \leq$$

$$\omega_1\left(f^{(n)}, h\right) \left(\frac{\|x - x_0\|_p^{n+1}}{(n+1)!h} + \frac{\|x - x_0\|_p^n}{2n!} + \frac{h\,\|x - x_0\|_p^{n-1}}{8\,(n-1)!} \right) < \infty, \tag{6.56}$$

$\forall\, x, x_0 \in M$.

Here $0 < \omega_1\left(f^{(n)}, h\right) < \infty$, by M being compact and $f^{(n)}$ being continuous on M.

One can rewrite (6.56) as

$$\left\| f(\cdot) - \sum_{j=0}^{n} \frac{f^{(j)}(x_0)(\cdot - x_0)^j}{j!} \right\|_\gamma \leq$$

$$\omega_1\left(f^{(n)}, h\right) \left(\frac{\|\cdot - x_0\|_p^{n+1}}{(n+1)!h} + \frac{\|\cdot - x_0\|_p^n}{2n!} + \frac{h\,\|\cdot - x_0\|_p^{n-1}}{8\,(n-1)!} \right), \ \forall\, x_0 \in M, \tag{6.57}$$

a pointwise functional inequality on M.

Here $(\cdot - x_0)^j$ maps M into $\left(\mathbb{R}^k\right)^j$ and it is continuous, also $f^{(j)}(x_0)$ maps $\left(\mathbb{R}^k\right)^j$ into X and it is continuous. Hence their composition $f^{(j)}(x_0)(\cdot - x_0)^j$ is continuous from M into X.

Clearly $f(\cdot) - \sum_{j=0}^{n} \frac{f^{(j)}(x_0)(\cdot - x_0)^j}{j!} \in C(M, X)$, hence $\left\| f(\cdot) - \sum_{j=0}^{n} \frac{f^{(j)}(x_0)(\cdot - x_0)^j}{j!} \right\|_\gamma \in C(M)$.

We mention the following fundamental approximation result.

Theorem 6.2. *Let M be a nonempty compact convex subset of the open subset O of $\left(\mathbb{R}^k, \|\cdot\|_p\right)$, $1 \leq p \leq +\infty$, and let $\left(X, \|\cdot\|_\gamma\right)$ be a general Banach space. For any $N \in \mathbb{N}$ let the linear operators $L_N : C(M, X) \to C(M, X)$ and the positive linear operators $\widetilde{L}_N : C(M) \to C(M)$, such that*

$$\|(L_N(f))(x_0)\|_\gamma \leq \left(\widetilde{L}_N\left(\|f\|_\gamma\right)\right)(x_0), \tag{6.58}$$

$\forall\, N \in \mathbb{N}$, $\forall\, f \in C(M, X)$, $\forall\, x_0 \in M$.

Furthermore assume that

$$L_N(cF) = c\widetilde{L}_N(F), \quad \forall\, F \in C(M), \forall\, c \in X. \tag{6.59}$$

Let $n \in \mathbb{N}$, here we deal with $f \in C^n(O, X)$, the space of n-times continuously Fréchet differentiable functions from O into X.

Here we study approximation to $f|_M$.

Let $x, x_0 \in M$, $r > 0$, and

$$\omega_1\left(f^{(n)}, h\right) := \sup_{\substack{x,y \in M: \\ \|x-y\|_p \leq h}} \left\|f^{(n)}(x) - f^{(n)}(y)\right\|. \tag{6.60}$$

Then

1)

$$\left\|(L_N(f))(x_0) - \sum_{j=0}^{n} \frac{1}{j!}\left(\widetilde{L}_N\left(f^{(j)}(x_0)(\cdot - x_0)^j\right)\right)(x_0)\right\|_\gamma \leq$$

$$\frac{\omega_1\left(f^{(n)}, r\left(\left(\widetilde{L}_N\left(\|\cdot - x_0\|_p^{n+1}\right)\right)(x_0)\right)^{\frac{1}{n+1}}\right)}{rn!}\left(\left(\widetilde{L}_N\left(\|\cdot - x_0\|_p^{n+1}\right)\right)(x_0)\right)^{\left(\frac{n}{n+1}\right)}$$

$$\left[\frac{1}{(n+1)} + \frac{r\left(\left(\widetilde{L}_N(1)\right)(x_0)\right)^{\frac{1}{n+1}}}{2} + \frac{nr^2\left(\left(\widetilde{L}_N(1)\right)(x_0)\right)^{\frac{2}{n+1}}}{8}\right], \tag{6.61}$$

2) additionally if $f^{(j)}(x_0) = 0$, $j = 0, 1, ..., n$, we have

$$\|(L_N(f))(x_0)\|_\gamma \leq$$

$$\frac{\omega_1\left(f^{(n)}, r\left(\left(\widetilde{L}_N\left(\|\cdot - x_0\|_p^{n+1}\right)\right)(x_0)\right)^{\frac{1}{n+1}}\right)}{rn!}\left(\left(\widetilde{L}_N\left(\|\cdot - x_0\|_p^{n+1}\right)\right)(x_0)\right)^{\left(\frac{n}{n+1}\right)}$$

$$\left[\frac{1}{(n+1)}+\frac{r\left(\left(\widetilde{L}_N(1)\right)(x_0)\right)^{\frac{1}{n+1}}}{2}+\frac{nr^2\left(\left(\widetilde{L}_N(1)\right)(x_0)\right)^{\frac{2}{n+1}}}{8}\right], \quad (6.62)$$

3)

$$\left\|(L_N(f))(x_0)-f(x_0)\right\|_\gamma \leq \|f(x_0)\|\left|\left(\widetilde{L}_N(1)\right)(x_0)-1\right|+$$

$$\sum_{j=1}^{n}\frac{1}{j!}\left\|\left(\widetilde{L}_N\left(f^{(j)}(x_0)(\cdot-x_0)^j\right)\right)(x_0)\right\|_\gamma+$$

$$\frac{\omega_1\left(f^{(n)},r\left(\left(\widetilde{L}_N\left(\|\cdot-x_0\|_p^{n+1}\right)\right)(x_0)\right)^{\frac{1}{n+1}}\right)}{rn!}\left(\left(\widetilde{L}_N\left(\|\cdot-x_0\|_p^{n+1}\right)\right)(x_0)\right)^{\left(\frac{n}{n+1}\right)}$$

$$\left[\frac{1}{(n+1)}+\frac{r\left(\left(\widetilde{L}_N(1)\right)(x_0)\right)^{\frac{1}{n+1}}}{2}+\frac{nr^2\left(\left(\widetilde{L}_N(1)\right)(x_0)\right)^{\frac{2}{n+1}}}{8}\right], \quad (6.63)$$

4)

$$\left\|\|L_N(f)-f\|_\gamma\right\|_{\infty,M} \leq \left\|\|f\|_\gamma\right\|_{\infty,M}\left\|\left(\widetilde{L}_N(1)\right)-1\right\|_{\infty,M}+$$

$$\sum_{j=1}^{n}\frac{1}{j!}\left\|\left\|\left(\widetilde{L}_N\left(f^{(j)}(x_0)(\cdot-x_0)^j\right)\right)(x_0)\right\|_\gamma\right\|_{\infty,x_0\in M}+$$

$$\frac{\omega_1\left(f^{(n)},r\left\|\left(\widetilde{L}_N\left(\|\cdot-x_0\|_p^{n+1}\right)\right)(x_0)\right\|_{\infty,x_0\in M}^{\frac{1}{n+1}}\right)}{rn!}$$

$$\left\|\left(\widetilde{L}_N\left(\|\cdot-x_0\|_p^{n+1}\right)\right)(x_0)\right\|_{\infty,x_0\in M}^{\left(\frac{n}{n+1}\right)}$$

$$\left[\frac{1}{(n+1)}+\frac{r\left\|\widetilde{L}_N(1)\right\|_{\infty,M}^{\frac{1}{n+1}}}{2}+\frac{nr^2\left\|\widetilde{L}_N(1)\right\|_{\infty,M}^{\frac{2}{n+1}}}{8}\right]. \quad (6.64)$$

We need

Lemma 6.1. *The function* $\left(\widetilde{L}_N\left(\|\cdot-x_0\|_p^m\right)\right)(x_0)$ *is continuous in* $x_0 \in M$, $m \in \mathbb{N}$.

Remark 6.4. We also have

$$\left\|\left(\widetilde{L}_N\left(\|\cdot-x_0\|_p^k\right)\right)(x_0)\right\|_{\infty,x_0\in M}\leq$$

$$\left\|\left(\widetilde{L}_N\left(\|\cdot-x_0\|_p^{n+1}\right)\right)(x_0)\right\|_{\infty,x_0\in M}^{\left(\frac{k}{n+1}\right)}\left\|\widetilde{L}_N(1)\right\|_{\infty,M}^{\left(\frac{n+1-k}{n+1}\right)}, \tag{6.65}$$

for all $k=1,...,n$.

Conclusion 6.1. Let $N\to\infty$ and $\widetilde{L}_N(1)\overset{u}{\to}1$, uniformly, and $\left\|\left(\widetilde{L}_N\left(\|\cdot-x_0\|_p^{n+1}\right)\right)(x_0)\right\|_{\infty,x_0\in M}\to 0$. Then by (6.64) we get, as $N\to\infty$, that $L_n(f)\overset{u}{\to}f$, uniformly in $\|\cdot\|_\gamma$ over M.

The last statement is also supported by

$$\left\|\left(L_N\left(f^{(j)}(x_0)(\cdot-x_0)^j\right)\right)(x_0)\right\|_\beta\leq\left\|f^{(j)}\right\|\left\|\left(\widetilde{L}_N\left(\|\cdot-x_0\|_p^j\right)\right)(x_0)\right\|_{\infty,x_0\in M}$$

and (6.65). Here notice that $\left\|\widetilde{L}_N(1)\right\|_{\infty,M}$ turns out to be bounded, and $\omega_1\left(f^{(n)},\cdot\right)$ is also bounded.

Conclusion 6.2. By (6.63) we get the following: as $N\to\infty$, assume that $\left(\widetilde{L}_N(1)\right)(x_0)\to 1$, and $\left(\left(\widetilde{L}_N\left(\|\cdot-x_0\|_p^{n+1}\right)\right)(x_0)\right)\to 0$. Then $(L_N(f))(x_0)\to f(x_0)$, as $N\to\infty$, pointwise. Notice that here $\left(\widetilde{L}_N(1)\right)(x_0)$ is bounded with respect to N, as well as $\omega_1\left(f^{(n)},\cdot\right)$ is bounded in general.

Corollary 6.1. *(to Theorem 6.2, case $n=1$, see (6.63)) Additionally assume that $\left(\widetilde{L}_N(1)\right)(x_0)=1$, $\forall\ x_0\in M$. Then*

1)

$$\|(L_N(f))(x_0)-f(x_0)\|_\gamma\leq\left\|\left(\widetilde{L}_N\left(f^{(1)}(x_0)(\cdot-x_0)\right)\right)(x_0)\right\|_\gamma+$$

$$\frac{1}{2r}\omega_1\left(f^{(1)},r\left(\left(\widetilde{L}_N\left(\|\cdot-x_0\|_p^2\right)\right)(x_0)\right)^{\frac{1}{2}}\right)\left(\left(\widetilde{L}_N\left(\|\cdot-x_0\|_p^2\right)\right)(x_0)\right)^{\frac{1}{2}}$$

$$\left[1+r+\frac{r^2}{4}\right], \tag{6.66}$$

2)

$$\|\|(L_N(f))-f\|\|_{\infty,M}\leq$$

$$\left\| \left\| \left(\widetilde{L}_N \left(f^{(1)}(x_0)(\cdot - x_0) \right) \right)(x_0) \right\|_\gamma \right\|_{\infty, x_0 \in M} +$$

$$\frac{1}{2r} \omega_1 \left(f^{(1)}, r \left\| \left(\widetilde{L}_N \left(\|\cdot - x_0\|_p^2 \right) \right)(x_0) \right\|_{\infty, x_0 \in M}^{\frac{1}{2}} \right) \left\| \left(\widetilde{L}_N \left(\|\cdot - x_0\|_p^2 \right) \right)(x_0) \right\|_{\infty, x_0 \in M}^{\frac{1}{2}}$$

$$\left[1 + r + \frac{r^2}{4} \right]. \tag{6.67}$$

We need

Definition 6.1. Let M be a convex and compaxt subset of $\left(\mathbb{R}^k, \|\cdot\|_p\right)$, $p \in [1, \infty]$, and $\left(X, \|\cdot\|_\gamma\right)$ be a general Banach space. Let $f \in C(M, X)$. We define the first modulus of continuity of f as

$$\omega_1(f, \delta) := \sup_{\substack{x, y \in M: \\ \|x - y\|_p \leq \delta}} \|f(x) - f(y)\|, \quad 0 < \delta \leq diam(M). \tag{6.68}$$

If $\delta > diam(M)$, then

$$\omega_1(f, \delta) = \omega_1(f, diam(M)). \tag{6.69}$$

Notice $\omega_1(f, \delta)$ is increasing in $\delta > 0$. For $f \in B(M, X)$ (bounded functions) $\omega_1(f, \delta)$ is defined similarly.

Lemma 6.2. *We have $\omega_1(f, \delta) \to 0$ as $\delta \downarrow 0$ iff $f \in C(M, X)$.*

Lemma 6.3. *(see [141], p. 208) Let M be a convex compact subset of $\left(\mathbb{R}^k, \|\cdot\|_p\right)$, $p \in [1, \infty]$, and $\left(X, \|\cdot\|_\gamma\right)$ be a general Banach space. Let $f \in B(M, X)$. Then*

$$\|f(x) - f(x_0)\|_\gamma \leq \omega_1(f, h) \left\lceil \frac{\|x - x_0\|_p}{h} \right\rceil$$

$$\leq \omega_1(f, h) \left(1 + \frac{\|x - x_0\|_p}{h} \right), \quad \forall\, x \in M, \tag{6.70}$$

where $\lceil \cdot \rceil$ is the ceiling of the number, $h > 0$.

We present the basic pointwise convergence result

Theorem 6.3. *Let $L_N : C(M, X) \to C(M, X)$ be linear operators, $\forall$ $N \in \mathbb{N}$; where M is a convex compact subset of $\left(\mathbb{R}^k, \|\cdot\|_p\right)$, $p \in [1, \infty]$,*

$\left(X, \|\cdot\|_{\gamma}\right)$ *is a general Banach space. Let* $\widetilde{L}_N : C(M) \to C(M)$ *be positive linear operators, such that*

$$\|(L_N(f))(x_0)\|_{\gamma} \leq \left(\widetilde{L}_N\left(\|f\|_{\gamma}\right)\right)(x_0), \tag{6.71}$$

$\forall\, N \in \mathbb{N}$, $\forall\, f \in C(M, X)$, $\forall\, x_0 \in M$.

Furthermore assume that

$$L_N(cg) = c\widetilde{L}_N(g), \quad \forall\, g \in C(M), \forall\, c \in X. \tag{6.72}$$

Then

$$\|(L_N(f))(x_0) - f(x_0)\|_{\gamma} \leq \|f(x_0)\|_{\gamma} \left|\left(\widetilde{L}_N(1)\right)(x_0) - 1\right| + \tag{6.73}$$

$$\left[\left(\widetilde{L}_N(1)\right)(x_0) + 1\right] \omega_1\left(f, \left(\widetilde{L}_N\left(\|\cdot - x_0\|_p\right)\right)(x_0)\right), \quad \forall\, f \in C(M, X).$$

By (6.73), as $\left(\widetilde{L}_N(1)\right)(x_0) \to 1$ *and* $\left(\widetilde{L}_N\left(\|\cdot - x_0\|_p\right)\right)(x_0) \to 0$, *then* $(L_N(f))(x_0) \overset{\|\cdot\|_{\gamma}}{\to} f(x_0)$, *as* $N \to \infty$, $\forall\, f \in C(M, X)$. *Here notice that* $\left(\widetilde{L}_N(1)\right)(x_0)$ *is bounded.*

Corollary 6.2. *All as in Theorem 6.3. Additionally assume that* $\left(\widetilde{L}_N(1)\right)(x_0) = 1$, $\forall\, x_0 \in M$. *Then*

$$\|(L_N(f))(x_0) - f(x_0)\|_{\gamma} \leq 2\omega_1\left(f, \left(\widetilde{L}_N\left(\|\cdot - x_0\|_p\right)\right)(x_0)\right), \forall\, f \in C(M, X). \tag{6.74}$$

Next we obtain the following uniform convergence result.

Theorem 6.4. *All as in Theorem 6.3. Then*

$$\left\|\|L_N(f) - f\|_{\gamma}\right\|_{\infty, M} \leq \left\|\|f\|_{\gamma}\right\|_{\infty, M} \left\|\widetilde{L}_N(1) - 1\right\|_{\infty, M} + \tag{6.75}$$

$$\left\|\widetilde{L}_N(1) + 1\right\|_{\infty, M} \omega_1\left(f, \left\|\left(\widetilde{L}_N\left(\|\cdot - x_0\|_p\right)\right)(x_0)\right\|_{\infty, x_0 \in M}\right), \forall\, f \in C(M, X).$$

Conclusion 6.3. By Lemma 6.1 $\left(\widetilde{L}_N\left(\|\cdot - x_0\|_p\right)\right)(x_0)$ is continuous in $x_0 \in M$. Thus $\left\|\left(\widetilde{L}_N\left(\|\cdot - x_0\|_p\right)\right)(x_0)\right\|_{\infty, x_0 \in M} < \infty$.

As $N \to \infty$, we assume that $\widetilde{L}_N(1) \overset{u}{\to} 1$, uniformly, and $\left\|\left(\widetilde{L}_N\left(\|\cdot - x_0\|_p\right)\right)(x_0)\right\|_{\infty, x_0 \in M} \to 0$, then $L_n(f) \overset{u}{\to} f$, uniformly on M, $\forall\, f \in C(M, X)$. Here $\left\|\widetilde{L}_N(1)\right\|$ is bounded.

Under convexity we have the following sharp general pointwise convergence result. Here M^0 denotes the interior of M.

Theorem 6.5. *All here as in Theorem 6.3. Additionally, assume that* $x_0 \in M^0$*, and the closed ball in* $\mathbb{R}^k : B\left(x_0, \left(\widetilde{L}_N\left(\|\cdot - x_0\|_p\right)\right)(x_0)\right) \subset M$*, where* $\left(\widetilde{L}_N\left(\|\cdot - x_0\|_p\right)\right)(x_0) \geq 0$*, and* $\|f(t) - f(x_0)\|_\gamma$ *is convex in* $t \in M$.

Then

$$\|(L_N(f))(x_0) - f(x_0)\|_\gamma \leq \|f(x_0)\|_\gamma \left|\left(\widetilde{L}_N(1)\right)(x_0) - 1\right| +$$

$$\omega_1\left(f, \left(\widetilde{L}_N\left(\|\cdot - x_0\|_p\right)\right)(x_0)\right), \quad \forall\, N \in \mathbb{N}. \tag{6.76}$$

Theorem 6.6. *All as in Theorem 6.5. Inequality (6.76) is sharp, in fact it is attained by* $f(t) = \overrightarrow{i}\,\|t - x_0\|_p$, $\overrightarrow{i}$ *is a unit vector of* $\left(X, \|\cdot\|_\gamma\right)$, $t \in M$.

6.4 Main Results

In this section we apply the results of Section 3 to the linear neural network operators A_n^s and $\widetilde{A}_n^s$, $n \in \mathbb{N}$, of Section 2. In fact $\widetilde{A}_n^s$ are positive linear neural network operators, with the property $\widetilde{A}_n^s(1) = 1$.

Also (6.44)–(6.46) are valid.

From now on $M = \prod_{i=1}^{N}[a_i, b_i] \subset \left(\mathbb{R}^N, \|\cdot\|_p\right)$, $N \in \mathbb{N}$, where $1 \leq p \leq +\infty$.

We present the following general approximation result.

Theorem 6.7. *Let* $\left(X, \|\cdot\|_\gamma\right)$ *is a general Banach space,* $x_0 \in \prod_{i=1}^{N}[a_i, b_i]$, $1 \leq p \leq +\infty$*. Let* $m \in \mathbb{N}$, $f \in C^m\left(\prod_{i=1}^{N}[a_i, b_i], X\right)$*, the space of m-times continuously Fréchet differentiable functions from* $\prod_{i=1}^{N}[a_i, b_i]$ *into* X.

Let $x, x_0 \in \prod_{i=1}^{N}[a_i, b_i]$, $r > 0$*, and*

$$\omega_1\left(f^{(m)}, h\right) := \sup_{\substack{x, y \in \prod_{i=1}^{N}[a_i, b_i]: \\ \|x - y\|_p \leq h}} \left\|f^{(m)}(x) - f^{(m)}(y)\right\|, \quad h > 0. \tag{6.77}$$

Then

1)

$$\left\| \left(A_n^s(f)\right)(x_0) - \sum_{j=0}^{n} \frac{1}{j!} \left(\widetilde{A}_n^s \left(f^{(j)}(x_0)(\cdot - x_0)^j\right)\right)(x_0) \right\|_\gamma \leq$$

$$\frac{\omega_1\left(f^{(m)}, r\left(\left(\widetilde{A}_n^s\left(\|\cdot - x_0\|_p^{m+1}\right)\right)(x_0)\right)^{\frac{1}{m+1}}\right)}{rm!} \left(\left(\widetilde{A}_n^s\left(\|\cdot - x_0\|_p^{m+1}\right)\right)(x_0)\right)^{\left(\frac{m}{m+1}\right)}$$

$$\left[\frac{1}{(m+1)} + \frac{r}{2} + \frac{mr^2}{8}\right], \tag{6.78}$$

2) additionally if $f^{(j)}(x_0) = 0$, $j = 0, 1, ..., m$, *we have*

$$\left\|\left(A_n^s(f)\right)(x_0)\right\|_\gamma \leq$$

$$\frac{\omega_1\left(f^{(m)}, r\left(\left(\widetilde{A}_n^s\left(\|\cdot - x_0\|_p^{m+1}\right)\right)(x_0)\right)^{\frac{1}{m+1}}\right)}{rm!} \left(\left(\widetilde{A}_n^s\left(\|\cdot - x_0\|_p^{m+1}\right)\right)(x_0)\right)^{\left(\frac{m}{m+1}\right)}$$

$$\left[\frac{1}{(m+1)} + \frac{r}{2} + \frac{mr^2}{8}\right], \tag{6.79}$$

3)

$$\left\|\left(A_n^s(f)\right)(x_0) - f(x_0)\right\|_\gamma \leq$$

$$\sum_{j=1}^{n} \frac{1}{j!} \left\| \left(\widetilde{A}_n^s\left(f^{(j)}(x_0)(\cdot - x_0)^j\right)\right)(x_0) \right\|_\gamma +$$

$$\frac{\omega_1\left(f^{(m)}, r\left(\left(\widetilde{A}_n^s\left(\|\cdot - x_0\|_p^{m+1}\right)\right)(x_0)\right)^{\frac{1}{m+1}}\right)}{rm!} \left(\left(\widetilde{A}_n^s\left(\|\cdot - x_0\|_p^{m+1}\right)\right)(x_0)\right)^{\left(\frac{m}{m+1}\right)}$$

$$\left[\frac{1}{(m+1)} + \frac{r}{2} + \frac{mr^2}{8}\right], \tag{6.80}$$

4)

$$\left\| \left\| A_n^s(f)(x_0) - f(x_0) \right\|_\gamma \right\|_{\infty, \prod_{i=1}^{N}[a_i, b_i]} \leq$$

$$\sum_{j=1}^{n} \frac{1}{j!} \left\| \left\| \left(\widetilde{A}_n^s\left(f^{(j)}(x_0)(\cdot - x_0)^j\right)\right)(x_0) \right\|_\gamma \right\|_{\infty, x_0 \in \prod_{i=1}^{N}[a_i, b_i]} +$$

$$\frac{\omega_1\left(f^{(m)}, r\left\|\left(\widetilde{A}_n^s\left(\|\cdot - x_0\|_p^{m+1}\right)\right)(x_0)\right\|_{\infty, x_0 \in \prod_{i=1}^N [a_i,b_i]}^{\frac{1}{m+1}}\right)}{rm!}$$

$$\left\|\left(\widetilde{A}_n^s\left(\|\cdot - x_0\|_p^{m+1}\right)\right)(x_0)\right\|_{\infty, x_0 \in \prod_{i=1}^N [a_i,b_i]}^{\left(\frac{m}{m+1}\right)}$$

$$\left[\frac{1}{(m+1)} + \frac{r}{2} + \frac{mr^2}{8}\right]. \tag{6.81}$$

Proof. A direct application of Theorem 6.2. □

We make

Remark 6.5. Clearly here $\omega_1\left(f^{(m)}, \cdot\right)$ is bounded. By Lemma 6.1 we get that $\left(\widetilde{A}_n^s\left(\|\cdot - x_0\|_p^{\lambda}\right)\right)(x_0)$ is continuous in $x_0 \in \prod_{i=1}^N [a_i, b_i]$, $\lambda \in \mathbb{N}$.

And, by (6.65) we obtain

$$\left\|\left(\widetilde{A}_n^s\left(\|\cdot - x_0\|_p^{k}\right)\right)(x_0)\right\|_{\infty, x_0 \in \prod_{i=1}^N [a_i,b_i]} \leq$$

$$\left\|\left(\widetilde{A}_n^s\left(\|\cdot - x_0\|_p^{m+1}\right)\right)(x_0)\right\|_{\infty, x_0 \in \prod_{i=1}^N [a_i,b_i]}^{\left(\frac{k}{m+1}\right)}, \tag{6.82}$$

for $k = 1, \ldots, m$.

We make

Remark 6.6. Here we take $p = \infty$. We observe that

$$\left(\widetilde{A}_n^s\left(\|\cdot - x_0\|_\infty^{m+1}\right)\right)(x_0) \overset{(6.43)}{=}$$

$$\frac{\sum_{k=\lceil na \rceil}^{\lfloor nb \rfloor} \left\|\frac{k}{n} - x_0\right\|_\infty^{m+1} Z_1(nx_0 - k)}{\sum_{k=\lceil na \rceil}^{\lfloor nb \rfloor} Z_1(nx_0 - k)} \overset{(6.39)}{\leq}$$

$$(\Delta(q))^N \left(\sum_{k=\lceil na \rceil}^{\lfloor nb \rfloor} \left\|\frac{k}{n} - x_0\right\|_\infty^{m+1} Z_1(nx_0 - k)\right) = \tag{6.83}$$

$$(\Delta(q))^N \left(\sum_{\left\{\begin{array}{c} k=\lceil na\rceil \\ :\left\|\frac{k}{n}-x_0\right\|_\infty \leq \frac{1}{n^\beta} \end{array}\right.}^{\lfloor nb\rfloor} \left\|\frac{k}{n}-x_0\right\|_\infty^{m+1} Z_1(nx_0-k) + \right.$$

$$\left. \sum_{\left\{\begin{array}{c} k=\lceil na\rceil \\ :\left\|\frac{k}{n}-x_0\right\|_\infty > \frac{1}{n^\beta} \end{array}\right.}^{\lfloor nb\rfloor} \left\|\frac{k}{n}-x_0\right\|_\infty^{m+1} Z_1(nx_0-k) \right) \overset{((6.35),(6.38))}{\leq}$$

$$(\Delta(q))^N \left[\frac{1}{n^{\beta(m+1)}} + \|b-a\|_\infty^{m+1} Te^{-2\lambda n^{(1-\beta)}} \right].$$

We have proved that

$$\left\| \left(\widetilde{A}_n^s \left(\|\cdot - x_0\|_p^{m+1} \right) \right)(x_0) \right\|_{\infty, x_0 \in \prod_{i=1}^N [a_i,b_i]} \leq \tag{6.84}$$

$$(\Delta(q))^N \left[\frac{1}{n^{\beta(m+1)}} + \|b-a\|_\infty^{m+1} Te^{-2\lambda n^{(1-\beta)}} \right] \to 0, \text{ as } n \to +\infty.$$

Here it is $0 < \beta < 1$, $T := 2\max\left\{q, \frac{1}{q}\right\} e^{4\lambda}$; $n \in \mathbb{N} : n^{1-\beta} > 2$, $x_0 \in \prod_{i=1}^N [a_i, b_i]$.

So, by Conclusion 6.1, see also (6.81) and (6.82), we derive that $A_n^s(f) \overset{u}{\to} f$, uniformly in $\|\cdot\|_\gamma$ over $\prod_{i=1}^N [a_i, b_i]$, as $n \to +\infty$; $p = +\infty$ case.

Pointwise convergence is obviously valid too.

We continue with

Corollary 6.3. *(to Theorem 6.7, case of $m = 1$, see (6.80)) Let $x_0 \in \prod_{i=1}^N [a_i, b_i]$, then*

1)

$$\|(A_n^s(f))(x_0) - f(x_0)\|_\gamma \leq \left\| \left(\widetilde{A}_n^s \left(f^{(1)}(x_0)(\cdot - x_0) \right) \right)(x_0) \right\|_\gamma +$$

$$\frac{1}{2r}\omega_1\left(f^{(1)}, r\left(\left(\widetilde{A}_n^s\left(\|\cdot - x_0\|_p^2\right)\right)(x_0)\right)^{\frac{1}{2}}\right)\left(\left(\widetilde{A}_n^s\left(\|\cdot - x_0\|_p^2\right)\right)(x_0)\right)^{\frac{1}{2}}$$

$$\left[1 + r + \frac{r^2}{4}\right], \tag{6.85}$$

and

2)

$$\left\|\|(A_n^s(f))(x_0) - f(x_0)\|_\gamma\right\|_{\infty, \prod_{i=1}^N [a_i,b_i]} \leq$$

$$\left\|\left\|\left(\widetilde{A}_n^s\left(f^{(1)}(x_0)(\cdot - x_0)\right)\right)(x_0)\right\|_\gamma\right\|_{\infty, x_0 \in \prod_{i=1}^N [a_i,b_i]} +$$

$$\frac{1}{2r}\omega_1\left(f^{(1)}, r\left\|\left(\widetilde{A}_n^s\left(\|\cdot - x_0\|_p^2\right)\right)(x_0)\right\|^{\frac{1}{2}}_{\infty, x_0 \in \prod_{i=1}^N [a_i,b_i]}\right)$$

$$\left\|\left(\widetilde{A}_n^s\left(\|\cdot - x_0\|_p^2\right)\right)(x_0)\right\|^{\frac{1}{2}}_{\infty, x_0 \in \prod_{i=1}^N [a_i,b_i]}\left[1 + r + \frac{r^2}{4}\right]. \tag{6.86}$$

Proof. Direct application to Corollary 6.1. □

We make

Remark 6.7. By (6.84) we get that

$$\left\|\left(\widetilde{A}_n^s\left(\|\cdot - x_0\|_p^2\right)\right)(x_0)\right\|_{\infty, x_0 \in \prod_{i=1}^N [a_i,b_i]} \leq \tag{6.87}$$

$$(\Delta(q))^N\left[\frac{1}{n^{2\beta}} + \|b - a\|_\infty^2 T e^{-2\lambda n^{(1-\beta)}}\right] \to 0, \text{ as } n \to +\infty.$$

Again by (6.86) we obtain $A_n^s(f) \to f$, as $n \to +\infty$, pointwise and uniformly, for $p = +\infty$ case.

Note 6.1. For $f \in C\left(\prod_{i=1}^N [a_i, b_i], X\right)$ the modulus of continuity $\omega_1(f, \delta)$ is defined by (6.68), (6.69).

By Lemma 6.2, we have that $\omega_1(f, \delta) \to 0$ as $\delta \downarrow 0$ iff $f \in C\left(\prod_{i=1}^N [a_i, b_i], X\right)$.

Also Lemma 6.3 now applies for an $f \in B\left(\prod_{i=1}^N [a_i, b_i], X\right)$ and (6.70) is valid.

We give

Theorem 6.8. *Let* $f \in C\left(\prod_{i=1}^{N}[a_i,b_i], X\right)$, *where* $1 \leq p \leq +\infty$.

Then

1)

$$\|(A_n^s(f))(x_0) - f(x_0)\|_\gamma \leq 2\omega_1\left(f, \left(\widetilde{A}_n^s\left(\|\cdot - x_0\|_p\right)\right)(x_0)\right), \tag{6.88}$$

and

2)

$$\left\|\|(A_n^s(f))(x_0) - f(x_0)\|_\gamma\right\|_{\infty, \prod_{i=1}^{N}[a_i,b_i]} \leq$$

$$2\omega_1\left(f, \left\|\left(\widetilde{A}_n^s\left(\|\cdot - x_0\|_p\right)\right)(x_0)\right\|_{\infty, x_0 \in \prod_{i=1}^{N}[a_i,b_i]}\right), \tag{6.89}$$

$\forall\, f \in C\left(\prod_{i=1}^{N}[a_i,b_i], X\right)$.

Proof. Direct application of (6.73), (6.74), (6.75). □

We make

Remark 6.8. Inequality (6.84) is clearly valid for $m = 0$. So we have

$$\left\|\|(A_n^s(f))(x_0) - f(x_0)\|_\gamma\right\|_{\infty, \prod_{i=1}^{N}[a_i,b_i]} \leq$$

$$2\omega_1\left(f, (\Delta(q))^N\left[\frac{1}{n^\beta} + \|b-a\|_\infty Te^{-2\lambda n^{(1-\beta)}}\right]\right) \to 0, \text{ as } n \to +\infty. \tag{6.90}$$

By (6.89) now we obtain $A_n^s(f) \to f$, as $n \to +\infty$, pointwise and uniformly, for $p = +\infty$ case.

Corollary 6.4. *It holds* $(0 < \beta < 1,\ p = +\infty)$

$$\left\|\|(A_n^s(f))(x_0) - f(x_0)\|_\gamma\right\|_{\infty, \prod_{i=1}^{N}[a_i,b_i]} \leq$$

$$2\omega_1\left(f, (\Delta(q))^N\left[\frac{1}{n^\beta} + \|b-a\|_\infty Te^{-2\lambda n^{(1-\beta)}}\right]\right), \tag{6.91}$$

$\forall\, f \in C\left(\prod_{i=1}^{N}[a_i,b_i], X\right)$.

Proof. By (6.89), (6.90), $p = +\infty$. $\square$

Under convexity we derive the following result.

Theorem 6.9. *Let* $f \in C\left(\prod_{i=1}^{N}[a_i, b_i], X\right)$, $1 \leq p \leq +\infty$. *Here* $x_0 \in \left(\prod_{i=1}^{N}[a_i, b_i]\right)^0$ *and* $\|f(t) - f(x_0)\|_\gamma$ *is convex in* $t \in \prod_{i=1}^{N}[a_i, b_i]$. *We assume that the closed ball in* $\mathbb{R}^N : B\left(x_0, \left(\widetilde{A}_n^s\left(\|\cdot - x_0\|_p\right)\right)(x_0)\right) \subset \prod_{i=1}^{N}[a_i, b_i]$.

Then

$$\|(A_n^s(f))(x_0) - f(x_0)\|_\gamma \leq \omega_1\left(f, \left(\widetilde{A}_n^s\left(\|\cdot - x_0\|_p\right)\right)(x_0)\right), \tag{6.92}$$

$\forall\, N \in \mathbb{N} : n^{1-\beta} > 2$, $0 < \beta < 1$.

Proof. By Theorem 6.5. Notice here that $\left(\widetilde{A}_n^s\left(\|\cdot - x_0\|_p\right)\right)(x_0) > 0$. $\square$

Theorem 6.10. *All as in Theorem 6.9. Inequality (6.92) is sharp, in fact it is attained by* $f(t) = \overrightarrow{i}\,\|t - x_0\|_p$, $\overrightarrow{i}$ *is a unit vector of* $\left(X, \|\cdot\|_\gamma\right)$, $t \in \prod_{i=1}^{N}[a_i, b_i]$.

Proof. By Theorem 6.6. $\square$

We finish with

Corollary 6.5. *All as in Theorem 6.9,* $p = +\infty$ *case.*

Then

$$\|(A_n^s(f))(x_0) - f(x_0)\|_\gamma \leq \omega_1\left(f, (\Delta(q))^N\left[\frac{1}{n^\beta} + \|b - a\|_\infty T e^{-2\lambda n^{(1-\beta)}}\right]\right), \tag{6.93}$$

$\forall\, N \in \mathbb{N} : n^{1-\beta} > 2$, $0 < \beta < 1$.

Clearly, $\lim_{n\to+\infty} A_n^s(f)(x_0) = f(x_0)$, *at* $x_0 \in \left(\prod_{i=1}^{N}[a_i, b_i]\right)^0$ *as in the assumptions.*

Bibliography

[140] G.A. Anastassiou, *A "K-Attainable" inequality related to the convergence of Positive Linear Operators*, J. Approximation Theory, 44, 380–383 (1985).

[141] G.A. Anastassiou, *Moments in probability and approximation theory*, Pitman Research Notes in Mathematics series / Longman Scientific & Technical, Essex, New York, UK, USA, 1993.

[142] G.A. Anastassiou, *Quantitative Approximations*, Chapmen & Hall/CRC, London, New York, 2001.

[143] G.A. Anastassiou, *Multivariate and abstract Approximation Theory for Banach space valued functions*, Demonstratio Mathematica, 2017; 50, 208–222.

[144] G.A. Anastassiou, *Intelligent Computations: Abstract, Fractional Calculus, Inequalities, Approximations*, Springer, Heidelberg, New York, 2018.

[145] G.A. Anastassiou, *Parametrized, Deformed and General Neural Networks*, Springer, Heidelberg, New York, 2023.

[146] G.A. Anastassiou, *Trigonometric and Hyperbolic Generated Approximation Theory*, World Scientific, Singapore, New York, 2025.

[147] H. Cartan, *Differential Calculus*, Herman, Paris, 1971.

[148] Z. Chen and F. Cao, *The approximation operators with sigmoidal functions*, Computers and Mathematics with Applications, 58 (2009), 758–765.

[149] D. Costarelli, R. Spigler, *Approximation results for neural network operators activated by sigmoidal functions*, Neural Networks 44 (2013), 101–106.

[150] D. Costarelli, R. Spigler, *Multivariate neural network operators with sigmoidal activation functions*, Neural Networks 48 (2013), 72–77.

[151] L.B. Rall, *Computational Solution of Nonlinear Operator Equations*, Joh Wiley & Sons, New York, 1969.

[152] O. Shisha and B. Mond, *The degree of convergence of sequences of linear positive operators*, Nat. Acad. Sci. U.S. 60 (1968), 1196–1200.

[153] O. Shisha and B. Mond, *The degree of Approximation to Periodic Functions by Linear Positive Operators*, J. Approximation Theory, 1, 335–339 (1968).

[154] S. Haykin, *Neural Networks: A Comprehensive Foundation* (2 ed.), Prentice Hall, New York, 1998.

[155] G.A. Anastassiou, *Neural Networks as Positive Linear Operators, Multivariate and Abstract case*, submitted, 2025.

[156] W. McCulloch and W. Pitts, *A logical calculus of the ideas immanent in nervous activity*, Bulletin of Mathematical Biophysics, 7 (1943), 115–133.

[157] T.M. Mitchell, *Machine Learning*, WCB-McGraw-Hill, New York, 1997.

[158] Yu, Dansheng; Cao, Feilong, *Construction and approximation rate for feed-forward neural network operators with sigmoidal functions*, J. Comput. Appl. Math. 453 (2025), paper No. 116150.

[159] Cen, Siyu; Jin, Bangti; Quan, Qimeng; Zhou, Zhi; *Hybrid neural-network FEM approximation of diffusion coeficient in elyptic and parabolic problems.* IMA J. Numer. Anal. 44 (2024), no. 5, 3059–3093.

[160] Coroianu Lucian; Costarelli, Danillo; Natale, Mariarosaria; Pantiş, Alexandra; *The approximation capabilities of Durrmeyer-type neural network operators.* J. Appl. Math. Comput. 70 (2024), no. 5, 4581–4599.

[161] Warin, Xavier; *The GroupMax neural network approximation of convex functions.* IEEE Trans. Neural Netw. Learn. Syst. 35 (2024), no. 8, 11608–11612.

[162] Fabra, Arnau; Guasch, Oriol; Baiges, Joan; Codina, Ramon; *Approximation of acoustic black holes with finite element mixed formulations and artificial neural network correction terms.* Finite Elem. Anal. Des. 241 (2024), paper No. 104236.

[163] Grohs, Philipp; Voigtlaender, Felix. *Proof of the theory-to-practice gap in deep learning via sampling complexity bounds for neural network approximation spaces.* Found. Comput. Math. 24 (2024), no. 4, 1085–1143.

[164] Basteri, Andrea; Trevisan, Dario; *Quantitative Gaussian approximation of randomly initialized deep neural networks.* Mach. Learn. 113 (2024), no. 9, 6373–6393.

[165] De Ryck, T.; Mishra, S. *Error analysis for deep neural network approximations of parametric hyperbolic conservation laws.* Math. Comp. 93 (2024), no. 350, 2643–2677.

[166] Liu, Jie; Zhang, Baoji; Lai, Yuyang; Fang, Liqiau. *Hull form optimization reserach based on multi-precision back-propagation neural network approximation model.* Internal. J. Numer. Methods Fluid 96 (2024), no. 8, 1445–1460.

[167] Yoo, Jihahm; Kim, Jaywon; Gim, Minjung; Lee, Haesung. *Error estimates of physics-informed neural networks for initial value problems.* J. Korean Soc. Ind. Appl. Math. 28 (2024), no. 1, 33–58.

Chapter 7

Multivariate Neural Networks in Infinite Domain as Positive Linear Operators and Measure Theory

Multivariate neural network operators of infinite domain are interpreted as positive linear operators and related general theory applies to them. These operators are induced by a symmetrized density function deriving from the parametrized and deformed hyperbolic tangent activation function. We are acting on the space of continuous and bounded functions on $\mathbb{R}^N$, $N \in \mathbb{N}$, to the reals. We study quantitatively the rate of convergence of these multivariate neural network operators to the unit operator. Our inequalities involve the modulus of continuity of the function under approximation or its Fréchet derivative. Connection to measure theory is shown. It follows [195].

7.1 Introduction

The author studied extensively the quantitative approximation of positive linear operators to the unit since 1985, see for example [168]–[170], [174], and [169] we use a lot in this work. He originated from the quantitative weak convergence of finite positive measures to the unit Dirac measure, having as a method the geometric moment theory, see [169], and he produced best upper bounds, leading to attained (i.e. sharp Jackson type inequalities), e.g. see [168], [169]. These studies have gone to all possible directions, univariate and multivariate, though in this work we stay only on the multivariate approach over infinite domain.

Also, since 1997, this author started the study of the quantitative convergence of neural network operators to the unit, and he has been written since then numerous of articles and books, e.g. see [170]–[172], [174]. Here we use a lot [172], [173].

A wide range of neural network operators over $\mathbb{R}^N$, $N \in \mathbb{N}$, being studied by the author all along and here are indeed by nature positive linear operators.

Here, for the first time in the literature, multivariate neural network operators of infinite domain are treated as positive linear operators.

So, methods of positive linear operators of infinite domain apply here to our infinite summation defined multivariate neural network operators, producing new and interesting pointwise and uniform results.

Convexity cases are also treated here producing sharp results. We also use Fréchet derivatives. Furthermore a strong connection of our operators to measure theory is established with respect to probability measures.

Of great inspiration always have been [180], [181]. For classical studies on neural network we recommend also, [176]–[178], [182]–[184].

For newer work on neural networks we also refer the reader to [185]–[194].

7.2 Basics

Initially we follow [172], pp. 455–460.

Our perturbed hyperbolic tangent activation function is

$$g_{q,\lambda}(x) := \frac{e^{\lambda x} - qe^{-\lambda x}}{e^{\lambda x} + qe^{-\lambda x}}, \quad \lambda, q > 0, \; x \in \mathbb{R}. \tag{7.1}$$

Above λ is the parameter and q is the deformation coefficient.

For more read Chapter 18 of [172]: "q-deformed and λ-Parametrized Hyperbolic Tangent based Banach space Valued Ordinary and Fractional Neural Network Approximation".

The chapters 17 and 18 of [172] motivate our current work.

The proposed "symmetrization method" aims to use half data feed to our multivariate neural networks.

We will employ the following density function

$$M_{q.\lambda}(x) := \frac{1}{4}\left(g_{q,\lambda}(x+1) - g_{q,\lambda}(x-1)\right) > 0, \tag{7.2}$$

$\forall x \in \mathbb{R}$; $q, \lambda > 0$.

We have that

$$M_{q,\lambda}(-x) = M_{\frac{1}{q},\lambda}(x), \quad \forall \, x \in \mathbb{R}; \; q, \lambda > 0, \tag{7.3}$$

and

$$M_{\frac{1}{q},\lambda}(-x) = M_{q,\lambda}(x), \quad \forall\, x \in \mathbb{R};\; q, \lambda > 0. \tag{7.4}$$

Adding (7.3) and (7.4) we obtain

$$M_{q,\lambda}(-x) + M_{\frac{1}{q},\lambda}(-x) = M_{q,\lambda}(x) + M_{\frac{1}{q},\lambda}(x), \tag{7.5}$$

a key to this work.

So that

$$\Phi(x) := \frac{M_{q,\lambda}(x) + M_{\frac{1}{q},\lambda}(x)}{2} \tag{7.6}$$

is an even function, symmetric with respect to the y-axis.

By (18.18) of [172], we have

$$\begin{aligned} &M_{q,\lambda}\left(\tfrac{\ln q}{2\lambda}\right) = \tfrac{\tanh(\lambda)}{2}, \\ &\text{and} \\ &M_{\frac{1}{q},\lambda}\left(-\tfrac{\ln q}{2\lambda}\right) = \tfrac{\tanh(\lambda)}{2}, \quad \lambda > 0, \end{aligned} \tag{7.7}$$

sharing the same maximum at symmetric points.

By Theorem 18.1, p. 458 of [172], we have that

$$\begin{aligned} &\sum_{i=-\infty}^{\infty} M_{q,\lambda}(x-i) = 1, \quad \forall\, x \in \mathbb{R},\ \lambda, q > 0, \\ &\text{and} \\ &\sum_{i=-\infty}^{\infty} M_{\frac{1}{q},\lambda}(x-i) = 1, \quad \forall\, x \in \mathbb{R},\ \lambda, q > 0. \end{aligned} \tag{7.8}$$

Consequently, we derive that

$$\sum_{i=-\infty}^{\infty} \Phi(x-i) = 1, \quad \forall\, x \in \mathbb{R}. \tag{7.9}$$

By Theorem 18.2, p. 459 of [172], we have that

$$\begin{aligned} &\int_{-\infty}^{\infty} M_{q,\lambda}(x)\,dx = 1, \\ &\text{and} \\ &\int_{-\infty}^{\infty} M_{\frac{1}{q},\lambda}(x)\,dx = 1, \end{aligned} \tag{7.10}$$

so that

$$\int_{-\infty}^{\infty} \Phi(x)\,dx = 1, \tag{7.11}$$

therefore Φ is a density function.

By Theorem 18.3, p. 459 of [172], we have:

Let $0 < \alpha < 1$, and $n \in \mathbb{N}$ with $n^{1-\alpha} > 2$; $q, \lambda > 0$. Then

$$\sum_{\begin{cases} k=-\infty \\ :|nx-k| \geq n^{1-\alpha} \end{cases}}^{\infty} M_{q,\lambda}(nx-k) < 2\max\left\{q, \frac{1}{q}\right\} e^{4\lambda} e^{-2\lambda n^{(1-\alpha)}} = Te^{-2\lambda n^{(1-\alpha)}}, \tag{7.12}$$

where $T := 2\max\left\{q, \frac{1}{q}\right\} e^{4\lambda}$.

Similarly, we get that

$$\sum_{k=-\infty}^{\infty} M_{\frac{1}{q},\lambda}(nx-k) < Te^{-2\lambda n^{(1-\alpha)}}. \tag{7.13}$$

Consequently we obtain that

$$\sum_{\begin{cases} k=-\infty \\ :|nx-k| \geq n^{1-\alpha} \end{cases}}^{\infty} \Phi(nx-k) < Te^{-2\lambda n^{(1-\alpha)}}, \tag{7.14}$$

where $T := 2\max\left\{q, \frac{1}{q}\right\} e^{4\lambda}$.

Remark 7.1. We introduce

$$Z(x_1, ..., x_N) := Z(x) := \prod_{i=1}^{N} \Phi(x_i), \quad x = (x_1, ..., x_N) \in \mathbb{R}^N, \ N \in \mathbb{N}. \tag{7.15}$$

It has the properties:

(i)

$$Z(x) > 0, \ \forall x \in \mathbb{R}^N,$$

(ii)

$$\sum_{k=-\infty}^{\infty} Z(x-k) := \sum_{k_1=-\infty}^{\infty} \sum_{k_2=-\infty}^{\infty} \cdots \sum_{k_N=-\infty}^{\infty} Z(x_1-k_1, ..., x_N-k_N) = 1, \tag{7.16}$$

where $k := (k_1, ..., k_n) \in \mathbb{Z}^N$, $\forall\, x \in \mathbb{R}^N$,

hence

(iii)

$$\sum_{k=-\infty}^{\infty} Z(nx-k) = 1, \tag{7.17}$$

$\forall\, x \in \mathbb{R}^N$; $n \in \mathbb{N}$,

and
(iv)

$$\int_{\mathbb{R}^N} Z(x)\,dx = 1, \tag{7.18}$$

that is Z is a multivariate density function.

Here denote $\|x\|_\infty := \max\{|x_1|, ..., |x_N|\}$, $x \in \mathbb{R}^N$, also set $\infty := (\infty, ..., \infty)$, $-\infty := (-\infty, ..., -\infty)$ upon the multivariate context,
(v)

$$\sum_{\substack{k=-\infty \\ \left\| \frac{k}{n} - x \right\|_\infty > \frac{1}{n^\beta}}}^{\infty} Z(nx-k) \overset{(7.14)}{<} Te^{-2\lambda n^{(1-\beta)}}, \tag{7.19}$$

where $T := 2\max\left\{q, \frac{1}{q}\right\} e^{4\lambda}$, $0 < \beta < 1$, $n \in \mathbb{N} : n^{1-\beta} > 2$, $x \in \mathbb{R}^N$.

Theorem 7.1. *([173]) Let* $0 < \beta < 1$, *and* $n \in \mathbb{N}$ *with* $n^{1-\beta} > 2$, $m \in \mathbb{N}$. *It holds*

$$\sum_{\substack{k=-\infty \\ \{k \in \mathbb{Z}^N : \|nx-k\|_\infty \geq n^{1-\beta}\}}}^{\infty} \|nx-k\|_\infty^m Z(nx-k) < \tag{7.20}$$

$$\frac{2m!}{\lambda^m}\left(q + \frac{1}{q}\right) e^{2\lambda} e^{-\lambda\left(n^{1-\beta}-1\right)}.$$

We need

Definition 7.1. The modulus of continuity here is defined by

$$\omega_1(f, \delta) := \sup_{\substack{x, y \in \mathbb{R}^N: \\ \|x-y\|_\infty < \delta}} |f(x) - f(y)|, \quad \delta > 0, \tag{7.21}$$

where $f : \mathbb{R}^N \to \mathbb{R}$ is a bounded and continuous function, denoted by $f \in C_B(\mathbb{R}^N)$, $N \in \mathbb{N}$. Similarly ω_1 is defined for $f \in C_U(\mathbb{R}^N)$ (uniformly continuous functions). We have that $f \in C_U(\mathbb{R}^N)$, iff $\omega_1(f, \delta) \to 0$ as $\delta \downarrow 0$.

Denote by $B(\mathbb{R}^N)$ the bounded functions on $\mathbb{R}^N$.
Next, we describe our neural network operators.

Definition 7.2. Let $f \in C_B(\mathbb{R}^N)$, we define

$$A_n(f,x) := A_n(f,x_1,...,x_N) := \sum_{k=-\infty}^{\infty} f\left(\frac{k}{n}\right) Z(nx-k) := \tag{7.22}$$

$$\sum_{k_1=-\infty}^{\infty} \sum_{k_2=-\infty}^{\infty} \cdots \sum_{k_N=-\infty}^{\infty} f\left(\frac{k_1}{n}, \frac{k_2}{n}, ..., \frac{k_N}{n}\right) \left(\prod_{i=1}^{N} \Phi(nx_i - k_i)\right),$$

$n \in \mathbb{N}$, $\forall\, x \in \mathbb{R}^N$, $N \in \mathbb{N}$, the multivariate quasi-interpolation neural network operator.

Also for $f \in C_B(\mathbb{R}^N)$ we define the multivariate Kantorovich type neural network operator

$$K_n(f,x) := K_n(f,x_1,...,x_N) := \sum_{k=-\infty}^{\infty} \left(n^N \int_{\frac{k}{n}}^{\frac{k+1}{n}} f(t)\,dt\right) Z(nx-k) := \tag{7.23}$$

$$\sum_{k_1=-\infty}^{\infty} \sum_{k_2=-\infty}^{\infty} \cdots \sum_{k_N=-\infty}^{\infty} \left(n^N \int_{\frac{k_1}{n}}^{\frac{k_1+1}{n}} \int_{\frac{k_2}{n}}^{\frac{k_2+1}{n}} \cdots \int_{\frac{k_N}{n}}^{\frac{k_N+1}{n}} f(t_1,...,t_N)\,dt_1...dt_N\right)$$

$$\left(\prod_{i=1}^{N} \Phi(nx_i - k_i)\right),$$

$n \in \mathbb{N}$, $\forall\, x \in \mathbb{R}^N$.

Again for $f \in C_B(\mathbb{R}^N)$, $N \in \mathbb{N}$, we define the multivariate neural network operator of quadrature type $Q_n(f,x)$, $n \in \mathbb{N}$, as follows. Let $\theta = (\theta_1,...,\theta_N) \in \mathbb{N}^N$, $r = (r_1,...,r_N) \in \mathbb{Z}_+^N$, $w_r = w_{r_1 r_2 ... r_N} \geq 0$, such that

$$\sum_{r=0}^{\theta} w_r = \sum_{r_1=0}^{\theta_1} \sum_{r_2=0}^{\theta_2} \cdots \sum_{r_N=0}^{\theta_N} w_{r_1 r_2 ... r_N} = 1; \quad k \in Z^N,$$

and

$$\delta_{nk}(f) := \delta_{n,k_1,k_2,...,k_N}(f) := \sum_{r=0}^{\theta} w_r f\left(\frac{k}{n} + \frac{r}{n\theta}\right) :=$$

$$\sum_{r_1=0}^{\theta_1} \sum_{r_2=0}^{\theta_2} \cdots \sum_{r_N=0}^{\theta_N} w_{r_1 r_2 ... r_N} f\left(\frac{k_1}{n} + \frac{r_1}{n\theta_1}, \frac{k_2}{n} + \frac{r_2}{n\theta_2}, ..., \frac{k_N}{n} + \frac{r_N}{n\theta_N}\right), \tag{7.24}$$

where $\frac{r}{\theta} := \left(\frac{r_1}{\theta_1}, \frac{r_2}{\theta_2}, ..., \frac{r_N}{\theta_N}\right)$.

We put

$$Q_n(f,x) := Q_n(f,x_1,...,x_N) := \sum_{k=-\infty}^{\infty} \delta_{nk}(f) Z(nx-k) := \tag{7.25}$$

$$\sum_{k_1=-\infty}^{\infty} \sum_{k_2=-\infty}^{\infty} \cdots \sum_{k_N=-\infty}^{\infty} \delta_{n,k_1,k_2,...,k_N}(f) \left(\prod_{i=1}^{N} \Phi(nx_i - k_i) \right), \quad \forall x \in \mathbb{R}^N.$$

Remark 7.2. We notice that

$$\int_{\frac{k}{n}}^{\frac{k+1}{n}} f(t)\,dt = \int_{\frac{k_1}{n}}^{\frac{k_1+1}{n}} \int_{\frac{k_2}{n}}^{\frac{k_2+1}{n}} \cdots \int_{\frac{k_N}{n}}^{\frac{k_N+1}{n}} f(t_1,t_2,...,t_N)\,dt_1 dt_2 ... dt_N =$$

$$\int_0^{\frac{1}{n}} \int_0^{\frac{1}{n}} \cdots \int_0^{\frac{1}{n}} f\left(t_1 + \frac{k_1}{n}, t_2 + \frac{k_2}{n}, ..., t_N + \frac{k_N}{n}\right) dt_1 ... dt_N$$

$$= \int_0^{\frac{1}{n}} f\left(t + \frac{k}{n}\right) dt. \tag{7.26}$$

Thus, it holds

$$K_n(f,x) = \sum_{k=-\infty}^{\infty} \left(n^N \int_0^{\frac{1}{n}} f\left(t + \frac{k}{n}\right) dt \right) Z(nx-k). \tag{7.27}$$

We do have that $A_n(1) = K_n(1) = Q_n(1) = 1$, $\forall\, n \in \mathbb{N}$.

Denote by Θ_n any of A_n, K_n, Q_n and they are positive linear operators mapping $C_B(\mathbb{R}^N)$ into $B(\mathbb{R}^N)$.

7.3 Auxilliary Results

Here let $x_0 \in \mathbb{R}^N$, $0 < \beta < 1$, $n \in \mathbb{N} : n^{1-\beta} > 2$ and $m \in \mathbb{N}$. It follows:

Proposition 7.1. *It holds that*

$$A_n\left(\|\cdot - x_0\|_\infty^m\right)(x_0) > 0,$$

and

$$0 < \left\| A_n\left(\|\cdot - x_0\|_\infty^m\right)(x_0) \right\|_\infty \le \tag{7.28}$$

$$\left(\frac{1}{n^\beta} + \frac{1}{n^m} \frac{2m!}{\lambda^m} \left(q + \frac{1}{q}\right) e^{2\lambda} e^{-\lambda\left(n^{1-\beta}-1\right)} \right) < \infty, \quad (\to 0, \text{ as } n \to +\infty).$$

Proof. We have that

$$0 < A_n \left(\|\cdot - x_0\|_\infty^m \right)(x_0) = \sum_{k=-\infty}^{\infty} \left\| \frac{k}{n} - x_0 \right\|_\infty^m Z(nx_0 - k) =$$

$$\sum_{\substack{k=-\infty \\ : \left\| \frac{k}{n} - x_0 \right\|_\infty < \frac{1}{n^\beta}}}^{\infty} \left\| \frac{k}{n} - x_0 \right\|_\infty^m Z(nx_0 - k) +$$

$$\sum_{\substack{k=-\infty \\ : \left\| \frac{k}{n} - x_0 \right\|_\infty \geq \frac{1}{n^\beta}}}^{\infty} \left\| \frac{k}{n} - x_0 \right\|_\infty^m Z(nx_0 - k) \leq \tag{7.29}$$

$$\frac{1}{n^\beta} + \frac{1}{n^m} \sum_{\substack{k=-\infty \\ : \|nx_0 - k\|_\infty \geq n^{1-\beta}}}^{\infty} \|nx_0 - k\|_\infty^m Z(nx_0 - k) \overset{(7.20)}{\leq}$$

$$\frac{1}{n^\beta} + \frac{1}{n^m} \frac{2m!}{\lambda^m} \left(q + \frac{1}{q} \right) e^{2\lambda} e^{-\lambda\left(n^{1-\beta} - 1\right)}.$$

Therefore it holds

$$0 < \left\| A_n \left(\|\cdot - x_0\|_\infty^m \right)(x_0) \right\|_\infty \leq \tag{7.30}$$

$$\frac{1}{n^\beta} + \frac{1}{n^m} \frac{2m!}{\lambda^m} \left(q + \frac{1}{q} \right) e^{2\lambda} e^{-\lambda\left(n^{1-\beta} - 1\right)} \to 0, \quad \text{as } n \to +\infty.$$

□

Proposition 7.2. *It holds*

$$0 < K_n \left(\|\cdot - x_0\|_\infty^m \right)(x_0) \leq \tag{7.31}$$

$$\left[\left(\frac{1}{n} + \frac{1}{n^\beta} \right)^m + \frac{2^{m-1}}{n^m} \left[T e^{-2\lambda n^{(1-\beta)}} + \frac{2m!}{\lambda^m} \left(q + \frac{1}{q} \right) e^{2\lambda} e^{-\lambda\left(n^{1-\beta} - 1\right)} \right] \right] < \infty,$$

$\forall\, x_0 \in \mathbb{R}^N$. *($\to 0$, as $n \to +\infty$).*

That is

$$0 < \left\| K_n \left(\|\cdot - x_0\|_\infty^m \right)(x_0) \right\|_\infty \leq \tag{7.32}$$

$$\left[\left(\frac{1}{n} + \frac{1}{n^\beta} \right)^m + \frac{2^{m-1}}{n^m} \left[T e^{-2\lambda n^{(1-\beta)}} + \frac{2m!}{\lambda^m} \left(q + \frac{1}{q} \right) e^{2\lambda} e^{-\lambda\left(n^{1-\beta} - 1\right)} \right] \right] < \infty,$$

($\to 0$, as $n \to +\infty$).

Proof. We have that

$$0 < K_n\left(\|\cdot - x_0\|_\infty^m\right)(x_0) = \sum_{k=-\infty}^{\infty}\left(n^N \int_0^{\frac{1}{n}} \left\|t + \frac{k}{n} - x_0\right\|_\infty^m dt\right) Z(nx_0 - k) \leq$$

$$\sum_{k=-\infty}^{\infty}\left(n^N \int_0^{\frac{1}{n}} \left(\left\|\frac{k}{n} - x_0\right\|_\infty + \|t\|_\infty\right)^m dt\right) Z(nx_0 - k) \leq$$

$$\sum_{k=-\infty}^{\infty} Z(nx_0 - k)\left(\left\|\frac{k}{n} - x_0\right\|_\infty + \frac{1}{n}\right)^m =$$

$$\sum_{\substack{k=-\infty \\ : \left\|\frac{k}{n} - x_0\right\|_\infty < \frac{1}{n^\beta}}}^{\infty} Z(nx_0 - k)\left(\left\|\frac{k}{n} - x_0\right\|_\infty + \frac{1}{n}\right)^m + \tag{7.33}$$

$$\sum_{\substack{k=-\infty \\ : \left\|\frac{k}{n} - x_0\right\|_\infty \geq \frac{1}{n^\beta}}}^{\infty} Z(nx_0 - k)\left(\left\|\frac{k}{n} - x_0\right\|_\infty + \frac{1}{n}\right)^m \leq$$

$$\left(\frac{1}{n^\beta} + \frac{1}{n}\right)^m + \frac{1}{n^m}\left(\sum_{\substack{k=-\infty \\ : \|nx_0 - k\|_\infty \geq n^{1-\beta}}}^{\infty} Z(nx_0 - k)\left(\|nx_0 - k\|_\infty + 1\right)^m\right) \leq$$

$$\left(\frac{1}{n} + \frac{1}{n^\beta}\right)^m + \frac{2^{m-1}}{n^m}\left(\sum_{\substack{k=-\infty \\ : \|nx_0 - k\|_\infty \geq n^{1-\beta}}}^{\infty} Z(nx_0 - k)\left(1 + \|nx_0 - k\|_\infty^m\right)\right) =$$

$$\left(\frac{1}{n} + \frac{1}{n^\beta}\right)^m + \frac{2^{m-1}}{n^m}\left[\sum_{\substack{k=-\infty \\ : \|nx_0 - k\|_\infty \geq n^{1-\beta}}}^{\infty} Z(nx_0 - k) + \right. \tag{7.34}$$

$$\left.\sum_{\left\{\substack{k=-\infty \\ :\left\|nx_0-k\right\|_\infty \geq n^{1-\beta}}\right.}^{\infty} Z\left(nx_0-k\right)\left\|nx_0-k\right\|_\infty^m\right] \overset{\text{(by (7.19), (7.20))}}{\leq}$$

$$\left(\frac{1}{n}+\frac{1}{n^\beta}\right)^m+\frac{2^{m-1}}{n^m}\left[Te^{-2\lambda n^{(1-\beta)}}+\frac{2m!}{\lambda^m}\left(q+\frac{1}{q}\right)e^{2\lambda}e^{-\lambda\left(n^{1-\beta}-1\right)}\right]<\infty.$$

□

Proposition 7.3. *It holds that*

$$0<Q_n\left(\left\|\cdot-x_0\right\|_\infty^m\right)\left(x_0\right) \leq \tag{7.35}$$

$$\left(\frac{1}{n}+\frac{1}{n^\beta}\right)^m+\frac{2^{m-1}}{n^m}\left[Te^{-2\lambda n^{(1-\beta)}}+\frac{2m!}{\lambda^m}\left(q+\frac{1}{q}\right)e^{2\lambda}e^{-\lambda\left(n^{1-\beta}-1\right)}\right]<\infty,$$

$\forall\, x_0 \in \mathbb{R}^N$ *($\rightarrow 0$, as $n \rightarrow +\infty$).*

Hence we get

$$0<\left\|Q_n\left(\left\|\cdot-x_0\right\|_\infty^m\right)\left(x_0\right)\right\|_\infty \leq \tag{7.36}$$

$$\left(\frac{1}{n}+\frac{1}{n^\beta}\right)^m+\frac{2^{m-1}}{n^m}\left[Te^{-2\lambda n^{(1-\beta)}}+\frac{2m!}{\lambda^m}\left(q+\frac{1}{q}\right)e^{2\lambda}e^{-\lambda\left(n^{1-\beta}-1\right)}\right]<\infty,$$

($\rightarrow 0$, as $n \rightarrow +\infty$).

Proof. We notice that

$$Q_n\left(\left\|\cdot-x_0\right\|_\infty^m\right)\left(x_0\right)=\sum_{k=-\infty}^{\infty}\left(\sum_{r=0}^{\theta} w_r\left\|\frac{k}{n}+\frac{r}{n\theta}-x_0\right\|_\infty^m\right) Z\left(nx_0-k\right) \leq$$

$$\sum_{k=-\infty}^{\infty}\left(\sum_{r=0}^{\theta} w_r\left(\left\|\frac{k}{n}-x_0\right\|_\infty+\left\|\frac{r}{n\theta}\right\|_\infty\right)^m\right) Z\left(nx_0-k\right) \leq \tag{7.37}$$

$$\sum_{k=-\infty}^{\infty}\left(\left\|\frac{k}{n}-x_0\right\|_\infty+\frac{1}{n}\right)^m Z\left(nx_0-k\right)$$

(as earlier in Proposition 7.2)

$$\leq\left(\frac{1}{n}+\frac{1}{n^\beta}\right)^m+\frac{2^{m-1}}{n^m}\left[Te^{-2\lambda n^{(1-\beta)}}+\frac{2m!}{\lambda^m}\left(q+\frac{1}{q}\right)e^{2\lambda}e^{-\lambda\left(n^{1-\beta}-1\right)}\right]<\infty. \tag{7.38}$$

□

7.4 Main Results

Next we present approximation results.

Theorem 7.2. *Let* $0 < \beta < 1$, $n \in \mathbb{N} : n^{1-\beta} > 2$, $f \in C_B\left(\mathbb{R}^N\right)$, $N \in \mathbb{N}$. *It holds*

$$\|A_n(f) - f\|_\infty \leq$$

$$2\omega_1\left(f, \left(\frac{1}{n^\beta} + \frac{2}{n\lambda}\left(q + \frac{1}{q}\right)e^{2\lambda}e^{-\lambda\left(n^{1-\beta}-1\right)}\right)\right) < \infty. \tag{7.39}$$

Given that f *is uniformly continuous we obtain that* $A_n(f) \to f$, *pointwise and uniformly, as* $n \to +\infty$.

Proof. We can write

$$A_n(f, x) = \int_{\mathbb{R}^N} f(t)\, d\mu_x(t), \tag{7.40}$$

where $\mu_x\left(\mathbb{R}^N\right) = 1$, μ_x is a probability measure on $\mathbb{R}^N$. The support of μ_x is discrete and countable, at the points $\frac{k}{n} = \left(\frac{k_1}{n}, \frac{k_2}{n}, ..., \frac{k_N}{n}\right) \in \mathbb{R}^N$, $k_i \in \mathbb{Z}$, $i = 1, ..., N$.

Each $\frac{k}{n}$ carries the weight $Z(nx - k) = \prod_{i=1}^{N} \Phi(nx_i - k_i)$.

Indeed here we have that

$$\sum_{k=-\infty}^{\infty} Z(nx - k) \overset{(7.17)}{=} \sum_{k_1=-\infty}^{\infty} \sum_{k_2=-\infty}^{\infty} \cdots \sum_{k_N=-\infty}^{\infty} \left(\prod_{i=1}^{N} \Phi(nx_i - k_i)\right) = 1. \tag{7.41}$$

Here it is $\omega_1(f, h) < \infty$, $h > 0$.

By Lemma 7.1.1, p. 208, [169], we get that

$$|f(t) - f(x)| \leq \omega_1(f, h)\left\lceil \frac{\|t - x\|_\infty}{h} \right\rceil, \tag{7.42}$$

where $\lceil \cdot \rceil$ is the ceiling function.

But it holds

$$\left\lceil \frac{\|t - x\|_\infty}{h} \right\rceil \leq 1 + \frac{\|t - x\|_\infty}{h}. \tag{7.43}$$

Thus we get

$$|f(t) - f(x)| \leq \omega_1(f, h)\left[1 + \frac{\|t - x\|_\infty}{h}\right], \quad \forall\, t \in \mathbb{R}^N. \tag{7.44}$$

The last can be written also as

$$|f(\cdot) - f(x)| \leq \omega_1(f,h)\left[1 + \frac{\|\cdot - x\|_\infty}{h}\right], \quad \text{over } \mathbb{R}^N. \tag{7.45}$$

We have that

$$A_n(f,x) - f(x) = \int_{\mathbb{R}^N} f(t)\, d\mu_x(t) - f(x) = \int_{\mathbb{R}^N} (f(t) - f(x))\, d\mu_x(t). \tag{7.46}$$

Therefore

$$|A_n(f,x) - f(x)| \leq \left|\int_{\mathbb{R}^N} (f(t) - f(x))\, d\mu_x(t)\right| \leq$$

$$\int_{\mathbb{R}^N} |f(t) - f(x)|\, d\mu_x(t) \leq \omega_1(f,h)\left[1 + \frac{\int_{\mathbb{R}^N} \|t - x\|_\infty\, d\mu_x(t)}{h}\right]. \tag{7.47}$$

Choosing $h := \int_{\mathbb{R}^N} \|t - x\|_\infty\, d\mu_x(t)$, we obtain

$$|A_n(f,x) - f(x)| \leq 2\omega_1\left(f, \int_{\mathbb{R}^N} \|t - x\|_\infty\, d\mu_x(t)\right) =$$

$$2\omega_1(f, A_n(\|\cdot - x\|_\infty)(x)). \tag{7.48}$$

But we have $(0 < \beta < 1)$

$$0 < A_n(\|\cdot - x\|_\infty)(x) = \sum_{k=-\infty}^{\infty} \left\|\frac{k}{n} - x\right\|_\infty Z(nx - k) =$$

$$\sum_{\begin{cases} k = -\infty \\ : \left\|\frac{k}{n} - x\right\|_\infty < \frac{1}{n^\beta} \end{cases}}^{\infty} \left\|\frac{k}{n} - x\right\|_\infty Z(nx - k) +$$

$$\sum_{\begin{cases} k = -\infty \\ : \left\|\frac{k}{n} - x\right\|_\infty \geq \frac{1}{n^\beta} \end{cases}}^{\infty} \left\|\frac{k}{n} - x\right\|_\infty Z(nx - k) \leq \tag{7.49}$$

$$\frac{1}{n^\beta} + \frac{1}{n} \sum_{\begin{cases} k = -\infty \\ : \|nx - k\|_\infty \geq n^{1-\beta} \end{cases}}^{\infty} \|nx - k\|_\infty Z(nx - k) \overset{(7.20)}{\leq}$$

$$\frac{1}{n^\beta} + \frac{1}{n}\frac{2}{\lambda}\left(q + \frac{1}{q}\right) e^{2\lambda} e^{-\lambda\left(n^{1-\beta} - 1\right)}.$$

So that it holds

$$A_n\left(\|\cdot - x\|_\infty\right)(x) \le \tag{7.50}$$

$$\frac{1}{n^\beta} + \frac{2}{n\lambda}\left(q + \frac{1}{q}\right) e^{2\lambda} e^{-\lambda\left(n^{1-\beta}-1\right)} (< +\infty) \to 0, \quad \text{as } n \to +\infty.$$

We have proved

$$\left|A_n(f, x) - f(x)\right| \le 2\omega_1\left(f, \left(\frac{1}{n^\beta} + \frac{2}{n\lambda}\left(q + \frac{1}{q}\right) e^{2\lambda} e^{-\lambda\left(n^{1-\beta}-1\right)}\right)\right) < \infty. \tag{7.51}$$

□

Next comes a convexity approximation result.

Theorem 7.3. *Let $f \in C_B\left(\mathbb{R}^N\right)$: $\left|f(\cdot) - f(x_0)\right|$ is convex, $x_0 \in \mathbb{R}^N$. Then*

$$\left|A_n(f, x_0) - f(x_0)\right| \le \tag{7.52}$$

$$\omega_1\left(f, \left(\frac{1}{n^\beta} + \frac{2}{n\lambda}\left(q + \frac{1}{q}\right) e^{2\lambda} e^{-\lambda\left(n^{1-\beta}-1\right)}\right)\right) < +\infty,$$

($\to 0$, as $n \to \infty$, given that f is uniformly continuous, i.e. $\lim_{n\to+\infty} A_n(f, x_0) = f(x_0)$).

Proof. Based on Lemma 8.1.1, p. 243, [169], we obtain:

Let $(V, \|\cdot\|)$ be a real normed vector space and $f : V \to \mathbb{R}$ such that $\left|f(t) - f(x_0)\right|$ is convex in $t \in V$; $x_0 \in V$. Then

$$\left|f(t) - f(x_0)\right| \le \frac{\omega_1(f, h)}{h} \left\|t - x_0\right\|, \quad \forall\, t \in V,\, h > 0. \tag{7.53}$$

That is, it holds here

$$\left|f(\cdot) - f(x_0)\right| \le \frac{\omega_1(f, h)}{h} \left\|\cdot - x_0\right\|_\infty, \text{ over } \mathbb{R}^N. \tag{7.54}$$

We have that

$$A_n(f, x_0) - f(x_0) = \int_{\mathbb{R}^N} \left(f(t) - f(x_0)\right) d\mu_{x_0}(t), \tag{7.55}$$

and

$$\left|A_n(f, x_0) - f(x_0)\right| \le \int_{\mathbb{R}^N} \left|f(t) - f(x_0)\right| d\mu_{x_0}(t) \le \tag{7.56}$$

$$\frac{\omega_1(f, h)}{h} \int_{\mathbb{R}^N} \left\|t - x_0\right\|_\infty d\mu_{x_0}(t)$$

(we take $h := \int_{\mathbb{R}^N} \|t - x_0\|_\infty \, d\mu_{x_0}(t) = A_n(\|\cdot - x_0\|_\infty)(x_0) > 0$)

$$= \omega_1\left(f, \int_{\mathbb{R}^N} \|t - x_0\|_\infty \, d\mu_{x_0}(t)\right)$$

$$= \omega_1(f, A_n(\|\cdot - x_0\|_\infty)(x_0)). \quad (7.57)$$

Thus, it holds

$$|A_n(f, x_0) - f(x_0)| \le \omega_1(f, A_n(\|\cdot - x_0\|_\infty)(x_0)). \quad (7.58)$$

Inequality (7.58) is sharp, namely it is attained by $f(t) = \|t - x_0\|_\infty$, $\forall\, t \in \mathbb{R}^N$, $x_0 \in \mathbb{R}^N$.

We have

$$|A_n(f, x_0) - f(x_0)| = A_n(\|\cdot - x_0\|_\infty)(x_0). \quad (7.59)$$

Next we observe

$$\omega_1(f, A_n(\|\cdot - x_0\|_\infty)(x_0)) = \omega_1(\|\cdot - x_0\|_\infty, A_n(\|\cdot - x_0\|_\infty)(x_0)) =$$

$$\sup_{\left\{\begin{array}{c} t_1, t_2 \in \mathbb{R}^N : \\ : \|t_1 - t_2\|_\infty \le A_n(\|\cdot - x_0\|_\infty)(x_0) \end{array}\right.} \left| \|t_1 - x_0\|_\infty - \|t_2 - x_0\|_\infty \right| \le \quad (7.60)$$

$$\sup_{\left\{\begin{array}{c} t_1, t_2 \in \mathbb{R}^N : \\ : \|t_1 - t_2\|_\infty \le A_n(\|\cdot - x_0\|_\infty)(x_0) \end{array}\right.} \|t_1 - t_2\|_\infty = A_n(\|\cdot - x_0\|_\infty)(x_0).$$

So that

$$\omega_1(f, A_n(\|\cdot - x_0\|_\infty)(x_0)) = A_n(\|\cdot - x_0\|_\infty)(x_0).$$

Thus (7.58) it is attained.

Continuing on (7.58) we have that

$$\omega_1(f, A_n(\|\cdot - x_0\|_\infty)(x_0)) \le \quad (7.61)$$

$$\omega_1\left(f, \left(\frac{1}{n^\beta} + \frac{2}{n\lambda}\left(q + \frac{1}{q}\right) e^{2\lambda} e^{-\lambda\left(n^{1-\beta} - 1\right)}\right)\right).$$

The theorem is proved. □

From [169], pp. 232–233 we mention

Theorem 7.4. *Let μ be a finite measure of mass $m > 0$ on $M \neq \varnothing$, which is a convex subset of the Banach space $(V, \|\cdot\|)$ and $x_0 \in M$ be given.*

Assume that

$$\left(\frac{1}{m}\int_M \|x - x_0\|^{n+1} \mu(dx)\right)^{\frac{1}{(n+1)}} := D_{n+1}(x_0) > 0, \quad n \in \mathbb{N}. \tag{7.62}$$

Let $f : M \to \mathbb{R}$ be a bounded continuous function, whose Fréchet derivatives $f^{(j)}$, $1 \leq j \leq n$ exist, are bounded and continuous functions. Assume that $\omega_1\left(f^{(n)}, rD_{n+1}(x_0)\right) \leq w$, where $r, w > 0$ are given. Then

$$\begin{aligned}\left|\int_M f d\mu - f(x_0)\right| &\leq |f(x_0)|\,|m-1| + \\ &\left|\sum_{j=1}^{n} \frac{1}{j!}\int_M f^{(j)}(x_0)(x - x_0)^j \mu(dx)\right| + \\ &\frac{mw}{rn!}\left[\frac{nr^2}{8} + \frac{r}{2} + \frac{1}{(n+1)}\right] D_{n+1}^n(x_0).\end{aligned} \tag{7.63}$$

Corollary 7.1. *When $\mu(M) = 1$ and $\left(\int_M \|x - x_0\|^2 \mu(dx)\right)^{\frac{1}{2}} := D_2(x_0) > 0$, it holds:*

$$\begin{aligned}\left|\int_M f d\mu - f(x_0)\right| - \left|\int_M f^{(1)}(x_0)(x - x_0)\mu(dx)\right| &\leq \\ (1.5625)\,\omega_1\left(f^{(1)}, \frac{1}{2}D_2(x_0)\right) D_2(x_0).\end{aligned} \tag{7.64}$$

Proposition 7.4. *Let μ be a finite measure of mass $m > 0$ on $M \neq \varnothing$, which is a convex subset of the Banach space $(V, \|\cdot\|)$ and $x_0 \in M$ be given. Assume that*

$$\left(\int_M \|x - x_0\|^2 \mu(dx)\right)^{\frac{1}{2}} := D_2(x_0) > 0 \tag{7.65}$$

and consider $r > 0$.

Let $f : M \to \mathbb{R}$ be a bounded continuous function, whose Fréchet derivative $f^{(1)}$ exists, is bounded and continuous.

Then

$$\begin{aligned}&\left|\int_M f d\mu - f(x_0)\right| - |f(x_0)|\,|m-1| - \\ &\quad\left|\int_M f^{(1)}(x_0)(x - x_0)\mu(dx)\right| \leq \\ &\frac{1}{8r}\left(2 + \sqrt{mr}\right)^2 \omega_1\left(f', rD_2(x_0)\right) D_2(x_0).\end{aligned} \tag{7.66}$$

Corollary 7.2. *Let μ be a finite measure of mass $m > 0$ on M, which is a convex subset of the Banach space $(V, \|\cdot\|)$ and $x_0 \in M$ be given. Assume that $\left(\int_M \|x - x_0\|^2 \mu(dx)\right)^{\frac{1}{2}} := D_2(x_0) > 0$. Consider $r > 0$, let $f : M \to \mathbb{R}$ be a bounded and continuous function, whose Fréchet derivative $f^{(1)}$ exists, is bounded and continuous. Then*

$$\left|\int_M f d\mu - f(x_0)\right| - |f(x_0)| |m - 1| -$$

$$\left|\int_M f^{(1)}(x_0)(x - x_0)\mu(dx)\right| \leq$$

$$\begin{cases} \frac{1}{8r}\left(2 + \sqrt{m}r\right)^2 \omega_1\left(f^{(1)}, rD_2(x_0)\right) D_2(x_0), & \text{if } r \leq \frac{2}{\sqrt{m}}, \\ \sqrt{m}\omega_1\left(f^{(1)}, rD_2(x_0)\right) D_2(x_0), & \text{if } r > \frac{2}{\sqrt{m}}. \end{cases} \tag{7.67}$$

By Theorem 7.4 we obtain:

Theorem 7.5. *Let $f \in C_B(\mathbb{R}^N)$, whose Fréchet derivatives $f^{(j)}$, $1 \leq j \leq m$, $m \in \mathbb{N}$, exist, are bounded and continuous, $r > 0$, $x_0 \in \mathbb{R}^N$.*

Denote

$$0 < \left(A_n\left(\|\cdot - x_0\|^{m+1}\right)(x_0)\right)^{\frac{1}{(m+1)}} := D_{m+1}(x_0). \tag{7.68}$$

Assume that

$$w := \omega_1\left(f^{(m)}, rD_{m+1}(x_0)\right) > 0. \tag{7.69}$$

Then

$$|A_n(f, x_0) - f(x_0)| \leq \left|\sum_{j=1}^{m} \frac{1}{j!} A_n\left(f^{(j)}(x_0)(\cdot - x_0)^j\right)(x_0)\right| + \tag{7.70}$$

$$\frac{w}{rm!}\left[\frac{mr^2}{8} + \frac{r}{2} + \frac{1}{(m+1)}\right] D_{m+1}^m(x_0).$$

By Corollary 7.1 we get:

Corollary 7.3. *Denote*

$$D_2(x_0) = \left(A_n\left(\|\cdot - x_0\|^2\right)(x_0)\right)^{\frac{1}{2}} \tag{7.71}$$

which is positive, $x_0 \in \mathbb{R}^N$.

Assume that

$$\omega_1\left(f^{(1)}, \frac{1}{2}D_2(x_0)\right) > 0. \tag{7.72}$$

Then

$$\left|A_n(f, x_0) - f(x_0)\right| - \left|A_n\left(f^{(1)}(x_0)(\cdot - x_0)\right)(x_0)\right| \leq$$
$$(1.5625)\,\omega_1\left(f^{(1)}, \frac{1}{2}D_2(x_0)\right) D_2(x_0). \tag{7.73}$$

By Proposition 7.4 we get:

Proposition 7.5. *Let $f \in C_B(\mathbb{R}^N)$, whose Fréchet derivative $f^{(1)}$ exists, is bounded and continuous, $r > 0$, $x_0 \in \mathbb{R}^N$.*

Denote

$$D_2(x_0) = \left(A_n\left(\|\cdot - x_0\|^2\right)(x_0)\right)^{\frac{1}{2}}, \tag{7.74}$$

which is positive.

Then

$$\left|A_n(f, x_0) - f(x_0)\right| - \left|A_n\left(f^{(1)}(x_0)(\cdot - x_0)\right)(x_0)\right| \leq$$
$$\frac{1}{8r}(2 + r)^2\,\omega_1\left(f', rD_2(x_0)\right) D_2(x_0). \tag{7.75}$$

By Corollary 7.2 we have:

Corollary 7.4. *Here all as in Proposition 7.5. Then*

$$\left|A_n(f, x_0) - f(x_0)\right| - \left|A_n\left(f^{(1)}(x_0)(\cdot - x_0)\right)(x_0)\right| \leq$$
$$\begin{cases} \frac{1}{8r}(2 + r)^2\,\omega_1\left(f^{(1)}, rD_2(x_0)\right) D_2(x_0), & \text{if } r \leq 2, \\ \omega_1\left(f^{(1)}, rD_2(x_0)\right) D_2(x_0), & \text{if } r > 2. \end{cases} \tag{7.76}$$

We make

Remark 7.3. We have that

$$D_{m+1}(x_0) = \left(A_n\left(\|\cdot - x_0\|^{m+1}\right)(x_0)\right)^{\frac{1}{(m+1)}} \leq$$
$$\left(\frac{1}{n^\beta} + \frac{1}{n^{m+1}}\frac{2(m+1)!}{\lambda^{m+1}}\left(q + \frac{1}{q}\right)e^{2\lambda}e^{-\lambda\left(n^{1-\beta}-1\right)}\right)^{\frac{1}{(m+1)}} \to 0, \text{ as } n \to +\infty. \tag{7.77}$$

And, furthermore we get

$$D_2(x_0) = \left(A_n\left(\|\cdot - x_0\|^2\right)(x_0)\right)^{\frac{1}{2}} \leq$$
$$\sqrt{\left(\frac{1}{n^\beta} + \frac{1}{n^2}\frac{4}{\lambda^2}\left(q + \frac{1}{q}\right)e^{2\lambda}e^{-\lambda\left(n^{1-\beta}-1\right)}\right)} \to 0, \text{ as } n \to +\infty. \tag{7.78}$$

We make

Remark 7.4. On Theorem 7.5:

Assume that $f^{(j)}(x_0) = 0$, $j = 1, ..., m$.

Then

$$|A_n(f, x_0) - f(x_0)| \leq$$

$$\frac{\omega_1\left(f^{(m)}, r\left(\frac{1}{n^\beta} + \frac{1}{n^{m+1}}\frac{2(m+1)!}{\lambda^{m+1}}\left(q+\frac{1}{q}\right)e^{2\lambda}e^{-\lambda\left(n^{1-\beta}-1\right)}\right)^{\frac{1}{(m+1)}}\right)}{rm!}$$

$$\left[\frac{mr^2}{8} + \frac{r}{2} + \frac{1}{(m+1)}\right] \tag{7.79}$$

$$\left(\frac{1}{n^\beta} + \frac{1}{n^{m+1}}\frac{2(m+1)!}{\lambda^{m+1}}\left(q+\frac{1}{q}\right)e^{2\lambda}e^{-\lambda\left(n^{1-\beta}-1\right)}\right)^{\frac{m}{(m+1)}} \to 0, \text{ as } n \to +\infty.$$

That is $\lim_{n\to+\infty} A_n(f, x_0) = f(x_0)$.

Case $m = 1$ in (7.79):

Assume $f^{(1)}(x_0) = 0$, then

$$|A_n(f, x_0) - f(x_0)| \leq$$

$$\frac{1}{r}\omega_1\left(f^{(1)}, r\left(\frac{1}{n^\beta} + \frac{1}{n^2}\frac{4}{\lambda^2}\left(q+\frac{1}{q}\right)e^{2\lambda}e^{-\lambda\left(n^{1-\beta}-1\right)}\right)^{\frac{1}{2}}\right)$$

$$\left[\frac{r^2}{8} + \frac{r}{2} + \frac{1}{2}\right] \tag{7.80}$$

$$\left(\frac{1}{n^\beta} + \frac{1}{n^2}\frac{4}{\lambda^2}\left(q+\frac{1}{q}\right)e^{2\lambda}e^{-\lambda\left(n^{1-\beta}-1\right)}\right)^{\frac{1}{2}} \to 0, \text{ as } n \to +\infty;$$

implying again $\lim_{n\to+\infty} A_n(f, x_0) = f(x_0)$.

We continue with general results and their applications.

Theorem 7.6. *Let $(L_n)_{n\in\mathbb{N}}$ be a sequence of positive linear operators from $C_B(\mathbb{R}^N)$ into $B(\mathbb{R}^N)$, such that $L_n(1) = 1$. Assume $0 < L_n(\|\cdot - x_0\|_\infty)(x_0) < +\infty$.*

Then

$$|L_n(f)(x_0) - f(x_0)| \leq 2\omega_1(f, L_n(\|\cdot - x_0\|_\infty)(x_0)) < +\infty. \tag{7.81}$$

Given that $\lim_{n\to+\infty} L_n(\|\cdot - x_0\|_\infty)(x_0) = 0$ and f is uniformly continuous, we get that $\lim_{n\to+\infty} L_n(f)(x_0) = f(x_0)$.

Proof. We have that $\omega_1(f,h)<\infty$, $h>0$.

We got by (7.45) that

$$|f(\cdot)-f(x_0)|\leq\omega_1(f,h)\left[1+\frac{\|\cdot-x_0\|_\infty}{h}\right],\quad \text{over } \mathbb{R}^N. \tag{7.82}$$

Hence we derive

$$|L_n(f)(x_0)-f(x_0)|=|L_n(f(\cdot)-f(x_0))(x_0)|\leq$$

$$L_n(|f(\cdot)-f(x_0)|)(x_0)\overset{(7.82)}{\leq}\omega_1(f,h)\left[1+\frac{L_n(\|\cdot-x_0\|_\infty)(x_0)}{h}\right] \tag{7.83}$$

(choose $h:=L_n(\|\cdot-x_0\|_\infty)(x_0)>0$)

$$=2\omega_1(f,L_n(\|\cdot-x_0\|_\infty)(x_0)).$$

□

Next we apply (7.81).

Theorem 7.7. *Let $n\in\mathbb{N}:n^{1-\beta}>2$, $0<\beta<1$, $x_0\in\mathbb{R}^N$. Then*

$$|K_n(f)(x_0)-f(x_0)|\leq2\omega_1(f,K_n(\|\cdot-x_0\|_\infty)(x_0))<+\infty, \tag{7.84}$$

and

$$|Q_n(f)(x_0)-f(x_0)|\leq2\omega_1(f,Q_n(\|\cdot-x_0\|_\infty)(x_0))<+\infty, \tag{7.85}$$

and

$$\left\{\begin{matrix}\|K_n(f)-f\|_\infty\\ \|Q_n(f)-f\|_\infty\end{matrix}\right\}\leq \tag{7.86}$$

$$2\omega_1\!\left(f,\left[\left(\frac{1}{n}+\frac{1}{n^\beta}\right)+\frac{1}{n}\left[Te^{-2\lambda n^{(1-\beta)}}+\frac{2}{\lambda}\left(q+\frac{1}{q}\right)e^{2\lambda}e^{-\lambda\left(n^{1-\beta}-1\right)}\right]\right]\right)<+\infty.$$

Given that $f\in C_U(\mathbb{R}^N)$ we derive that $K_n(f)\to f$, $Q_n(f)\to f$, as $n\to+\infty$, pointwise and uniformly.

Proof. Along with (7.81) we use of Propositions 7.2, 7.3. □

We continue with the convexity case.

Remark 7.5. Let $f\in C_B(\mathbb{R}^N)$ and $|f(\cdot)-f(x_0)|$ is convex over $\mathbb{R}^N$. Then, by (7.54), we have

$$|f(\cdot)-f(x_0)|\leq\frac{\omega_1(f,h)}{h}\|\cdot-x_0\|_\infty, \tag{7.87}$$

over $\mathbb{R}^N$, $x_0\in\mathbb{R}^N$.

Acting as in Theorem 7.6 we derive:

Theorem 7.8. *Let* $(L_n)_{n\in\mathbb{N}}$ *positive linear operators from* $C_B(\mathbb{R}^N)$ *into* $B(\mathbb{R}^N)$: $L_n(1) = 1$. *Let* $f \in C_B(\mathbb{R}^N)$ *and the convex function* $|f(\cdot) - f(x_0)|$ *over* $\mathbb{R}^N$, $x_0 \in \mathbb{R}^N$. *Assume that* $0 < L_n(\|\cdot - x_0\|_\infty)(x_0) < +\infty$.

Then

$$|L_n(f)(x_0) - f(x_0)| \leq \omega_1(f, L_n(\|\cdot - x_0\|_\infty)(x_0)). \tag{7.88}$$

The last (7.88) is attained by $f(t) = \|t - x_0\|_\infty$, $\forall\, t \in \mathbb{R}^N$, $x_0 \in \mathbb{R}^N$, *that is sharp. Given that* f *is uniformly continuous and* $L_n(\|\cdot - x_0\|_\infty)(x_0) \to 0$, *as* $n \to \infty$, *then* $L_n(f)(x_0) \to f(x_0)$, *as* $n \to \infty$.

Next we apply Theorem 7.8.

Theorem 7.9. *Let* $f \in C_B(\mathbb{R}^N)$: $|f(\cdot) - f(x_0)|$ *is convex,* $x_0 \in \mathbb{R}^N$. *Then*

$$\left\{ \begin{array}{l} |K_n(f, x_0) - f(x_0)| \\ |Q_n(f, x_0) - f(x_0)| \end{array} \right\} \leq \tag{7.89}$$

$$\omega_1\left(f, \left[\left(\frac{1}{n} + \frac{1}{n^\beta}\right) + \frac{1}{n}\left[Te^{-2\lambda n^{(1-\beta)}} + \frac{2}{\lambda}\left(q + \frac{1}{q}\right)e^{2\lambda}e^{-\lambda\left(n^{1-\beta}-1\right)}\right]\right]\right) < +\infty.$$

Given that $f \in C_U(\mathbb{R}^N)$, *we obtain that* $K_n(f)(x_0) \to f(x_0)$, $Q_n(f)(x_0) \to f(x_0)$, *as* $n \to +\infty$.

Proof. By (7.88) and Propositions 7.2, 7.3. □

We make

Remark 7.6. We consider $(\mathbb{R}^N, \|\cdot\|_\infty)$ and $(\mathbb{R}^N)^j$ denote the j-fold product space $\mathbb{R}^N \times \ldots \times \mathbb{R}^N$ endowed with the max-norm $\|x\|_{(\mathbb{R}^N)^j} := \max_{1\leq k\leq j} \|x_k\|_\infty$, where $x := (x_1, \ldots, x_j) \in (\mathbb{R}^N)^j$. Then the space $L_j := L_j\left((\mathbb{R}^N)^j; \mathbb{R}\right)$ of all real-valued multilinear continuous functions $g : (\mathbb{R}^N)^j \to \mathbb{R}$ is a Banach space with norm

$$\|g\| := \|g\|_{L_j} := \sup_{\left(\|x\|_{(\mathbb{R}^N)^j}=1\right)} |g(x)| = \sup \frac{|g(x)|}{\|x_1\|_\infty \cdots \|x_j\|_\infty}. \tag{7.90}$$

Let $x_0 \in \mathbb{R}^N$ be fixed, and $f : \mathbb{R}^N \to \mathbb{R}$ be a continuous and bounded function whose Fréchet derivatives

$$f^{(j)} : \mathbb{R}^N \to L_j = L_j\left((\mathbb{R}^N)^j; \mathbb{R}\right)$$

exist and are continuous and bounded for all $1 \leq j \leq m$, $m \in \mathbb{N}$. Call $(x - x_0)^j := (x - x_0, ..., x - x_0) \in (\mathbb{R}^N)^j = \underbrace{\mathbb{R}^N \times ... \times \mathbb{R}^N}_{-j-}$.

Then, by Taylor's formula, see [179], p. 124 and [175], we get

$$f(x) = \sum_{j=0}^{m} \frac{f^{(j)}(x_0)(x - x_0)^j}{j!} + R_m(x, x_0), \text{ all } x \in \mathbb{R}^N, \tag{7.91}$$

where

$$R_m(x, x_0) := \int_0^1 \frac{(1-u)^{m-1}}{(m-1)!} \left(f^{(m)}(x_0 + u(x - x_0)) - f^{(m)}(x_0) \right) (x - x_0)^m du. \tag{7.92}$$

Here we assume that $f^{(j)}(x_0) = 0$, $j = 1, ..., m$, then

$$f(x) - f(x_0) = \int_0^1 \frac{(1-u)^{m-1}}{(m-1)!} \left(f^{(m)}(x_0 + u(x - x_0)) - f^{(m)}(x_0) \right) (x - x_0)^m du. \tag{7.93}$$

Consider

$$w := \omega_1\left(f^{(m)}, h\right) = \sup_{\substack{x, y \in \mathbb{R}^N \\ : \|x - y\|_\infty \leq h}} \left\| f^{(m)}(x) - f^{(m)}(y) \right\|, \quad h > 0. \tag{7.94}$$

Notice that $w < +\infty$.

We obtain that

$$\left\| \left(f^{(m)}(x_0 + u(x - x_0)) - f^{(m)}(x_0) \right) (x - x_0)^m \right\| \leq$$

$$\left\| f^{(m)}(x_0 + u(x - x_0)) - f^{(m)}(x_0) \right\| \|x - x_0\|^m \leq \tag{7.95}$$

$$w \|x - x_0\|^m \left\lceil \frac{u \|x - x_0\|}{h} \right\rceil,$$

by Lemma 7.1.1, p. 208, [169].

Therefore for all $x \in \mathbb{R}^N$:

$$|f(x) - f(x_0)| \leq w \|x - x_0\|^m \int_0^1 \left\lceil \frac{u \|x - x_0\|_\infty}{h} \right\rceil \frac{(1-u)^{m-1}}{(m-1)!} du \tag{7.96}$$

$$= w \Phi_m(\|x - x_0\|_\infty),$$

by a change of variable, where

$$\Phi_m(t) := \int_0^{|t|} \left\lceil \frac{s}{h} \right\rceil \frac{(|t|-s)^{m-1}}{(m-1)!} ds = \frac{1}{m!} \left(\sum_{j=0}^{\infty} (|t| - j \cdot h)_+^m \right), \quad (7.97)$$

$t \in \mathbb{R}$; $(x)_+ = \max\{x, 0\}$.

For a full list of properties of the spline function Φ_m see Remark 7.1.3 of [169], p. 210.

By (7.1.18) of [169], p. 210, we have

$$\Phi_m(x) \leq \left(\frac{|x|^{m+1}}{(m+1)!h} + \frac{|x|^m}{2m!} + \frac{h|x|^{m-1}}{8(m-1)!} \right), \quad \forall x \in \mathbb{R}. \quad (7.98)$$

Consequently it holds ($x \in \mathbb{R}^N$)

$$|f(x) - f(x_0)| \leq \omega_1\left(f^{(m)}, h\right) \left(\frac{\|x - x_0\|_\infty^{m+1}}{(m+1)!h} + \frac{\|x - x_0\|_\infty^m}{2m!} + \frac{h\|x - x_0\|_\infty^{m-1}}{8(m-1)!} \right). \quad (7.99)$$

So that

$$|f(\cdot) - f(x_0)| \leq \omega_1\left(f^{(m)}, h\right) \left(\frac{\|\cdot - x_0\|_\infty^{m+1}}{(m+1)!h} + \frac{\|\cdot - x_0\|_\infty^m}{2m!} + \frac{h\|\cdot - x_0\|_\infty^{m-1}}{8(m-1)!} \right), \quad (7.100)$$

over $\mathbb{R}^N$.

Let $(L_n)_{n\in\mathbb{N}}$ be a sequence of positive linear operators from $C_B(\mathbb{R}^N)$ into $B(\mathbb{R}^N)$ such that $L_n(1) = 1$, $\forall\, n \in \mathbb{N}$.

By $|f| \leq |f|$, iff $-|f| \leq f \leq |f|$, then $-L_n(|f|) \leq L_n(f) \leq L_n(|f|)$, and $|L_n(f)| \leq L_n(|f|)$.

Hence

$$|L_n(f(\cdot) - f(x_0))(x_0)| \leq L_n(|f(\cdot) - f(x_0)|)(x_0) \overset{(7.100)}{\leq} \omega_1\left(f^{(m)}, h\right)$$

$$\left[\frac{L_n\left(\|\cdot - x_0\|_\infty^{m+1}\right)(x_0)}{(m+1)!h} + \frac{L_n\left(\|\cdot - x_0\|_\infty^m\right)(x_0)}{2m!} + \frac{hL_n\left(\|\cdot - x_0\|_\infty^{m-1}\right)(x_0)}{8(m-1)!} \right] =: (\xi). \quad (7.101)$$

We need the following Hölder's type inequality for positive linear operators.

Theorem 7.10. *Let L be a positive linear operator from $C_B(\mathbb{R}^N)$ into $B(\mathbb{R}^N)$, and $f, g \in C_B(\mathbb{R}^N)$, furthermore let $p, q > 1 : \frac{1}{p} + \frac{1}{q} = 1$. Assume that $L((|f(\cdot)|^p))(s_*)$, $L((|g(\cdot)|^q))(s_*) > 0$ for some $s_* \in \mathbb{R}^N$. Then*

$$L(|f(\cdot)g(\cdot)|)(s_*) \leq (L((|f(\cdot)|^p))(s_*))^{\frac{1}{p}} (L((|g(\cdot)|^q))(s_*))^{\frac{1}{q}}. \quad (7.102)$$

Proof. Let $a, b \geq 0$, $p, q > 1 : \frac{1}{p} + \frac{1}{q} = 1$. The Young's inequality says

$$ab \leq \frac{a^p}{p} + \frac{b^q}{q}. \tag{7.103}$$

Then

$$\frac{|f(s)|}{\left(L\left(\left(|f(\cdot)|^p\right)\right)(s_*)\right)^{\frac{1}{p}}} \cdot \frac{|g(s)|}{\left(L\left(\left(|g(\cdot)|^q\right)\right)(s_*)\right)^{\frac{1}{q}}} \leq$$

$$\frac{|f(s)|^p}{p\left(L\left(\left(|f(\cdot)|^p\right)\right)(s_*)\right)} + \frac{|g(s)|^q}{q\left(L\left(\left(|g(\cdot)|^q\right)\right)(s_*)\right)}, \quad \forall\, s \in \mathbb{R}^N. \tag{7.104}$$

Hence it holds

$$\frac{L\left(|f(\cdot)\, g(\cdot)|\right)(s_*)}{\left(L\left(\left(|f(\cdot)|^p\right)\right)(s_*)\right)^{\frac{1}{p}}\left(L\left(\left(|g(\cdot)|^q\right)\right)(s_*)\right)^{\frac{1}{q}}} \leq$$

$$\frac{L\left(\left(|f(\cdot)|^p\right)\right)(s_*)}{p\left(L\left(\left(|f(\cdot)|^p\right)\right)(s_*)\right)} + \frac{L\left(\left(|g(\cdot)|^q\right)\right)(s_*)}{q\left(L\left(\left(|g(\cdot)|^q\right)\right)(s_*)\right)} = \frac{1}{p} + \frac{1}{q} = 1, \tag{7.105}$$

for $s_* \in \mathbb{R}^N$, proving the claim. $\square$

We continue with

Remark 7.7. Let $g = 1$ and $L_n : L_n(1) = 1$, then by (7.102) we get:

$$L_n\left(\|\cdot - x_0\|_\infty^m\right)(x_0) \leq \left(L_n\left(\|\cdot - x_0\|_\infty^{m+1}\right)(x_0)\right)^{\left(\frac{m}{m+1}\right)}, \tag{7.106}$$

and

$$L_n\left(\|\cdot - x_0\|_\infty^{(m-1)}\right)(x_0) \leq \left(L_n\left(\|\cdot - x_0\|_\infty^{m+1}\right)(x_0)\right)^{\left(\frac{m-1}{m+1}\right)}, \tag{7.107}$$

where $m \in \mathbb{N}$ and $L_n\left(\|\cdot - x_0\|_\infty^{m+1}\right)(x_0) > 0$.

Continuing from (7.101):

$$(\xi) \leq \omega_1\left(f^{(m)}, h\right)\left[\frac{L_n\left(\|\cdot - x_0\|_\infty^{m+1}\right)(x_0)}{(m+1)!h} + \right. \tag{7.108}$$

$$\left. \frac{\left(L_n\left(\|\cdot - x_0\|_\infty^{m+1}\right)(x_0)\right)^{\frac{m}{m+1}}}{2m!} + \frac{h\left(L_n\left(\|\cdot - x_0\|_\infty^{m+1}\right)(x_0)\right)^{\frac{m-1}{m+1}}}{8(m-1)!}\right]$$

(assume $h := \left(L_n\left(\|\cdot - x_0\|_\infty^{m+1}\right)(x_0)\right)^{\frac{1}{m+1}} > 0$)

$$= \omega_1\left(f^{(m)}, \left(L_n\left(\|\cdot - x_0\|_\infty^{m+1}\right)(x_0)\right)^{\frac{1}{m+1}}\right)$$

$$\left[\frac{h^m}{(m+1)!} + \frac{h^m}{2m!} + \frac{h^m}{8(m-1)!}\right] \tag{7.109}$$

$$= \omega_1\left(f^{(m)}, \left(L_n\left(\|\cdot - x_0\|_\infty^{m+1}\right)(x_0)\right)^{\frac{1}{m+1}}\right)$$

$$\left(L_n\left(\|\cdot - x_0\|_\infty^{m+1}\right)(x_0)\right)^{\frac{m}{m+1}}\left[\frac{1}{(m+1)!} + \frac{1}{2m!} + \frac{1}{8(m-1)!}\right]. \tag{7.110}$$

We have proved that

Theorem 7.11. *Let $f \in C_B\left(\mathbb{R}^N\right)$, $x_0 \in \mathbb{R}^N$, Fréchet derivatives $f^{(j)}$ continuous and bounded with $f^{(j)}(x_0) = 0$, $j = 1, ..., m$. Then*

$$|L_n(f)(x_0) - f(x_0)| \leq \omega_1\left(f^{(m)}, \left(L_n\left(\|\cdot - x_0\|_\infty^{m+1}\right)(x_0)\right)^{\frac{1}{m+1}}\right) \tag{7.111}$$

$$\left(L_n\left(\|\cdot - x_0\|_\infty^{m+1}\right)(x_0)\right)^{\frac{m}{m+1}}\left[\frac{1}{(m+1)!} + \frac{1}{2m!} + \frac{1}{8(m-1)!}\right], \quad \forall\, n \in \mathbb{N}.$$

Here $L_n(1) = 1$, L_n positive linear operators from $C_b\left(\mathbb{R}^N\right)$ into $B\left(\mathbb{R}^N\right)$, under the assumption $L_n\left(\|\cdot - x_0\|_\infty^{m+1}\right)(x_0) > 0$.

Let $L_n\left(\|\cdot - x_0\|_\infty^{m+1}\right)(x_0) \to 0$, as $n \to +\infty$, then $\lim_{n\to+\infty} L_n(f)(x_0) = f(x_0)$.

Next we apply Theorem 7.11 to the operators Θ_n.

Theorem 7.12. *Let $f \in C_B\left(\mathbb{R}^N\right)$, $x_0 \in \mathbb{R}^N$, Fréchet derivatives $f^{(j)}$ continuous and bounded with $f^{(j)}(x_0) = 0$, $j = 1, ..., m$. Then*

$$|\Theta_n(f)(x_0) - f(x_0)| \leq \omega_1\left(f^{(m)}, \left(\Theta_n\left(\|\cdot - x_0\|_\infty^{m+1}\right)(x_0)\right)^{\frac{1}{m+1}}\right) \tag{7.112}$$

$$\left(\Theta_n\left(\|\cdot - x_0\|_\infty^{m+1}\right)(x_0)\right)^{\frac{m}{m+1}}\left[\frac{1}{(m+1)!} + \frac{1}{2m!} + \frac{1}{8(m-1)!}\right],$$

$\forall\, n \in \mathbb{N} : n^{1-\beta} > 2$, $0 < \beta < 1$.

We have that $\lim_{n\to+\infty} \Theta_n(f)(x_0) = f(x_0)$.

Proof. Here Θ_n any of A_n, K_n, Q_n.

By Theorem 7.11, and Propositions 7.1–7.3 we get $\Theta_n \left(\|\cdot - x_0\|_\infty^{m+1} \right) (x_0) \to 0$, as $n \to +\infty$. □

Corollary 7.5. *All as in Theorem 7.12. Set*

$$\varphi_1 (n) := \frac{1}{n^\beta} + \frac{1}{n^{m+1}} \frac{2(m+1)!}{\lambda^{m+1}} \left(q + \frac{1}{q} \right) e^{2\lambda} e^{-\lambda \left(n^{1-\beta} - 1 \right)}. \tag{7.113}$$

Then

$$|A_n (f)(x_0) - f(x_0)| \leq \omega_1 \left(f^{(m)}, (\varphi_1 (n))^{\frac{1}{m+1}} \right)$$

$$(\varphi_1 (n))^{\frac{m}{m+1}} \left[\frac{1}{(m+1)!} + \frac{1}{2m!} + \frac{1}{8(m-1)!} \right], \tag{7.114}$$

$\forall\, n \in \mathbb{N} : n^{1-\beta} > 2$, $0 < \beta < 1$.

We have that $\lim_{n \to +\infty} A_n (f)(x_0) = f(x_0)$.

Proof. By Theorem 7.12 and Proposition 7.1. □

We finish with

Corollary 7.6. *All as in Theorem 7.12. Set*

$$\varphi_2 (n) := \left[\left(\frac{1}{n} + \frac{1}{n^\beta} \right)^m + \right.$$

$$\left. \frac{2^m}{n^{m+1}} \left[T e^{-2\lambda n^{(1-\beta)}} + \frac{2(m+1)!}{\lambda^{m+1}} \left(q + \frac{1}{q} \right) e^{2\lambda} e^{-\lambda \left(n^{1-\beta} - 1 \right)} \right] \right]. \tag{7.115}$$

Then

$$\left\{ \begin{array}{l} |K_n (f)(x_0) - f(x_0)| \\ |Q_n (f)(x_0) - f(x_0)| \end{array} \right\} \leq \omega_1 \left(f^{(m)}, (\varphi_2 (n))^{\frac{1}{m+1}} \right)$$

$$(\varphi_2 (n))^{\frac{m}{m+1}} \left[\frac{1}{(m+1)!} + \frac{1}{2m!} + \frac{1}{8(m-1)!} \right], \tag{7.116}$$

$\forall\, n \in \mathbb{N} : n^{1-\beta} > 2$, $0 < \beta < 1$.

We have that $\lim_{n \to +\infty} K_n (f)(x_0) = \lim_{n \to +\infty} Q_n (f)(x_0) = f(x_0)$.

Proof. By Theorem 7.12 and Propositions 7.2, 7.3. □

Bibliography

[168] G.A. Anastassiou, *A "K-Attainable" inequality related to the convergence of Positive Linear Operators*, J. Approximation Theory, 44, 380–383 (1985).

[169] G.A. Anastassiou, *Moments in probability and approximation theory*, Pitman Research Notes in Mathematics series / Longman Scientific & Technical, Essex, New York, UK, USA, 1993.

[170] G.A. Anastassiou, *Quantitative Approximations*, Chapmen & Hall/CRC, London, New York, 2001.

[171] G.A. Anastassiou, *Intelligent Computations: Abstract Fractional Calculus, Inequalities, Approximations*, Springer, Heidelberg, New York, 2018.

[172] G.A. Anastassiou, *Parametrized, Deformed and General Neural Networks*, Springer, Heidelberg, New York, 2023.

[173] G.A. Anastassiou, *Multivariate smooth symmetrized and perturbed hyperbolic tangent neural network approximation over infinite domains*, 2024, 12, 3777, `https://doi.org/10.3390/math12233777`.

[174] G.A. Anastassiou, *Trigonometric and Hyperbolic Generated Approximation Theory*, World Scientific, Singapore, New York, 2025.

[175] H. Cartan, *Differential Calculus*, Herman, Paris, 1971.

[176] Z. Chen and F. Cao, *The approximation operators with sigmoidal functions*, Computers and Mathematics with Applications, 58 (2009), 758–765.

[177] D. Costarelli, R. Spigler, *Approximation results for neural network operators activated by sigmoidal functions*, Neural Networks 44 (2013), 101–106.

[178] D. Costarelli, R. Spigler, *Multivariate neural network operators with sigmoidal activation functions*, Neural Networks 48 (2013), 72–77.

[179] L.B. Rall, *Computational Solution of Nonlinear Operator Equations*, John Wiley & Sons, New York, 1969.

[180] O. Shisha and B. Mond, *The degree of convergence of sequences of linear positive operators*, Nat. Acad. Sci. U.S. 60 (1968), 1196–1200.

[181] O. Shisha and B. Mond, *The degree of Approximation to Periodic Functions by Linear Positive Operators*, J. Approximation Theory, 1, 335–339 (1968).

[182] S. Haykin, *Neural Networks: A Comprehensive Foundation* (2 ed.), Prentice Hall, New York, 1998.

[183] W. McCulloch and W. Pitts, *A logical calculus of the ideas immanent in nervous activity*, Bulletin of Mathematical Biophysics, 7 (1943), 115–133.

[184] T.M. Mitchell, *Machine Learning*, WCB-McGraw-Hill, New York, 1997.

[185] Yu, Dansheng; Cao, Feilong, *Construction and approximation rate for feed-forward neural network operators with sigmoidal functions*, J. Comput. Appl. Math. 453 (2025), paper No. 116150.

[186] Cen, Siyu; Jin, Bangti; Quan, Qimeng; Zhou, Zhi; *Hybrid neural-network FEM approximation of diffusion coeficient in elyptic and parabolic problems.* IMA J. Numer. Anal. 44 (2024), no. 5, 3059–3093.

[187] Coroianu Lucian; Costarelli, Danillo; Natale, Mariarosaria; Pantiş, Alexandra; *The approximation capabilities of Durrmeyer-type neural network operators.* J. Appl. Math. Comput. 70 (2024), no. 5, 4581–4599.

[188] Warin, Xavier; *The GroupMax neural network approximation of convex functions.* IEEE Trans. Neural Netw. Learn. Syst. 35 (2024), no. 8, 11608–11612.

[189] Fabra, Arnau; Guasch, Oriol; Baiges, Joan; Codina, Ramon; *Approximation of acoustic black holes with finite element mixed formulations and artificial neural network correction terms.* Finite Elem. Anal. Des. 241 (2024), paper No. 104236.

[190] Grohs, Philipp; Voigtlaender, Felix. *Proof of the theory-to-practice gap in deep learning via sampling complexity bounds for neural network approximation spaces.* Found. Comput. Math. 24 (2024), no. 4, 1085–1143.

[191] Basteri, Andrea; Trevisan, Dario; *Quantitative Gaussian approximation of randomly initialized deep neural networks.* Mach. Learn. 113 (2024), no. 9, 6373–6393.

[192] De Ryck, T.; Mishra, S. *Error analysis for deep neural network approximations of parametric hyperbolic conservation laws.* Math. Comp. 93 (2024), no. 350, 2643–2677.

[193] Liu, Jie; Zhang, Baoji; Lai, Yuyang; Fang, Liqiau. *Hull form optimization reserach based on multi-precision back-propagation neural network approximation model.* Internal. J. Numer. Methods Fluid 96 (2024), no. 8, 1445–1460.

[194] Yoo, Jihahm; Kim, Jaywon; Gim, Minjung; Lee, Haesung. *Error estimates of physics-informed neural networks for initial value problems.* J. Korean Soc. Ind. Appl. Math. 28 (2024), no. 1, 33–58.

[195] G.A. Anastassiou, *Multivariate Neural Networks over infinite domain as Positive Linear Operators and a touch of measure theory*, Moroccan Journal of Pure and Applied Analysis, accepted, 2025.

Chapter 8

Multivariate Generalized Logistic Neural Networks in Infinite Domain as Positive Linear Operators with Measure Theory

Multivariate neural network operators of infinite domain are interpreted as positive linear operators and related general theory applies to them. These operators are induced by a symmetrized density function deriving from the parametrized and deformed generalized logistic activation function. We are acting on the space of continuous and bounded functions on $\mathbb{R}^N$, $N \in \mathbb{N}$, to the reals. We study quantitatively the rate of convergence of these multivariate neural network operators to the unit operator. Our inequalities involve the modulus of continuity of the function under approximation or its Fréchet derivative. Connection to measure theory is studied in detail. It follows [223].

8.1 Introduction

The author studied extensively the quantitative approximation of positive linear operators to the unit since 1985, see for example [196]–[198], [202], and [197] we use a lot in this work. He originated from the quantitative weak convergence of finite positive measures to the unit Dirac measure, having as a method the geometric moment theory, see [197], and he produced best upper bounds, leading to attained (i.e. sharp Jackson type inequalities), e.g. see [196], [197]. These studies have gone to all possible directions, univariate and multivariate, though in this work we stay only on the multivariate approach on infinite domain.

Also, since 1997, this author started the study of the quantitative convergence of neural network operators to the unit, and he has been written since then numerous of articles and books, e.g. see [198]–[200], [202]. Here we use a lot [200], and we are acting based on [201].

A wide range of neural network operators over $\mathbb{R}^N$, $N \in \mathbb{N}$, being studied by the author all along and here are indeed by nature positive linear operators.

Here, for the first time in the literature, multivariate generalized logistic neural network operators of infinite domain are treated as positive linear operators.

So, methods of positive linear operators of infinite domain apply here to our infinite summation defined multivariate generalized logistic neural network operators, producing new and interesting pointwise and uniform results.

Convexity cases are also treated here producing sharp results. We also use Fréchet derivatives. Furthermore a strong connection of our operators to measure theory is established with respect to probability measures.

Of great inspiration always have been [208], [209]. For classical studies on neural network we recommend also, [204]–[206], [210]–[212].

For newer work on neural networks we also refer the reader to [213]–[222].

8.2 Basics

Initially we follow [200], pp. 395–417.

Our activation function here is the q-deformed and λ-parametrized function

$$\varphi_{q,\lambda}(x) = \frac{1}{1 + qA^{-\lambda x}}, \quad x \in \mathbb{R}, \; q, \lambda > 0, \; A > 1. \tag{8.1}$$

This is the A-generalized logistic function.

For more read Chapter 16 of [200]: "Banach space valued ordinary and fractional neural network approximation based on q-deformed and λ-parametrized A-generalized logistic function".

The chapters 15, 16 of [200] motivate our current work.

The proposed "symmetrization technique" aims to use half data feed to our neural networks.

We will employ the following density function

$$G_{q,\lambda}(x) := \frac{1}{2}\left(\varphi_{q,\lambda}(x+1) - \varphi_{q,\lambda}(x-1)\right), \quad x \in \mathbb{R}, \; q, \lambda > 0. \tag{8.2}$$

We have that

$$G_{q,\lambda}(-x) = G_{\frac{1}{q},\lambda}(x), \tag{8.3}$$

and

$$G_{\frac{1}{q},\lambda}(-x) = G_{q,\lambda}(x), \ \ \forall\, x \in \mathbb{R}. \tag{8.4}$$

Adding (8.3) and (8.4) we obtain

$$G_{q,\lambda}(-x) + G_{\frac{1}{q},\lambda}(-x) = G_{q,\lambda}(x) + G_{\frac{1}{q},\lambda}(x), \ \ \forall\, x \in \mathbb{R}, \tag{8.5}$$

the key to this work.

So that

$$W(x) := \frac{G_{q,\lambda}(x) + G_{\frac{1}{q},\lambda}(x)}{2} \tag{8.6}$$

is an even function, symmetric with respect to the y-axis.

The global maximum of $G_{q,\lambda}$ is given by (16.18), p. 401 of [200] as

$$G_{q,\lambda}\left(\frac{\log_A q}{\lambda}\right) = \frac{A^\lambda - 1}{2(A^\lambda + 1)}. \tag{8.7}$$

And, the global max of $G_{\frac{1}{q},\lambda}$ is

$$G_{\frac{1}{q},\lambda}\left(\frac{\log_A \frac{1}{q}}{\lambda}\right) = G_{\frac{1}{q},\lambda}\left(\frac{-\log_A q}{\lambda}\right) = \frac{A^\lambda - 1}{2(A^\lambda + 1)}, \tag{8.8}$$

both sharing the same maximum at symmetric points.

By Theorem 16.1, p. 401 of [200], we have that

$$\sum_{i=-\infty}^{\infty} G_{q,\lambda}(x - i) = 1, \ \ \forall\, x \in \mathbb{R},\ \lambda, q > 0,\ A > 1, \tag{8.9}$$

and

$$\sum_{i=-\infty}^{\infty} G_{\frac{1}{q},\lambda}(x - i) = 1, \ \ \forall\, x \in \mathbb{R},\ \lambda, q > 0,\ A > 1. \tag{8.10}$$

Consequently, we derive that

$$\sum_{i=-\infty}^{\infty} W(x - i) = 1, \ \ \forall\, x \in \mathbb{R}. \tag{8.11}$$

By Theorem 16.2, p. 402 of [200], we have that

$$\int_{-\infty}^{\infty} G_{q,\lambda}(x)\, dx = 1, \quad \lambda, q > 0,\ A > 1, \tag{8.12}$$

similarly it holds

$$\int_{-\infty}^{\infty} G_{\frac{1}{q},\lambda}(x)\,dx = 1, \tag{8.13}$$

so that

$$\int_{-\infty}^{\infty} W(x)\,dx = 1, \tag{8.14}$$

therefore $W(x)$ is a density function.

By Theorem 16.3, p. 402 of [200], we have:

Let $0 < \alpha < 1$, and $n \in \mathbb{N}$ with $n^{1-\alpha} > 2$, then

$$\sum_{\substack{k=-\infty \\ : |nx-k| \geq n^{1-\alpha}}}^{\infty} G_{q,\lambda}(nx-k) < 2\max\left\{q, \frac{1}{q}\right\} \frac{1}{A^{\lambda(n^{1-\alpha}-2)}} = \gamma A^{-\lambda(n^{1-\alpha}-2)}, \tag{8.15}$$

where $\lambda, q > 0$, $A > 1$; $\gamma := 2\max\left\{q, \frac{1}{q}\right\}$.

Similarly, we get that

$$\sum_{\substack{k=-\infty \\ : |nx-k| \geq n^{1-\alpha}}}^{\infty} G_{\frac{1}{q},\lambda}(nx-k) < \gamma A^{-\lambda(n^{1-\alpha}-2)}. \tag{8.16}$$

Consequently we obtain that

$$\sum_{\substack{k=-\infty \\ : |nx-k| \geq n^{1-\alpha}}}^{\infty} W(nx-k) < \gamma A^{-\lambda(n^{1-\alpha}-2)}, \tag{8.17}$$

where $\gamma := 2\max\left\{q, \frac{1}{q}\right\}$.

Next, we move on to the multivariate case, see Chapter 15 of [200], pp. 365–394, as a model of action.

We make

Remark 8.1. We introduce

$$Z(x_1, ..., x_N) := Z(x) := \prod_{i=1}^{N} W(x_i) = \frac{1}{2^N} \prod_{i=1}^{N} \left(G_{q,\lambda} + G_{\frac{1}{q},\lambda}\right)(x_i), \tag{8.18}$$

$x = (x_1, ..., x_N) \in \mathbb{R}^N$, $\lambda, q > 0$, $N \in \mathbb{N}$.

Properties:

(i)

$$Z(x) > 0, \ \forall x \in \mathbb{R}^N, \tag{8.19}$$

(ii)

$$\sum_{k=-\infty}^{\infty} Z(x-k) := \sum_{k_1=-\infty}^{\infty} \sum_{k_2=-\infty}^{\infty} \cdots \sum_{k_N=-\infty}^{\infty} Z(x_1-k_1, ..., x_N-k_N) = 1, \tag{8.20}$$

$k := (k_1, ..., k_n) \in \mathbb{Z}^N$, $\forall\ x \in \mathbb{R}^N$,

hence

(iii)

$$\sum_{k=-\infty}^{\infty} Z(nx-k) = 1, \ \forall x \in \mathbb{R}^N; n \in \mathbb{N}, \tag{8.21}$$

(iv)

$$\int_{\mathbb{R}^N} Z(x)\, dx = 1, \tag{8.22}$$

that is Z is a multivariate density function.

Here denote $\|x\|_\infty := \max\{|x_1|, ..., |x_N|\}$, $x \in \mathbb{R}^N$, $\infty := (\infty, ..., \infty)$, $-\infty := (-\infty, ..., -\infty)$.

It is clear that

(v)

$$\sum_{\begin{cases} k = -\infty \\ \left\|\frac{k}{n} - x\right\|_\infty > \frac{1}{n^\beta} \end{cases}}^{\infty} Z(nx-k) < \gamma A^{-\lambda\left(n^{1-\beta}-2\right)}, \tag{8.23}$$

$0 < \beta < 1$, $n \in \mathbb{N} : n^{1-\beta} > 2$, $\gamma := 2\max\left\{q, \frac{1}{q}\right\}$, $x \in \mathbb{R}^N$.

We make

Remark 8.2. Let $x \in [1, \infty)$, and we apply the mean value theorem. We have that

$$G_{q,\lambda}(x) = \frac{1}{2}\left(\varphi_{q,\lambda}(x+1) - \varphi_{q,\lambda}(q-1)\right)$$

$$\frac{1}{2}\varphi'_{q,\lambda}(\xi)\, 2 = \varphi'_{q,\lambda}(\xi) = \frac{q\lambda \ln A}{\left(1+qA^{-\lambda\xi}\right)^2 A^{\lambda\xi}}, \tag{8.24}$$

where $0 \leq x-1 < \xi < x+1$.

We obtain that

$$G_{q,\lambda}(x) < (q\lambda \ln A) A^{-\lambda\xi} < (q\lambda \ln A) A^{-\lambda(x-1)}, \quad x \geq 1. \tag{8.25}$$

Similarly it holds

$$G_{\frac{1}{q},\lambda}(x) < \left(\frac{1}{q}\lambda \ln A\right) A^{-\lambda(x-1)}, \quad x \geq 1. \tag{8.26}$$

And, finally it holds

$$W(x) < \frac{1}{2}\left(q + \frac{1}{q}\right)(\lambda \ln A) A^{-\lambda(x-1)}, \quad x \geq 1, \tag{8.27}$$

where $q, \lambda > 0$, $A > 1$.

We also make

Remark 8.3. We would like to estimate $(n, m \in \mathbb{N})$:

$$\sum_{\begin{cases} k=-\infty \\ : |nx-k| \geq n^{1-\beta} \end{cases}}^{\infty} W(|nx-k|) |nx-k|^m \overset{(8.27)}{\leq}$$

$$\frac{1}{2}\left(q + \frac{1}{q}\right)(\lambda \ln A) \sum_{\begin{cases} k=-\infty \\ : |nx-k| \geq n^{1-\beta} \end{cases}}^{\infty} |nx-k|^m A^{-\lambda(|nx-k|-1)} = \tag{8.28}$$

$$\frac{1}{2}\left(q + \frac{1}{q}\right)(\lambda \ln A) \sum_{\begin{cases} k=-\infty \\ : |nx-k| \geq n^{1-\beta} \end{cases}}^{\infty} |nx-k|^m e^{-\lambda(\ln A)(|nx-k|-1)} =$$

$$\frac{1}{2}\left(q + \frac{1}{q}\right)\lambda(\ln A) A^{\lambda} \sum_{\begin{cases} k=-\infty \\ : |nx-k| \geq n^{1-\beta} \end{cases}}^{\infty} |nx-k|^m e^{-\lambda(\ln A)|nx-k|} =: (*).$$

For convenience set $\mu := \lambda \ln A > 0$.

Notice that

$$e^{\frac{\mu|nx-k|}{2}} = \sum_{\theta=0}^{\infty} \frac{\left(\frac{\mu|nx-k|}{2}\right)^{\theta}}{\theta!} \geq \left(\frac{\mu|nx-k|}{2}\right)^m \frac{1}{m!}. \tag{8.29}$$

Therefore we have

$$\left(\frac{\mu\,|nx-k|}{2}\right)^m \leq m!e^{\frac{\mu|nx-k|}{2}}, \quad \text{or} \tag{8.30}$$

$$(\mu\,|nx-k|)^m \leq 2^m m!e^{\frac{\mu|nx-k|}{2}}, \text{ or} \tag{8.31}$$

$$|nx-k|^m \leq \frac{2^m}{\mu^m} m!e^{\frac{\mu|nx-k|}{2}}. \tag{8.32}$$

Hence it holds

$$(*) \leq \frac{1}{2}\left(q+\frac{1}{q}\right)\lambda\,(\ln A)\,A^\lambda$$

$$\sum_{\begin{cases} k=-\infty \\ : |nx-k| \geq n^{1-\beta} \end{cases}}^{\infty} \frac{2^m}{\lambda^m (\ln A)^m} m!e^{\frac{\lambda \ln A|nx-k|}{2}} e^{-\lambda \ln A|nx-k|} =$$

$$\frac{1}{2}\left(q+\frac{1}{q}\right)\lambda\,(\ln A)\,A^\lambda \frac{2^m}{\lambda^m (\ln A)^m} m! \sum_{\begin{cases} k=-\infty \\ : |nx-k| \geq n^{1-\beta} \end{cases}}^{\infty} e^{-\frac{\lambda \ln A|nx-k|}{2}} = \tag{8.33}$$

$$\frac{1}{2}\left(q+\frac{1}{q}\right)\frac{A^\lambda 2^m m!}{\lambda^{m-1}(\ln A)^{m-1}} \left(\sum_{\begin{cases} k=-\infty \\ : |nx-k| \geq n^{1-\beta} \end{cases}}^{\infty} e^{-\frac{\lambda \ln A}{2}|nx-k|} \right) \leq$$

$$\left(q+\frac{1}{q}\right)\frac{A^\lambda 2^m m!}{\lambda^{m-1}(\ln A)^{m-1}} \left(\int_{n^{1-\beta}-1}^{\infty} e^{-\frac{\lambda \ln A}{2}x} dx\right) =$$

$$\frac{2}{\lambda \ln A}\left(q+\frac{1}{q}\right)\frac{A^\lambda 2^m m!}{\lambda^{m-1}(\ln A)^{m-1}} \left(\int_{n^{1-\beta}-1}^{\infty} e^{-\frac{\lambda \ln A}{2}x} d\left(\frac{\lambda \ln A}{2}\right)x\right) = \tag{8.34}$$

$$\frac{\left(q+\frac{1}{q}\right)A^\lambda 2^{m+1} m!}{\lambda^m (\ln A)^m}\left(\int_{n^{1-\beta}-1}^{\infty} e^{-y} dy\right) =$$

$$\frac{\left(q+\frac{1}{q}\right)A^{\lambda}2^{m+1}m!}{\lambda^{m}(\ln A)^{m}}\left(-e^{-y}|_{n^{1-\beta}-1}^{\infty}\right)=$$

$$\frac{\left(q+\frac{1}{q}\right)A^{\lambda}2^{m+1}m!}{\lambda^{m}(\ln A)^{m}}\left(e^{-y}|_{\infty}^{n^{1-\beta}-1}\right)= \tag{8.35}$$

$$\frac{\left(q+\frac{1}{q}\right)A^{\lambda}2^{m+1}m!}{\lambda^{m}(\ln A)^{m}}\left(e^{-\frac{\lambda\ln A}{2}x}|_{\infty}^{n^{1-\beta}-1}\right)=$$

$$\frac{\left(q+\frac{1}{q}\right)A^{\lambda}2^{m+1}m!}{\lambda^{m}(\ln A)^{m}}\left(e^{-\frac{\lambda\ln A}{2}\left(n^{1-\beta}-1\right)}\right).$$

Thus, we have proved:

$$\sum_{\substack{k=-\infty\\ :|nx-k|\geq n^{1-\beta}}}^{\infty} W(|nx-k|)\,|nx-k|^{m}\leq$$

$$\frac{\left(q+\frac{1}{q}\right)2^{m+1}m!}{\lambda^{m}(\ln A)^{m}}A^{-\frac{\lambda}{2}\left(n^{1-\beta}-3\right)}, \tag{8.36}$$

where $m,n\in\mathbb{N}$; $q,\lambda>0$, $A>1$, $0<\beta<1$; see also [201].

We also make

Remark 8.4. We also estimate ($\gamma:=2\max\left\{q,\frac{1}{q}\right\}$)

$$\sum_{\substack{k=-\infty\\ :|nx-k|\geq n^{1-\beta}}}^{\infty} W(|nx-k|)\,(1+|nx-k|)^{m}\leq$$

$$2^{m-1}\sum_{\substack{k=-\infty\\ :|nx-k|\geq n^{1-\beta}}}^{\infty} W(|nx-k|)\,(1+|nx-k|^{m})= \tag{8.37}$$

$$2^{m-1}\left\{\sum_{\substack{k=-\infty\\ :|nx-k|\geq n^{1-\beta}}}^{\infty} W(|nx-k|)+\right.$$

$$\left.\sum_{\begin{cases} k=-\infty \\ : |nx-k| \geq n^{1-\beta} \end{cases}}^{\infty} W(|nx-k|)\,|nx-k|^{m}\right\} \overset{\text{(by (8.17), (8.36))}}{\leq}$$

$$2^{m-1}\left\{\gamma A^{-\lambda\left(n^{1-\beta}-2\right)}+\left(\frac{\left(q+\frac{1}{q}\right)2^{m+1}m!}{\lambda^{m}(\ln A)^{m}}\right)A^{-\frac{\lambda}{2}\left(n^{1-\beta}-3\right)}\right\}.$$

We have proved that

$$\sum_{\begin{cases} k=-\infty \\ : |nx-k| \geq n^{1-\beta} \end{cases}}^{\infty} W(|nx-k|)\,(1+|nx-k|)^{m} \leq$$

$$2^{m-1}\left\{\gamma A^{-\lambda\left(n^{1-\beta}-2\right)}+\left(\frac{\left(q+\frac{1}{q}\right)2^{m+1}m!}{\lambda^{m}(\ln A)^{m}}\right)A^{-\frac{\lambda}{2}\left(n^{1-\beta}-3\right)}\right\}, \qquad (8.38)$$

where $m, n \in \mathbb{N}$; $q, \lambda > 0$, $A > 1$, $0 < \beta < 1$, $\gamma = 2\max\left\{q, \frac{1}{q}\right\}$.

We give

Theorem 8.1. *Let* $0 < \beta < 1$, *and* $n \in \mathbb{N}$ *with* $n^{1-\beta} > 2$, $m \in \mathbb{N}$; $A > 1$, $q, \lambda > 0$, $\gamma = 2\max\left\{q, \frac{1}{q}\right\}$. *Then*

$$\sum_{\begin{cases} k=-\infty \\ \left\{k\in\mathbb{Z}^{N} : \|nx-k\|_{\infty} \geq n^{1-\beta}\right\} \end{cases}}^{\infty} \|nx-k\|_{\infty}^{m} Z(nx-k) <$$

$$\frac{\left(q+\frac{1}{q}\right)2^{m+1}m!}{\lambda^{m}(\ln A)^{m}}A^{-\frac{\lambda}{2}\left(n^{1-\beta}-3\right)}. \qquad (8.39)$$

Proof. The condition $k \in \mathbb{Z}^{N} : \|nx-k\|_{\infty} \geq n^{1-\beta}$, implies that there exists at least one $k_r \in \mathbb{Z} : |nx_r - k_r| \geq n^{1-\beta}$, where $r \in \{1, ..., N\}$.

Indeed, it is

$$\left\{k \in \mathbb{Z}^{N} : \|nx-k\|_{\infty} \geq n^{1-\beta}\right\} \subset$$

$$\cup_{r=1}^{N}\left\{\mathbb{Z}^{N-1}\cup\left\{k_r\in\mathbb{Z}:|nx_r-k_r|\geq n^{1-\beta}\right\}\right\}\subset$$
$$\mathbb{Z}^{N-1}\cup\left\{k_{r^*}\in\mathbb{Z}:|nx_{r^*}-k_{r^*}|\geq n^{1-\beta}\right\}, \tag{8.40}$$

for some $k^*\in\{1,...,N\}$.

Next, we observe that

$$\sum_{\begin{cases} k=-\infty \\ \left\{k\in\mathbb{Z}^N:\|nx-k\|_\infty\geq n^{1-\beta}\right\}\end{cases}}^{\infty}\|nx-k\|_\infty^m Z(nx-k)\leq$$

$$\sum_{\begin{cases} k_{r^*}=-\infty \\ \left\{k_{r^*}\in\mathbb{Z}^N:|nx_{r^*}-k_{r^*}|\geq n^{1-\beta}\right\}\end{cases}}^{\infty}|nx_{r^*}-k_{r^*}|^m W(|nx_{r^*}-k_{r^*}|)\overset{(8.36)}{<}$$

$$\frac{\left(q+\frac{1}{q}\right)2^{m+1}m!}{\lambda^m(\ln A)^m}A^{-\frac{\lambda}{2}\left(n^{1-\beta}-3\right)}. \tag{8.41}$$

□

We also give

Theorem 8.2. *All as in Theorem 8.1. Then*

$$\sum_{k=-\infty}^{\infty}Z(nx-k)\left(1+\|nx-k\|_\infty\right)^m\leq$$

$$2^{m-1}\left(\gamma A^{-\lambda\left(n^{1-\beta}-2\right)}+\frac{\left(q+\frac{1}{q}\right)2^{m+1}m!}{\lambda^m(\ln A)^m}A^{-\frac{\lambda}{2}\left(n^{1-\beta}-3\right)}\right). \tag{8.42}$$

Proof. We have

$$\sum_{\begin{cases} k=-\infty \\ :\|nx-k\|_\infty\geq n^{1-\beta}\end{cases}}^{\infty}Z(nx-k)\left(1+\|nx-k\|_\infty\right)^m\leq$$

$$2^{m-1}\left(\sum_{\begin{cases} k=-\infty \\ :\|nx-k\|_\infty\geq n^{1-\beta}\end{cases}}^{\infty}Z(nx-k)\left(1+\|nx-k\|_\infty^m\right)\right)= \tag{8.43}$$

$$2^{m-1}\left(\sum_{\begin{cases} k=-\infty \\ :\|nx-k\|_\infty \geq n^{1-\beta}\end{cases}}^{\infty} Z(nx-k)+\right.$$

$$\left.\sum_{\begin{cases} k=-\infty \\ :\|nx-k\|_\infty \geq n^{1-\beta}\end{cases}}^{\infty} Z(nx-k)\|nx-k\|_\infty^m\right) \overset{(\text{by } (8.23),\,(8.39))}{\leq}$$

$$2^{m-1}\left(\gamma A^{-\lambda\left(n^{1-\beta}-2\right)}+\frac{\left(q+\frac{1}{q}\right)2^{m+1}m!}{\lambda^m(\ln A)^m}A^{-\frac{\lambda}{2}\left(n^{1-\beta}-3\right)}\right).$$

□

We need

Definition 8.1. The modulus of continuity here is defined by

$$\omega_1(f,\delta):=\sup_{\substack{x,y\in\mathbb{R}^N: \\ \|x-y\|_\infty<\delta}}|f(x)-f(y)|,\quad \delta>0, \tag{8.44}$$

where $f:\mathbb{R}^N\to\mathbb{R}$ is a bounded and continuous function, denoted by $f\in C_B\left(\mathbb{R}^N\right)$, $N\in\mathbb{N}$. Similarly ω_1 is defined for $f\in C_U\left(\mathbb{R}^N\right)$ (uniformly continuous functions). We have that $f\in C_U\left(\mathbb{R}^N\right)$, iff $\omega_1(f,\delta)\to 0$ as $\delta\downarrow 0$, see also [199].

Denote by $B\left(\mathbb{R}^N\right)$ the bounded functions on $\mathbb{R}^N$.

Next, we describe our neural network operators.

Definition 8.2. Let $f\in C_B\left(\mathbb{R}^N\right)$, we define

$$\widetilde{A}_n(f,x):=\widetilde{A}_n(f,x_1,...,x_N):=\sum_{k=-\infty}^{\infty} f\left(\frac{k}{n}\right)Z(nx-k):= \tag{8.45}$$

$$\sum_{k_1=-\infty}^{\infty}\sum_{k_2=-\infty}^{\infty}...\sum_{k_N=-\infty}^{\infty} f\left(\frac{k_1}{n},\frac{k_2}{n},...,\frac{k_N}{n}\right)\left(\prod_{i=1}^{N}W(nx_i-k_i)\right),$$

$n \in \mathbb{N}$, $\forall\, x \in \mathbb{R}^N$, $N \in \mathbb{N}$, the multivariate quasi-interpolation neural network operator.

Also for $f \in C_B(\mathbb{R}^N)$ we define the multivariate Kantorovich type neural network operator

$$\widetilde{K}_n(f,x) := \widetilde{K}_n(f,x_1,...,x_N) := \sum_{k=-\infty}^{\infty}\left(n^N \int_{\frac{k}{n}}^{\frac{k+1}{n}} f(t)\,dt\right) Z(nx-k) := \tag{8.46}$$

$$\sum_{k_1=-\infty}^{\infty}\sum_{k_2=-\infty}^{\infty}\cdots\sum_{k_N=-\infty}^{\infty}\left(n^N \int_{\frac{k_1}{n}}^{\frac{k_1+1}{n}}\int_{\frac{k_2}{n}}^{\frac{k_2+1}{n}}\cdots\int_{\frac{k_N}{n}}^{\frac{k_N+1}{n}} f(t_1,...,t_N)\,dt_1...dt_N\right)$$

$$\left(\prod_{i=1}^{N} W(nx_i-k_i)\right),$$

$n \in \mathbb{N}$, $\forall\, x \in \mathbb{R}^N$.

Again for $f \in C_B(\mathbb{R}^N)$, $N \in \mathbb{N}$, we define the multivariate neural network operator of quadrature type $\widetilde{Q}_n(f,x)$, $n \in \mathbb{N}$, as follows. Let $\theta = (\theta_1,...,\theta_N) \in \mathbb{N}^N$, $r = (r_1,...,r_N) \in \mathbb{Z}_+^N$, $w_r = w_{r_1r_2...r_N} \geq 0$, such that

$$\sum_{r=0}^{\theta} w_r = \sum_{r_1=0}^{\theta_1}\sum_{r_2=0}^{\theta_2}\cdots\sum_{r_N=0}^{\theta_N} w_{r_1r_2...r_N} = 1; \quad k \in Z^N,$$

and

$$\delta_{nk}(f) := \delta_{n,k_1,k_2,...,k_N}(f) := \sum_{r=0}^{\theta} w_r f\left(\frac{k}{n}+\frac{r}{n\theta}\right) :=$$

$$\sum_{r_1=0}^{\theta_1}\sum_{r_2=0}^{\theta_2}\cdots\sum_{r_N=0}^{\theta_N} w_{r_1r_2...r_N} f\left(\frac{k_1}{n}+\frac{r_1}{n\theta_1}, \frac{k_2}{n}+\frac{r_2}{n\theta_2},...,\frac{k_N}{n}+\frac{r_N}{n\theta_N}\right), \tag{8.47}$$

where $\frac{r}{\theta} := \left(\frac{r_1}{\theta_1},\frac{r_2}{\theta_2},...,\frac{r_N}{\theta_N}\right)$.

We put

$$\widetilde{Q}_n(f,x) := \widetilde{Q}_n(f,x_1,...,x_N) := \sum_{k=-\infty}^{\infty} \delta_{nk}(f)\, Z(nx-k) := \tag{8.48}$$

$$\sum_{k_1=-\infty}^{\infty}\sum_{k_2=-\infty}^{\infty}\cdots\sum_{k_N=-\infty}^{\infty} \delta_{n,k_1,k_2,...,k_N}(f)\left(\prod_{i=1}^{N} W(nx_i-k_i)\right), \quad \forall x \in \mathbb{R}^N.$$

Remark 8.5. We notice that

$$\int_{\frac{k}{n}}^{\frac{k+1}{n}} f(t)\,dt = \int_{\frac{k_1}{n}}^{\frac{k_1+1}{n}} \int_{\frac{k_2}{n}}^{\frac{k_2+1}{n}} \dots \int_{\frac{k_N}{n}}^{\frac{k_N+1}{n}} f(t_1, t_2, ..., t_N)\, dt_1 dt_2 ... dt_N =$$

$$\int_0^{\frac{1}{n}} \int_0^{\frac{1}{n}} \dots \int_0^{\frac{1}{n}} f\left(t_1 + \frac{k_1}{n}, t_2 + \frac{k_2}{n}, ..., t_N + \frac{k_N}{n}\right) dt_1 ... dt_N$$

$$= \int_0^{\frac{1}{n}} f\left(t + \frac{k}{n}\right) dt. \tag{8.49}$$

Thus, it holds

$$\widetilde{K}_n(f, x) = \sum_{k=-\infty}^{\infty} \left(n^N \int_0^{\frac{1}{n}} f\left(t + \frac{k}{n}\right) dt\right) Z(nx - k). \tag{8.50}$$

We do have that $\widetilde{A}_n(1) = \widetilde{K}_n(1) = \widetilde{Q}_n(1) = 1$, $\forall\ n \in \mathbb{N}$.

Denote by $\widetilde{\Theta}_n$ any of $\widetilde{A}_n$, $\widetilde{K}_n$, $\widetilde{Q}_n$ and they are positive linear operators mapping $C_B\left(\mathbb{R}^N\right)$ into $B\left(\mathbb{R}^N\right)$.

8.3 Auxilliary Results

Here let $x_0 \in \mathbb{R}^N$, $0 < \beta < 1$, $n \in \mathbb{N} : n^{1-\beta} > 2$ and $m \in \mathbb{N}$. It follows:

Proposition 8.1. *It holds that*

$$\widetilde{A}_n\left(\|\cdot - x_0\|_\infty^m\right)(x_0) > 0,$$

and

$$0 < \left\|\widetilde{A}_n\left(\|\cdot - x_0\|_\infty^m\right)(x_0)\right\|_\infty \leq \tag{8.51}$$

$$\left(\frac{1}{n^\beta} + \frac{1}{n^m} \frac{\left(q + \frac{1}{q}\right) 2^{m+1} m!}{\lambda^m (\ln A)^m} A^{-\frac{\lambda}{2}\left(n^{1-\beta} - 3\right)}\right) < \infty, \quad (\to 0,\ as\ n \to +\infty).$$

Proof. We have that

$$0 < \widetilde{A}_n\left(\|\cdot - x_0\|_\infty^m\right)(x_0) = \sum_{k=-\infty}^{\infty} \left\|\frac{k}{n} - x_0\right\|_\infty^m Z(nx_0 - k) =$$

$$\sum_{\substack{k = -\infty \\ : \left\|\frac{k}{n} - x_0\right\|_\infty < \frac{1}{n^\beta}}}^{\infty} \left\|\frac{k}{n} - x_0\right\|_\infty^m Z(nx_0 - k) +$$

$$\sum_{\left\{\begin{array}{c} k=-\infty \\ : \left\|\frac{k}{n}-x_0\right\|_\infty \geq \frac{1}{n^\beta} \end{array}\right.}^{\infty} \left\|\frac{k}{n}-x_0\right\|_\infty^m Z\left(nx_0-k\right) \leq \tag{8.52}$$

$$\frac{1}{n^\beta}+\frac{1}{n^m} \sum_{\left\{\begin{array}{c} k=-\infty \\ : \left\|nx_0-k\right\|_\infty \geq n^{1-\beta} \end{array}\right.}^{\infty} \left\|nx_0-k\right\|_\infty^m Z\left(nx_0-k\right) \overset{(8.39)}{\leq}$$

$$\frac{1}{n^\beta}+\frac{1}{n^m}\frac{\left(q+\frac{1}{q}\right)2^{m+1}m!}{\lambda^m\left(\ln A\right)^m}A^{-\frac{\lambda}{2}\left(n^{1-\beta}-3\right)}.$$

Therefore it holds

$$0<\left\|\widetilde{A}_n\left(\left\|\cdot-x_0\right\|_\infty^m\right)\left(x_0\right)\right\|_\infty \leq \tag{8.53}$$

$$\frac{1}{n^\beta}+\frac{1}{n^m}\frac{\left(q+\frac{1}{q}\right)2^{m+1}m!}{\lambda^m\left(\ln A\right)^m}A^{-\frac{\lambda}{2}\left(n^{1-\beta}-3\right)} \to 0, \quad \text{as } n\to+\infty.$$

□

Proposition 8.2. *It holds*

$$0<\widetilde{K}_n\left(\left\|\cdot-x_0\right\|_\infty^m\right)\left(x_0\right) \leq \tag{8.54}$$

$$\left[\left(\frac{1}{n}+\frac{1}{n^\beta}\right)^m+\frac{2^{m-1}}{n^m}\left[\gamma A^{-\lambda\left(n^{1-\beta}-2\right)}+\frac{\left(q+\frac{1}{q}\right)2^{m+1}m!}{\lambda^m\left(\ln A\right)^m}A^{-\frac{\lambda}{2}\left(n^{1-\beta}-3\right)}\right]\right]<\infty,$$

$\forall\ x_0 \in \mathbb{R}^N$ *($\to 0$, as $n \to +\infty$).*

That is

$$0<\left\|\widetilde{K}_n\left(\left\|\cdot-x_0\right\|_\infty^m\right)\left(x_0\right)\right\|_\infty \leq \tag{8.55}$$

$$\left[\left(\frac{1}{n}+\frac{1}{n^\beta}\right)^m+\frac{2^{m-1}}{n^m}\left[\gamma A^{-\lambda\left(n^{1-\beta}-2\right)}+\frac{\left(q+\frac{1}{q}\right)2^{m+1}m!}{\lambda^m\left(\ln A\right)^m}A^{-\frac{\lambda}{2}\left(n^{1-\beta}-3\right)}\right]\right]<\infty,$$

($\to 0$, as $n \to +\infty$).

Proof. We have that

$$0<\widetilde{K}_n\left(\|\cdot-x_0\|_\infty^m\right)(x_0)=\sum_{k=-\infty}^{\infty}\left(n^N\int_0^{\frac{1}{n}}\left\|t+\frac{k}{n}-x_0\right\|_\infty^m dt\right)Z(nx_0-k)\leq$$

$$\sum_{k=-\infty}^{\infty}\left(n^N\int_0^{\frac{1}{n}}\left(\left\|\frac{k}{n}-x_0\right\|_\infty+\|t\|_\infty\right)^m dt\right)Z(nx_0-k)\leq$$

$$\sum_{k=-\infty}^{\infty}Z(nx_0-k)\left(\left\|\frac{k}{n}-x_0\right\|_\infty+\frac{1}{n}\right)^m=$$

$$\sum_{\left\{\begin{array}{c}k=-\infty\\ :\left\|\frac{k}{n}-x_0\right\|_\infty<\frac{1}{n^\beta}\end{array}\right.}^{\infty}Z(nx_0-k)\left(\left\|\frac{k}{n}-x_0\right\|_\infty+\frac{1}{n}\right)^m+\tag{8.56}$$

$$\sum_{\left\{\begin{array}{c}k=-\infty\\ :\left\|\frac{k}{n}-x_0\right\|_\infty\geq\frac{1}{n^\beta}\end{array}\right.}^{\infty}Z(nx_0-k)\left(\left\|\frac{k}{n}-x_0\right\|_\infty+\frac{1}{n}\right)^m\leq$$

$$\left(\frac{1}{n^\beta}+\frac{1}{n}\right)^m+\frac{1}{n^m}\left(\sum_{\left\{\begin{array}{c}k=-\infty\\ :\|nx_0-k\|_\infty\geq n^{1-\beta}\end{array}\right.}^{\infty}Z(nx_0-k)\left(\|nx_0-k\|_\infty+1\right)^m\right)\overset{(8.42)}{\leq}$$

$$\left(\frac{1}{n^\beta}+\frac{1}{n}\right)^m+\frac{2^{m-1}}{n^m}\left(\gamma A^{-\lambda\left(n^{1-\beta}-2\right)}+\frac{\left(q+\frac{1}{q}\right)2^{m+1}m!}{\lambda^m(\ln A)^m}A^{-\frac{\lambda}{2}\left(n^{1-\beta}-3\right)}\right).\tag{8.57}$$

□

Proposition 8.3. *It holds that*

$$0<\widetilde{Q}_n\left(\|\cdot-x_0\|_\infty^m\right)(x_0)\leq\tag{8.58}$$

$$\left(\frac{1}{n}+\frac{1}{n^\beta}\right)^m+\frac{2^{m-1}}{n^m}\left[\gamma A^{-\lambda\left(n^{1-\beta}-2\right)}+\frac{\left(q+\frac{1}{q}\right)2^{m+1}m!}{\lambda^m(\ln A)^m}A^{-\frac{\lambda}{2}\left(n^{1-\beta}-3\right)}\right]<\infty,$$

$\forall\ x_0\in\mathbb{R}^N$, *($\to 0$, as $n\to+\infty$).*

Hence we get

$$0 < \left\| \widetilde{Q}_n \left(\|\cdot - x_0\|_\infty^m \right) (x_0) \right\|_\infty \leq \tag{8.59}$$

$$\left(\frac{1}{n} + \frac{1}{n^\beta} \right)^m + \frac{2^{m-1}}{n^m} \left[\gamma A^{-\lambda\left(n^{1-\beta}-2\right)} + \frac{\left(q+\frac{1}{q}\right) 2^{m+1} m!}{\lambda^m (\ln A)^m} A^{-\frac{\lambda}{2}\left(n^{1-\beta}-3\right)} \right] < \infty,$$

($\to 0$, as $n \to +\infty$).

Proof. We notice that

$$\widetilde{Q}_n \left(\|\cdot - x_0\|_\infty^m \right) (x_0) = \sum_{k=-\infty}^{\infty} \left(\sum_{r=0}^{\theta} w_r \left\| \frac{k}{n} + \frac{r}{n\theta} - x_0 \right\|_\infty^m \right) Z(nx_0 - k) \leq$$

$$\sum_{k=-\infty}^{\infty} \left(\sum_{r=0}^{\theta} w_r \left(\left\| \frac{k}{n} - x_0 \right\|_\infty + \left\| \frac{r}{n\theta} \right\|_\infty \right)^m \right) Z(nx_0 - k) \leq \tag{8.60}$$

$$\sum_{k=-\infty}^{\infty} \left(\left\| \frac{k}{n} - x_0 \right\|_\infty + \frac{1}{n} \right)^m Z(nx_0 - k)$$

(as earlier in Proposition 8.2)

$$\leq \left(\frac{1}{n} + \frac{1}{n^\beta} \right)^m + \frac{2^{m-1}}{n^m} \left[\gamma A^{-\lambda\left(n^{1-\beta}-2\right)} + \frac{\left(q+\frac{1}{q}\right) 2^{m+1} m!}{\lambda^m (\ln A)^m} A^{-\frac{\lambda}{2}\left(n^{1-\beta}-3\right)} \right] < \infty. \tag{8.61}$$

□

8.4 Main Results

Next we present approximation results.

Theorem 8.3. *Let $0 < \beta < 1$, $n \in \mathbb{N} : n^{1-\beta} > 2$, $f \in C_B\left(\mathbb{R}^N\right)$, $N \in \mathbb{N}$. It holds*

$$\left\| \widetilde{A}_n (f) - f \right\|_\infty \leq$$

$$2\omega_1 \left(f, \left(\frac{1}{n^\beta} + \frac{1}{n} \frac{\left(q+\frac{1}{q}\right) 4}{\lambda \ln A} A^{-\frac{\lambda}{2}\left(n^{1-\beta}-3\right)} \right) \right) < \infty. \tag{8.62}$$

Given that f is uniformly continuous we obtain that $\widetilde{A}_n (f) \to f$, pointwise and uniformly, as $n \to +\infty$.

Proof. We can write

$$\widetilde{A}_n(f,x) = \int_{\mathbb{R}^N} f(t)\, d\mu_x(t), \tag{8.63}$$

where $\mu_x(\mathbb{R}^N) = 1$, μ_x is a probability measure on $\mathbb{R}^N$. The support of μ_x is discrete and countable, at the points $\frac{k}{n} = \left(\frac{k_1}{n}, \frac{k_2}{n}, ..., \frac{k_N}{n}\right) \in \mathbb{R}^N$, $k_i \in \mathbb{Z}$, $i = 1, ..., N$.

Each $\frac{k}{n}$ carries the weight $Z(nx - k) = \prod_{i=1}^{N} W(nx_i - k_i)$.

Indeed here we have that

$$\sum_{k=-\infty}^{\infty} Z(nx-k) = \sum_{k_1=-\infty}^{\infty} \sum_{k_2=-\infty}^{\infty} \cdots \sum_{k_N=-\infty}^{\infty} \left(\prod_{i=1}^{N} W(nx_i - k_i)\right) = 1. \tag{8.64}$$

Here it is $\omega_1(f,h) < \infty$, $h > 0$.

By Lemma 7.1.1, p. 208, [197], we get that

$$|f(t) - f(x)| \leq \omega_1(f,h) \left\lceil \frac{\|t-x\|_\infty}{h} \right\rceil, \tag{8.65}$$

where $\lceil \cdot \rceil$ is the ceiling function.

But it holds

$$\left\lceil \frac{\|t-x\|_\infty}{h} \right\rceil \leq 1 + \frac{\|t-x\|_\infty}{h}. \tag{8.66}$$

Thus we get

$$|f(t) - f(x)| \leq \omega_1(f,h) \left[1 + \frac{\|t-x\|_\infty}{h}\right], \quad \forall\, t \in \mathbb{R}^N. \tag{8.67}$$

The last can be written also as

$$|f(\cdot) - f(x)| \leq \omega_1(f,h) \left[1 + \frac{\|\cdot - x\|_\infty}{h}\right], \quad \text{over } \mathbb{R}^N. \tag{8.68}$$

We have that

$$\widetilde{A}_n(f,x) - f(x) = \int_{\mathbb{R}^N} f(t)\, d\mu_x(t) - f(x) = \int_{\mathbb{R}^N} (f(t) - f(x))\, d\mu_x(t). \tag{8.69}$$

Therefore

$$\left|\widetilde{A}_n(f,x) - f(x)\right| \leq \left|\int_{\mathbb{R}^N} (f(t) - f(x))\, d\mu_x(t)\right| \leq$$

$$\int_{\mathbb{R}^N} |f(t) - f(x)|\, d\mu_x(t) \leq \omega_1(f,h) \left[1 + \frac{\int_{\mathbb{R}^N} \|t-x\|_\infty\, d\mu_x(t)}{h}\right]. \tag{8.70}$$

Choosing $h := \int_{\mathbb{R}^N} \|t - x\|_\infty d\mu_x(t)$, we obtain

$$\left|\widetilde{A}_n(f,x) - f(x)\right| \leq 2\omega_1\left(f, \int_{\mathbb{R}^N} \|t - x\|_\infty d\mu_x(t)\right) =$$

$$2\omega_1\left(f, \widetilde{A}_n(\|\cdot - x\|_\infty)(x)\right). \tag{8.71}$$

But we have $(0 < \beta < 1)$

$$0 < \widetilde{A}_n(\|\cdot - x\|_\infty)(x) = \sum_{k=-\infty}^{\infty} \left\|\frac{k}{n} - x\right\|_\infty Z(nx - k) =$$

$$\sum_{\substack{k=-\infty \\ : \left\|\frac{k}{n} - x\right\|_\infty < \frac{1}{n^\beta}}}^{\infty} \left\|\frac{k}{n} - x\right\|_\infty Z(nx-k) +$$

$$\sum_{\substack{k=-\infty \\ : \left\|\frac{k}{n} - x\right\|_\infty \geq \frac{1}{n^\beta}}}^{\infty} \left\|\frac{k}{n} - x\right\|_\infty Z(nx-k) \leq \tag{8.72}$$

$$\frac{1}{n^\beta} + \frac{1}{n} \sum_{\substack{k=-\infty \\ : \|nx - k\|_\infty \geq n^{1-\beta}}}^{\infty} \|nx - k\|_\infty Z(nx-k) \overset{(8.39)}{\leq}$$

$$\frac{1}{n^\beta} + \frac{1}{n}\frac{\left(q + \frac{1}{q}\right)4}{\lambda \ln A} A^{-\frac{\lambda}{2}\left(n^{1-\beta} - 3\right)}.$$

So that it holds

$$\widetilde{A}_n(\|\cdot - x\|_\infty)(x) \leq \tag{8.73}$$

$$\frac{1}{n^\beta} + \frac{1}{n}\frac{\left(q + \frac{1}{q}\right)4}{\lambda \ln A} A^{-\frac{\lambda}{2}\left(n^{1-\beta} - 3\right)} (< +\infty) \to 0, \quad \text{as } n \to +\infty.$$

We have proved

$$\left|\widetilde{A}_n(f,x) - f(x)\right| \leq 2\omega_1\left(f, \left(\frac{1}{n^\beta} + \frac{1}{n}\frac{\left(q + \frac{1}{q}\right)4}{\lambda \ln A} A^{-\frac{\lambda}{2}\left(n^{1-\beta} - 3\right)}\right)\right) < \infty. \tag{8.74}$$

□

Next comes a convexity approximation result.

Theorem 8.4. *Let $f \in C_B(\mathbb{R}^N) : |f(\cdot) - f(x_0)|$ is convex, $x_0 \in \mathbb{R}^N$. Then*

$$\left|\widetilde{A}_n(f, x_0) - f(x_0)\right| \leq \tag{8.75}$$

$$\omega_1\left(f, \left(\frac{1}{n^\beta} + \frac{1}{n}\frac{\left(q + \frac{1}{q}\right)4}{\lambda \ln A} A^{-\frac{\lambda}{2}\left(n^{1-\beta}-3\right)}\right)\right) < +\infty,$$

($\to 0$, as $n \to \infty$, given that f is uniformly continuous, i.e. $\lim_{n\to+\infty} \widetilde{A}_n(f, x_0) = f(x_0)$).

Proof. Based on Lemma 8.1.1, p. 243, [197], we obtain:

Let $(V, \|\cdot\|)$ be a real normed vector space and $f : V \to \mathbb{R}$ such that $|f(t) - f(x_0)|$ is convex in $t \in V$; $x_0 \in V$. Then

$$|f(t) - f(x_0)| \leq \frac{\omega_1(f, h)}{h}\|t - x_0\|, \quad \forall\, t \in V,\ h > 0. \tag{8.76}$$

That is, it holds here

$$|f(\cdot) - f(x_0)| \leq \frac{\omega_1(f, h)}{h}\|\cdot - x_0\|_\infty, \text{ over } \mathbb{R}^N. \tag{8.77}$$

We have that

$$\widetilde{A}_n(f, x_0) - f(x_0) = \int_{\mathbb{R}^N} (f(t) - f(x_0))\, d\mu_{x_0}(t), \tag{8.78}$$

and

$$\left|\widetilde{A}_n(f, x_0) - f(x_0)\right| \leq \int_{\mathbb{R}^N} |f(t) - f(x_0)|\, d\mu_{x_0}(t) \leq \tag{8.79}$$

$$\frac{\omega_1(f, h)}{h}\int_{\mathbb{R}^N} \|t - x_0\|_\infty\, d\mu_{x_0}(t)$$

(we take $h := \int_{\mathbb{R}^N} \|t - x_0\|_\infty\, d\mu_{x_0}(t) = \widetilde{A}_n(\|\cdot - x_0\|_\infty)(x_0) > 0$)

$$= \omega_1\left(f, \int_{\mathbb{R}^N} \|t - x_0\|_\infty\, d\mu_{x_0}(t)\right)$$

$$= \omega_1\left(f, \widetilde{A}_n(\|\cdot - x_0\|_\infty)(x_0)\right). \tag{8.80}$$

Thus, it holds

$$\left|\widetilde{A}_n(f, x_0) - f(x_0)\right| \leq \omega_1\left(f, \widetilde{A}_n(\|\cdot - x_0\|_\infty)(x_0)\right). \tag{8.81}$$

Inequality (8.81) is sharp, namely it is attained by $f(t) = \|t - x_0\|_\infty$, $\forall$ $t \in \mathbb{R}^N$, $x_0 \in \mathbb{R}^N$.

We have

$$\left|\widetilde{A}_n(f, x_0) - f(x_0)\right| = \widetilde{A}_n\left(\|\cdot - x_0\|_\infty\right)(x_0). \tag{8.82}$$

Next we observe

$$\omega_1\left(f, \widetilde{A}_n\left(\|\cdot - x_0\|_\infty\right)(x_0)\right) = \omega_1\left(\|\cdot - x_0\|_\infty, \widetilde{A}_n\left(\|\cdot - x_0\|_\infty\right)(x_0)\right) =$$

$$\sup_{\left\{\begin{array}{l} t_1, t_2 \in \mathbb{R}^N : \\ : \|t_1 - t_2\|_\infty \leq \widetilde{A}_n\left(\|\cdot - x_0\|_\infty\right)(x_0) \end{array}\right.} \left|\|t_1 - x_0\|_\infty - \|t_2 - x_0\|_\infty\right| \leq \tag{8.83}$$

$$\sup_{\left\{\begin{array}{l} t_1, t_2 \in \mathbb{R}^N : \\ : \|t_1 - t_2\|_\infty \leq \widetilde{A}_n\left(\|\cdot - x_0\|_\infty\right)(x_0) \end{array}\right.} \|t_1 - t_2\|_\infty = \widetilde{A}_n\left(\|\cdot - x_0\|_\infty\right)(x_0).$$

So that

$$\omega_1\left(f, \widetilde{A}_n\left(\|\cdot - x_0\|_\infty\right)(x_0)\right) = \widetilde{A}_n\left(\|\cdot - x_0\|_\infty\right)(x_0).$$

Thus (8.81) it is attained.

Continuing on (8.81) we have that

$$\omega_1\left(f, \widetilde{A}_n\left(\|\cdot - x_0\|_\infty\right)(x_0)\right) \leq \tag{8.84}$$

$$\omega_1\left(f, \left(\frac{1}{n^\beta} + \frac{1}{n}\frac{\left(q + \frac{1}{q}\right)4}{\lambda \ln A} A^{-\frac{\lambda}{2}\left(n^{1-\beta} - 3\right)}\right)\right).$$

The theorem is proved. □

From [197], pp. 232–233 we mention

Theorem 8.5. *Let μ be a finite measure of mass $m > 0$ on $M \neq \varnothing$, which is a convex subset of the Banach space $(V, \|\cdot\|)$ and $x_0 \in M$ be given.*

Assume that

$$\left(\frac{1}{m}\int_M \|x - x_0\|^{n+1} \mu(dx)\right)^{\frac{1}{(n+1)}} := D_{n+1}(x_0) > 0, \quad n \in \mathbb{N}. \tag{8.85}$$

Let $f : M \to \mathbb{R}$ be a bounded continuous function, whose Fréchet derivatives $f^{(j)}$, $1 \leq j \leq n$ exist, are bounded and continuous functions. Assume that $\omega_1\left(f^{(n)}, rD_{n+1}(x_0)\right) \leq w$, where $r, w > 0$ are given. Then

$$\left|\int_M f d\mu - f(x_0)\right| \leq |f(x_0)|\,|m - 1| + \tag{8.86}$$

$$\left|\sum_{j=1}^{n}\frac{1}{j!}\int_{M}f^{(j)}(x_{0})(x-x_{0})^{j}\mu(dx)\right|+$$

$$\frac{mw}{rn!}\left[\frac{nr^{2}}{8}+\frac{r}{2}+\frac{1}{(n+1)}\right]D_{n+1}^{n}(x_{0}).$$

Corollary 8.1. *When* $\mu(M)=1$ *and* $\left(\int_{M}\|x-x_{0}\|^{2}\mu(dx)\right)^{\frac{1}{2}}:=D_{2}(x_{0})>0$, *it holds:*

$$\left|\int_{M}fd\mu-f(x_{0})\right|-\left|\int_{M}f^{(1)}(x_{0})(x-x_{0})\mu(dx)\right|\leq \tag{8.87}$$

$$(1.5625)\,\omega_{1}\left(f^{(1)},\frac{1}{2}D_{2}(x_{0})\right)D_{2}(x_{0}).$$

Proposition 8.4. *Let* μ *be a finite measure of mass* $m>0$ *on* $M\neq\varnothing$, *which is a convex subset of the Banach space* $(V,\|\cdot\|)$ *and* $x_{0}\in M$ *be given. Assume that*

$$\left(\int_{M}\|x-x_{0}\|^{2}\mu(dx)\right)^{\frac{1}{2}}:=D_{2}(x_{0})>0 \tag{8.88}$$

and consider $r>0$.

Let $f:M\to\mathbb{R}$ *be a bounded continuous function, whose Fréchet derivative* $f^{(1)}$ *exists, is bounded and continuous.*

Then

$$\left|\int_{M}fd\mu-f(x_{0})\right|-|f(x_{0})|\,|m-1|-$$

$$\left|\int_{M}f^{(1)}(x_{0})(x-x_{0})\mu(dx)\right|\leq \tag{8.89}$$

$$\frac{1}{8r}\left(2+\sqrt{m}r\right)^{2}\omega_{1}\left(f',rD_{2}(x_{0})\right)D_{2}(x_{0}).$$

Corollary 8.2. *Let* μ *be a finite measure of mass* $m>0$ *on* M, *which is a convex subset of the Banach space* $(V,\|\cdot\|)$ *and* $x_{0}\in M$ *be given. Assume that* $\left(\int_{M}\|x-x_{0}\|^{2}\mu(dx)\right)^{\frac{1}{2}}:=D_{2}(x_{0})>0$. *Consider* $r>0$, *let* $f:M\to\mathbb{R}$ *be a bounded and continuous function, whose Fréchet derivative* $f^{(1)}$ *exists, is bounded and continuous. Then*

$$\left|\int_{M}fd\mu-f(x_{0})\right|-|f(x_{0})|\,|m-1|-$$

$$\left|\int_M f^{(1)}(x_0)(x-x_0)\mu(dx)\right| \leq$$

$$\begin{cases} \frac{1}{8r}\left(2+\sqrt{m}r\right)^2 \omega_1\left(f^{(1)}, rD_2(x_0)\right) D_2(x_0), & \text{if } r \leq \frac{2}{\sqrt{m}}, \\ \sqrt{m}\omega_1\left(f^{(1)}, rD_2(x_0)\right) D_2(x_0), & \text{if } r > \frac{2}{\sqrt{m}}. \end{cases} \tag{8.90}$$

By Theorem 8.5 we obtain:

Theorem 8.6. *Let $f \in C_B\left(\mathbb{R}^N\right)$, whose Fréchet derivatives $f^{(j)}$, $1 \leq j \leq m$, $m \in \mathbb{N}$, exist, are bounded and continuous, $r > 0$, $x_0 \in \mathbb{R}^N$.*

Denote

$$0 < \left(\widetilde{A}_n\left(\|\cdot - x_0\|^{m+1}\right)(x_0)\right)^{\frac{1}{(m+1)}} := D_{m+1}(x_0). \tag{8.91}$$

Assume that

$$w := \omega_1\left(f^{(m)}, rD_{m+1}(x_0)\right) > 0. \tag{8.92}$$

Then

$$\left|\widetilde{A}_n(f, x_0) - f(x_0)\right| \leq \left|\sum_{j=1}^{m} \frac{1}{j!}\widetilde{A}_n\left(f^{(j)}(x_0)(\cdot - x_0)^j\right)(x_0)\right| + \tag{8.93}$$

$$\frac{w}{rm!}\left[\frac{mr^2}{8} + \frac{r}{2} + \frac{1}{(m+1)}\right] D_{m+1}^m(x_0).$$

By Corollary 8.1 we get:

Corollary 8.3. *Denote*

$$D_2(x_0) = \left(\widetilde{A}_n\left(\|\cdot - x_0\|^2\right)(x_0)\right)^{\frac{1}{2}} \tag{8.94}$$

which is positive, $x_0 \in \mathbb{R}^N$.

Assume that

$$\omega_1\left(f^{(1)}, \frac{1}{2}D_2(x_0)\right) > 0. \tag{8.95}$$

Then

$$\left|\widetilde{A}_n(f, x_0) - f(x_0)\right| - \left|\widetilde{A}_n\left(f^{(1)}(x_0)(\cdot - x_0)\right)(x_0)\right| \leq$$

$$(1.5625)\,\omega_1\left(f^{(1)}, \frac{1}{2}D_2(x_0)\right) D_2(x_0). \tag{8.96}$$

By Proposition 8.4 we get:

Proposition 8.5. *Let $f \in C_B\left(\mathbb{R}^N\right)$, whose Fréchet derivative $f^{(1)}$ exists, is bounded and continuous, $r > 0$, $x_0 \in \mathbb{R}^N$.*

Denote

$$D_2\left(x_0\right) = \left(\widetilde{A}_n\left(\left\|\cdot - x_0\right\|^2\right)\left(x_0\right)\right)^{\frac{1}{2}}, \tag{8.97}$$

which is positive.

Then

$$\left|\widetilde{A}_n\left(f, x_0\right) - f\left(x_0\right)\right| - \left|\widetilde{A}_n\left(f^{(1)}\left(x_0\right)\left(\cdot - x_0\right)\right)\left(x_0\right)\right| \leq$$

$$\frac{1}{8r}\left(2+r\right)^2 \omega_1\left(f', rD_2\left(x_0\right)\right) D_2\left(x_0\right). \tag{8.98}$$

By Corollary 8.2 we have:

Corollary 8.4. *Here all as in Proposition 8.5. Then*

$$\left|\widetilde{A}_n\left(f, x_0\right) - f\left(x_0\right)\right| - \left|\widetilde{A}_n\left(f^{(1)}\left(x_0\right)\left(\cdot - x_0\right)\right)\left(x_0\right)\right| \leq$$

$$\begin{cases} \frac{1}{8r}\left(2+r\right)^2 \omega_1\left(f^{(1)}, rD_2\left(x_0\right)\right) D_2\left(x_0\right), & \text{if } r \leq 2, \\ \omega_1\left(f^{(1)}, rD_2\left(x_0\right)\right) D_2\left(x_0\right), & \text{if } r > 2. \end{cases} \tag{8.99}$$

We make

Remark 8.6. We have that

$$D_{m+1}\left(x_0\right) = \left(\widetilde{A}_n\left(\left\|\cdot - x_0\right\|^{m+1}\right)\left(x_0\right)\right)^{\frac{1}{(m+1)}} \overset{(8.51)}{\leq}$$

$$\left(\frac{1}{n^\beta} + \frac{1}{n^{m+1}} \frac{\left(q+\frac{1}{q}\right) 2^{m+2}\left(m+1\right)!}{\lambda^{m+1}\left(\ln A\right)^{m+1}} A^{-\frac{\lambda}{2}\left(n^{1-\beta}-3\right)}\right)^{\frac{1}{(m+1)}} \to 0, \quad \text{as } n \to +\infty. \tag{8.100}$$

And, furthermore we get

$$D_2\left(x_0\right) = \left(\widetilde{A}_n\left(\left\|\cdot - x_0\right\|^2\right)\left(x_0\right)\right)^{\frac{1}{2}} \leq$$

$$\sqrt{\left(\frac{1}{n^\beta} + \frac{1}{n^2} \frac{\left(q+\frac{1}{q}\right) 16}{\lambda^2\left(\ln A\right)^2} A^{-\frac{\lambda}{2}\left(n^{1-\beta}-3\right)}\right)} \to 0, \quad \text{as } n \to +\infty. \tag{8.101}$$

We make

Remark 8.7. On Theorem 8.6:

Assume that $f^{(j)}(x_0) = 0$, $j = 1, ..., m$.

Then

$$\left|\widetilde{A}_n(f, x_0) - f(x_0)\right| \leq$$

$$\frac{\omega_1\left(f^{(m)}, r\left(\frac{1}{n^\beta} + \frac{1}{n^{m+1}}\frac{\left(q+\frac{1}{q}\right)2^{m+2}(m+1)!}{\lambda^{m+1}(\ln A)^{m+1}}A^{-\frac{\lambda}{2}\left(n^{1-\beta}-3\right)}\right)^{\frac{1}{(m+1)}}\right)}{rm!}$$

$$\left[\frac{mr^2}{8} + \frac{r}{2} + \frac{1}{(m+1)}\right] \tag{8.102}$$

$$\left(\frac{1}{n^\beta} + \frac{1}{n^{m+1}}\frac{\left(q+\frac{1}{q}\right)2^{m+2}(m+1)!}{\lambda^{m+1}(\ln A)^{m+1}}A^{-\frac{\lambda}{2}\left(n^{1-\beta}-3\right)}\right)^{\frac{m}{(m+1)}} \to 0, \quad \text{as } n \to +\infty.$$

That is $\lim_{n\to+\infty} \widetilde{A}_n(f, x_0) = f(x_0)$.

Case $m = 1$ in (8.102):

Assume $f^{(1)}(x_0) = 0$, then

$$\left|\widetilde{A}_n(f, x_0) - f(x_0)\right| \leq$$

$$\frac{1}{r}\omega_1\left(f^{(1)}, r\left(\frac{1}{n^\beta} + \frac{1}{n^2}\frac{\left(q+\frac{1}{q}\right)16}{\lambda^2(\ln A)^2}A^{-\frac{\lambda}{2}\left(n^{1-\beta}-3\right)}\right)^{\frac{1}{2}}\right)$$

$$\left[\frac{r^2}{8} + \frac{r}{2} + \frac{1}{2}\right] \tag{8.103}$$

$$\left(\frac{1}{n^\beta} + \frac{1}{n^2}\frac{\left(q+\frac{1}{q}\right)16}{\lambda^2(\ln A)^2}A^{-\frac{\lambda}{2}\left(n^{1-\beta}-3\right)}\right)^{\frac{1}{2}} \to 0, \quad \text{as } n \to +\infty;$$

implying again $\lim_{n\to+\infty} \widetilde{A}_n(f, x_0) = f(x_0)$.

We continue with general results and their applications.

Theorem 8.7. *Let* $(L_n)_{n\in\mathbb{N}}$ *be a sequence of positive linear operators from* $C_B(\mathbb{R}^N)$ *into* $B(\mathbb{R}^N)$, *such that* $L_n(1) = 1$. *Assume* $0 < L_n(\|\cdot - x_0\|_\infty)(x_0) < +\infty$.

Then

$$|L_n(f)(x_0) - f(x_0)| \leq 2\omega_1(f, L_n(\|\cdot - x_0\|_\infty)(x_0)) < +\infty. \tag{8.104}$$

Given that $\lim_{n\to+\infty} L_n(\|\cdot - x_0\|_\infty)(x_0) = 0$ *and* f *is uniformly continuous, we get that* $\lim_{n\to+\infty} L_n(f)(x_0) = f(x_0)$.

Proof. We have that $\omega_1(f,h) < \infty$, $h > 0$.

We got by (8.77) that

$$|f(\cdot) - f(x_0)| \leq \omega_1(f,h)\left[1 + \frac{\|\cdot - x_0\|_\infty}{h}\right], \quad \text{over } \mathbb{R}^N. \tag{8.105}$$

Hence we derive

$$|L_n(f)(x_0) - f(x_0)| = |L_n(f(\cdot) - f(x_0))(x_0)| \leq$$

$$L_n(|f(\cdot) - f(x_0)|)(x_0) \overset{(8.105)}{\leq} \omega_1(f,h)\left[1 + \frac{L_n(\|\cdot - x_0\|_\infty)(x_0)}{h}\right] \tag{8.106}$$

(choose $h := L_n(\|\cdot - x_0\|_\infty)(x_0) > 0$)

$$= 2\omega_1(f, L_n(\|\cdot - x_0\|_\infty)(x_0)).$$

□

Next we apply (8.104).

Theorem 8.8. *Let* $n \in \mathbb{N} : n^{1-\beta} > 2$, $0 < \beta < 1$, $x_0 \in \mathbb{R}^N$. *Then*

$$\left|\widetilde{K}_n(f)(x_0) - f(x_0)\right| \leq 2\omega_1\left(f, \widetilde{K}_n(\|\cdot - x_0\|_\infty)(x_0)\right) < +\infty, \tag{8.107}$$

and

$$\left|\widetilde{Q}_n(f)(x_0) - f(x_0)\right| \leq 2\omega_1\left(f, \widetilde{Q}_n(\|\cdot - x_0\|_\infty)(x_0)\right) < +\infty, \tag{8.108}$$

and

$$\left\{\begin{array}{l}\left\|\widetilde{K}_n(f) - f\right\|_\infty \\ \left\|\widetilde{Q}_n(f) - f\right\|_\infty\end{array}\right\} \leq \tag{8.109}$$

$$2\omega_1\left(f, \left[\left(\frac{1}{n} + \frac{1}{n^\beta}\right) + \frac{1}{n}\left[\gamma A^{-\lambda\left(n^{1-\beta}-2\right)} + \frac{\left(q + \frac{1}{q}\right)4}{\lambda \ln A} A^{-\frac{\lambda}{2}\left(n^{1-\beta}-3\right)}\right]\right]\right) < +\infty.$$

Given that $f \in C_U(\mathbb{R}^N)$ *we derive that* $\widetilde{K}_n(f) \to f$, $\widetilde{Q}_n(f) \to f$, *as* $n \to +\infty$, *pointwise and uniformly.*

Proof. Along with (8.104) we use of Propositions 8.2, 8.3. □

We continue with the convexity case.

Remark 8.8. Let $f \in C_B(\mathbb{R}^N)$ and $|f(\cdot) - f(x_0)|$ is convex over $\mathbb{R}^N$. Then, by (8.77), we have

$$|f(\cdot) - f(x_0)| \leq \frac{\omega_1(f,h)}{h} \|\cdot - x_0\|_\infty, \tag{8.110}$$

over $\mathbb{R}^N$, $x_0 \in \mathbb{R}^N$.

Acting as in Theorem 8.7 we derive:

Theorem 8.9. *Let $(L_n)_{n\in\mathbb{N}}$ positive linear operators from $C_B(\mathbb{R}^N)$ into $B(\mathbb{R}^N)$: $L_n(1) = 1$. Let $f \in C_B(\mathbb{R}^N)$ and the convex function $|f(\cdot) - f(x_0)|$ over $\mathbb{R}^N$, $x_0 \in \mathbb{R}^N$. Assume that $0 < L_n(\|\cdot - x_0\|_\infty)(x_0) < +\infty$.*

Then

$$|L_n(f)(x_0) - f(x_0)| \leq \omega_1(f, L_n(\|\cdot - x_0\|_\infty)(x_0)). \tag{8.111}$$

The last (8.111) is attained by $f(t) = \|t - x_0\|_\infty$, $\forall\, t \in \mathbb{R}^N$, $x_0 \in \mathbb{R}^N$, that is sharp. Given that f is uniformly continuous and $L_n(\|\cdot - x_0\|_\infty)(x_0) \to 0$, as $n \to \infty$, then $L_n(f)(x_0) \to f(x_0)$, as $n \to \infty$.

Next we apply Theorem 8.9.

Theorem 8.10. *Let $f \in C_B(\mathbb{R}^N)$: $|f(\cdot) - f(x_0)|$ is convex, $x_0 \in \mathbb{R}^N$. Then*

$$\left\{ \begin{vmatrix} \widetilde{K}_n(f,x_0) - f(x_0) \\ \widetilde{Q}_n(f,x_0) - f(x_0) \end{vmatrix} \right\} \leq \tag{8.112}$$

$$\omega_1\left(f, \left[\left(\frac{1}{n} + \frac{1}{n^\beta}\right) + \frac{1}{n}\left[\gamma A^{-\lambda(n^{1-\beta}-2)} + \frac{\left(q+\frac{1}{q}\right)4}{\lambda \ln A} A^{-\frac{\lambda}{2}(n^{1-\beta}-3)}\right]\right]\right) < +\infty.$$

Given that $f \in C_U(\mathbb{R}^N)$, we obtain that $\widetilde{K}_n(f)(x_0) \to f(x_0)$, $\widetilde{Q}_n(f)(x_0) \to f(x_0)$, as $n \to +\infty$.

Proof. By (8.111) and Propositions 8.2, 8.3. □

We make

Remark 8.9. We consider $\left(\mathbb{R}^N, \|\cdot\|_\infty\right)$ and $\left(\mathbb{R}^N\right)^j$ denote the j-fold product space $\mathbb{R}^N \times ... \times \mathbb{R}^N$ endowed with the max-norm $\|x\|_{(\mathbb{R}^N)^j} := \max_{1\leq k\leq j} \|x_k\|_\infty$, where $x := (x_1, ..., x_j) \in \left(\mathbb{R}^N\right)^j$. Then the space $L_j := L_j\left(\left(\mathbb{R}^N\right)^j ; \mathbb{R}\right)$ of all real-valued multilinear continuous functions $g : \left(\mathbb{R}^N\right)^j \to \mathbb{R}$ is a Banach space with norm

$$\|g\| := \|g\|_{L_j} := \sup_{\left(\|x\|_{(\mathbb{R}^N)^j}=1\right)} |g(x)| = \sup \frac{|g(x)|}{\|x_1\|_\infty ... \|x_j\|_\infty}. \tag{8.113}$$

Let $x_0 \in \mathbb{R}^N$ be fixed, and $f : \mathbb{R}^N \to \mathbb{R}$ be a continuous and bounded function whose Fréchet derivatives

$$f^{(j)} : \mathbb{R}^N \to L_j = L_j\left(\left(\mathbb{R}^N\right)^j ; \mathbb{R}\right)$$

exist and are continuous and bounded for all $1 \leq j \leq m$, $m \in \mathbb{N}$. Call $(x - x_0)^j := (x - x_0, ..., x - x_0) \in \left(\mathbb{R}^N\right)^j = \underbrace{\mathbb{R}^N \times ... \times \mathbb{R}^N}_{-j-}$.

Then, by Taylor's formula, see [207], p. 124 and [203], we get

$$f(x) = \sum_{j=0}^{m} \frac{f^{(j)}(x_0)(x - x_0)^j}{j!} + R_m(x, x_0), \text{ all } x \in \mathbb{R}^N, \tag{8.114}$$

where

$$R_m(x, x_0) := \int_0^1 \frac{(1-u)^{m-1}}{(m-1)!} \left(f^{(m)}(x_0 + u(x - x_0)) - f^{(m)}(x_0)\right)(x - x_0)^m du. \tag{8.115}$$

Here we assume that $f^{(j)}(x_0) = 0$, $j = 1, ..., m$, then

$$f(x) - f(x_0) = \int_0^1 \frac{(1-u)^{m-1}}{(m-1)!} \left(f^{(m)}(x_0 + u(x - x_0)) - f^{(m)}(x_0)\right)(x - x_0)^m du. \tag{8.116}$$

Consider

$$w := \omega_1\left(f^{(m)}, h\right) = \sup_{\begin{cases} x, y \in \mathbb{R}^N \\ : \|x - y\|_\infty \leq h \end{cases}} \left\|f^{(m)}(x) - f^{(m)}(y)\right\|, \quad h > 0. \tag{8.117}$$

Notice that $w < +\infty$.

We obtain that

$$\left\| \left(f^{(m)}\left(x_0 + u\left(x - x_0 \right) \right) - f^{(m)}\left(x_0 \right) \right) \left(x - x_0 \right)^m \right\| \leq$$

$$\left\| f^{(m)}\left(x_0 + u\left(x - x_0 \right) \right) - f^{(m)}\left(x_0 \right) \right\| \left\| x - x_0 \right\|^m \leq \tag{8.118}$$

$$w \left\| x - x_0 \right\|^m \left\lceil \frac{u \left\| x - x_0 \right\|}{h} \right\rceil,$$

by Lemma 7.1.1, p. 208, [197].

Therefore for all $x \in \mathbb{R}^N$:

$$\left| f(x) - f(x_0) \right| \leq w \left\| x - x_0 \right\|^m \int_0^1 \left\lceil \frac{u \left\| x - x_0 \right\|_\infty}{h} \right\rceil \frac{(1-u)^{m-1}}{(m-1)!} du \tag{8.119}$$

$$= w \Phi_m \left(\left\| x - x_0 \right\|_\infty \right),$$

by a change of variable, where

$$\Phi_m(t) := \int_0^{|t|} \left\lceil \frac{s}{h} \right\rceil \frac{(|t| - s)^{m-1}}{(m-1)!} ds = \frac{1}{m!} \left(\sum_{j=0}^{\infty} \left(|t| - j \cdot h \right)_+^m \right), \tag{8.120}$$

$t \in \mathbb{R}$; $(x)_+ = \max\{x, 0\}$.

For a full list of properties of the spline function Φ_m see Remark 7.1.3 of [197], p. 210.

By (7.1.18) of [197], p. 210, we have

$$\Phi_m(x) \leq \left(\frac{|x|^{m+1}}{(m+1)!h} + \frac{|x|^m}{2m!} + \frac{h|x|^{m-1}}{8(m-1)!} \right), \quad \forall\, x \in \mathbb{R}. \tag{8.121}$$

Consequently it holds ($x \in \mathbb{R}^N$)

$$\left| f(x) - f(x_0) \right| \leq \omega_1 \left(f^{(m)}, h \right) \left(\frac{\left\| x - x_0 \right\|_\infty^{m+1}}{(m+1)!h} + \frac{\left\| x - x_0 \right\|_\infty^m}{2m!} + \frac{h \left\| x - x_0 \right\|_\infty^{m-1}}{8(m-1)!} \right). \tag{8.122}$$

So that

$$\left| f(\cdot) - f(x_0) \right| \leq \omega_1 \left(f^{(m)}, h \right) \left(\frac{\left\| \cdot - x_0 \right\|_\infty^{m+1}}{(m+1)!h} + \frac{\left\| \cdot - x_0 \right\|_\infty^m}{2m!} + \frac{h \left\| \cdot - x_0 \right\|_\infty^{m-1}}{8(m-1)!} \right), \tag{8.123}$$

over $\mathbb{R}^N$.

Let $(L_n)_{n \in \mathbb{N}}$ be a sequence of positive linear operators from $C_B\left(\mathbb{R}^N\right)$ into $B\left(\mathbb{R}^N\right)$ such that $L_n(1) = 1$, $\forall\, n \in \mathbb{N}$.

By $|f| \leq |f|$, iff $-|f| \leq f \leq |f|$, then $-L_n(|f|) \leq L_n(f) \leq L_n(|f|)$, and $|L_n(f)| \leq L_n(|f|)$.

Hence

$$|L_n(f(\cdot) - f(x_0))(x_0)| \leq L_n(|f(\cdot) - f(x_0)|)(x_0) \overset{(8.123)}{\leq} \omega_1\left(f^{(m)}, h\right)$$

$$\left[\frac{L_n\left(\|\cdot - x_0\|_\infty^{m+1}\right)(x_0)}{(m+1)!h} + \frac{L_n\left(\|\cdot - x_0\|_\infty^{m}\right)(x_0)}{2m!} + \frac{hL_n\left(\|\cdot - x_0\|_\infty^{m-1}\right)(x_0)}{8(m-1)!}\right] =: (\xi). \tag{8.124}$$

We need the following Hölder's type inequality for positive linear operators.

Theorem 8.11. *Let L be a positive linear operator from $C_B\left(\mathbb{R}^N\right)$ into $B\left(\mathbb{R}^N\right)$, and $f, g \in C_B\left(\mathbb{R}^N\right)$, furthermore let $p, q > 1 : \frac{1}{p} + \frac{1}{q} = 1$. Assume that $L((|f(\cdot)|^p))(s_*)$, $L((|g(\cdot)|^q))(s_*) > 0$ for some $s_* \in \mathbb{R}^N$. Then*

$$L(|f(\cdot)g(\cdot)|)(s_*) \leq (L((|f(\cdot)|^p))(s_*))^{\frac{1}{p}} (L((|g(\cdot)|^q))(s_*))^{\frac{1}{q}}. \tag{8.125}$$

Proof. Let $a, b \geq 0$, $p, q > 1 : \frac{1}{p} + \frac{1}{q} = 1$. The Young's inequality says

$$ab \leq \frac{a^p}{p} + \frac{b^q}{q}. \tag{8.126}$$

Then

$$\frac{|f(s)|}{(L((|f(\cdot)|^p))(s_*))^{\frac{1}{p}}} \cdot \frac{|g(s)|}{(L((|g(\cdot)|^q))(s_*))^{\frac{1}{q}}} \leq$$

$$\frac{|f(s)|^p}{p(L((|f(\cdot)|^p))(s_*))} + \frac{|g(s)|^q}{q(L((|g(\cdot)|^q))(s_*))}, \quad \forall\, s \in \mathbb{R}^N. \tag{8.127}$$

Hence it holds

$$\frac{L(|f(\cdot)g(\cdot)|)(s_*)}{(L((|f(\cdot)|^p))(s_*))^{\frac{1}{p}} (L((|g(\cdot)|^q))(s_*))^{\frac{1}{q}}} \leq$$

$$\frac{L((|f(\cdot)|^p))(s_*)}{p(L((|f(\cdot)|^p))(s_*))} + \frac{L((|g(\cdot)|^q))(s_*)}{q(L((|g(\cdot)|^q))(s_*))} = \frac{1}{p} + \frac{1}{q} = 1, \tag{8.128}$$

for $s_* \in \mathbb{R}^N$, proving the claim. □

We continue with

Remark 8.10. Let $g = 1$ and $L_n : L_n(1) = 1$, then by (8.125) we get:

$$L_n\left(\|\cdot - x_0\|_\infty^m\right)(x_0) \le \left(L_n\left(\|\cdot - x_0\|_\infty^{m+1}\right)(x_0)\right)^{\left(\frac{m}{m+1}\right)}, \tag{8.129}$$

and

$$L_n\left(\|\cdot - x_0\|_\infty^{(m-1)}\right)(x_0) \le \left(L_n\left(\|\cdot - x_0\|_\infty^{m+1}\right)(x_0)\right)^{\left(\frac{m-1}{m+1}\right)}, \tag{8.130}$$

where $m \in \mathbb{N}$ and $L_n\left(\|\cdot - x_0\|_\infty^{m+1}\right)(x_0) > 0$.

Continuing from (8.124):

$$(\xi) \le \omega_1\left(f^{(m)}, h\right)\left[\frac{L_n\left(\|\cdot - x_0\|_\infty^{m+1}\right)(x_0)}{(m+1)!h} + \right. \tag{8.131}$$

$$\left. \frac{\left(L_n\left(\|\cdot - x_0\|_\infty^{m+1}\right)(x_0)\right)^{\frac{m}{m+1}}}{2m!} + \frac{h\left(L_n\left(\|\cdot - x_0\|_\infty^{m+1}\right)(x_0)\right)^{\frac{m-1}{m+1}}}{8(m-1)!}\right]$$

(assume $h := \left(L_n\left(\|\cdot - x_0\|_\infty^{m+1}\right)(x_0)\right)^{\frac{1}{m+1}} > 0$)

$$= \omega_1\left(f^{(m)}, \left(L_n\left(\|\cdot - x_0\|_\infty^{m+1}\right)(x_0)\right)^{\frac{1}{m+1}}\right)$$

$$\left[\frac{h^m}{(m+1)!} + \frac{h^m}{2m!} + \frac{h^m}{8(m-1)!}\right] \tag{8.132}$$

$$= \omega_1\left(f^{(m)}, \left(L_n\left(\|\cdot - x_0\|_\infty^{m+1}\right)(x_0)\right)^{\frac{1}{m+1}}\right)$$

$$\left(L_n\left(\|\cdot - x_0\|_\infty^{m+1}\right)(x_0)\right)^{\frac{m}{m+1}}\left[\frac{1}{(m+1)!} + \frac{1}{2m!} + \frac{1}{8(m-1)!}\right]. \tag{8.133}$$

We have proved that

Theorem 8.12. *Let $f \in C_B\left(\mathbb{R}^N\right)$, $x_0 \in \mathbb{R}^N$, Fréchet derivatives $f^{(j)}$ continuous and bounded with $f^{(j)}(x_0) = 0$, $j = 1, ..., m$. Then*

$$|L_n(f)(x_0) - f(x_0)| \le \omega_1\left(f^{(m)}, \left(L_n\left(\|\cdot - x_0\|_\infty^{m+1}\right)(x_0)\right)^{\frac{1}{m+1}}\right) \tag{8.134}$$

$$\left(L_n\left(\|\cdot - x_0\|_\infty^{m+1}\right)(x_0)\right)^{\frac{m}{m+1}}\left[\frac{1}{(m+1)!}+\frac{1}{2m!}+\frac{1}{8(m-1)!}\right], \quad \forall\, n\in\mathbb{N}.$$

Here $L_n(1)=1$, L_n *positive linear operators from* $C_B(\mathbb{R}^N)$ *into* $B(\mathbb{R}^N)$, *under the assumption* $L_n\left(\|\cdot - x_0\|_\infty^{m+1}\right)(x_0)>0$.

Let $L_n\left(\|\cdot - x_0\|_\infty^{m+1}\right)(x_0)\to 0$, *as* $n\to+\infty$, *then* $\lim_{n\to+\infty} L_n(f)(x_0)=f(x_0)$.

Next we apply Theorem 8.12 to the operators $\widetilde{\Theta}_n$.

Theorem 8.13. *Let* $f\in C_B(\mathbb{R}^N)$, $x_0\in\mathbb{R}^N$, *Fréchet derivatives* $f^{(j)}$ *continuous and bounded with* $f^{(j)}(x_0)=0$, $j=1,\ldots,m$. *Then*

$$\left|\widetilde{\Theta}_n(f)(x_0)-f(x_0)\right|\le\omega_1\left(f^{(m)},\left(\widetilde{\Theta}_n\left(\|\cdot - x_0\|_\infty^{m+1}\right)(x_0)\right)^{\frac{1}{m+1}}\right) \tag{8.135}$$

$$\left(\widetilde{\Theta}_n\left(\|\cdot - x_0\|_\infty^{m+1}\right)(x_0)\right)^{\frac{m}{m+1}}\left[\frac{1}{(m+1)!}+\frac{1}{2m!}+\frac{1}{8(m-1)!}\right],$$

$\forall\, n\in\mathbb{N}: n^{1-\beta}>2$, $0<\beta<1$.

We have that $\lim_{n\to+\infty}\widetilde{\Theta}_n(f)(x_0)=f(x_0)$.

Proof. Here $\widetilde{\Theta}_n$ any of $\widetilde{A}_n$, $\widetilde{K}_n$, $\widetilde{Q}_n$.

By Theorem 8.12, and Propositions 8.1–8.3 we get $\widetilde{\Theta}_n\left(\|\cdot - x_0\|_\infty^{m+1}\right)(x_0)\to 0$, as $n\to+\infty$. □

Corollary 8.5. *All as in Theorem 8.13. Set*

$$\widetilde{\varphi}_1(n):=\frac{1}{n^\beta}+\frac{1}{n^{m+1}}\frac{\left(q+\frac{1}{q}\right)2^{m+2}(m+1)!}{\lambda^{m+1}(\ln A)^{m+1}}A^{-\frac{\lambda}{2}\left(n^{1-\beta}-3\right)}. \tag{8.136}$$

Then

$$\left|\widetilde{A}_n(f)(x_0)-f(x_0)\right|\le\omega_1\left(f^{(m)},\left(\widetilde{\varphi}_1(n)\right)^{\frac{1}{m+1}}\right)$$

$$\left(\widetilde{\varphi}_1(n)\right)^{\frac{m}{m+1}}\left[\frac{1}{(m+1)!}+\frac{1}{2m!}+\frac{1}{8(m-1)!}\right], \tag{8.137}$$

$\forall\, n\in\mathbb{N}: n^{1-\beta}>2$, $0<\beta<1$.

We have that $\lim_{n\to+\infty}\widetilde{A}_n(f)(x_0)=f(x_0)$.

Proof. By Theorem 8.13 and Proposition 8.1. □

We finish with

Corollary 8.6. *All as in Theorem 8.13. Set*

$$\widetilde{\varphi}_2(n) := \left[\left(\frac{1}{n} + \frac{1}{n^\beta} \right)^{m+1} + \right.$$

$$\left. \frac{2^m}{n^{m+1}} \left(\frac{\left(q + \frac{1}{q}\right) 2^{m+2} (m+1)!}{\lambda^{m+1} (\ln A)^{m+1}} A^{-\frac{\lambda}{2}\left(n^{1-\beta}-3\right)} \right) \right]. \tag{8.138}$$

Then

$$\left\{ \begin{array}{l} \left| \widetilde{K}_n(f)(x_0) - f(x_0) \right| \\ \left| \widetilde{Q}_n(f)(x_0) - f(x_0) \right| \end{array} \right\} \leq \omega_1 \left(f^{(m)}, (\widetilde{\varphi}_2(n))^{\frac{1}{m+1}} \right)$$

$$(\widetilde{\varphi}_2(n))^{\frac{m}{m+1}} \left[\frac{1}{(m+1)!} + \frac{1}{2m!} + \frac{1}{8(m-1)!} \right], \tag{8.139}$$

$\forall\, n \in \mathbb{N} : n^{1-\beta} > 2,\ 0 < \beta < 1.$

We have that $\lim_{n \to +\infty} \widetilde{K}_n(f)(x_0) = \lim_{n \to +\infty} \widetilde{Q}_n(f)(x_0) = f(x_0)$.

Proof. By Theorem 8.13 and Propositions 8.2, 8.3. □

Bibliography

[196] G.A. Anastassiou, *A "K-Attainable" inequality related to the convergence of Positive Linear Operators*, J. Approximation Theory, 44, 380–383 (1985).

[197] G.A. Anastassiou, *Moments in probability and approximation theory*, Pitman Research Notes in Mathematics series / Longman Scientific & Technical, Essex, New York, UK, USA, 1993.

[198] G.A. Anastassiou, *Quantitative Approximations*, Chapmen & Hall/CRC, London, New York, 2001.

[199] G.A. Anastassiou, *Intelligent Computations: Abstract Fractional Calculus, Inequalities, Approximations*, Springer, Heidelberg, New York, 2018.

[200] G.A. Anastassiou, *Parametrized, Deformed and General Neural Networks*, Springer, Heidelberg, New York, 2023.

[201] G.A. Anastassiou, *Multivariate smooth symmetrized and perturbed hyperbolic tangent neural network approximation over infinite domains*, Mathematics, 2024, 12, 3777, `https://doi.org/10.3390/math12233777`.

[202] G.A. Anastassiou, *Trigonometric and Hyperbolic Generated Approximation Theory*, World Scientific, Singapore, New York, 2025.

[203] H. Cartan, *Differential Calculus*, Herman, Paris, 1971.

[204] Z. Chen and F. Cao, *The approximation operators with sigmoidal functions*, Computers and Mathematics with Applications, 58 (2009), 758–765.

[205] D. Costarelli, R. Spigler, *Approximation results for neural network operators activated by sigmoidal functions*, Neural Networks 44 (2013), 101–106.

[206] D. Costarelli, R. Spigler, *Multivariate neural network operators with sigmoidal activation functions*, Neural Networks 48 (2013), 72–77.

[207] L.B. Rall, *Computational Solution of Nonlinear Operator Equations*, John Wiley & Sons, New York, 1969.

[208] O. Shisha and B. Mond, *The degree of convergence of sequences of linear positive operators*, Nat. Acad. Sci. U.S. 60 (1968), 1196–1200.

[209] O. Shisha and B. Mond, *The degree of Approximation to Periodic Functions by Linear Positive Operators*, J. Approximation Theory, 1, 335–339 (1968).

[210] S. Haykin, *Neural Networks: A Comprehensive Foundation* (2 ed.), Prentice Hall, New York, 1998.

[211] W. McCulloch and W. Pitts, *A logical calculus of the ideas immanent in nervous activity*, Bulletin of Mathematical Biophysics, 7 (1943), 115–133.

[212] T.M. Mitchell, *Machine Learning*, WCB-McGraw-Hill, New York, 1997.

[213] Yu, Dansheng; Cao, Feilong, *Construction and approximation rate for feed-forward neural network operators with sigmoidal functions*, J. Comput. Appl. Math. 453 (2025), paper No. 116150.

[214] Cen, Siyu; Jin, Bangti; Quan, Qimeng; Zhou, Zhi; *Hybrid neural-network FEM approximation of diffusion coeficient in elyptic and parabolic problems.* IMA J. Numer. Anal. 44 (2024), no. 5, 3059–3093.

[215] Coroianu Lucian; Costarelli, Danillo; Natale, Mariarosaria; Pantiş, Alexandra; *The approximation capabilities of Durrmeyer-type neural network operators.* J. Appl. Math. Comput. 70 (2024), no. 5, 4581–4599.

[216] Warin, Xavier; *The GroupMax neural network approximation of convex functions.* IEEE Trans. Neural Netw. Learn. Syst. 35 (2024), no. 8, 11608–11612.

[217] Fabra, Arnau; Guasch, Oriol; Baiges, Joan; Codina, Ramon; *Approximation of acoustic black holes with finite element mixed formulations and artificial neural network correction terms.* Finite Elem. Anal. Des. 241 (2024), paper No. 104236.

[218] Grohs, Philipp; Voigtlaender, Felix. *Proof of the theory-to-practice gap in deep learning via sampling complexity bounds for neural network approximation spaces.* Found. Comput. Math. 24 (2024), no. 4, 1085–1143.

[219] Basteri, Andrea; Trevisan, Dario; *Quantitative Gaussian approximation of randomly initialized deep neural networks.* Mach. Learn. 113 (2024), no. 9, 6373–6393.

[220] De Ryck, T.; Mishra, S. *Error analysis for deep neural network approximations of parametric hyperbolic conservation laws.* Math. Comp. 93 (2024), no. 350, 2643–2677.

[221] Liu, Jie; Zhang, Baoji; Lai, Yuyang; Fang, Liqiau. *Hull form optimization reserach based on multi-precision back-propagation neural network approximation model.* Internal. J. Numer. Methods Fluid 96 (2024), no. 8, 1445–1460.

[222] Yoo, Jihahm; Kim, Jaywon; Gim, Minjung; Lee, Haesung. *Error estimates of physics-informed neural networks for initial value problems.* J. Korean Soc. Ind. Appl. Math. 28 (2024), no. 1, 33–58.

[223] G.A. Anastassiou, *Multivariate Generalized Logistic Neural Networks on infinite domain as Positive Linear Operators and a taste of measure theory*, submitted, 2025.

Chapter 9

Multivariate Fuzzy-Random Symmetrized and Perturbed Neural Network Strong Approximation

In this work we estimate the degree of approximation of multivariate pointwise and uniform convergences in the p-mean to the Fuzzy-Random unit operator of multivariate fuzzy-random quasi-interpolation symmetrized and perturbed activation functions based neural network operators. These symmetrized multivariate fuzzy-random operators arise in a natural way among multivariate fuzzy-random symmetrized neural networks. The rates are given through multivariate probabilistic-Jackson type inequalities involving the multivariate fuzzy-random modulus of continuity of the engaged multivariate fuzzy-random function. It follows [234].

9.1 About Fuzzy-Random Functions

See also [225], Ch. 22, pp. 497–501.

We start with

Definition 9.1. (see [231]) Let $\mu : \mathbb{R} \to [0, 1]$ with the following properties:

(i) is normal, i.e., $\exists\ x_0 \in \mathbb{R} : \mu(x_0) = 1$.

(ii) $\mu(\lambda x + (1 - \lambda) y) \geq \min\{\mu(x), \mu(y)\}$, $\forall\ x, y \in \mathbb{R}$, $\forall\ \lambda \in [0, 1]$ (μ is called a convex fuzzy subset).

(iii) μ is upper semicontinuous on $\mathbb{R}$, i.e., $\forall\ x_0 \in \mathbb{R}$ and $\forall\ \varepsilon > 0$, $\exists$ neighborhood $V(x_0) : \mu(x) \leq \mu(x_0) + \varepsilon$, $\forall\ x \in V(x_0)$.

(iv) the set $\overline{\operatorname{supp}(\mu)}$ is compact in $\mathbb{R}$ (where $\operatorname{supp}(\mu) := \{x \in \mathbb{R}; \mu(x) > 0\}$).

We call μ a fuzzy real number. Denote the set of all μ with $\mathbb{R}_{\mathcal{F}}$.

E.g., $\chi_{\{x_0\}} \in \mathbb{R}_{\mathcal{F}}$, for any $x_0 \in \mathbb{R}$, where $\chi_{\{x_0\}}$ is the characteristic function at x_0.

For $0 < r \leq 1$ and $\mu \in \mathbb{R}_{\mathcal{F}}$ define $[\mu]^r := \{x \in \mathbb{R} : \mu(x) \geq r\}$ and $[\mu]^0 := \overline{\{x \in \mathbb{R} : \mu(x) > 0\}}$.

Then it is well known that for each $r \in [0,1]$, $[\mu]^r$ is a closed and bounded interval of $\mathbb{R}$. For $u, v \in \mathbb{R}_{\mathcal{F}}$ and $\lambda \in \mathbb{R}$, we define uniquely the sum $u \oplus v$ and the product $\lambda \odot u$ by

$$[u \oplus v]^r = [u]^r + [v]^r, \quad [\lambda \odot u]^r = \lambda [u]^r, \quad \forall\, r \in [0,1],$$

where $[u]^r + [v]^r$ means the usual addition of two intervals (as subsets of $\mathbb{R}$) and $\lambda [u]^r$ means the usual product between a scalar and a subset of $\mathbb{R}$ (see, e.g., [231]). Notice $1 \odot u = u$ and it holds $u \oplus v = v \oplus u$, $\lambda \odot u = u \odot \lambda$. If $0 \leq r_1 \leq r_2 \leq 1$ then $[u]^{r_2} \subseteq [u]^{r_1}$. Actually $[u]^r = \left[u_-^{(r)}, u_+^{(r)}\right]$, where $u_-^{(r)} < u_+^{(r)}$, $u_-^{(r)}, u_+^{(r)} \in \mathbb{R}$, $\forall\, r \in [0,1]$.

Define

$$D : \mathbb{R}_{\mathcal{F}} \times \mathbb{R}_{\mathcal{F}} \to \mathbb{R}_+ \cup \{0\}$$

by

$$D(u,v) := \sup_{r \in [0,1]} \max \left\{ \left| u_-^{(r)} - v_-^{(r)} \right|, \left| u_+^{(r)} - v_+^{(r)} \right| \right\},$$

where $[v]^r = \left[v_-^{(r)}, v_+^{(r)}\right]$; $u, v \in \mathbb{R}_{\mathcal{F}}$. We have that D is a metric on $\mathbb{R}_{\mathcal{F}}$. Then $(\mathbb{R}_{\mathcal{F}}, D)$ is a complete metric space, see [231], with the properties

$$\begin{aligned} D(u \oplus w, v \oplus w) &= D(u,v), \quad \forall\, u, v, w \in \mathbb{R}_{\mathcal{F}}, \\ D(k \odot u, k \odot v) &= |k|\, D(u,v), \quad \forall\, u, v \in \mathbb{R}_{\mathcal{F}}, \forall\, k \in \mathbb{R}, \\ D(u \oplus v, w \oplus e) &\leq D(u,w) + D(v,e), \quad \forall\, u, v, w, e \in \mathbb{R}_{\mathcal{F}}. \end{aligned} \tag{9.1}$$

Let (M, d) metric space and $f, g : M \to \mathbb{R}_{\mathcal{F}}$ be fuzzy real number valued functions. The distance between f, g is defined by

$$D^*(f,g) := \sup_{x \in M} D(f(x), g(x)).$$

On $\mathbb{R}_{\mathcal{F}}$ we define a partial order by "$\leq$": $u, v \in \mathbb{R}_{\mathcal{F}}$, $u \leq v$ iff $u_-^{(r)} \leq v_-^{(r)}$ and $u_+^{(r)} \leq v_+^{(r)}$, $\forall\, r \in [0,1]$.

$\overset{*}{\sum}$ denotes the fuzzy summation, $\widetilde{o} := \chi_{\{0\}} \in \mathbb{R}_{\mathcal{F}}$ the neutral element with respect to $\oplus$. For more see also [232], [233].

We need

Definition 9.2. (see also [230], Definition 13.16, p. 654) Let $(X, \mathcal{B}, P)$ be a probability space. A fuzzy-random variable is a $\mathcal{B}$-measurable mapping $g : X \to \mathbb{R}_{\mathcal{F}}$ (i.e., for any open set $U \subseteq \mathbb{R}_{\mathcal{F}}$, in the topology of $\mathbb{R}_{\mathcal{F}}$ generated by the metric D, we have

$$g^{-1}(U) = \{s \in X; g(s) \in U\} \in \mathcal{B}). \tag{9.2}$$

The set of all fuzzy-random variables is denoted by $\mathcal{L}_{\mathcal{F}}(X, \mathcal{B}, P)$. Let $g_n, g \in \mathcal{L}_{\mathcal{F}}(X, \mathcal{B}, P)$, $n \in \mathbb{N}$ and $0 < p < +\infty$. We say $g_n(s) \overset{\text{"}p\text{-mean"}}{\underset{n \to +\infty}{\to}} g(s)$ if

$$\lim_{n \to +\infty} \int_X D(g_n(s), g(s))^p P(ds) = 0. \tag{9.3}$$

Remark 9.1. (see [230], p. 654) If $f, g \in \mathcal{L}_{\mathcal{F}}(X, \mathcal{B}, P)$, let us denote $F : X \to \mathbb{R}_+ \cup \{0\}$ by $F(s) = D(f(s), g(s))$, $s \in X$. Here, F is $\mathcal{B}$-measurable, because $F = G \circ H$, where $G(u, v) = D(u, v)$ is continuous on $\mathbb{R}_{\mathcal{F}} \times \mathbb{R}_{\mathcal{F}}$, and $H : X \to \mathbb{R}_{\mathcal{F}} \times \mathbb{R}_{\mathcal{F}}$, $H(s) = (f(s), g(s))$, $s \in X$, is $\mathcal{B}$-measurable. This shows that the above convergence in q-mean makes sense.

Definition 9.3. (see [230], p. 654, Definition 13.17) Let $(T, \mathcal{T})$ be a topological space. A mapping $f : T \to \mathcal{L}_{\mathcal{F}}(X, \mathcal{B}, P)$ will be called fuzzy-random function (or fuzzy-stochastic process) on T. We denote $f(t)(s) = f(t, s)$, $t \in T$, $s \in X$.

Remark 9.2. (see [230], p. 655) Any usual fuzzy real function $f : T \to \mathbb{R}_{\mathcal{F}}$ can be identified with the degenerate fuzzy-random function $f(t, s) = f(t)$, $\forall\, t \in T$, $s \in X$.

Remark 9.3. (see [230], p. 655) Fuzzy-random functions that coincide with probability one for each $t \in T$ will be consider equivalent.

Remark 9.4. (see [230], p. 655) Let $f, g : T \to \mathcal{L}_{\mathcal{F}}(X, \mathcal{B}, P)$. Then $f \oplus g$ and $k \odot f$ are defined pointwise, i.e.,

$$(f \oplus g)(t, s) = f(t, s) \oplus g(t, s),$$
$$(k \odot f)(t, s) = k \odot f(t, s), \quad t \in T, s \in X, \ k \in \mathbb{R}.$$

Definition 9.4. (see also Definition 13.18, pp. 655–656, [230]) For a fuzzy-random function $f : W \subseteq \mathbb{R}^N \to \mathcal{L}_{\mathcal{F}}(X, \mathcal{B}, P)$, $N \in \mathbb{N}$, we define the (first) fuzzy-random modulus of continuity

$$\Omega_1^{(\mathcal{F})}(f, \delta)_{L^p} =$$

$$\sup\left\{\left(\int_X D^p\left(f\left(x,s\right),f\left(y,s\right)\right)P\left(ds\right)\right)^{\frac{1}{p}}:x,y\in W,\ \left\|x-y\right\|_\infty\le\delta\right\},$$

$0<\delta$, $1\le p<\infty$.

Definition 9.5. ([224]) Here $1\le p<+\infty$. Let $f:W\subseteq\mathbb{R}^N\to\mathcal{L}_{\mathcal{F}}\left(X,\mathcal{B},P\right)$, $N\in\mathbb{N}$, be a fuzzy random function. We call f a (p-mean) uniformly continuous fuzzy random function over W, iff $\forall\ \varepsilon>0\ \exists\ \delta>0$: whenever $\left\|x-y\right\|_\infty\le\delta$, $x,y\in W$, implies that

$$\int_X\left(D\left(f\left(x,s\right),f\left(y,s\right)\right)\right)^p P\left(ds\right)\le\varepsilon.$$

We denote it as $f\in C_{FR}^{U_p}\left(W\right)$.

Proposition 9.1. *([224]) Let $f\in C_{FR}^{U_p}\left(W\right)$, where $W\subseteq\mathbb{R}^N$ is convex. Then $\Omega_1^{(\mathcal{F})}\left(f,\delta\right)_{L^p}<\infty$, any $\delta>0$.*

Proposition 9.2. *([224]) Let $f,g:W\subseteq\mathbb{R}^N\to\mathcal{L}_{\mathcal{F}}\left(X,\mathcal{B},P\right)$, $N\in\mathbb{N}$, be fuzzy random functions. It holds*

(i) $\Omega_1^{(\mathcal{F})}\left(f,\delta\right)_{L^p}$ is nonnegative and nondecreasing in $\delta>0$.

(ii) $\lim_{\delta\downarrow 0}\Omega_1^{(\mathcal{F})}\left(f,\delta\right)_{L^p}=\Omega_1^{(\mathcal{F})}\left(f,0\right)_{L^p}=0$, iff $f\in C_{FR}^{U_p}\left(W\right)$.

We need also

Proposition 9.3. *([224]) Let f,g be fuzzy random variables from $\mathcal{S}$ into $\mathbb{R}_{\mathcal{F}}$. Then*

(i) Let $c\in\mathbb{R}$, then $c\odot f$ is a fuzzy random variable.

(ii) $f\oplus g$ is a fuzzy random variable.

9.2 About Symmetrized and Perturbed Neural Networks

9.2.1 *About Symmetrized and q-Deformed and λ-Parametrized Hyperbolic Tangent Function $g_{q,\lambda}$*

Initially we follow [227], pp. 455–460.

Our perturbed hyperbolic tangent activation function here to be used is

$$g_{q,\lambda}\left(x\right):=\frac{e^{\lambda x}-qe^{-\lambda x}}{e^{\lambda x}+qe^{-\lambda x}},\quad\lambda,q>0,\ x\in\mathbb{R}.\tag{9.4}$$

Above λ is the parameter and q is the deformation coefficient, typically it is $0<\lambda,q\le 1$.

For more read Chapter 18 of [227]: "q-Deformed and λ-Parametrized Hyperbolic Tangent based Banach space Valued Ordinary and Fractional Neural Network Approximation".

The chapters 17 and 18 of [227] motivate our current work.

The proposed "symmetrization method" aims to use half data feed to our multivariate neural networks.

We will employ the following density function

$$M_{q.\lambda}(x) := \frac{1}{4}\left(g_{q,\lambda}(x+1) - g_{q,\lambda}(x-1)\right) > 0, \tag{9.5}$$

$\forall x \in \mathbb{R}$; $q, \lambda > 0$.

We have that

$$M_{q,\lambda}(-x) = M_{\frac{1}{q},\lambda}(x), \quad \forall\, x \in \mathbb{R};\ q, \lambda > 0, \tag{9.6}$$

and

$$M_{\frac{1}{q},\lambda}(-x) = M_{q,\lambda}(x), \quad \forall\, x \in \mathbb{R};\ q, \lambda > 0. \tag{9.7}$$

Adding (9.6) and (9.7) we obtain

$$M_{q,\lambda}(-x) + M_{\frac{1}{q},\lambda}(-x) = M_{q,\lambda}(x) + M_{\frac{1}{q},\lambda}(x), \tag{9.8}$$

a key to this work.

So that

$$\Phi(x) := \frac{M_{q,\lambda}(x) + M_{\frac{1}{q},\lambda}(x)}{2} > 0, \tag{9.9}$$

is an even function, symmetric with respect to the y-axis.

By (18.18) of [227], we have

$$\begin{gathered} M_{q,\lambda}\left(\frac{\ln q}{2\lambda}\right) = \frac{\tanh(\lambda)}{2}, \\ \text{and} \\ M_{\frac{1}{q},\lambda}\left(-\frac{\ln q}{2\lambda}\right) = \frac{\tanh(\lambda)}{2}, \quad \lambda > 0, \end{gathered} \tag{9.10}$$

sharing the same maximum at symmetric points.

By Theorem 18.1, p. 458 of [227], we have that

$$\begin{gathered} \sum_{i=-\infty}^{\infty} M_{q,\lambda}(x-i) = 1, \quad \forall\, x \in \mathbb{R},\ \lambda, q > 0, \\ \text{and} \\ \sum_{i=-\infty}^{\infty} M_{\frac{1}{q},\lambda}(x-i) = 1, \quad \forall\, x \in \mathbb{R},\ \lambda, q > 0. \end{gathered} \tag{9.11}$$

Consequently, we derive that

$$\sum_{i=-\infty}^{\infty} \Phi (x-i) = 1, \ \ \forall \, x \in \mathbb{R}. \tag{9.12}$$

By Theorem 18.2, p. 459 of [227], we have that

$$\begin{gathered} \int_{-\infty}^{\infty} M_{q,\lambda} (x) \, dx = 1, \\ \text{and} \\ \int_{-\infty}^{\infty} M_{\frac{1}{q},\lambda} (x) \, dx = 1, \end{gathered} \tag{9.13}$$

so that

$$\int_{-\infty}^{\infty} \Phi (x) \, dx = 1, \tag{9.14}$$

therefore Φ is a density function.

By Theorem 18.3, p. 459 of [227], we have:

Let $0 < \alpha < 1$, and $n \in \mathbb{N}$ with $n^{1-\alpha} > 2$; $q, \lambda > 0$. Then

$$\sum_{\substack{k=-\infty \\ : |nx-k| \geq n^{1-\alpha}}}^{\infty} M_{q,\lambda} (nx-k) < 2 \max \left\{ q, \frac{1}{q} \right\} e^{4\lambda} e^{-2\lambda n^{(1-\alpha)}} = Te^{-2\lambda n^{(1-\alpha)}}, \tag{9.15}$$

where $T := 2 \max \left\{ q, \frac{1}{q} \right\} e^{4\lambda}$.

Similarly, we get that

$$\sum_{k=-\infty}^{\infty} M_{\frac{1}{q},\lambda} (nx-k) < Te^{-2\lambda n^{(1-\alpha)}}. \tag{9.16}$$

Consequently we obtain that

$$\sum_{\substack{k=-\infty \\ : |nx-k| \geq n^{1-\alpha}}}^{\infty} \Phi (nx-k) < Te^{-2\lambda n^{(1-\alpha)}} =: c_1 (n, \alpha), \tag{9.17}$$

where $T := 2 \max \left\{ q, \frac{1}{q} \right\} e^{4\lambda}$.

Here $\lceil \cdot \rceil$ denotes the ceiling of the number, and $\lfloor \cdot \rfloor$ its integral part.

We mention

Theorem 18.4 (p. 459, [227]) Let $x \in [a, b] \subset \mathbb{R}$ and $n \in \mathbb{N}$ so that $\lceil na \rceil \leq \lfloor nb \rfloor$. For $q, \lambda > 0$, we consider $\lambda_q > z_0 > 0$, such that $M_{q,\lambda} (z_0) = M_{q,\lambda} (0)$, and $\lambda_q > 1$. Then

$$\frac{1}{\sum_{k=\lceil na \rceil}^{\lfloor nb \rfloor} M_{q,\lambda} (nx-k)} < \max \left\{ \frac{1}{M_{q,\lambda} (\lambda_q)}, \frac{1}{M_{\frac{1}{q},\lambda} \left(\lambda_{\frac{1}{q}} \right)} \right\} =: \Delta (q). \tag{9.18}$$

Similarly, we consider $\lambda_{\frac{1}{q}} > z_1 > 0$, such that $M_{\frac{1}{q},\lambda}(z_1) = M_{\frac{1}{q},\lambda}(0)$, and $\lambda_{\frac{1}{q}} > 1$. Thus

$$\frac{1}{\sum_{k=\lceil na \rceil}^{\lfloor nb \rfloor} M_{\frac{1}{q},\lambda}(nx-k)} < \max\left\{ \frac{1}{M_{\frac{1}{q},\lambda}\left(\lambda_{\frac{1}{q}}\right)}, \frac{1}{M_{q,\lambda}(\lambda_q)} \right\} = \Delta(q). \quad (9.19)$$

Hence

$$\sum_{k=\lceil na \rceil}^{\lfloor nb \rfloor} M_{q,\lambda}(nx-k) > \frac{1}{\Delta(q)}, \quad (9.20)$$

and

$$\sum_{k=\lceil na \rceil}^{\lfloor nb \rfloor} M_{\frac{1}{q},\lambda}(nx-k) > \frac{1}{\Delta(q)}. \quad (9.21)$$

Consequently it holds

$$\sum_{k=\lceil na \rceil}^{\lfloor nb \rfloor} \frac{\left(M_{q,\lambda}(nx-k) + M_{\frac{1}{q},\lambda}(nx-k)\right)}{2} > \frac{2}{2\Delta(q)} = \frac{1}{\Delta(q)}, \quad (9.22)$$

so that

$$\frac{1}{\sum_{k=\lceil na \rceil}^{\lfloor nb \rfloor} \frac{\left(M_{q,\lambda}(nx-k) + M_{\frac{1}{q},\lambda}(nx-k)\right)}{2}} < \Delta(q), \quad (9.23)$$

that is

$$\frac{1}{\sum_{k=\lceil na \rceil}^{\lfloor nb \rfloor} \Phi(nx-k)} < \Delta(q). \quad (9.24)$$

We have proved

Theorem 9.1. *Let $x \in [a,b] \subset \mathbb{R}$ and $n \in \mathbb{N}$ so that $\lceil na \rceil \leq \lfloor nb \rfloor$. For $q, \lambda > 0$, we consider $\lambda_q > z_0 > 0$, such that $M_{q,\lambda}(z_0) = M_{q,\lambda}(0)$, and $\lambda_q > 1$. Also consider $\lambda_{\frac{1}{q}} > z_1 > 0$, such that $M_{\frac{1}{q},\lambda}(z_1) = M_{\frac{1}{q},\lambda}(0)$, and $\lambda_{\frac{1}{q}} > 1$. Then*

$$\frac{1}{\sum_{k=\lceil na \rceil}^{\lfloor nb \rfloor} \Phi(nx-k)} < \Delta(q). \quad (9.25)$$

We make

Remark 9.5.

I) By Remark 18.5, p. 460 of [227], we have

$$\lim_{n\to\infty} \sum_{k=\lceil na \rceil}^{\lfloor nb \rfloor} M_{q,\lambda}(nx_1 - k) \neq 1, \text{ for some } x_1 \in [a,b], \tag{9.26}$$

and

$$\lim_{n\to\infty} \sum_{k=\lceil na \rceil}^{\lfloor nb \rfloor} M_{\frac{1}{q},\lambda}(nx_2 - k) \neq 1, \text{ for some } x_2 \in [a,b]. \tag{9.27}$$

Therefore it holds

$$\lim_{n\to\infty} \sum_{k=\lceil na \rceil}^{\lfloor nb \rfloor} \frac{\left(M_{q,\lambda}(nx_1 - k) + M_{\frac{1}{q},\lambda}(nx_2 - k)\right)}{2} \neq 1. \tag{9.28}$$

Hence

$$\lim_{n\to\infty} \sum_{k=\lceil na \rceil}^{\lfloor nb \rfloor} \frac{\left(M_{q,\lambda}(nx_1 - k) + M_{\frac{1}{q},\lambda}(nx_1 - k)\right)}{2} \neq 1, \tag{9.29}$$

even if

$$\lim_{n\to\infty} \sum_{k=\lceil na \rceil}^{\lfloor nb \rfloor} M_{\frac{1}{q},\lambda}(nx_1 - k) = 1, \tag{9.30}$$

because then

$$\lim_{n\to\infty} \sum_{k=\lceil na \rceil}^{\lfloor nb \rfloor} \frac{M_{q,\lambda}(nx_1 - k)}{2} + \frac{1}{2} \neq 1, \tag{9.31}$$

equivalently

$$\lim_{n\to\infty} \sum_{k=\lceil na \rceil}^{\lfloor nb \rfloor} \frac{M_{q,\lambda}(nx_1 - k)}{2} \neq \frac{1}{2}, \tag{9.32}$$

true by

$$\lim_{n\to\infty} \sum_{k=\lceil na \rceil}^{\lfloor nb \rfloor} M_{q,\lambda}(nx_1 - k) \neq 1. \tag{9.33}$$

II) Let $[a,b] \subset \mathbb{R}$. For large n we always have $\lceil na \rceil \leq \lfloor nb \rfloor$. Also $a \leq \frac{k}{n} \leq b$, iff $\lceil na \rceil \leq k \leq \lfloor nb \rfloor$. So in general it holds

$$\sum_{k=\lceil na \rceil}^{\lfloor nb \rfloor} \Phi(nx - k) \leq 1. \tag{9.34}$$

Next, we move on to the multivariate case, see Chapter 17 of [227], pp. 419–452, as a model of action.

We make

Remark 9.6. We introduce

$$Z_1(x_1,...,x_N) := Z_1(x) := \prod_{i=1}^{N} \Phi(x_i) = \frac{1}{2^N} \prod_{i=1}^{N} \left(M_{q,\lambda} + M_{\frac{1}{q},\lambda} \right)(x_i), \tag{9.35}$$

$x = (x_1,...,x_N) \in \mathbb{R}^N$, $\lambda, q > 0$, $N \in \mathbb{N}$.

Properties:

(i)

$$Z_1(x) > 0, \ \forall \ x \in \mathbb{R}^N, \tag{9.36}$$

(ii)

$$\sum_{k=-\infty}^{\infty} Z_1(x-k) := \sum_{k_1=-\infty}^{\infty} \sum_{k_2=-\infty}^{\infty} ... \sum_{k_N=-\infty}^{\infty} Z_1(x_1-k_1,...,x_N-k_N) = 1, \tag{9.37}$$

$k := (k_1,...,k_n) \in \mathbb{Z}^N$, $\forall \ x \in \mathbb{R}^N$,

hence

(iii)

$$\sum_{k=-\infty}^{\infty} Z_1(nx-k) = 1, \tag{9.38}$$

$\forall \ x \in \mathbb{R}^N$, $n \in \mathbb{N}$,

(iv)

$$\int_{\mathbb{R}^N} Z_1(x)\, dx = 1, \tag{9.39}$$

that is Z_1 is a multivariate density function.

Here denote $\|x\|_\infty := \max\{|x_1|,...,|x_N|\}$, $x \in \mathbb{R}^N$, $\infty := (\infty,...,\infty)$, $-\infty := (-\infty,...,-\infty)$, $\lceil na \rceil := (\lceil na_1 \rceil,...,\lceil na_N \rceil)$, $\lfloor nb \rfloor := (\lfloor nb_1 \rfloor,...,\lfloor nb_N \rfloor)$, $a := (a_1,...,a_N)$, $b := (b_1,...,b_N)$.

We obviously see that

$$\sum_{k=\lceil na \rceil}^{\lfloor nb \rfloor} Z_1(nx-k) = \sum_{k=\lceil na \rceil}^{\lfloor nb \rfloor} \left(\prod_{i=1}^{N} \Phi(nx_i - k_i) \right) = \prod_{i=1}^{N} \left(\sum_{k_i=\lceil na_i \rceil}^{\lfloor nb_i \rfloor} \Phi(nx_i - k_i) \right). \tag{9.40}$$

(v) We derive that

$$\sum_{\substack{k=\lceil na\rceil \\ \left\|\frac{k}{n}-x\right\|_{\infty}>\frac{1}{n^{\beta}}}}^{\lfloor nb\rfloor} Z_1(nx-k) < Te^{-2\lambda n^{(1-\beta)}} = c_1(n,\beta), \text{ where } 0<\beta<1, \tag{9.41}$$

with $n\in\mathbb{N}: n^{1-\beta}>2$, $x\in\prod_{i=1}^{N}[a_i,b_i]$; $T:=2\max\left\{q,\frac{1}{q}\right\}e^{4\lambda}$.

(vi) It holds

$$0<\frac{1}{\sum_{k=\lceil na\rceil}^{\lfloor nb\rfloor} Z_1(nx-k)}<(\Delta(q))^N, \tag{9.42}$$

$\forall\, x\in\left(\prod_{i=1}^{N}[a_i,b_i]\right)$, $n\in\mathbb{N}$.

It is clear that

(vii)

$$\sum_{\substack{k=-\infty \\ \left\|\frac{k}{n}-x\right\|_{\infty}>\frac{1}{n^{\beta}}}}^{\infty} Z_1(nx-k) < Te^{-2\lambda n^{(1-\beta)}} = c_1(n,\beta), \tag{9.43}$$

where $0<\beta<1$, $n\in\mathbb{N}: n^{1-\beta}>2$, $x\in\mathbb{R}^N$.

Furthermore it holds

$$\lim_{n\to\infty}\sum_{k=\lceil na\rceil}^{\lfloor nb\rfloor} Z_1(nx-k)\neq 1, \tag{9.44}$$

for at least some $x\in\left(\prod_{i=1}^{N}[a_i,b_i]\right)$.

Let $f\in C\left(\prod_{i=1}^{N}[a_i,b_i]\right)$, $x=(x_1,...,x_N)\in\prod_{i=1}^{N}[a_i,b_i]$, $n\in\mathbb{N}$: $\lceil na_i\rceil\leq\lfloor nb_i\rfloor$, $i=1,...,N$.

We introduce and define the following multivariate linear normalized symmetrized real neural network operator, let $x:=(x_1,...,x_N)\in\prod_{i=1}^{N}[a_i,b_i]$:

$$A_n^s(f,x_1,...,x_N):=A_n^s(f,x):=\frac{\sum_{k=\lceil na\rceil}^{\lfloor nb\rfloor} f\left(\frac{k}{n}\right)Z_1(nx-k)}{\sum_{k=\lceil na\rceil}^{\lfloor nb\rfloor} Z_1(nx-k)}= \tag{9.45}$$

$$\frac{\sum_{k_1=\lceil na_1\rceil}^{\lfloor nb_1\rfloor}\sum_{k_2=\lceil na_2\rceil}^{\lfloor nb_2\rfloor}\cdots\sum_{k_N=\lceil na_N\rceil}^{\lfloor nb_N\rfloor} f\left(\frac{k_1}{n},...,\frac{k_N}{n}\right)\left(\prod_{i=1}^{N}\left(M_{q,\lambda}(nx_i-k_i)+M_{\frac{1}{q},\lambda}(nx_i-k_i)\right)\right)}{\prod_{i=1}^{N}\left(\sum_{k_i=\lceil na_i\rceil}^{\lfloor nb_i\rfloor}\left(M_{q,\lambda}(nx_i-k_i)+M_{\frac{1}{q},\lambda}(nx_i-k_i)\right)\right)}.$$

For large enough $n\in\mathbb{N}$ we always obtain $\lceil na_i\rceil\leq\lfloor nb_i\rfloor$, $i=1,...,N$. Also $a_i\leq\frac{k_i}{n}\leq b_i$, iff $\lceil na_i\rceil\leq k_i\leq\lfloor nb_i\rfloor$, $i=1,...,N$.

We need

Definition 9.6. Let M be a convex and compact subset of $\left(\mathbb{R}^k, \|\cdot\|_\infty\right)$, $f \in C(M)$. We define the first modulus of continuity of f as

$$\omega_1(f,\delta) := \sup_{\substack{x, y \in M: \\ \|x-y\|_\infty \le \delta}} |f(x) - f(y)|, \quad 0 < \delta \le diam(M). \tag{9.46}$$

If $\delta > diam(M)$, then

$$\omega_1(f,\delta) = \omega_1(f, diam(M)). \tag{9.47}$$

We have $\omega_1(f,\delta) \to 0$ as $\delta \downarrow 0$ iff $f \in C(M)$.

We need also

Theorem 9.2. *([228]) It holds*

$$\|A_n^s(f) - f\|_{\infty,\prod_{i=1}^N [a_i,b_i]} \le 2\omega_1\left(f, (\Delta(q))^N \left[\frac{1}{n^\beta} + \|b-a\|_\infty T e^{-2\lambda n^{(1-\beta)}}\right]\right), \tag{9.48}$$

$\forall\ f \in C\left(\prod_{i=1}^N [a_i,b_i]\right)$, $0 < \beta < 1$.

By (9.48), we obtain $A_n^s(f) \to f$, as $n \to +\infty$, pointwise and uniformly.

9.2.2 About Symmetrized and q-Deformed and λ-Parametrized A-Generalized Logistic Function

Initially we follow [227], pp. 395–471.

Our activation function here to be used is the q-deformed and λ-parametrized function

$$\varphi_{q,\lambda}(x) = \frac{1}{1 + qA^{-\lambda x}}, \quad x \in \mathbb{R},\ q, \lambda > 0,\ A > 1. \tag{9.49}$$

This is the A-generalized logistic function.

For more read Chapter 16 of [227]: "Banach space valued ordinary and fractional neural network approximation based on q-deformed and λ-parametrized A-generalized logistic function".

The chapters 15, 16 of [227] motivate our current work.

The proposed "symmetrization technique" aims to use half data feed to our neural networks.

We will employ the following density function

$$G_{q,\lambda}(x) := \frac{1}{2}\left(\varphi_{q,\lambda}(x+1) - \varphi_{q,\lambda}(x-1)\right), \quad x \in \mathbb{R},\ q, \lambda > 0. \tag{9.50}$$

We have that

$$G_{q,\lambda}(-x) = G_{\frac{1}{q},\lambda}(x), \tag{9.51}$$

and

$$G_{\frac{1}{q},\lambda}(-x) = G_{q,\lambda}(x), \quad \forall\, x \in \mathbb{R}. \tag{9.52}$$

Adding (9.51) and (9.52) we obtain

$$G_{q,\lambda}(-x) + G_{\frac{1}{q},\lambda}(-x) = G_{q,\lambda}(x) + G_{\frac{1}{q},\lambda}(x), \quad \forall\, x \in \mathbb{R}, \tag{9.53}$$

the key to this work.

So that

$$W(x) := \frac{G_{q,\lambda}(x) + G_{\frac{1}{q},\lambda}(x)}{2} > 0, \tag{9.54}$$

is an even function, symmetric with respect to the y-axis.

The global maximum of $G_{q,\lambda}$ is given by (16.18), p. 401 of [227] as

$$G_{q,\lambda}\left(\frac{\log_A q}{\lambda}\right) = \frac{A^{\lambda}-1}{2(A^{\lambda}+1)}. \tag{9.55}$$

And, the global max of $G_{\frac{1}{q},\lambda}$ is

$$G_{\frac{1}{q},\lambda}\left(\frac{\log_A \frac{1}{q}}{\lambda}\right) = G_{\frac{1}{q},\lambda}\left(\frac{-\log_A q}{\lambda}\right) = \frac{A^{\lambda}-1}{2(A^{\lambda}+1)}, \tag{9.56}$$

both sharing the same maximum at symmetric points.

By Theorem 16.1, p. 401 of [227], we have that

$$\sum_{i=-\infty}^{\infty} G_{q,\lambda}(x-i) = 1, \quad \forall\, x \in \mathbb{R},\ \lambda, q > 0,\ A > 1, \tag{9.57}$$

and

$$\sum_{i=-\infty}^{\infty} G_{\frac{1}{q},\lambda}(x-i) = 1, \quad \forall\, x \in \mathbb{R},\ \lambda, q > 0,\ A > 1. \tag{9.58}$$

Consequently, we derive that

$$\sum_{i=-\infty}^{\infty} W(x-i) = 1, \quad \forall\, x \in \mathbb{R}. \tag{9.59}$$

By Theorem 16.2, p. 402 of [227], we have that

$$\int_{-\infty}^{\infty} G_{q,\lambda}(x)\, dx = 1, \quad \lambda, q > 0,\ A > 1, \tag{9.60}$$

similarly it holds

$$\int_{-\infty}^{\infty} G_{\frac{1}{q},\lambda}(x)\,dx = 1, \tag{9.61}$$

so that

$$\int_{-\infty}^{\infty} W(x)\,dx = 1, \tag{9.62}$$

therefore W is a density function.

By Theorem 16.3, p. 402 of [227], we have:

Let $0 < \alpha < 1$, and $n \in \mathbb{N}$ with $n^{1-\alpha} > 2$. Then

$$\sum_{\begin{cases} k=-\infty \\ : |nx-k| \geq n^{1-\alpha} \end{cases}}^{\infty} G_{q,\lambda}(nx-k) < 2\max\left\{q, \frac{1}{q}\right\} \frac{1}{A^{\lambda\left(n^{1-\alpha}-2\right)}} = \widetilde{\gamma} A^{-\lambda\left(n^{1-\alpha}-2\right)}, \tag{9.63}$$

where $\lambda, q > 0$, $A > 1$; $\widetilde{\gamma} := 2\max\left\{q, \frac{1}{q}\right\}$.

Similarly, we get that

$$\sum_{\begin{cases} k=-\infty \\ : |nx-k| \geq n^{1-\alpha} \end{cases}}^{\infty} G_{\frac{1}{q},\lambda}(nx-k) < \widetilde{\gamma} A^{-\lambda\left(n^{1-\alpha}-2\right)}. \tag{9.64}$$

Consequently we obtain that

$$\sum_{\begin{cases} k=-\infty \\ : |nx-k| \geq n^{1-\alpha} \end{cases}}^{\infty} W(nx-k) < \widetilde{\gamma} A^{-\lambda\left(n^{1-\alpha}-2\right)} =: c_2(n,\alpha), \tag{9.65}$$

where $\widetilde{\gamma} := 2\max\left\{q, \frac{1}{q}\right\}$.

Here $\lceil \cdot \rceil$ denotes the ceiling of the number, and $\lfloor \cdot \rfloor$ its integral part.

We mention

Theorem 16.4 (p. 402, [227]) Let $x \in [a,b] \subset \mathbb{R}$ and $n \in \mathbb{N}$ so that $\lceil na \rceil \leq \lfloor nb \rfloor$. For $q > 0$, $\lambda > 0$, $A > 1$, we consider the number $\lambda_q > z_0 > 0$ with $G_{q,\lambda}(z_0) = G_{q,\lambda}(0)$, and $\lambda_q > 1$. Then

$$\frac{1}{\sum_{k=\lceil na \rceil}^{\lfloor nb \rfloor} G_{q,\lambda}(nx-k)} < \max\left\{\frac{1}{G_{q,\lambda}(\lambda_q)}, \frac{1}{G_{\frac{1}{q},\lambda}\left(\lambda_{\frac{1}{q}}\right)}\right\} =: K(q). \tag{9.66}$$

Similarly, we consider $\lambda_{\frac{1}{q}} > z_1 > 0$, such that $G_{\frac{1}{q},\lambda}(z_1) = G_{\frac{1}{q},\lambda}(0)$, and $\lambda_{\frac{1}{q}} > 1$. Thus

$$\frac{1}{\sum_{k=\lceil na \rceil}^{\lfloor nb \rfloor} G_{\frac{1}{q},\lambda}(nx-k)} < \max\left\{\frac{1}{G_{\frac{1}{q},\lambda}\left(\lambda_{\frac{1}{q}}\right)}, \frac{1}{G_{q,\lambda}(\lambda_q)}\right\} = K(q). \quad (9.67)$$

Hence

$$\sum_{k=\lceil na \rceil}^{\lfloor nb \rfloor} G_{q,\lambda}(nx-k) > \frac{1}{K(q)}, \quad (9.68)$$

and

$$\sum_{k=\lceil na \rceil}^{\lfloor nb \rfloor} G_{\frac{1}{q},\lambda}(nx-k) > \frac{1}{K(q)}. \quad (9.69)$$

Consequently it holds

$$\sum_{k=\lceil na \rceil}^{\lfloor nb \rfloor} \frac{\left(G_{q,\lambda}(nx-k) + G_{\frac{1}{q},\lambda}(nx-k)\right)}{2} > \frac{2}{2K(q)} = \frac{1}{K(q)}, \quad (9.70)$$

so that

$$\frac{1}{\sum_{k=\lceil na \rceil}^{\lfloor nb \rfloor} \frac{\left(G_{q,\lambda}(nx-k)+G_{\frac{1}{q},\lambda}(nx-k)\right)}{2}} < K(q), \quad (9.71)$$

that is

$$\frac{1}{\sum_{k=\lceil na \rceil}^{\lfloor nb \rfloor} W(nx-k)} < K(q). \quad (9.72)$$

We have proved

Theorem 9.3. *Let $x \in [a,b] \subset \mathbb{R}$ and $n \in \mathbb{N}$ so that $\lceil na \rceil \leq \lfloor nb \rfloor$. For $q, \lambda > 0$, $A > 1$, we consider $\lambda_q > z_0 > 0$ with $G_{q,\lambda}(z_0) = G_{q,\lambda}(0)$, and $\lambda_q > 1$. Also consider $\lambda_{\frac{1}{q}} > z_1 > 0$, such that $G_{\frac{1}{q},\lambda}(z_1) = G_{\frac{1}{q},\lambda}(0)$, and $\lambda_{\frac{1}{q}} > 1$. Then*

$$\frac{1}{\sum_{k=\lceil na \rceil}^{\lfloor nb \rfloor} W(nx-k)} < K(q). \quad (9.73)$$

We make

Remark 9.7.
I) By Remark 16.5, p. 402 of [227], we have

$$\lim_{n\to\infty} \sum_{k=\lceil na\rceil}^{\lfloor nb\rfloor} G_{q,\lambda}(nx_1 - k) \neq 1, \text{ for some } x_1 \in [a,b], \tag{9.74}$$

and

$$\lim_{n\to\infty} \sum_{k=\lceil na\rceil}^{\lfloor nb\rfloor} G_{\frac{1}{q},\lambda}(nx_2 - k) \neq 1, \text{ for some } x_2 \in [a,b]. \tag{9.75}$$

Therefore it holds

$$\lim_{n\to\infty} \sum_{k=\lceil na\rceil}^{\lfloor nb\rfloor} \frac{\left(G_{q,\lambda}(nx_1 - k) + G_{\frac{1}{q},\lambda}(nx_2 - k)\right)}{2} \neq 1. \tag{9.76}$$

Hence

$$\lim_{n\to\infty} \sum_{k=\lceil na\rceil}^{\lfloor nb\rfloor} \frac{\left(G_{q,\lambda}(nx_1 - k) + G_{\frac{1}{q},\lambda}(nx_1 - k)\right)}{2} \neq 1, \tag{9.77}$$

even if

$$\lim_{n\to\infty} \sum_{k=\lceil na\rceil}^{\lfloor nb\rfloor} G_{\frac{1}{q},\lambda}(nx_1 - k) = 1, \tag{9.78}$$

because then

$$\lim_{n\to\infty} \sum_{k=\lceil na\rceil}^{\lfloor nb\rfloor} \frac{G_{q,\lambda}(nx_1 - k)}{2} + \frac{1}{2} \neq 1, \tag{9.79}$$

equivalently

$$\lim_{n\to\infty} \sum_{k=\lceil na\rceil}^{\lfloor nb\rfloor} \frac{G_{q,\lambda}(nx_1 - k)}{2} \neq \frac{1}{2}, \tag{9.80}$$

true by

$$\lim_{n\to\infty} \sum_{k=\lceil na\rceil}^{\lfloor nb\rfloor} G_{q,\lambda}(nx_1 - k) \neq 1. \tag{9.81}$$

II) Let $[a,b] \subset \mathbb{R}$. For large n we always have $\lceil na\rceil \leq \lfloor nb\rfloor$. Also $a \leq \frac{k}{n} \leq b$, iff $\lceil na\rceil \leq k \leq \lfloor nb\rfloor$. So in general it holds

$$\sum_{k=\lceil na\rceil}^{\lfloor nb\rfloor} W(nx - k) \leq 1. \tag{9.82}$$

Next, we move on to the multivariate case, see Chapter 15 of [227], pp. 365–394, as a model of action.

We make

Remark 9.8. We introduce

$$\widetilde{Z}(x_1, ..., x_N) := \widetilde{Z}(x) := \prod_{i=1}^{N} W(x_i) = \frac{1}{2^N} \prod_{i=1}^{N} \left(G_{q,\lambda} + G_{\frac{1}{q},\lambda} \right)(x_i), \tag{9.83}$$

$x = (x_1, ..., x_N) \in \mathbb{R}^N$, $\lambda, q > 0$, $N \in \mathbb{N}$.

Properties:

(i)

$$\widetilde{Z}(x) > 0, \ \forall \ x \in \mathbb{R}^N, \tag{9.84}$$

(ii)

$$\sum_{k=-\infty}^{\infty} \widetilde{Z}(x-k) := \sum_{k_1=-\infty}^{\infty} \sum_{k_2=-\infty}^{\infty} ... \sum_{k_N=-\infty}^{\infty} \widetilde{Z}(x_1 - k_1, ..., x_N - k_N) = 1, \tag{9.85}$$

$k := (k_1, ..., k_n) \in \mathbb{Z}^N$, $\forall \ x \in \mathbb{R}^N$,

hence

(iii)

$$\sum_{k=-\infty}^{\infty} \widetilde{Z}(nx-k) = 1, \tag{9.86}$$

$\forall \ x \in \mathbb{R}^N$, $n \in \mathbb{N}$,

(iv)

$$\int_{\mathbb{R}^N} \widetilde{Z}(x)\, dx = 1, \tag{9.87}$$

that is $\widetilde{Z}$ is a multivariate density function.

Here denote $\|x\|_{\infty} := \max\{|x_1|, ..., |x_N|\}$, $x \in \mathbb{R}^N$, $\infty := (\infty, ..., \infty)$, $-\infty := (-\infty, ..., -\infty)$, $\lceil na \rceil := (\lceil na_1 \rceil, ..., \lceil na_N \rceil)$, $\lfloor nb \rfloor := (\lfloor nb_1 \rfloor, ..., \lfloor nb_N \rfloor)$, $a := (a_1, ..., a_N)$, $b := (b_1, ..., b_N)$.

We obviously see that

$$\sum_{k=\lceil na \rceil}^{\lfloor nb \rfloor} \widetilde{Z}(nx-k) = \sum_{k=\lceil na \rceil}^{\lfloor nb \rfloor} \left(\prod_{i=1}^{N} W(nx_i - k_i) \right) = \prod_{i=1}^{N} \left(\sum_{k_i=\lceil na_i \rceil}^{\lfloor nb_i \rfloor} W(nx_i - k_i) \right). \tag{9.88}$$

(v) We derive that

$$\sum_{\begin{cases} k = \lceil na \rceil \\ \left\| \frac{k}{n} - x \right\|_{\infty} > \frac{1}{n^{\beta}} \end{cases}}^{\lfloor nb \rfloor} \widetilde{Z}(nx - k) < \widetilde{\gamma} A^{-\lambda\left(n^{1-\beta}-2\right)} = c_2(n, \beta), \text{ where } 0 < \beta < 1, \tag{9.89}$$

$0 < \beta < 1$, $n \in \mathbb{N} : n^{1-\beta} > 2$, $x \in \prod_{i=1}^{N} [a_i, b_i]$.

(vi) It holds

$$0 < \frac{1}{\sum_{k=\lceil na \rceil}^{\lfloor nb \rfloor} \widetilde{Z}(nx - k)} < (K(q))^N, \tag{9.90}$$

$\forall\, x \in \left(\prod_{i=1}^{N} [a_i, b_i]\right)$, $n \in \mathbb{N}$.

It is clear that

(vii)

$$\sum_{\begin{cases} k = -\infty \\ \left\| \frac{k}{n} - x \right\|_{\infty} > \frac{1}{n^{\beta}} \end{cases}}^{\infty} \widetilde{Z}(nx - k) < \widetilde{\gamma} A^{-\lambda\left(n^{1-\beta}-2\right)} = c_2(n, \beta), \tag{9.91}$$

$0 < \beta < 1$, $n \in \mathbb{N} : n^{1-\beta} > 2$, $x \in \prod_{i=1}^{N} [a_i, b_i]$.

Furthermore it holds

$$\lim_{n \to \infty} \sum_{k=\lceil na \rceil}^{\lfloor nb \rfloor} \widetilde{Z}(nx - k) \neq 1, \tag{9.92}$$

for at least some $x \in \left(\prod_{i=1}^{N} [a_i, b_i]\right)$.

Let $f \in C\left(\prod_{i=1}^{N} [a_i, b_i]\right)$, $x = (x_1, ..., x_N) \in \prod_{i=1}^{N} [a_i, b_i]$, $n \in \mathbb{N}$: $\lceil na_i \rceil \leq \lfloor nb_i \rfloor$, $i = 1, ..., N$.

We introduce and define the following multivariate linear normalized symmetrized real neural network operator, let $x := (x_1, ..., x_N) \in \prod_{i=1}^{N} [a_i, b_i]$:

$$\psi_n^s(f, x_1, ..., x_N) := \psi_n^s(f, x) := \frac{\sum_{k=\lceil na \rceil}^{\lfloor nb \rfloor} f\left(\frac{k}{n}\right) \widetilde{Z}(nx - k)}{\sum_{k=\lceil na \rceil}^{\lfloor nb \rfloor} \widetilde{Z}(nx - k)} = \tag{9.93}$$

$$\frac{\sum_{k_1=\lceil na_1 \rceil}^{\lfloor nb_1 \rfloor} \sum_{k_2=\lceil na_2 \rceil}^{\lfloor nb_2 \rfloor} \cdots \sum_{k_N=\lceil na_N \rceil}^{\lfloor nb_N \rfloor} f\left(\frac{k_1}{n}, \ldots, \frac{k_N}{n}\right) \left(\prod_{i=1}^{N} \left(G_{q,\lambda}(nx_i - k_i) + G_{\frac{1}{q},\lambda}(nx_i - k_i)\right)\right)}{\prod_{i=1}^{N} \left(\sum_{k_i=\lceil na_i \rceil}^{\lfloor nb_i \rfloor} \left(G_{q,\lambda}(nx_i - k_i) + G_{\frac{1}{q},\lambda}(nx_i - k_i)\right)\right)}.$$

We need

Theorem 9.4. *([229]) It holds*

$$\|\psi_n^s(f) - f\|_{\infty,\prod_{i=1}^N [a_i,b_i]} \leq 2\omega_1\left(f,(K(q))^N \left[\frac{1}{n^\beta} + \|b-a\|_\infty \widetilde{\gamma} A^{-\lambda\left(n^{1-\beta}-2\right)}\right]\right), \tag{9.94}$$

$\forall\ f \in C\left(\prod_{i=1}^N [a_i,b_i]\right)$, $0 < \beta < 1$.

By (9.94), we obtain $\psi_n^s(f) \to f$, as $n \to +\infty$, pointwise and uniformly.

Notation 9.1. We set

$$_1Z_q := Z_1,\ _2Z_q := \widetilde{Z}, \tag{9.95}$$

and

$$\Phi_1^*(q) := \Delta(q),\ \Phi_2^*(q) := K(q). \tag{9.96}$$

9.3 Main Result

About p-Mean Basic Approximation by Fuzzy-Random Symmetrized Perturbed Quasi-Interpolation Neural Network Operators

All terms and assumptions here as in Sections 1, 2.

Let $f \in C_{\mathcal{FR}}^{U_p}\left(\prod_{i=1}^N [a_i,b_i]\right)$, $1 \leq p < +\infty$, $n, N \in \mathbb{N}$, $0 < \beta^* < 1$, $\overrightarrow{x} \in \left(\prod_{i=1}^N [a_i,b_i]\right)$, $(X, \mathcal{B}, P)$ probability space, $s \in X$; $j = 1, 2$.

We define the following multivariate fuzzy random symmetrized perturbed quasi-interpolation linear neural network operators

$$\left({}_j^sF_n^{\mathcal{FR}}(f)\right)(\overrightarrow{x}, s) := \sum_{\overrightarrow{k}=\lceil na\rceil}^{\lfloor nb\rfloor *} f\left(\frac{\overrightarrow{k}}{n}, s\right) \odot \frac{{}_jZ_q\left(n\overrightarrow{x} - \overrightarrow{k}\right)}{\sum_{\overrightarrow{k}=\lceil na\rceil}^{\lfloor nb\rfloor} {}_jZ_q\left(n\overrightarrow{x} - \overrightarrow{k}\right)}, \tag{9.97}$$

for $j = 1, 2$.

We present

Theorem 9.5. *Let $f \in C_{\mathcal{FR}}^{U_p}\left(\prod_{i=1}^N [a_i,b_i]\right)$, $0 < \beta^* < 1$, $\overrightarrow{x} \in \left(\prod_{i=1}^N [a_i,b_i]\right)$, $n, N \in \mathbb{N}$, with $n^{1-\beta^*} > 2$, $1 \leq p < +\infty$. Assume that $\int_X (D^*(f(\cdot,s), \widetilde{o}))^p P(ds) < \infty$; $j = 1, 2$. Then*

1)

$$\left(\int_X D^p\left(\left({}_j^sF_n^{\mathcal{FR}}(f)\right)(\overrightarrow{x},s),f(\overrightarrow{x},s)\right)P(ds)\right)^{\frac{1}{p}} \leq \tag{9.98}$$

$$\left(\Phi_j^*(q)\right)^N\left\{\Omega_1\left(f,\frac{1}{n^{\beta^*}}\right)_{L^p}+2c_j(n,\beta^*)\left(\int_X\left(D^*(f(\cdot,s),\widetilde{o})\right)^pP(ds)\right)^{\frac{1}{p}}\right\}=:{}^s\mu_j^{(\mathcal{FR})},$$

2)

$$\left\|\left(\int_X D^p\left(\left({}_j^sFn^{\mathcal{FR}}(f)\right)(\overrightarrow{x},s),f(\overrightarrow{x},s)\right)P(ds)\right)^{\frac{1}{p}}\right\|_{\infty,\left(\prod_{i=1}^N[a_i,b_i]\right)}\leq {}^s\mu_j^{(\mathcal{FR})}, \tag{9.99}$$

where $\left(\Phi_j^*(q)\right)^N$ *as in (9.96), (9.42), (9.90) and* $c_j(n,\beta^*)$ *as in (9.43), (9.91).*

Proof. We notice that

$$D\left(f\left(\frac{\overrightarrow{k}}{n},s\right),f(\overrightarrow{x},s)\right)\leq D\left(f\left(\frac{\overrightarrow{k}}{n},s\right),\widetilde{o}\right)+D\left(f(\overrightarrow{x},s),\widetilde{o}\right) \tag{9.100}$$

$$\leq 2D^*\left(f(\cdot,s),\widetilde{o}\right).$$

Hence

$$D^p\left(f\left(\frac{\overrightarrow{k}}{n},s\right),f(\overrightarrow{x},s)\right)\leq 2^pD^{*p}\left(f(\cdot,s),\widetilde{o}\right), \tag{9.101}$$

and

$$\left(\int_X D^p\left(f\left(\frac{\overrightarrow{k}}{n},s\right),f(\overrightarrow{x},s)\right)P(ds)\right)^{\frac{1}{p}}\leq 2\left(\int_X\left(D^*(f(\cdot,s),\widetilde{o})\right)^pP(ds)\right)^{\frac{1}{p}}. \tag{9.102}$$

We observe that

$$D\left(\left({}_j^sF_n^{\mathcal{FR}}(f)\right)(\overrightarrow{x},s),f(\overrightarrow{x},s)\right)= \tag{9.103}$$

$$D\left(\sum_{\overrightarrow{k}=\lceil na\rceil}^{\lfloor nb\rfloor *}f\left(\frac{\overrightarrow{k}}{n},s\right)\odot\frac{{}_jZ_q(nx-k)}{\sum_{\overrightarrow{k}=\lceil na\rceil}^{\lfloor nb\rfloor}{}_jZ_q(nx-k)},f(\overrightarrow{x},s)\odot 1\right)=$$

$$D\left(\sum_{\vec{k}=\lceil na\rceil}^{\lfloor nb\rfloor *} f\left(\frac{\vec{k}}{n},s\right)\odot\frac{{}_{j}Z_q(nx-k)}{\sum\limits_{\vec{k}=\lceil na\rceil}^{\lfloor nb\rfloor}{}_{j}Z_q(nx-k)},\ f(\vec{x},s)\odot\frac{\sum\limits_{\vec{k}=\lceil na\rceil}^{\lfloor nb\rfloor}{}_{j}Z_q(nx-k)}{\sum\limits_{\vec{k}=\lceil na\rceil}^{\lfloor nb\rfloor}{}_{j}Z_q(nx-k)}\right)= \tag{9.104}$$

$$D\left(\sum_{\vec{k}=\lceil na\rceil}^{\lfloor nb\rfloor *} f\left(\frac{\vec{k}}{n},s\right)\odot\frac{{}_{j}Z_q(nx-k)}{\sum\limits_{\vec{k}=\lceil na\rceil}^{\lfloor nb\rfloor}{}_{j}Z_q(nx-k)},\ \sum_{\vec{k}=\lceil na\rceil}^{\lfloor nb\rfloor *} f(\vec{x},s)\odot\frac{{}_{j}Z_q(nx-k)}{\sum\limits_{\vec{k}=\lceil na\rceil}^{\lfloor nb\rfloor}{}_{j}Z_q(nx-k)}\right)$$

$$\leq\sum_{\vec{k}=\lceil na\rceil}^{\lfloor nb\rfloor}\left(\frac{{}_{j}Z_q(nx-k)}{\sum\limits_{\vec{k}=\lceil na\rceil}^{\lfloor nb\rfloor}{}_{j}Z_q(nx-k)}\right)D\left(f\left(\frac{\vec{k}}{n},s\right),f(\vec{x},s)\right). \tag{9.105}$$

So that

$$D\left(\left({}_{j}^{s}F_n^{\mathcal{FR}}(f)\right)(\vec{x},s),f(\vec{x},s)\right)\leq$$

$$\sum_{\vec{k}=\lceil na\rceil}^{\lfloor nb\rfloor}\left(\frac{{}_{j}Z_q(nx-k)}{\sum\limits_{\vec{k}=\lceil na\rceil}^{\lfloor nb\rfloor}{}_{j}Z_q(nx-k)}\right)D\left(f\left(\frac{\vec{k}}{n},s\right),f(\vec{x},s)\right)= \tag{9.106}$$

$$\sum_{\substack{\vec{k}=\lceil na\rceil\\ \left\|\frac{\vec{k}}{n}-\vec{x}\right\|_{\infty}\leq\frac{1}{n^{\beta *}}}}^{\lfloor nb\rfloor}\left(\frac{{}_{j}Z_q(nx-k)}{\sum\limits_{\vec{k}=\lceil na\rceil}^{\lfloor nb\rfloor}{}_{j}Z_q(nx-k)}\right)D\left(f\left(\frac{\vec{k}}{n},s\right),f(\vec{x},s)\right)+$$

$$\sum_{\substack{\vec{k}=\lceil na\rceil\\ \left\|\frac{\vec{k}}{n}-\vec{x}\right\|_{\infty}>\frac{1}{n^{\beta *}}}}^{\lfloor nb\rfloor}\left(\frac{{}_{j}Z_q(nx-k)}{\sum\limits_{\vec{k}=\lceil na\rceil}^{\lfloor nb\rfloor}{}_{j}Z_q(nx-k)}\right)D\left(f\left(\frac{\vec{k}}{n},s\right),f(\vec{x},s)\right).$$

Hence it holds

$$\left(\int_X D^p\left(\left({}_j^sF_n^{\mathcal{FR}}(f)\right)(\overrightarrow{x},s), f(\overrightarrow{x},s)\right)P(ds)\right)^{\frac{1}{p}} \leq \tag{9.107}$$

$$\sum_{\substack{\overrightarrow{k}=\lceil na\rceil \\ \left\|\frac{\overrightarrow{k}}{n}-\overrightarrow{x}\right\|_\infty \leq \frac{1}{n^{\beta^*}}}}^{\lfloor nb\rfloor}\left(\frac{{}_jZ_q(nx-k)}{\sum_{\overrightarrow{k}=\lceil na\rceil}^{\lfloor nb\rfloor}{}_jZ_q(nx-k)}\right)\left(\int_X D^p\left(f\left(\frac{\overrightarrow{k}}{n},s\right), f(\overrightarrow{x},s)\right)P(ds)\right)^{\frac{1}{p}} +$$

$$\sum_{\substack{\overrightarrow{k}=\lceil na\rceil \\ \left\|\frac{\overrightarrow{k}}{n}-\overrightarrow{x}\right\|_\infty > \frac{1}{n^{\beta^*}}}}^{\lfloor nb\rfloor}\left(\frac{{}_jZ_q(nx-k)}{\sum_{\overrightarrow{k}=\lceil na\rceil}^{\lfloor nb\rfloor}{}_jZ_q(nx-k)}\right)\left(\int_X D^p\left(f\left(\frac{\overrightarrow{k}}{n},s\right), f(\overrightarrow{x},s)\right)P(ds)\right)^{\frac{1}{p}} \leq$$

$$\left(\frac{1}{\sum_{\overrightarrow{k}=\lceil na\rceil}^{\lfloor nb\rfloor}{}_jZ_q(nx-k)}\right)\cdot\left\{\Omega_1^{(\mathcal{F})}\left(f,\frac{1}{n^{\beta^*}}\right)_{L^p} + \right. \tag{9.108}$$

$$\left.2\left(\int_X \left(D^*\left(f(\cdot,s),\widetilde{o}\right)\right)^p P(ds)\right)^{\frac{1}{p}}\left(\sum_{\substack{\overrightarrow{k}=\lceil na\rceil \\ \left\|\frac{\overrightarrow{k}}{n}-\overrightarrow{x}\right\|_\infty > \frac{1}{n^{\beta^*}}}}^{\lfloor nb\rfloor}{}_jZ_q(nx-k)\right)\right\}$$

(by (9.41), (9.42), and (9.89), (9.90), (9.96)).

$$\leq\left(\Phi_j^*(q)\right)^N\left\{\Omega_1^{(\mathcal{F})}\left(f,\frac{1}{n^{\beta^*}}\right)_{L^p} + 2c_j(n,\beta^*)\left(\int_X \left(D^*\left(f(\cdot,s),\widetilde{o}\right)\right)^p P(ds)\right)^{\frac{1}{p}}\right\}. \tag{9.109}$$

We have proved claim. □

Conclusion 9.1. By Theorem 9.5 we obtain the pointwise and uniform convergences with rates in the p-mean and D-metric of the symmetrized operator ${}_j^sF_n^{\mathcal{FR}}$ to the unit operator for $f\in C_{\mathcal{FR}}^{U_p}\left(\prod_{i=1}^N[a_i,b_i]\right)$, $j=1,2$.

Bibliography

[224] G.A. Anastassiou, *Multivariate Fuzzy-Random Quasi-interpolation neural network approximation operators*, J. Fuzzy Mathematics, Vol. 22, No. 1, 2014, 167–184.

[225] G.A. Anastassiou, *Intelligent Systems II: Complete Approximation by Neural Network Operators*, Springer, Heidelberg, New York, 2016.

[226] G.A. Anastassiou, *General sigmoid based Banach space valued neural network approximation*, J. of Computational Analysis and Applications, 31 (4) (2023), 520–534.

[227] G.A. Anastassiou, *Parametrized Deformed and General Neural Networks*, Chapter 15, Springer, Heidelberg, New York, 2023.

[228] G.A. Anastassiou, *Generalized Logistic Neural Networks as Positive Linear Operators, Multivariate and finite domain case*, submitted, 2025.

[229] G.A. Anastassiou, *Neural Networks Positive Linear Operators, Multivariate and abstract case*, submitted, 2025.

[230] S. Gal, *Approximation Theory in Fuzzy Setting*, Chapter 13 in Handbook of Analytic-Computational Methods in Applied Mathematics, pp. 617–666, edited by G. Anastassiou, Chapman & Hall/CRC, 2000, Boca Raton, New York.

[231] Wu Congxin, Gong Zengtai, *On Henstock integral of interval-valued functions and fuzzy valued functions*, Fuzzy Sets and Systems, Vol. 115, No. 3, 2000, 377–391.

[232] C. Wu, Z. Gong, *On Henstock integral of fuzzy-number-valued functions (I)*, Fuzzy Sets and Systems, 120, No. 3, (2001), 523–532.

[233] C. Wu, M. Ma, *On embedding problem of fuzzy number space: Part 1*, Fuzzy Sets and Systems, 44 (1991), 33–38.

[234] G.A. Anastassiou, *Multivariate Fuzzy-Random Symmetrized and Perturbed Neural Network enhanced Approximation*, submitted, 2025.

Chapter 10

Multivariate Generalized Logistic Neural Networks as Positive Linear Operators in Finite Domain

Essential multivariate-abstract neural network operators are interpreted as positive linear operators and the related general theory applies to them. These operators are induced by a symmetrized density function deriving from the parametrized and deformed generalized logistic activation function. We are acting on the space of continuous functions on a box in $\mathbb{R}^N$, $N \in \mathbb{N}$, with values to a general Banach space. We study quantitatively the rate of convergence of these neural network operators to the unit operator. Our inequalities involve the modulus of continuity of the function under approximation or its Fréchet derivative. Convexity of function case is studied in detail. It follows [262].

10.1 Introduction

The author studied extensively the quantitative approximation of positive linear operators to the unit since 1985, see for example [235]–[241], which motivated this work. He originated from the quantitative weak convergence of finite positive measures to the unit Dirac measure, having as a method the geometric moment theory, see [236], and he produced best upper bounds, leading to attained (i.e. sharp Jackson type inequalities), e.g. see [235], [236]. These studies have gone to all possible directions, univariate and multivariate, though in this work we treat only the multivariate-abstract case of finite domain.

Also, since 1997, this author started the study of the quantitative convergence of neural network operators to the unit, and he has been written since then numerous of articles and books, e.g. see [237], [239]–[241], which are what we use here.

The wide range of neural network operators being treated by the author all along, are indeed by nature positive linear operators.

Here, for the first time in the literature, multivariate-abstract neural network operators are treated as positive linear operators.

So, important methods of positive linear operators apply here to our finite summation defined neural network operators, producing new and interesting results: pointwise and uniform. Via the Riesz representation theorem neural networks are connected to measure theory. Convexity of functions produces optimal and sharper results.

The use of perturbed generalized logistic function as an activation function is well established and supported in [240], Chapters 15, 16. Generalized logistic activation function is greatly behaving and one of the most commonly used activation functions on this subject.

The author's symmetrization method presented here aims that our operators converge at high speed to the unit operator using half feed of data.

This chapter establishes a bridge for the first time connecting Neural Networks Approximation (part of AI) to Positive Linear Operators (a great part of Functional Analysis).

So it is a theoretical work intended by nature mainly to related experts.

Of great inspiration always have been [247], [248]. For classical studies on neural network we recommend also, [243]–[245], and [249]–[251].

For newer work on neural networks we also refer the reader to [252]–[261].

10.2 Basics

Initially we follow [240], pp. 395–471.

Our activation function here is the q-deformed and λ-parametrized function

$$\varphi_{q,\lambda}(x) = \frac{1}{1 + qA^{-\lambda x}}, \quad x \in \mathbb{R}, \ q, \lambda > 0, \ A > 1. \tag{10.1}$$

This is the A-generalized logistic function.

For more read Chapter 16 of [240]: "Banach space valued ordinary and fractional neural network approximation based on q-deformed and λ-parametrized A-generalized logistic function".

The chapters 15, 16 of [240] motivate our current work.

The proposed "symmetrization technique" aims to use half data to feed into our neural networks.

We will employ the following density function

$$G_{q,\lambda}(x) := \frac{1}{2}\left(\varphi_{q,\lambda}(x+1) - \varphi_{q,\lambda}(x-1)\right), \quad x \in \mathbb{R},\ q,\lambda > 0. \tag{10.2}$$

We have that

$$G_{q,\lambda}(-x) = G_{\frac{1}{q},\lambda}(x), \tag{10.3}$$

and

$$G_{\frac{1}{q},\lambda}(-x) = G_{q,\lambda}(x), \quad \forall\, x \in \mathbb{R}. \tag{10.4}$$

Adding (10.3) and (10.4) we obtain

$$G_{q,\lambda}(-x) + G_{\frac{1}{q},\lambda}(-x) = G_{q,\lambda}(x) + G_{\frac{1}{q},\lambda}(x), \quad \forall\, x \in \mathbb{R}, \tag{10.5}$$

the key to this work.

So that

$$W(x) := \frac{G_{q,\lambda}(x) + G_{\frac{1}{q},\lambda}(x)}{2} \tag{10.6}$$

is an even function, symmetric with respect to the y-axis.

The global maximum of $G_{q,\lambda}$ is given by (16.18), p. 401 of [240] as

$$G_{q,\lambda}\left(\frac{\log_A q}{\lambda}\right) = \frac{A^{\lambda}-1}{2\left(A^{\lambda}+1\right)}. \tag{10.7}$$

And, the global max of $G_{\frac{1}{q},\lambda}$ is

$$G_{\frac{1}{q},\lambda}\left(\frac{\log_A \frac{1}{q}}{\lambda}\right) = G_{\frac{1}{q},\lambda}\left(\frac{-\log_A q}{\lambda}\right) = \frac{A^{\lambda}-1}{2\left(A^{\lambda}+1\right)}, \tag{10.8}$$

both sharing the same maximum at symmetric points.

By Theorem 16.1, p. 401 of [240], we have that

$$\sum_{i=-\infty}^{\infty} G_{q,\lambda}(x-i) = 1, \quad \forall\, x \in \mathbb{R},\ \lambda, q > 0,\ A > 1, \tag{10.9}$$

and

$$\sum_{i=-\infty}^{\infty} G_{\frac{1}{q},\lambda}(x-i) = 1, \quad \forall\, x \in \mathbb{R},\ \lambda, q > 0,\ A > 1. \tag{10.10}$$

Consequently, we derive that

$$\sum_{i=-\infty}^{\infty} W(x-i) = 1, \quad \forall\, x \in \mathbb{R}. \tag{10.11}$$

By Theorem 16.2, p. 402 of [240], we have that

$$\int_{-\infty}^{\infty} G_{q,\lambda}(x)\,dx = 1, \quad \lambda, q > 0, \ A > 1, \tag{10.12}$$

similarly it holds

$$\int_{-\infty}^{\infty} G_{\frac{1}{q},\lambda}(x)\,dx = 1, \tag{10.13}$$

so that

$$\int_{-\infty}^{\infty} W(x)\,dx = 1, \tag{10.14}$$

therefore W is a density function.

By Theorem 16.3, p. 402 of [240], we have:

Let $0 < \alpha < 1$, and $n \in \mathbb{N}$ with $n^{1-\alpha} > 2$. Then

$$\sum_{\begin{cases} k = -\infty \\ : |nx - k| \geq n^{1-\alpha} \end{cases}}^{\infty} G_{q,\lambda}(nx - k) < 2\max\left\{q, \frac{1}{q}\right\} \frac{1}{A^{\lambda\left(n^{1-\alpha}-2\right)}} = \widetilde{\gamma} A^{-\lambda\left(n^{1-\alpha}-2\right)}, \tag{10.15}$$

where $\lambda, q > 0$, $A > 1$; $\widetilde{\gamma} := 2\max\left\{q, \frac{1}{q}\right\}$.

Similarly, we get that

$$\sum_{\begin{cases} k = -\infty \\ : |nx - k| \geq n^{1-\alpha} \end{cases}}^{\infty} G_{\frac{1}{q},\lambda}(nx - k) < \widetilde{\gamma} A^{-\lambda\left(n^{1-\alpha}-2\right)}. \tag{10.16}$$

Consequently we obtain that

$$\sum_{\begin{cases} k = -\infty \\ : |nx - k| \geq n^{1-\alpha} \end{cases}}^{\infty} W(nx - k) < \widetilde{\gamma} A^{-\lambda\left(n^{1-\alpha}-2\right)}, \tag{10.17}$$

where $\widetilde{\gamma} := 2\max\left\{q, \frac{1}{q}\right\}$.

Here $\lceil\cdot\rceil$ denotes the ceiling of the number, and $\lfloor\cdot\rfloor$ its integral part.

We mention

Theorem 16.4 (p. 402, [240]) Let $x \in [a,b] \subset \mathbb{R}$ and $n \in \mathbb{N}$ so that $\lceil na \rceil \leq \lfloor nb \rfloor$. For $q > 0$, $\lambda > 0$, $A > 1$, we consider the number $\lambda_q > z_0 > 0$ with $G_{q,\lambda}(z_0) = G_{q,\lambda}(0)$, and $\lambda_q > 1$. Then

$$\frac{1}{\sum_{k=\lceil na \rceil}^{\lfloor nb \rfloor} G_{q,\lambda}(nx - k)} < \max\left\{\frac{1}{G_{q,\lambda}(\lambda_q)}, \frac{1}{G_{\frac{1}{q},\lambda}\left(\lambda_{\frac{1}{q}}\right)}\right\} =: K(q). \tag{10.18}$$

Similarly, we consider $\lambda_{\frac{1}{q}} > z_1 > 0$, such that $G_{\frac{1}{q},\lambda}(z_1) = G_{\frac{1}{q},\lambda}(0)$, and $\lambda_{\frac{1}{q}} > 1$. Thus

$$\frac{1}{\sum_{k=\lceil na \rceil}^{\lfloor nb \rfloor} G_{\frac{1}{q},\lambda}(nx-k)} < \max\left\{\frac{1}{G_{\frac{1}{q},\lambda}\left(\lambda_{\frac{1}{q}}\right)}, \frac{1}{G_{q,\lambda}(\lambda_q)}\right\} = K(q). \quad (10.19)$$

Hence

$$\sum_{k=\lceil na \rceil}^{\lfloor nb \rfloor} G_{q,\lambda}(nx-k) > \frac{1}{K(q)}, \quad (10.20)$$

and

$$\sum_{k=\lceil na \rceil}^{\lfloor nb \rfloor} G_{\frac{1}{q},\lambda}(nx-k) > \frac{1}{K(q)}. \quad (10.21)$$

Consequently it holds

$$\sum_{k=\lceil na \rceil}^{\lfloor nb \rfloor} \frac{\left(G_{q,\lambda}(nx-k) + G_{\frac{1}{q},\lambda}(nx-k)\right)}{2} > \frac{2}{2K(q)} = \frac{1}{K(q)}, \quad (10.22)$$

so that

$$\frac{1}{\sum_{k=\lceil na \rceil}^{\lfloor nb \rfloor} \frac{\left(G_{q,\lambda}(nx-k) + G_{\frac{1}{q},\lambda}(nx-k)\right)}{2}} < K(q), \quad (10.23)$$

that is

$$\frac{1}{\sum_{k=\lceil na \rceil}^{\lfloor nb \rfloor} W(nx-k)} < K(q). \quad (10.24)$$

We have proved

Theorem 10.1. *Let $x \in [a,b] \subset \mathbb{R}$ and $n \in \mathbb{N}$ so that $\lceil na \rceil \leq \lfloor nb \rfloor$. For $q, \lambda > 0$, $A > 1$, we consider $\lambda_q > z_0 > 0$ with $G_{q,\lambda}(z_0) = G_{q,\lambda}(0)$, and $\lambda_q > 1$. Also consider $\lambda_{\frac{1}{q}} > z_1 > 0$, such that $G_{\frac{1}{q},\lambda}(z_1) = G_{\frac{1}{q},\lambda}(0)$, and $\lambda_{\frac{1}{q}} > 1$. Then*

$$\frac{1}{\sum_{k=\lceil na \rceil}^{\lfloor nb \rfloor} W(nx-k)} < K(q). \quad (10.25)$$

We make

Remark 10.1.

I) By Remark 16.5, p. 402 of [240], we have

$$\lim_{n\to\infty}\sum_{k=\lceil na\rceil}^{\lfloor nb\rfloor} G_{q,\lambda}(nx_1-k)\neq 1,\ \text{ for some } x_1\in[a,b], \tag{10.26}$$

and

$$\lim_{n\to\infty}\sum_{k=\lceil na\rceil}^{\lfloor nb\rfloor} G_{\frac{1}{q},\lambda}(nx_2-k)\neq 1,\ \text{ for some } x_2\in[a,b]. \tag{10.27}$$

Therefore it holds

$$\lim_{n\to\infty}\sum_{k=\lceil na\rceil}^{\lfloor nb\rfloor} \frac{\left(G_{q,\lambda}(nx_1-k)+G_{\frac{1}{q},\lambda}(nx_2-k)\right)}{2}\neq 1. \tag{10.28}$$

Hence

$$\lim_{n\to\infty}\sum_{k=\lceil na\rceil}^{\lfloor nb\rfloor} \frac{\left(G_{q,\lambda}(nx_1-k)+G_{\frac{1}{q},\lambda}(nx_1-k)\right)}{2}\neq 1, \tag{10.29}$$

even if

$$\lim_{n\to\infty}\sum_{k=\lceil na\rceil}^{\lfloor nb\rfloor} G_{\frac{1}{q},\lambda}(nx_1-k)=1, \tag{10.30}$$

because then

$$\lim_{n\to\infty}\sum_{k=\lceil na\rceil}^{\lfloor nb\rfloor} \frac{G_{q,\lambda}(nx_1-k)}{2}+\frac{1}{2}\neq 1, \tag{10.31}$$

equivalently

$$\lim_{n\to\infty}\sum_{k=\lceil na\rceil}^{\lfloor nb\rfloor} \frac{G_{q,\lambda}(nx_1-k)}{2}\neq \frac{1}{2}, \tag{10.32}$$

true by

$$\lim_{n\to\infty}\sum_{k=\lceil na\rceil}^{\lfloor nb\rfloor} G_{q,\lambda}(nx_1-k)\neq 1. \tag{10.33}$$

II) Let $[a,b]\subset\mathbb{R}$. For large n we always have $\lceil na\rceil\leq\lfloor nb\rfloor$. Also $a\leq\frac{k}{n}\leq b$, iff $\lceil na\rceil\leq k\leq\lfloor nb\rfloor$. So in general it holds

$$\sum_{k=\lceil na\rceil}^{\lfloor nb\rfloor} W(nx-k)\leq 1. \tag{10.34}$$

Next, we move on to the multivariate case, see Chapter 15 of [240], pp. 365–394, as a model of action.

We make

Remark 10.2. We introduce

$$\widetilde{Z}(x_1,...,x_N) := \widetilde{Z}(x) := \prod_{i=1}^{N} W(x_i) = \frac{1}{2^N}\prod_{i=1}^{N}\left(G_{q,\lambda} + G_{\frac{1}{q},\lambda}\right)(x_i), \tag{10.35}$$

$x = (x_1,...,x_N) \in \mathbb{R}^N$, $\lambda, q > 0$, $N \in \mathbb{N}$.

Properties:

(i)

$$\widetilde{Z}(x) > 0, \ \forall\, x \in \mathbb{R}^N, \tag{10.36}$$

(ii)

$$\sum_{k=-\infty}^{\infty} \widetilde{Z}(x-k) := \sum_{k_1=-\infty}^{\infty}\sum_{k_2=-\infty}^{\infty} ... \sum_{k_N=-\infty}^{\infty} \widetilde{Z}(x_1-k_1,...,x_N-k_N) = 1, \tag{10.37}$$

$k := (k_1,...,k_n) \in \mathbb{Z}^N$, $\forall\, x \in \mathbb{R}^N$,

hence

(iii)

$$\sum_{k=-\infty}^{\infty} \widetilde{Z}(nx-k) = 1, \tag{10.38}$$

$\forall\, x \in \mathbb{R}^N$, $n \in \mathbb{N}$,

(iv)

$$\int_{\mathbb{R}^N} \widetilde{Z}(x)\,dx = 1, \tag{10.39}$$

that is $\widetilde{Z}$ is a multivariate density function.

Here denote $\|x\|_{\infty} := \max\{|x_1|,...,|x_N|\}$, $x \in \mathbb{R}^N$, $\infty := (\infty,...,\infty)$, $-\infty := (-\infty,...,-\infty)$, $\lceil na \rceil := (\lceil na_1 \rceil,...,\lceil na_N \rceil)$, $\lfloor nb \rfloor := (\lfloor nb_1 \rfloor,...,\lfloor nb_N \rfloor)$, $a := (a_1,...,a_N)$, $b := (b_1,...,b_N)$.

We obviously see that

$$\sum_{k=\lceil na \rceil}^{\lfloor nb \rfloor} \widetilde{Z}(nx-k) = \sum_{k=\lceil na \rceil}^{\lfloor nb \rfloor}\left(\prod_{i=1}^{N} W(nx_i-k_i)\right) = \prod_{i=1}^{N}\left(\sum_{k_i=\lceil na_i \rceil}^{\lfloor nb_i \rfloor} W(nx_i-k_i)\right). \tag{10.40}$$

(v) We derive that

$$\sum_{\substack{k=\lceil na\rceil \\ \left\|\frac{k}{n}-x\right\|_{\infty}>\frac{1}{n^{\beta}}}}^{\lfloor nb\rfloor} \widetilde{Z}(nx-k)<\widetilde{\gamma} A^{-\lambda\left(n^{1-\beta}-2\right)}, \text{ where } 0<\beta<1, \tag{10.41}$$

$0<\beta<1$, $n \in \mathbb{N}: n^{1-\beta}>2$, $x \in \prod_{i=1}^{N}\left[a_{i}, b_{i}\right]$.

(vi) It holds

$$0<\frac{1}{\sum_{k=\lceil na\rceil}^{\lfloor nb\rfloor} \widetilde{Z}(nx-k)}<(K(q))^{N}, \tag{10.42}$$

$\forall\, x \in\left(\prod_{i=1}^{N}\left[a_{i}, b_{i}\right]\right)$, $n \in \mathbb{N}$.

It is clear that

(vii)

$$\sum_{\substack{k=-\infty \\ \left\|\frac{k}{n}-x\right\|_{\infty}>\frac{1}{n^{\beta}}}}^{\infty} \widetilde{Z}(nx-k)<\widetilde{\gamma} A^{-\lambda\left(n^{1-\beta}-2\right)}, \tag{10.43}$$

$0<\beta<1$, $n \in \mathbb{N}: n^{1-\beta}>2$, $x \in \prod_{i=1}^{N}\left[a_{i}, b_{i}\right]$.

Furthermore it holds

$$\lim_{n \rightarrow \infty} \sum_{k=\lceil na\rceil}^{\lfloor nb\rfloor} \widetilde{Z}(nx-k) \neq 1, \tag{10.44}$$

for at least some $x \in\left(\prod_{i=1}^{N}\left[a_{i}, b_{i}\right]\right)$.

Here $\left(X,\|\cdot\|_{\gamma}\right)$ is a Banach space.

Let $f \in C\left(\prod_{i=1}^{N}\left[a_{i}, b_{i}\right], X\right)$, $x=\left(x_{1}, \ldots, x_{N}\right) \in \prod_{i=1}^{N}\left[a_{i}, b_{i}\right]$, $n \in \mathbb{N}$: $\lceil na_{i}\rceil \leq\lfloor nb_{i}\rfloor$, $i=1, \ldots, N$.

We introduce and define the following multivariate linear normalized symmetrized neural network operator, let $x:=\left(x_{1}, \ldots, x_{N}\right) \in \prod_{i=1}^{N}\left[a_{i}, b_{i}\right]$:

$$\Psi_{n}^{s}\left(f, x_{1}, \ldots, x_{N}\right):=\Psi_{n}^{s}(f, x):=\frac{\sum_{k=\lceil na\rceil}^{\lfloor nb\rfloor} f\left(\frac{k}{n}\right) \widetilde{Z}(nx-k)}{\sum_{k=\lceil na\rceil}^{\lfloor nb\rfloor} \widetilde{Z}(nx-k)}=$$

$$\frac{\sum_{k_{1}=\lceil na_{1}\rceil}^{\lfloor nb_{1}\rfloor} \sum_{k_{2}=\lceil na_{2}\rceil}^{\lfloor nb_{2}\rfloor} \cdots \sum_{k_{N}=\lceil na_{N}\rceil}^{\lfloor nb_{N}\rfloor} f\left(\frac{k_{1}}{n}, \ldots, \frac{k_{N}}{n}\right)\left(\prod_{i=1}^{N}\left(G_{q, \lambda}\left(nx_{i}-k_{i}\right)+G_{\frac{1}{q}, \lambda}\left(nx_{i}-k_{i}\right)\right)\right)}{\prod_{i=1}^{N}\left(\sum_{k_{i}=\lceil na_{i}\rceil}^{\lfloor nb_{i}\rfloor}\left(G_{q, \lambda}\left(nx_{i}-k_{i}\right)+G_{\frac{1}{q}, \lambda}\left(nx_{i}-k_{i}\right)\right)\right)}. \tag{10.45}$$

For large enough $n \in \mathbb{N}$ we always obtain $\lceil na_i \rceil \leq \lfloor nb_i \rfloor$, $i = 1, ..., N$. Also $a_i \leq \frac{k_i}{n} \leq b_i$, iff $\lceil na_i \rceil \leq k_i \leq \lfloor nb_i \rfloor$, $i = 1, ..., N$.

When $g \in C\left(\prod_{i=1}^{N}[a_i, b_i]\right)$ we define the companion operator

$$\widetilde{\Psi}_n^s(g, x) := \frac{\sum_{k=\lceil na \rceil}^{\lfloor nb \rfloor} g\left(\frac{k}{n}\right) \widetilde{Z}(nx - k)}{\sum_{k=\lceil na \rceil}^{\lfloor nb \rfloor} \widetilde{Z}(nx - k)}. \tag{10.46}$$

Clearly $\widetilde{\Psi}_n^s$ is a positive linear operator. We have that

$$\widetilde{\Psi}_n^s(1, x) = 1, \ \ \forall\, x \in \prod_{i=1}^{N}[a_i, b_i].$$

Notice that $\Psi_n^s(f) \in C\left(\prod_{i=1}^{N}[a_i, b_i], X\right)$ and $\widetilde{\Psi}_n^s(g) \in C\left(\prod_{i=1}^{N}[a_i, b_i]\right)$.

Furthermore it holds

$$\|\Psi_n^s(f, x)\|_\gamma \leq \widetilde{\Psi}_n^s\left(\|f\|_\gamma, x\right), \tag{10.47}$$

$\forall\, x \in \prod_{i=1}^{N}[a_i, b_i]$,

and

$$\Psi_n^s(cg, x) = c\widetilde{\Psi}_n^s(g, x), \ \ \forall\, x \in \prod_{i=1}^{N}[a_i, b_i], \tag{10.48}$$

and

$$\Psi_n^s(c) = c, \ \ \forall\, c \in X. \tag{10.49}$$

Set also

$${}^*\Psi_n^s(f, x) := \sum_{k=\lceil na \rceil}^{\lfloor nb \rfloor} f\left(\frac{k}{n}\right) \widetilde{Z}(nx - k). \tag{10.50}$$

Consequently we derive (by (42))

$$|\Psi_n^s(f, x) - f(x)| \leq (K(q))^N \left|{}^*\Psi_n^s(f, x) - f(x) \sum_{k=\lceil na \rceil}^{\lfloor nb \rfloor} \widetilde{Z}(nx - k)\right|, \tag{10.51}$$

$\forall\, x \in \left(\prod_{i=1}^{N}[a_i, b_i]\right)$.

We will estimate the right-hand side of (10.51), when $X = (\mathbb{C}, |\cdot|)$, the complex numbers.

For the last and others we need

Definition 10.1. ([239], p. 274) Let M be a convex and compact subset of $\left(\mathbb{R}^N, \|\cdot\|_p\right)$, $p \in [1, \infty]$, and $\left(X, \|\cdot\|_\gamma\right)$ be a Banach space. Let $f \in C(M, X)$. We define the first modulus of continuity of f as

$$\omega_1(f, \delta) := \sup_{\substack{x, y \in M: \\ \|x-y\|_p \leq \delta}} \|f(x) - f(y)\|_\gamma, \quad 0 < \delta \leq diam(M). \tag{10.52}$$

If $\delta > diam(M)$, then

$$\omega_1(f, \delta) = \omega_1(f, diam(M)). \tag{10.53}$$

Notice $\omega_1(f, \delta)$ is increasing in $\delta > 0$. For $f \in C_B(M, X)$ (continuous and bounded functions) $\omega_1(f, \delta)$ is defined similarly.

Lemma 10.1. *([239], p. 274) We have $\omega_1(f, \delta) \to 0$ as $\delta \downarrow 0$, iff $f \in C(M, X)$, where M is a convex compact subset of $\left(\mathbb{R}^N, \|\cdot\|_p\right)$, $p \in [1, \infty]$.*

10.3 Abstract Korovkin Inequalities

All here in this section come from [238] and [239], Ch. 10, pp. 261–281.

We make

Remark 10.3. Let $\left(\mathbb{R}^k, \|\cdot\|_p\right)$, $k \in \mathbb{N}$; where $\|\cdot\|_p$ is the L_p-norm, $1 \leq p \leq +\infty$. $\mathbb{R}^k$ is a Banach space, and $\left(\mathbb{R}^k\right)^j$ denotes the j-fold product space $\mathbb{R}^k \times ... \times \mathbb{R}^k$ endowed with the max-norm $\|x\|_{(\mathbb{R}^k)^j} := \max_{1 \leq \lambda \leq j} \|x_\lambda\|_p$, where $x := (x_1, ..., x_j) \in \left(\mathbb{R}^k\right)^j$.

Let $\left(X, \|\cdot\|_\gamma\right)$ be a general Banach space. Then the space $L_j := L_j\left(\left(\mathbb{R}^k\right)^j; X\right)$ of all j-multilinear continuous maps $g : \left(\mathbb{R}^k\right)^j \to X$, $j = 1, ..., n$, is a Banach space with norm

$$\|g\| := \|g\|_{L_j} := \sup_{\left(\|x\|_{(\mathbb{R}^k)^j} = 1\right)} \|g(x)\|_\beta = \sup \frac{\|g(x)\|_\beta}{\|x_1\|_p \cdots \|x_j\|_p}. \tag{10.54}$$

Let M be a non-empty convex and compact subset of $\mathbb{R}^k$ and $x_0 \in M$ is fixed.

Let O an open subset of $\mathbb{R}^k : M \subset O$. Let $f : O \to X$ be a continuous function whose Fréchet derivatives (see [246]) $f^{(j)} : O \to L_j = L_j\left(\left(\mathbb{R}^k\right)^j ; X\right)$ exist and are continuous for $1 \leq j \leq n$, $n \in \mathbb{N}$. Call $(x - x_0)^j := (x - x_0, ..., x - x_0) \in \left(\mathbb{R}^k\right)^j$, $x \in M$.

Here we deal with $f|_M$.

Then by Taylor's formula ([242]), ([246], p. 124), we get

$$f(x) = \sum_{j=0}^{n} \frac{f^{(j)}(x_0)(x - x_0)^j}{j!} + R_n(x, x_0), \quad \text{all } x \in M, \tag{10.55}$$

where the remainder is the Riemann integral

$$R_n(x, x_0) := \int_0^1 \frac{(1-u)^{n-1}}{(n-1)!} \left(f^{(n)}(x_0 + u(x - x_0)) - f^{(n)}(x_0)\right)(x - x_0)^n \, du, \tag{10.56}$$

here we set $f^{(0)}(x_0)(x - x_0)^0 = f(x_0)$.

Considering

$$w := \omega_1\left(f^{(n)}, h\right) := \sup_{\substack{x,y \in M: \\ \|x-y\|_p \leq h}} \left\| f^{(n)}(x) - f^{(n)}(y) \right\|, \tag{10.57}$$

we obtain

$$\left\| \left(f^{(n)}(x_0 + u(x - x_0)) - f^{(n)}(x_0)\right)(x - x_0)^n \right\|_\gamma \leq$$

$$\left\| f^{(n)}(x_0 + u(x - x_0)) - f^{(n)}(x_0) \right\| \cdot \|x - x_0\|_p^n \leq$$

$$w \cdot \|x - x_0\|_p^n \cdot \left\lceil \frac{u \|x - x_0\|_p}{h} \right\rceil, \tag{10.58}$$

by Lemma 7.1.1, [236], p. 208, $\lceil \cdot \rceil$ is the ceiling.

Therefore for all $x \in M$ (see [236], pp. 121–122):

$$\|R_n(x, x_0)\|_\beta \leq w \|x - x_0\|_p^n \int_0^1 \left\lceil \frac{u \|x - x_0\|_p}{h} \right\rceil \frac{(1-u)^{n-1}}{(n-1)!} du$$

$$= w \Phi_n\left(\|x - x_0\|_p\right) \tag{10.59}$$

by a change of variable, where

$$\Phi_n(t) := \int_0^{|t|} \left\lceil \frac{s}{h} \right\rceil \frac{(|t| - s)^{n-1}}{(n-1)!} ds = \frac{1}{n!} \left(\sum_{j=0}^{\infty} (|t| - jh)_+^n \right), \quad \forall\, t \in \mathbb{R}, \tag{10.60}$$

is a (polynomial) spline function, see [236], pp. 210–211. Also from there we get

$$\Phi_n(t) \leq \left(\frac{|t|^{n+1}}{(n+1)!h} + \frac{|t|^n}{2n!} + \frac{h|t|^{n-1}}{8(n-1)!} \right), \quad \forall\, t \in \mathbb{R}, \tag{10.61}$$

with equality true only at $t = 0$. Therefore it holds

$$\|R_n(x, x_0)\|_\gamma \leq w\left(\frac{\|x - x_0\|_p^{n+1}}{(n+1)!h} + \frac{\|x - x_0\|_p^n}{2n!} + \frac{h\|x - x_0\|_p^{n-1}}{8(n-1)!} \right), \forall\, x \in M. \tag{10.62}$$

We have found that

$$\left\| f(x) - \sum_{j=0}^{n} \frac{f^{(j)}(x_0)(x - x_0)^j}{j!} \right\|_\gamma \leq$$

$$\omega_1\left(f^{(n)}, h\right) \left(\frac{\|x - x_0\|_p^{n+1}}{(n+1)!h} + \frac{\|x - x_0\|_p^n}{2n!} + \frac{h\|x - x_0\|_p^{n-1}}{8(n-1)!} \right) < \infty, \tag{10.63}$$

$\forall\, x, x_0 \in M$.

Here $0 < \omega_1\left(f^{(n)}, h\right) < \infty$, by M being compact and $f^{(n)}$ being continuous on M.

One can rewrite (10.63) as

$$\left\| f(\cdot) - \sum_{j=0}^{n} \frac{f^{(j)}(x_0)(\cdot - x_0)^j}{j!} \right\|_\gamma \leq$$

$$\omega_1\left(f^{(n)}, h\right) \left(\frac{\|\cdot - x_0\|_p^{n+1}}{(n+1)!h} + \frac{\|\cdot - x_0\|_p^n}{2n!} + \frac{h\|\cdot - x_0\|_p^{n-1}}{8(n-1)!} \right), \quad \forall\, x_0 \in M, \tag{10.64}$$

a pointwise functional inequality on M.

Here $(\cdot - x_0)^j$ maps M into $\left(\mathbb{R}^k\right)^j$ and it is continuous, also $f^{(j)}(x_0)$ maps $\left(\mathbb{R}^k\right)^j$ into X and it is continuous. Hence their composition $f^{(j)}(x_0)(\cdot - x_0)^j$ is continuous from M into X.

Clearly $f(\cdot) - \sum_{j=0}^{n} \frac{f^{(j)}(x_0)(\cdot - x_0)^j}{j!} \in C(M, X)$, hence $\left\| f(\cdot) - \sum_{j=0}^{n} \frac{f^{(j)}(x_0)(\cdot - x_0)^j}{j!} \right\|_\gamma \in C(M)$.

We mention the following fundamental approximation result.

Theorem 10.2. *Let M be a nonempty compact convex subset of the open subset O of $\left(\mathbb{R}^k, \|\cdot\|_p\right)$, $1 \leq p \leq +\infty$, and let $\left(X, \|\cdot\|_\gamma\right)$ be a general Banach space. For any $N \in \mathbb{N}$ let the linear operators $L_N : C(M, X) \to C(M, X)$ and the positive linear operators $\widetilde{L}_N : C(M) \to C(M)$, such that*

$$\|(L_N(f))(x_0)\|_\gamma \leq \left(\widetilde{L}_N\left(\|f\|_\gamma\right)\right)(x_0), \tag{10.65}$$

$\forall\ N \in \mathbb{N}$, $\forall\ f \in C(M, X)$, $\forall\ x_0 \in M$.

Furthermore assume that

$$L_N(cF) = c\widetilde{L}_N(F), \quad \forall\ F \in C(M), \forall\ c \in X. \tag{10.66}$$

Let $n \in \mathbb{N}$, here we deal with $f \in C^n(O, X)$, the space of n-times continuously Fréchet differentiable functions from O into X.

Here we study approximation to $f|_M$.

Let $x, x_0 \in M$, $r > 0$, and

$$\omega_1\left(f^{(n)}, h\right) := \sup_{\substack{x,y \in M: \\ \|x-y\|_p \leq h}} \left\|f^{(n)}(x) - f^{(n)}(y)\right\|. \tag{10.67}$$

Then

1)

$$\left\|(L_N(f))(x_0) - \sum_{j=0}^{n} \frac{1}{j!}\left(\widetilde{L}_N\left(f^{(j)}(x_0)(\cdot - x_0)^j\right)\right)(x_0)\right\|_\gamma \leq$$

$$\frac{\omega_1\left(f^{(n)}, r\left(\left(\widetilde{L}_N\left(\|\cdot - x_0\|_p^{n+1}\right)\right)(x_0)\right)^{\frac{1}{n+1}}\right)}{rn!}\left(\left(\widetilde{L}_N\left(\|\cdot - x_0\|_p^{n+1}\right)\right)(x_0)\right)^{\left(\frac{n}{n+1}\right)}$$

$$\left[\frac{1}{(n+1)} + \frac{r\left(\left(\widetilde{L}_N(1)\right)(x_0)\right)^{\frac{1}{n+1}}}{2} + \frac{nr^2\left(\left(\widetilde{L}_N(1)\right)(x_0)\right)^{\frac{2}{n+1}}}{8}\right], \tag{10.68}$$

2) additionally if $f^{(j)}(x_0) = 0$, $j = 0, 1, ..., n$, we have

$$\|(L_N(f))(x_0)\|_\gamma \leq$$

$$\frac{\omega_1\left(f^{(n)}, r\left(\left(\widetilde{L}_N\left(\|\cdot - x_0\|_p^{n+1}\right)\right)(x_0)\right)^{\frac{1}{n+1}}\right)}{rn!}\left(\left(\widetilde{L}_N\left(\|\cdot - x_0\|_p^{n+1}\right)\right)(x_0)\right)^{\left(\frac{n}{n+1}\right)}$$

$$\left[\frac{1}{(n+1)}+\frac{r\left(\left(\widetilde{L}_N(1)\right)(x_0)\right)^{\frac{1}{n+1}}}{2}+\frac{nr^2\left(\left(\widetilde{L}_N(1)\right)(x_0)\right)^{\frac{2}{n+1}}}{8}\right], \quad (10.69)$$

3)

$$\left\|(L_N(f))(x_0)-f(x_0)\right\|_\gamma \leq \|f(x_0)\|\left|\left(\widetilde{L}_N(1)\right)(x_0)-1\right|+$$

$$\sum_{j=1}^{n}\frac{1}{j!}\left\|\left(\widetilde{L}_N\left(f^{(j)}(x_0)(\cdot-x_0)^j\right)\right)(x_0)\right\|_\gamma+$$

$$\frac{\omega_1\left(f^{(n)},r\left(\left(\widetilde{L}_N\left(\|\cdot-x_0\|_p^{n+1}\right)\right)(x_0)\right)^{\frac{1}{n+1}}\right)}{rn!}\left(\left(\widetilde{L}_N\left(\|\cdot-x_0\|_p^{n+1}\right)\right)(x_0)\right)^{\left(\frac{n}{n+1}\right)}$$

$$\left[\frac{1}{(n+1)}+\frac{r\left(\left(\widetilde{L}_N(1)\right)(x_0)\right)^{\frac{1}{n+1}}}{2}+\frac{nr^2\left(\left(\widetilde{L}_N(1)\right)(x_0)\right)^{\frac{2}{n+1}}}{8}\right], \quad (10.70)$$

4)

$$\left\|\|L_N(f)-f\|_\gamma\right\|_{\infty,M} \leq \left\|\|f\|_\gamma\right\|_{\infty,M}\left\|\left(\widetilde{L}_N(1)\right)-1\right\|_{\infty,M}+$$

$$\sum_{j=1}^{n}\frac{1}{j!}\left\|\left\|\left(\widetilde{L}_N\left(f^{(j)}(x_0)(\cdot-x_0)^j\right)\right)(x_0)\right\|_\gamma\right\|_{\infty,x_0\in M}+$$

$$\frac{\omega_1\left(f^{(n)},r\left\|\left(\widetilde{L}_N\left(\|\cdot-x_0\|_p^{n+1}\right)\right)(x_0)\right\|_{\infty,x_0\in M}^{\frac{1}{n+1}}\right)}{rn!}$$

$$\left\|\left(\widetilde{L}_N\left(\|\cdot-x_0\|_p^{n+1}\right)\right)(x_0)\right\|_{\infty,x_0\in M}^{\left(\frac{n}{n+1}\right)}$$

$$\left[\frac{1}{(n+1)}+\frac{r\left\|\widetilde{L}_N(1)\right\|_{\infty,M}^{\frac{1}{n+1}}}{2}+\frac{nr^2\left\|\widetilde{L}_N(1)\right\|_{\infty,M}^{\frac{2}{n+1}}}{8}\right]. \quad (10.71)$$

We need

Lemma 10.2. *The function* $\left(\widetilde{L}_N\left(\|\cdot-x_0\|_p^m\right)\right)(x_0)$ *is continuous in* $x_0 \in M$, $m \in \mathbb{N}$.

Remark 10.4. We also have

$$\left\|\left(\widetilde{L}_N\left(\|\cdot - x_0\|_p^k\right)\right)(x_0)\right\|_{\infty, x_0\in M} \leq$$

$$\left\|\left(\widetilde{L}_N\left(\|\cdot - x_0\|_p^{n+1}\right)\right)(x_0)\right\|_{\infty, x_0\in M}^{\left(\frac{k}{n+1}\right)} \left\|\widetilde{L}_N(1)\right\|_{\infty, M}^{\left(\frac{n+1-k}{n+1}\right)}, \tag{10.72}$$

for all $k = 1, ..., n$.

Conclusion 10.1. Let $N \to \infty$ and $\widetilde{L}_N(1) \overset{u}{\to} 1$, uniformly, and $\left\|\left(\widetilde{L}_N\left(\|\cdot - x_0\|_p^{n+1}\right)\right)(x_0)\right\|_{\infty, x_0\in M} \to 0$. Then by (10.71) we get, as $N \to \infty$, that $L_n(f) \overset{u}{\to} f$, uniformly in $\|\cdot\|_\gamma$ over M.

The last statement is also supported by

$$\left\|\left(L_N\left(f^{(j)}(x_0)(\cdot - x_0)^j\right)\right)(x_0)\right\|_\beta \leq \left\|f^{(j)}\right\| \left\|\left(\widetilde{L}_N\left(\|\cdot - x_0\|_p^j\right)\right)(x_0)\right\|_{\infty, x_0\in M}$$

and (10.72). Here notice that $\left\|\widetilde{L}_N(1)\right\|_{\infty, M}$ turns out to be bounded, and $\omega_1\left(f^{(n)}, \cdot\right)$ is also bounded.

Conclusion 10.2. By (10.70) we get the following: as $N \to \infty$, assume that $\left(\widetilde{L}_N(1)\right)(x_0) \to 1$, and $\left(\left(\widetilde{L}_N\left(\|\cdot - x_0\|_p^{n+1}\right)\right)(x_0)\right) \to 0$. Then $(L_N(f))(x_0) \to f(x_0)$, as $N \to \infty$, pointwise. Notice that here $\left(\widetilde{L}_N(1)\right)(x_0)$ is bounded with respect to N, as well as $\omega_1\left(f^{(n)}, \cdot\right)$ is bounded in general.

Corollary 10.1. *(to Theorem 10.2, case $n = 1$, see (10.70)) Additionally assume that $\left(\widetilde{L}_N(1)\right)(x_0) = 1$, $\forall\, x_0 \in M$. Then*

1)

$$\|(L_N(f))(x_0) - f(x_0)\|_\gamma \leq \left\|\left(\widetilde{L}_N\left(f^{(1)}(x_0)(\cdot - x_0)\right)\right)(x_0)\right\|_\gamma +$$

$$\frac{1}{2r}\omega_1\left(f^{(1)}, r\left(\left(\widetilde{L}_N\left(\|\cdot - x_0\|_p^2\right)\right)(x_0)\right)^{\frac{1}{2}}\right)\left(\left(\widetilde{L}_N\left(\|\cdot - x_0\|_p^2\right)\right)(x_0)\right)^{\frac{1}{2}}$$

$$\left[1 + r + \frac{r^2}{4}\right], \tag{10.73}$$

2)

$$\|\|(L_N(f)) - f\|\|_{\infty, M} \leq$$

$$\left\| \left\| \left(\widetilde{L}_N \left(f^{(1)} (x_0) (\cdot - x_0) \right) \right) (x_0) \right\|_\gamma \right\|_{\infty, x_0 \in M} +$$

$$\frac{1}{2r} \omega_1 \left(f^{(1)}, r \left\| \left(\widetilde{L}_N \left(\|\cdot - x_0\|_p^2 \right) \right) (x_0) \right\|_{\infty, x_0 \in M}^{\frac{1}{2}} \right) \left\| \left(\widetilde{L}_N \left(\|\cdot - x_0\|_p^2 \right) \right) (x_0) \right\|_{\infty, x_0 \in M}^{\frac{1}{2}}$$

$$\left[1 + r + \frac{r^2}{4} \right]. \tag{10.74}$$

We need

Lemma 10.3. *(see [236], p. 208) Let M be a convex compact subset of $\left(\mathbb{R}^k, \|\cdot\|_p\right)$, $p \in [1, \infty]$, and $\left(X, \|\cdot\|_\gamma\right)$ be a general Banach space. Let $f \in B(M, X)$ (bounded functions). Then*

$$\|f(x) - f(x_0)\|_\gamma \leq \omega_1 (f, h) \left\lceil \frac{\|x - x_0\|_p}{h} \right\rceil$$

$$\leq \omega_1 (f, h) \left(1 + \frac{\|x - x_0\|_p}{h} \right), \quad \forall\, x \in M, \tag{10.75}$$

where $\lceil \cdot \rceil$ is the ceiling of the number, $h > 0$.

We present the basic pointwise convergence result

Theorem 10.3. *Let $L_N : C(M, X) \to C(M, X)$ be linear operators, $\forall\, N \in \mathbb{N}$; where M is a convex compact subset of $\left(\mathbb{R}^k, \|\cdot\|_p\right)$, $p \in [1, \infty]$, $\left(X, \|\cdot\|_\gamma\right)$ is a general Banach space. Let $\widetilde{L}_N : C(M) \to C(M)$ be positive linear operators, such that*

$$\|(L_N(f))(x_0)\|_\gamma \leq \left(\widetilde{L}_N \left(\|f\|_\gamma \right) \right) (x_0), \tag{10.76}$$

$\forall\, N \in \mathbb{N}$, $\forall\, f \in C(M, X)$, $\forall\, x_0 \in M$.

Furthermore assume that

$$L_N(cg) = c\widetilde{L}_N(g), \quad \forall\, g \in C(M), \forall\, c \in X. \tag{10.77}$$

Then

$$\|(L_N(f))(x_0) - f(x_0)\|_\gamma \leq \|f(x_0)\|_\gamma \left| \left(\widetilde{L}_N(1) \right)(x_0) - 1 \right| + \tag{10.78}$$

$$\left[\left(\widetilde{L}_N(1) \right)(x_0) + 1 \right] \omega_1 \left(f, \left(\widetilde{L}_N \left(\|\cdot - x_0\|_p \right) \right)(x_0) \right), \quad \forall\, f \in C(M, X).$$

By (78), as $\left(\widetilde{L}_N(1)\right)(x_0) \to 1$ and $\left(\widetilde{L}_N\left(\|\cdot - x_0\|_p\right)\right)(x_0) \to 0$, then $(L_N(f))(x_0) \overset{\|\cdot\|_\gamma}{\to} f(x_0)$, as $N \to \infty$, $\forall\, f \in C(M, X)$. Here notice that $\left(\widetilde{L}_N(1)\right)(x_0)$ is bounded.

Corollary 10.2. *All as in Theorem 10.3. Additionally assume that* $\left(\widetilde{L}_N(1)\right)(x_0) = 1$, $\forall\, x_0 \in M$. *Then*

$$\|(L_N(f))(x_0) - f(x_0)\|_\gamma \leq 2\omega_1\left(f, \left(\widetilde{L}_N\left(\|\cdot - x_0\|_p\right)\right)(x_0)\right), \forall\, f \in C(M, X). \tag{10.79}$$

Next we obtain the following uniform convergence result.

Theorem 10.4. *All as in Theorem 10.3. Then*

$$\left\| \|L_N(f) - f\|_\gamma \right\|_{\infty,M} \leq \left\| \|f\|_\gamma \right\|_{\infty,M} \left\| \widetilde{L}_N(1) - 1 \right\|_{\infty,M} + \tag{10.80}$$

$$\left\| \widetilde{L}_N(1) + 1 \right\|_{\infty,M} \omega_1\left(f, \left\| \left(\widetilde{L}_N\left(\|\cdot - x_0\|_p\right)\right)(x_0) \right\|_{\infty, x_0 \in M}\right), \forall\, f \in C(M, X).$$

Conclusion 10.3. By Lemma 10.2 $\left(\widetilde{L}_N\left(\|\cdot - x_0\|_p\right)\right)(x_0)$ is continuous in $x_0 \in M$. Thus $\left\| \left(\widetilde{L}_N\left(\|\cdot - x_0\|_p\right)\right)(x_0) \right\|_{\infty, x_0 \in M} < \infty$.

As $N \to \infty$, we assume that $\widetilde{L}_N(1) \overset{u}{\to} 1$, uniformly, and $\left\| \left(\widetilde{L}_N\left(\|\cdot - x_0\|_p\right)\right)(x_0) \right\|_{\infty, x_0 \in M} \to 0$, then $L_n(f) \overset{u}{\to} f$, uniformly on M, $\forall\, f \in C(M, X)$. Here $\left\| \widetilde{L}_N(1) \right\|$ is bounded.

Under convexity we have the following sharp general pointwise convergence result. Here M^0 denotes the interior of M.

Theorem 10.5. *All here as in Theorem 10.3. Additionally, assume that* $x_0 \in M^0$, *and the closed ball in* $\mathbb{R}^k : B\left(x_0, \left(\widetilde{L}_N\left(\|\cdot - x_0\|_p\right)\right)(x_0)\right) \subset M$, *where* $\left(\widetilde{L}_N\left(\|\cdot - x_0\|_p\right)\right)(x_0) \geq 0$, *and* $\|f(t) - f(x_0)\|_\gamma$ *is convex in* $t \in M$.

Then

$$\|(L_N(f))(x_0) - f(x_0)\|_\gamma \leq \|f(x_0)\|_\gamma \left| \left(\widetilde{L}_N(1)\right)(x_0) - 1 \right| +$$

$$\omega_1\left(f, \left(\widetilde{L}_N\left(\|\cdot - x_0\|_p\right)\right)(x_0)\right), \quad \forall\, N \in \mathbb{N}. \tag{10.81}$$

Theorem 10.6. *All as in Theorem 10.5. Inequality (10.81) is sharp, in fact it is attained by* $f(t) = \overrightarrow{i}\, \|t - x_0\|_p$, $\overrightarrow{i}$ *is a unit vector of* $\left(X, \|\cdot\|_\gamma\right)$, $t \in M$.

10.4 Main Results

In this section we apply the results of Section 3 to the linear neural network operators Ψ_n^s and $\widetilde{\Psi}_n^s$, $n \in \mathbb{N}$, of Section 2. In fact $\widetilde{\Psi}_n^s$ are positive linear neural network operators, with the property $\widetilde{\Psi}_n^s(1) = 1$.

Also (10.47)–(10.51) are valid.

From now on $M = \prod_{i=1}^{N} [a_i, b_i] \subset \left(\mathbb{R}^N, \|\cdot\|_p\right)$, $N \in \mathbb{N}$, where $1 \leq p \leq +\infty$.

We present the following general approximation result.

Theorem 10.7. *Let* $\left(X, \|\cdot\|_\gamma\right)$ *is a general Banach space,* $x_0 \in \prod_{i=1}^{N} [a_i, b_i]$, $1 \leq p \leq +\infty$. *Let* $m \in \mathbb{N}$, $f \in C^m\left(\prod_{i=1}^{N} [a_i, b_i], X\right)$, *the space of* m*-times continuously Fréchet differentiable functions from* $\prod_{i=1}^{N} [a_i, b_i]$ *into* X.

Let $x, x_0 \in \prod_{i=1}^{N} [a_i, b_i]$, $r > 0$, *and*

$$\omega_1\left(f^{(m)}, h\right) := \sup_{\substack{x,y \in \prod_{i=1}^{N} [a_i,b_i]: \\ \|x-y\|_p \leq h}} \left\| f^{(m)}(x) - f^{(m)}(y) \right\|, \quad h > 0. \tag{10.82}$$

Then

1)

$$\left\| \left(\Psi_n^s(f)\right)(x_0) - \sum_{j=0}^{n} \frac{1}{j!} \left(\widetilde{\Psi}_n^s\left(f^{(j)}(x_0)(\cdot - x_0)^j\right)\right)(x_0) \right\|_\gamma \leq$$

$$\frac{\omega_1\left(f^{(m)}, r\left(\left(\widetilde{\Psi}_n^s\left(\|\cdot - x_0\|_p^{m+1}\right)\right)(x_0)\right)^{\frac{1}{m+1}}\right)}{rm!} \left(\left(\widetilde{\Psi}_n^s\left(\|\cdot - x_0\|_p^{m+1}\right)\right)(x_0)\right)^{\left(\frac{m}{m+1}\right)}$$

$$\left[\frac{1}{(m+1)} + \frac{r}{2} + \frac{mr^2}{8}\right], \tag{10.83}$$

2) additionally if $f^{(j)}(x_0) = 0$, $j = 0, 1, ..., m$, *we have*

$$\left\|\left(\Psi_n^s(f)\right)(x_0)\right\|_\gamma \leq$$

$$\frac{\omega_1\left(f^{(m)}, r\left(\left(\widetilde{\Psi}_n^s\left(\|\cdot - x_0\|_p^{m+1}\right)\right)(x_0)\right)^{\frac{1}{m+1}}\right)}{rm!} \left(\left(\widetilde{\Psi}_n^s\left(\|\cdot - x_0\|_p^{m+1}\right)\right)(x_0)\right)^{\left(\frac{m}{m+1}\right)}$$

$$\left[\frac{1}{(m+1)} + \frac{r}{2} + \frac{mr^2}{8}\right], \tag{10.84}$$

3)

$$\left\| \left(\Psi_n^s (f) \right) (x_0) - f(x_0) \right\|_\gamma \leq$$

$$\sum_{j=1}^{n} \frac{1}{j!} \left\| \left(\widetilde{\Psi}_n^s \left(f^{(j)} (x_0) (\cdot - x_0)^j \right) \right) (x_0) \right\|_\gamma +$$

$$\frac{\omega_1\left(f^{(m)}, r \left(\left(\widetilde{\Psi}_n^s \left(\| \cdot - x_0 \|_p^{m+1} \right) \right) (x_0) \right)^{\frac{1}{m+1}} \right)}{rm!} \left(\left(\widetilde{\Psi}_n^s \left(\| \cdot - x_0 \|_p^{m+1} \right) \right) (x_0) \right)^{\left(\frac{m}{m+1} \right)}$$

$$\left[\frac{1}{(m+1)} + \frac{r}{2} + \frac{mr^2}{8}\right], \tag{10.85}$$

4)

$$\left\| \left\| \Psi_n^s (f) (x_0) - f(x_0) \right\|_\gamma \right\|_{\infty, \prod_{i=1}^{N} [a_i, b_i]} \leq$$

$$\sum_{j=1}^{n} \frac{1}{j!} \left\| \left\| \left(\widetilde{\Psi}_n^s \left(f^{(j)} (x_0) (\cdot - x_0)^j \right) \right) (x_0) \right\|_\gamma \right\|_{\infty, x_0 \in \prod_{i=1}^{N} [a_i, b_i]} +$$

$$\frac{\omega_1 \left(f^{(m)}, r \left\| \left(\widetilde{\Psi}_n^s \left(\| \cdot - x_0 \|_p^{m+1} \right) \right) (x_0) \right\|_{\infty, x_0 \in \prod_{i=1}^{N} [a_i, b_i]}^{\frac{1}{m+1}} \right)}{rm!}$$

$$\left\| \left(\widetilde{\Psi}_n^s \left(\| \cdot - x_0 \|_p^{m+1} \right) \right) (x_0) \right\|_{\infty, x_0 \in \prod_{i=1}^{N} [a_i, b_i]}^{\left(\frac{m}{m+1} \right)}$$

$$\left[\frac{1}{(m+1)} + \frac{r}{2} + \frac{mr^2}{8}\right]. \tag{10.86}$$

Proof. A direct application of Theorem 10.2. □

We make

Remark 10.5. Clearly here $\omega_1\left(f^{(m)},\cdot\right)$ is bounded. By Lemma 10.2 we get that $\left(\widetilde{\Psi}_n^s\left(\|\cdot-x_0\|_p^\lambda\right)\right)(x_0)$ is continuous in $x_0\in\prod_{i=1}^N[a_i,b_i]$, $\lambda\in\mathbb{N}$.

And, by (10.72) we obtain

$$\left\|\left(\widetilde{\Psi}_n^s\left(\|\cdot-x_0\|_p^k\right)\right)(x_0)\right\|_{\infty,x_0\in\prod_{i=1}^N[a_i,b_i]}\leq$$

$$\left\|\left(\widetilde{\Psi}_n^s\left(\|\cdot-x_0\|_p^{m+1}\right)\right)(x_0)\right\|_{\infty,x_0\in\prod_{i=1}^N[a_i,b_i]}^{\left(\frac{k}{m+1}\right)},\tag{10.87}$$

for $k=1,...,m$.

We make

Remark 10.6. Here we take $p=\infty$. We observe that

$$\left(\widetilde{\Psi}_n^s\left(\|\cdot-x_0\|_\infty^{m+1}\right)\right)(x_0)\overset{(10.46)}{=}$$

$$\frac{\sum_{k=\lceil na\rceil}^{\lfloor nb\rfloor}\left\|\frac{k}{n}-x_0\right\|_\infty^{m+1}\widetilde{Z}(nx_0-k)}{\sum_{k=\lceil na\rceil}^{\lfloor nb\rfloor}\widetilde{Z}(nx_0-k)}\overset{(10.42)}{\leq}$$

$$(K(q))^N\left(\sum_{k=\lceil na\rceil}^{\lfloor nb\rfloor}\left\|\frac{k}{n}-x_0\right\|_\infty^{m+1}\widetilde{Z}(nx_0-k)\right)=\tag{10.88}$$

$$(K(q))^N\left(\sum_{\substack{k=\lceil na\rceil\\ :\left\|\frac{k}{n}-x_0\right\|_\infty\leq\frac{1}{n^\beta}}}^{\lfloor nb\rfloor}\left\|\frac{k}{n}-x_0\right\|_\infty^{m+1}\widetilde{Z}(nx_0-k)+\right.$$

$$\left.\sum_{\substack{k=\lceil na\rceil\\ :\left\|\frac{k}{n}-x_0\right\|_\infty>\frac{1}{n^\beta}}}^{\lfloor nb\rfloor}\left\|\frac{k}{n}-x_0\right\|_\infty^{m+1}\widetilde{Z}(nx_0-k)\right)\overset{((10.38),(10.43))}{\leq}$$

$$(K(q))^N \left[\frac{1}{n^{\beta(m+1)}} + \|b-a\|_\infty^{m+1} \widetilde{\gamma} A^{-\lambda\left(n^{1-\beta}-2\right)} \right].$$

We have proved that

$$\left\| \left(\widetilde{\Psi}_n^s \left(\|\cdot - x_0\|_p^{m+1} \right) \right) (x_0) \right\|_{\infty, x_0 \in \prod_{i=1}^N [a_i, b_i]} \leq \tag{10.89}$$

$$(K(q))^N \left[\frac{1}{n^{\beta(m+1)}} + \|b-a\|_\infty^{m+1} \widetilde{\gamma} A^{-\lambda\left(n^{1-\beta}-2\right)} \right] \to 0, \text{ as } n \to +\infty.$$

Here it is $0 < \beta < 1$, $\widetilde{\gamma} := 2 \max\left\{q, \frac{1}{q}\right\}$, $A > 1$; $n \in \mathbb{N} : n^{1-\beta} > 2$, $x_0 \in \prod_{i=1}^N [a_i, b_i]$.

So, by Conclusion 10.1, see also (10.86) and (10.87), we derive that $\Psi_n^s(f) \overset{u}{\to} f$, uniformly in $\|\cdot\|_\gamma$ over $\prod_{i=1}^N [a_i, b_i]$, as $n \to +\infty$; $p = +\infty$ case.

Pointwise convergence is obviously valid too.

We continue with

Corollary 10.3. *(to Theorem 10.7, case of $m = 1$, see (10.85)) Let $x_0 \in \prod_{i=1}^N [a_i, b_i]$, then*

1)

$$\|(\Psi_n^s(f))(x_0) - f(x_0)\|_\gamma \leq \left\| \left(\widetilde{\Psi}_n^s \left(f^{(1)}(x_0)(\cdot - x_0) \right) \right) (x_0) \right\|_\gamma +$$

$$\frac{1}{2r} \omega_1 \left(f^{(1)}, r \left(\left(\widetilde{\Psi}_n^s \left(\|\cdot - x_0\|_p^2 \right) \right) (x_0) \right)^{\frac{1}{2}} \right) \left(\left(\widetilde{\Psi}_n^s \left(\|\cdot - x_0\|_p^2 \right) \right) (x_0) \right)^{\frac{1}{2}}$$

$$\left[1 + r + \frac{r^2}{4} \right], \tag{10.90}$$

and

2)

$$\left\| \|(\Psi_n^s(f))(x_0) - f(x_0)\|_\gamma \right\|_{\infty, \prod_{i=1}^N [a_i, b_i]} \leq$$

$$\left\| \left\| \left(\widetilde{\Psi}_n^s \left(f^{(1)}(x_0)(\cdot - x_0) \right) \right) (x_0) \right\|_\gamma \right\|_{\infty, x_0 \in \prod_{i=1}^N [a_i, b_i]} +$$

$$\frac{1}{2r} \omega_1 \left(f^{(1)}, r \left\| \left(\widetilde{\Psi}_n^s \left(\|\cdot - x_0\|_p^2 \right) \right) (x_0) \right\|_{\infty, x_0 \in \prod_{i=1}^N [a_i, b_i]}^{\frac{1}{2}} \right)$$

$$\left\| \left(\widetilde{\Psi}_n^s \left(\|\cdot - x_0\|_p^2 \right) \right) (x_0) \right\|_{\infty, x_0 \in \prod_{i=1}^N [a_i, b_i]}^{\frac{1}{2}} \left[1 + r + \frac{r^2}{4} \right]. \tag{10.91}$$

Proof. Direct application to Corollary 10.1. □

We make

Remark 10.7. By (10.89) we get that

$$\left\| \left(\widetilde{\Psi}_n^s \left(\|\cdot - x_0\|_p^2 \right) \right) (x_0) \right\|_{\infty, x_0 \in \prod_{i=1}^N [a_i, b_i]} \leq \tag{10.92}$$

$$(K(q))^N \left[\frac{1}{n^{2\beta}} + \|b - a\|_\infty^2 \widetilde{\gamma} A^{-\lambda\left(n^{1-\beta}-2\right)} \right] \to 0, \text{ as } n \to +\infty.$$

Again by (10.91) we obtain $\Psi_n^s(f) \to f$, as $n \to +\infty$, pointwise and uniformly, for $p = +\infty$ case.

Note 10.1. For $f \in C\left(\prod_{i=1}^N [a_i, b_i], X \right)$ the modulus of continuity $\omega_1(f, \delta)$ is defined by (10.52), (10.53).

By Lemma 10.1, we have that $\omega_1(f, \delta) \to 0$ as $\delta \downarrow 0$ iff $f \in C\left(\prod_{i=1}^N [a_i, b_i], X \right)$.

Also Lemma 10.3 now applies for an $f \in B\left(\prod_{i=1}^N [a_i, b_i], X \right)$ and (10.75) is valid.

We give

Theorem 10.8. *Let* $f \in C\left(\prod_{i=1}^N [a_i, b_i], X \right)$*, where* $1 \leq p \leq +\infty$.

Then

1)

$$\|(\Psi_n^s(f))(x_0) - f(x_0)\|_\gamma \leq 2\omega_1\left(f, \left(\widetilde{\Psi}_n^s \left(\|\cdot - x_0\|_p \right) \right)(x_0) \right), \tag{10.93}$$

and

2)

$$\left\| \|(\Psi_n^s(f))(x_0) - f(x_0)\|_\gamma \right\|_{\infty, \prod_{i=1}^N [a_i, b_i]} \leq$$

$$2\omega_1\left(f, \left\| \left(\widetilde{\Psi}_n^s \left(\|\cdot - x_0\|_p \right) \right)(x_0) \right\|_{\infty, x_0 \in \prod_{i=1}^N [a_i, b_i]} \right), \tag{10.94}$$

$\forall\, f \in C\left(\prod_{i=1}^N [a_i, b_i], X \right)$.

Proof. Direct application of (10.78), (10.79), (10.80). □

We make

Remark 10.8. Inequality (10.89) is clearly valid for $m = 0$. So we have

$$\left\| \left\| \left(\Psi_n^s (f) \right) (x_0) - f (x_0) \right\|_\gamma \right\|_{\infty, \prod_{i=1}^N [a_i, b_i]} \leq$$

$$2\omega_1 \left(f, (K (q))^N \left[\frac{1}{n^\beta} + \|b - a\|_\infty \widetilde{\gamma} A^{-\lambda\left(n^{1-\beta}-2\right)} \right] \right) \to 0, \text{ as } n \to +\infty. \tag{10.95}$$

By (10.94) now we obtain $\Psi_n^s (f) \to f$, as $n \to +\infty$, pointwise and uniformly, for $p = +\infty$ case.

Corollary 10.4. *It holds* $(0 < \beta < 1,\ p = +\infty)$

$$\left\| \left\| \left(\Psi_n^s (f) \right) (x_0) - f (x_0) \right\|_\gamma \right\|_{\infty, \prod_{i=1}^N [a_i, b_i]} \leq$$

$$2\omega_1 \left(f, (K (q))^N \left[\frac{1}{n^\beta} + \|b - a\|_\infty \widetilde{\gamma} A^{-\lambda\left(n^{1-\beta}-2\right)} \right] \right), \tag{10.96}$$

$\forall\ f \in C \left(\prod_{i=1}^N [a_i, b_i], X \right)$.

Proof. By (10.94), (10.95), $p = +\infty$. □

Under convexity we derive the following result.

Theorem 10.9. *Let* $f \in C \left(\prod_{i=1}^N [a_i, b_i], X \right)$, $1 \leq p \leq +\infty$. *Here* $x_0 \in \left(\prod_{i=1}^N [a_i, b_i] \right)^0$ *and* $\|f (t) - f (x_0)\|_\gamma$ *is convex in* $t \in \prod_{i=1}^N [a_i, b_i]$. *We assume that the closed ball in* $\mathbb{R}^N$: $B \left(x_0, \left(\widetilde{\Psi}_n^s \left(\|\cdot - x_0\|_p \right) \right) (x_0) \right) \subset \prod_{i=1}^N [a_i, b_i]$.

Then

$$\|(\Psi_n^s (f)) (x_0) - f (x_0)\|_\gamma \leq \omega_1 \left(f, \left(\widetilde{\Psi}_n^s \left(\|\cdot - x_0\|_p \right) \right) (x_0) \right), \tag{10.97}$$

$\forall\ N \in \mathbb{N} : n^{1-\beta} > 2,\ 0 < \beta < 1$.

Proof. By Theorem 10.5. Notice here that $\left(\widetilde{\Psi}_n^s \left(\|\cdot - x_0\|_p \right) \right) (x_0) > 0$. □

Theorem 10.10. *All as in Theorem 10.9. Inequality (10.97) is sharp, in fact it is attained by* $f(t) = \overrightarrow{i} \|t - x_0\|_p$, $\overrightarrow{i}$ *is a unit vector of* $\left(X, \|\cdot\|_\gamma\right)$, $t \in \prod_{i=1}^{N} [a_i, b_i]$.

Proof. By Theorem 10.6. □

We finish with

Corollary 10.5. *All as in Theorem 10.9,* $p = +\infty$ *case.*
Then

$$\|(\Psi_n^s(f))(x_0) - f(x_0)\|_\gamma \leq \omega_1\left(f, (K(q))^N \left[\frac{1}{n^\beta} + \|b - a\|_\infty \widetilde{\gamma} A^{-\lambda\left(n^{1-\beta}-2\right)}\right]\right), \tag{10.98}$$

$\forall\, N \in \mathbb{N} : n^{1-\beta} > 2,\ 0 < \beta < 1.$

Clearly, $\lim_{n \to +\infty} \Psi_n^s(f)(x_0) = f(x_0)$, *at* $x_0 \in \left(\prod_{i=1}^{N} [a_i, b_i]\right)^0$ *as in the assumptions.*

Bibliography

[235] G.A. Anastassiou, *A "K-Attainable" inequality related to the convergence of Positive Linear Operators*, J. Approximation Theory, 44, 380–383 (1985).

[236] G.A. Anastassiou, *Moments in probability and approximation theory*, Pitman Research Notes in Mathematics series / Longman Scientific & Technical, Essex, New York, UK, USA, 1993.

[237] G.A. Anastassiou, *Quantitative Approximations*, Chapmen & Hall/CRC, London, New York, 2001.

[238] G.A. Anastassiou, *Multivariate and abstract Approximation Theory for Banach space valued functions*, Demonstratio Mathematica, 2017; 50, 208–222.

[239] G.A. Anastassiou, *Intelligent Computations: Abstract, Fractional Calculus, Inequalities, Approximations*, Springer, Heidelberg, New York, 2018.

[240] G.A. Anastassiou, *Parametrized, Deformed and General Neural Networks*, Springer, Heidelberg, New York, 2023.

[241] G.A. Anastassiou, *Trigonometric and Hyperbolic Generated Approximation Theory*, World Scientific, Singapore, New York, 2025.

[242] H. Cartan, *Differential Calculus*, Herman, Paris, 1971.

[243] Z. Chen and F. Cao, *The approximation operators with sigmoidal functions*, Computers and Mathematics with Applications, 58 (2009), 758–765.

[244] D. Costarelli, R. Spigler, *Approximation results for neural network operators activated by sigmoidal functions*, Neural Networks 44 (2013), 101–106.

[245] D. Costarelli, R. Spigler, *Multivariate neural network operators with sigmoidal activation functions*, Neural Networks 48 (2013), 72–77.

[246] L.B. Rall, *Computational Solution of Nonlinear Operator Equations*, Joh Wiley & Sons, New York, 1969.

[247] O. Shisha and B. Mond, *The degree of convergence of sequences of linear positive operators*, Nat. Acad. Sci. U.S. 60 (1968), 1196–1200.

[248] O. Shisha and B. Mond, *The degree of Approximation to Periodic Functions by Linear Positive Operators*, J. Approximation Theory, 1, 335–339 (1968).

[249] S. Haykin, *Neural Networks: A Comprehensive Foundation* (2 ed.), Prentice Hall, New York, 1998.

[250] W. McCulloch and W. Pitts, *A logical calculus of the ideas immanent in nervous activity*, Bulletin of Mathematical Biophysics, 7 (1943), 115–133.

[251] T.M. Mitchell, *Machine Learning*, WCB-McGraw-Hill, New York, 1997.

[252] Yu, Dansheng; Cao, Feilong, *Construction and approximation rate for feed-forward neural network operators with sigmoidal functions*, J. Comput. Appl. Math. 453 (2025), paper No. 116150.

[253] Cen, Siyu; Jin, Bangti; Quan, Qimeng; Zhou, Zhi; *Hybrid neural-network FEM approximation of diffusion coeficient in elyptic and parabolic problems.* IMA J. Numer. Anal. 44 (2024), no. 5, 3059–3093.

[254] Coroianu Lucian; Costarelli, Danillo; Natale, Mariarosaria; Pantiş, Alexandra; *The approximation capabilities of Durrmeyer-type neural network operators.* J. Appl. Math. Comput. 70 (2024), no. 5, 4581–4599.

[255] Warin, Xavier; *The GroupMax neural network approximation of convex functions.* IEEE Trans. Neural Netw. Learn. Syst. 35 (2024), no. 8, 11608–11612.

[256] Fabra, Arnau; Guasch, Oriol; Baiges, Joan; Codina, Ramon; *Approximation of acoustic black holes with finite element mixed formulations and artificial neural network correction terms.* Finite Elem. Anal. Des. 241 (2024), paper No. 104236.

[257] Grohs, Philipp; Voigtlaender, Felix. *Proof of the theory-to-practice gap in deep learning via sampling complexity bounds for neural network approximation spaces.* Found. Comput. Math. 24 (2024), no. 4, 1085–1143.

[258] Basteri, Andrea; Trevisan, Dario; *Quantitative Gaussian approximation of randomly initialized deep neural networks.* Mach. Learn. 113 (2024), no. 9, 6373–6393.

[259] De Ryck, T.; Mishra, S. *Error analysis for deep neural network approximations of parametric hyperbolic conservation laws.* Math. Comp. 93 (2024), no. 350, 2643–2677.

[260] Liu, Jie; Zhang, Baoji; Lai, Yuyang; Fang, Liqiau. *Hull form optimization reserach based on multi-precision back-propagation neural network approximation model.* Internal. J. Numer. Methods Fluid 96 (2024), no. 8, 1445–1460.

[261] Yoo, Jihahm; Kim, Jaywon; Gim, Minjung; Lee, Haesung. *Error estimates of physics-informed neural networks for initial value problems.* J. Korean Soc. Ind. Appl. Math. 28 (2024), no. 1, 33–58.

[262] G.A. Anastassiou, *Generalized Logistic Neural Networks as Positive Linear Operators, Multivariate and finite domain case*, submitted, 2025.

Chapter 11

Approximation by Symmetrized and Disturbed Hyperbolic Tangent Energized Convolutions as Positive Linear Operators

In this work, the univariate symmetrized and perturbed hyperbolic tangent activated convolution type operators of three kinds are studied further. Here this is done with the method of positive linear operators. Their new approximation properties are established by the quantitative convergence to the unit operator using the modulus of continuity. It also studies the related simultaneous approximation, as well as the iterated approximation. It follows [279].

11.1 Introduction

The author studied extensively the quantitative approximation of positive linear operators to the unit since 1985, see for example [263]–[265], [268]. He originated from the quantitative weak convergence of finite positive measures to the unit Dirac measure, having as a method the geometric moment theory, see [264], and he produced best upper bounds, leading to attained (i.e. sharp Jackson type inequalities), e.g. see [263], [264]. These studies have gone to all possible directions, univariate and multivariate, though in this work we stay only on the univariate approach over an infinite domain.

Our convolution operators here have as a kernel the symmetrized and perturbed hyperbolic tangent activation function, which is used very commonly in the study of neural networks, and they can be interpreted as positive linear operators.

So here our proving methods come from the theory of positive linear operators.

Thus in Section 2, we discuss about the symmetrized and pertubed hyperbolic tangent activation function.

In Section 3 we describe our activated convolution type operators and we present their properties, such as differentiation and iteration, along with positive linear operators results to be applied.

In Section 4, we derive some auxilliary results which are estimates to our operators, when applied to polynomial type functions and to be used into our main results.

In Section 5, we present our main explicit results under the lens of positive linear operators theory. We treat also the simultaneous and iterated approximation cases under the same spirit.

We are greatly inspired by our earlier works [4], [269], [271].

Furthermore, general motivation comes from the great works [273]–[278].

11.2 About q-Deformed and λ-Parametrized Hyperbolic Tangent Function $g_{q,\lambda}$

Here, all this initial background comes from Chapter 18, [267].

We use $g_{q,\lambda}$, see (11.1), and exhibit that it is a sigmoid function and we will present several of its properties related to the approximation by neural network operators.

So, let us consider the hyperbolic tangent activation function

$$g_{q,\lambda}(x) := \frac{e^{\lambda x} - qe^{-\lambda x}}{e^{\lambda x} + qe^{-\lambda x}}, \quad \lambda, q > 0, \ x \in \mathbb{R}. \tag{11.1}$$

We have that

$$g_{q,\lambda}(0) = \frac{1-q}{1+q}.$$

We notice also that

$$g_{q,\lambda}(-x) = \frac{e^{-\lambda x} - qe^{\lambda x}}{e^{-\lambda x} + qe^{\lambda x}} = \frac{\frac{1}{q}e^{-\lambda x} - e^{\lambda x}}{\frac{1}{q}e^{-\lambda x} + e^{\lambda x}} = -\frac{\left(e^{\lambda x} - \frac{1}{q}e^{-\lambda x}\right)}{e^{\lambda x} + \frac{1}{q}e^{-\lambda x}} = -g_{\frac{1}{q},\lambda}(x). \tag{11.2}$$

That is

$$g_{q,\lambda}(-x) = -g_{\frac{1}{q},\lambda}(x), \quad \forall\, x \in \mathbb{R}, \tag{11.3}$$

and

$$g_{\frac{1}{q},\lambda}(x) = -g_{q,\lambda}(-x),$$

hence

$$g'_{\frac{1}{q},\lambda}(x) = g'_{q,\lambda}(-x). \tag{11.4}$$

It is

$$g_{q,\lambda}(x) = \frac{e^{2\lambda x} - q}{e^{2\lambda x} + q} = \frac{1 - \frac{q}{e^{2lx}}}{1 + \frac{q}{e^{2\lambda x}}} \underset{(x \to +\infty)}{\to} 1,$$

i.e.

$$g_{q,\lambda}(+\infty) = 1, \tag{11.5}$$

Furthermore

$$g_{q,\lambda}(x) = \frac{e^{2\lambda x} - q}{e^{2\lambda x} + q} \underset{(x \to -\infty)}{\to} \frac{-q}{q} = -1,$$

i.e.

$$g_{q,\lambda}(-\infty) = -1. \tag{11.6}$$

We find that

$$g'_{q,\lambda}(x) = \frac{4q\lambda e^{2\lambda x}}{\left(e^{2\lambda x} + q\right)^2} > 0, \tag{11.7}$$

therefore $g_{q,\lambda}$ is strictly increasing.

Next we obtain ($x \in \mathbb{R}$)

$$g''_{q,\lambda}(x) = 8q\lambda^2 e^{2\lambda x}\left(\frac{q - e^{2\lambda x}}{\left(e^{2\lambda x} + q\right)^3}\right) \in C(\mathbb{R}). \tag{11.8}$$

We observe that

$$q - e^{2\lambda x} \gtrless 0 \Leftrightarrow q \gtrless e^{2\lambda x} \Leftrightarrow \ln q \gtrless 2\lambda x \Leftrightarrow x \lessgtr \frac{\ln q}{2\lambda}.$$

So, in case of $x < \frac{\ln q}{2\lambda}$, we have that $g_{q,\lambda}$ is strictly concave up, with $g''_{q,\lambda}\left(\frac{\ln q}{2\lambda}\right) = 0$.

And in case of $x > \frac{\ln q}{2\lambda}$, we have that $g_{q,\lambda}$ is strictly concave down.

Clearly, $g_{q,\lambda}$ is a shifted sigmoid function with $g_{q,\lambda}(0) = \frac{1-q}{1+q}$, and $g_{q,\lambda}(-x) = -g_{q^{-1},\lambda}(x)$, (a semi-odd function).

By $1 > -1$, $x + 1 > x - 1$, we consider the function

$$M_{q.\lambda}(x) := \frac{1}{4}\left(g_{q,\lambda}(x+1) - g_{q,\lambda}(x-1)\right) > 0, \tag{11.9}$$

$\forall x \in \mathbb{R}$; $q, \lambda > 0$. Notice that $M_{q,\lambda}(\pm\infty) = 0$, so the x-axis is horizontal asymptote.

We have that

$$M_{q,\lambda}(-x) = \frac{1}{4}\left(g_{q,\lambda}(-x+1) - g_{q,\lambda}(-x-1)\right) =$$

$$\frac{1}{4}\left(g_{q,\lambda}(-(x-1)) - g_{q,\lambda}(-(x+1))\right) =$$

$$\frac{1}{4}\left(-g_{\frac{1}{q},\lambda}(x-1) + g_{\frac{1}{q},\lambda}(x+1)\right) = \tag{11.10}$$

$$\frac{1}{4}\left(g_{\frac{1}{q},\lambda}(x+1) - g_{\frac{1}{q},\lambda}(x-1)\right) = M_{\frac{1}{q},\lambda}(x), \quad \forall\, x \in \mathbb{R}.$$

Thus

$$M_{q,\lambda}(-x) = M_{\frac{1}{q},\lambda}(x), \quad \forall\, x \in \mathbb{R};\ q, \lambda > 0, \tag{11.11}$$

a deformed symmetry.

Next, we have that

$$M'_{q,\lambda}(x) = \frac{1}{4}\left(g'_{q,\lambda}(x+1) - g'_{q,\lambda}(x-1)\right), \quad \forall\, x \in \mathbb{R}. \tag{11.12}$$

Let $x < \frac{\ln q}{2\lambda} - 1$, then $x-1 < x+1 < \frac{\ln q}{2\lambda}$ and $g'_{q,\lambda}(x+1) > g'_{q,\lambda}(x-1)$ (by $g_{q,\lambda}$ being strictly concave up for $x < \frac{\ln q}{2\lambda}$), that is $M'_{q,\lambda}(x) > 0$. Hence $M_{q,\lambda}$ is strictly increasing over $\left(-\infty, \frac{\ln q}{2\lambda} - 1\right)$.

Let now $x - 1 > \frac{\ln q}{2\lambda}$, then $x+1 > x-1 > \frac{\ln q}{2\lambda}$, and $g'_{q,\lambda}(x+1) < g'_{q,\lambda}(x-1)$, that is $M'_{q,\lambda}(x) < 0$.

Therefore $M_{q,\lambda}$ is strictly decreasing over $\left(\frac{\ln q}{2\lambda} + 1, +\infty\right)$.

Let us next consider, $\frac{\ln q}{2\lambda} - 1 \leq x \leq \frac{\ln q}{2\lambda} + 1$. We have that

$$M''_{q,\lambda}(x) = \frac{1}{4}\left(g''_{q,\lambda}(x+1) - g''_{q,\lambda}(x-1)\right) =$$

$$2q\lambda^2\left[e^{2\lambda(x+1)}\left(\frac{q - e^{2\lambda(x+1)}}{\left(e^{2\lambda(x+1)} + q\right)^3}\right) - e^{2\lambda(x-1)}\left(\frac{q - e^{2\lambda(x-1)}}{\left(e^{2\lambda(x-1)} + q\right)^3}\right)\right]. \tag{11.13}$$

By $\frac{\ln q}{2\lambda} - 1 \leq x \Leftrightarrow \frac{\ln q}{2\lambda} \leq x+1 \Leftrightarrow \ln q \leq 2\lambda(x+1) \Leftrightarrow q \leq e^{2\lambda(x+1)} \Leftrightarrow q - e^{2\lambda(x+1)} \leq 0$.

By $x \leq \frac{\ln q}{2\lambda} + 1 \Leftrightarrow x - 1 \leq \frac{\ln q}{2\lambda} \Leftrightarrow 2\lambda(x-1) \leq \ln q \Leftrightarrow e^{2\lambda(x-1)} \leq q \Leftrightarrow q - e^{2\lambda\beta(x-1)} \geq 0$.

Clearly by (11.13) we get that $M''_{q,\lambda}(x) \leq 0$, for $x \in \left[\frac{\ln q}{2\lambda} - 1, \frac{\ln q}{2\lambda} + 1\right]$.

More precisely $M_{q,\lambda}$ is concave down over $\left[\frac{\ln q}{2\lambda}-1, \frac{\ln q}{2\lambda}+1\right]$, and strictly concave down over $\left(\frac{\ln q}{2\lambda}-1, \frac{\ln q}{2\lambda}+1\right)$.

Consequently $M_{q,\lambda}$ has a bell-type shape over $\mathbb{R}$.

Of course it holds $M''_{q,\lambda}\left(\frac{\ln q}{2\lambda}\right)<0$.

At $x=\frac{\ln q}{2\lambda}$, we have

$$M'_{q,\lambda}(x)=\frac{1}{4}\left(g'_{q,\lambda}(x+1)-g'_{q,\lambda}(x-1)\right)=$$

$$q\lambda\left(\frac{e^{2\lambda(x+1)}}{\left(e^{2\lambda(x+1)}+q\right)^2}-\frac{e^{2\lambda(x-1)}}{\left(e^{2\lambda(x-1)}+q\right)^2}\right). \tag{11.14}$$

Thus

$$M'_{q,\lambda}\left(\frac{\ln q}{2\lambda}\right)=q\lambda\left(\frac{e^{2\lambda\left(\frac{\ln q}{2\lambda}+1\right)}}{\left(e^{2\lambda\left(\frac{\ln q}{2\lambda}+1\right)}+q\right)^2}-\frac{e^{2\lambda\left(\frac{\ln q}{2\lambda}-1\right)}}{\left(e^{2\lambda\left(\frac{\ln q}{2\lambda}-1\right)}+q\right)^2}\right)=$$

$$\lambda\left(\frac{e^{2\lambda}\left(e^{-2\lambda}+1\right)^2-e^{-2\lambda}\left(e^{2\lambda}+1\right)^2}{\left(e^{2\lambda}+1\right)^2\left(e^{-2\lambda}+1\right)^2}\right)=0. \tag{11.15}$$

That is, $\frac{\ln q}{2\lambda}$ is the only critical number of $M_{q,\lambda}$ over $\mathbb{R}$. Hence at $x=\frac{\ln q}{2\lambda}$, $M_{q,\lambda}$ achieves its global maximum, which is

$$M_{q,\lambda}\left(\frac{\ln q}{2\lambda}\right)=\frac{1}{4}\left[g_{q,\lambda}\left(\frac{\ln q}{2\lambda}+1\right)-g_{q,\lambda}\left(\frac{\ln q}{2\lambda}-1\right)\right]= \tag{11.16}$$

$$\frac{1}{4}\left[\left(\frac{e^{\lambda}-e^{-\lambda}}{e^{\lambda}+e^{-\lambda}}\right)-\left(\frac{e^{-\lambda}-e^{\lambda}}{e^{-\lambda}+e^{\lambda}}\right)\right]=$$

$$\frac{1}{4}\left[\frac{2\left(e^{\lambda}-e^{-\lambda}\right)}{e^{\lambda}+e^{-\lambda}}\right]=\frac{1}{2}\left(\frac{e^{\lambda}-e^{-\lambda}}{e^{\lambda}+e^{-\lambda}}\right)=\frac{\tanh(\lambda)}{2}.$$

Conclusion: The maximum value of $M_{q,\lambda}$ is

$$M_{q,\lambda}\left(\frac{\ln q}{2\lambda}\right)=\frac{\tanh(\lambda)}{2},\quad \lambda>0. \tag{11.17}$$

We mention

Theorem 11.1. *([267], Ch. 18, p. 458) We have that*

$$\sum_{i=-\infty}^{\infty}M_{q,\lambda}(x-i)=1,\quad \forall\, x\in\mathbb{R},\ \forall\ \lambda,q>0. \tag{11.18}$$

Also it holds

Theorem 11.2. *([267], Ch. 18, p. 459) It holds*

$$\int_{-\infty}^{\infty} M_{q,\lambda}(x)\,dx = 1, \quad \lambda, q > 0. \tag{11.19}$$

So that $M_{q,\lambda}$ is a density function on $\mathbb{R}$; $\lambda, q > 0$.

Similarly we get that

$$\int_{-\infty}^{\infty} M_{\frac{1}{q},\lambda}(x)\,dx = 1, \quad \lambda, q > 0, \tag{11.20}$$

so that $M_{\frac{1}{q},\lambda}$ is a density function.

Furthermore, we observe the symmetry

$$\left(M_{q,\lambda} + M_{\frac{1}{q},\lambda}\right)(-x) = \left(M_{q,\lambda} + M_{\frac{1}{q},\lambda}\right)(x), \quad \forall\, x \in \mathbb{R}. \tag{11.21}$$

Furthermore

$$\varphi = \frac{M_{q,\lambda} + M_{\frac{1}{q},\lambda}}{2} > 0 \tag{11.22}$$

is a new density function over $\mathbb{R}$, i.e.

$$\int_{-\infty}^{\infty} \varphi(x)\,dx = 1,$$

and φ is an even function.

Clearly, then

$$\int_{-\infty}^{\infty} \varphi(nx - u)\,du = 1, \quad \forall\, n \in \mathbb{N},\, x \in \mathbb{R}. \tag{11.23}$$

11.3 Basics

We give

Definition 11.1. Let $f \in C_B(\mathbb{R})$ (continuous and bounded functions on $\mathbb{R}$), $n \in \mathbb{N}$. We define the following basic activated hyperbolic tangent perturbed convolution type operators

$$A_n(f)(x) := \int_{-\infty}^{\infty} f\left(\frac{u}{n}\right) \varphi(nx - u)\,du, \quad \forall\, x \in \mathbb{R}. \tag{11.24}$$

In this work we examine the quantitative convergence of A_n to the unit operator.

We study similarly the activated Kantorovich type operators,

$$A_n^*(f)(x) := n\int_{-\infty}^{\infty}\left(\int_{\frac{u}{n}}^{\frac{u+1}{n}} f(t)\,dt\right)\varphi(nx-u)\,du$$

$$= n\int_{-\infty}^{\infty}\left(\int_{0}^{\frac{1}{n}} f\left(t+\frac{u}{n}\right)dt\right)\varphi(nx-u)\,du, \tag{11.25}$$

where $f \in C_B(\mathbb{R})$, $n \in \mathbb{N}$, $\forall\ x \in \mathbb{R}$.
And the activated Quadrature operators

$$\overline{A_n}(f)(x) := \int_{-\infty}^{\infty}\left(\sum_{i=1}^{r} w_i f\left(\frac{u}{n}+\frac{i}{nr}\right)\right)\varphi(nx-u)\,du, \tag{11.26}$$

where $w_i \geq 0$, $\sum\limits_{i=1}^{r} w_i = 1$; $f \in C_B(\mathbb{R})$, $n \in \mathbb{N}$, $\forall\ x \in \mathbb{R}$.

An essential property follows:

Theorem 11.3. *([269]) Let* $0 < \alpha < 1$, $n \in \mathbb{N} : n^{1-\alpha} > 2$. *Then*

$$\int_{\{u\in\mathbb{R}:|nx-u|\geq n^{1-\alpha}\}} \varphi(nx-u)\,du < \frac{\left(q+\frac{1}{q}\right)}{e^{2\lambda\left(n^{1-\alpha}-1\right)}}, \quad q,\lambda > 0. \tag{11.27}$$

The first modulus of continuity here is

$$\omega_1(f,\delta) := \sup_{\substack{x,y\in\mathbb{R},\\ |x-y|\leq\delta}} |f(x)-f(y)|, \quad \delta > 0. \tag{11.28}$$

We need

Proposition 11.1. *([269]) It holds* ($k \in \mathbb{N}$)

$$\int_{-\infty}^{\infty} |z|^k \varphi(z)\,dz \leq \left[\frac{\tanh(\lambda)}{(k+1)} + \frac{\left(q+\frac{1}{q}\right)e^{2\lambda}k!}{(2\lambda)^k}\right] < \infty. \tag{11.29}$$

We make

Remark 11.1. Given $f \in C_B(\mathbb{R})$, by [7], we obtain that $A_n(f)$, $A_n^*(f)$, $\overline{A}_n(f) \in C_B(\mathbb{R})$. Clearly here A_n, A_n^*, $\overline{A}_n$ are positive linear operators from $C_B(\mathbb{R})$ into itself, with the property $A_n(1) = A_n^*(1) = \overline{A}_n(1) = 1$, $\forall$ $n \in \mathbb{N}$.

Let $i \in \mathbb{N}$ be fixed. Assume that $f \in C^{(i)}(\mathbb{R})$, with $f^{(j)} \in C_B(\mathbb{R})$, for $j = 0, 1, ..., i$. We derive from [269] that

$$(A_n(f))^{(j)}(x) = A_n\left(f^{(j)}\right)(x),$$

$$(A_n^*(f))^{(j)}(x) = A_n^*\left(f^{(j)}\right)(x), \tag{11.30}$$

$$\left(\overline{A_n}(f)\right)^{(j)}(x) = \overline{A_n}\left(f^{(j)}\right)(x), \quad \forall\, x \in \mathbb{R},$$

for all $j = 1, ..., i$.

Call Ψ_n any of the A_n, A_n^* and $\overline{A}_n$, $\forall\, n \in \mathbb{N}$.
We also make

Remark 11.2. Furthermore it holds

$$|A_n(f)(x)| \le \|f\|_\infty \int_{-\infty}^{\infty} \varphi(z)\, dz = \|f\|_\infty,$$

i.e.

$$\|A_n(f)\|_\infty \le \|f\|_\infty. \tag{11.31}$$

So A_n is a bounded positive linear operator.
Clearly it holds

$$\left\|A_n^2(f)\right\|_\infty = \|A_n(A_n(f))\|_\infty \le \|A_n(f)\|_\infty \le \|f\|_\infty. \tag{11.32}$$

And for $k \in \mathbb{N}$ we obtain

$$\left\|A_n^k(f)\right\|_\infty \le \left\|A_n^{k-1}(f)\right\|_\infty \le \left\|A_n^{k-2}(f)\right\|_\infty \le ... \le \|f\|_\infty, \tag{11.33}$$

so the contraction property valid and A_n^k is a bounded linear operator.
Let $r \in \mathbb{N}$. We observe that

$$A_n^r f - f = \left(A_n^r f - A_n^{r-1} f\right) + \left(A_n^{r-1} f - A_n^{r-2} f\right) + \left(A_n^{r-2} f - A_n^{r-3} f\right)$$

$$+ ... + \left(A_n^2 f - A_n f\right) + \left(A_n f - f\right). \tag{11.34}$$

Then

$$\|A_n^r f - f\|_\infty \le \left\|A_n^r f - A_n^{r-1} f\right\|_\infty + \left\|A_n^{r-1} f - A_n^{r-2} f\right\|_\infty + \left\|A_n^{r-2} f - A_n^{r-3} f\right\|_\infty$$

$$+ ... + \left\|A_n^2 f - A_n f\right\|_\infty + \|A_n f - f\|_\infty =$$

$$\left\|A_n^{r-1}(A_n f - f)\right\|_\infty + \left\|A_n^{r-2}(A_n f - f)\right\|_\infty + ... + \|A_n(A_n f - f)\|_\infty +$$

$$\|A_n f - f\|_\infty \le r\, \|A_n f - f\|_\infty.$$

Therefore

$$\|A_n^r f - f\|_\infty \leq r \|A_n f - f\|_\infty . \tag{11.35}$$

Let now $m_1, m_2, ..., m_r \in \mathbb{N} : m_1 \leq m_2 \leq ... \leq m_r$, and A_{m_i} as above, then

$$A_{m_r}\left(A_{m_{r-1}}\left(...A_{m_2}\left(A_{m_1}f\right)\right)\right) - f = ... =$$

$$A_{m_r}\left(A_{m_{r-1}}\left(...A_{m_2}\right)\right)\left(A_{m_1}f - f\right) + A_{m_r}\left(A_{m_{r-1}}\left(...A_{m_3}\right)\right)\left(A_{m_2}f - f\right) + \tag{11.36}$$

$$A_{m_r}\left(A_{m_{r-1}}\left(...A_{m_4}\right)\right)\left(A_{m_3}f - f\right) + ... + A_{m_r}\left(A_{m_{r-1}}f - f\right) + A_{m_r}f - f.$$

Consequently it holds, as in [267], Chapter 2,

$$\left\|A_{m_r}\left(A_{m_{r-1}}\left(...A_{m_2}\left(A_{m_1}f\right)\right)\right) - f\right\|_\infty \leq \sum_{i=1}^{r} \|A_{m_i}f - f\|_\infty . \tag{11.37}$$

All of (11.31)–(11.37) are also true for A_n^* and $\overline{A}_n$, $\forall\, n \in \mathbb{N}$.

We need the following Hölder's type inequality for positive linear operators.

Theorem 11.4. *([270]) Let L be a positive linear operator from $C(\mathbb{R})$ into $C_B(\mathbb{R})$, and $f, g \in C(\mathbb{R})$, furthermore let $p, q > 1 : \frac{1}{p} + \frac{1}{q} = 1$. Assume that $L\left(\left(|f(\cdot)|^p\right)\right)(s_*)$, $L\left(\left(|g(\cdot)|^q\right)\right)(s_*) > 0$ for some $s_* \in \mathbb{R}$. Then*

$$L\left(|f(\cdot)\, g(\cdot)|\right)(s_*) \leq \left(L\left(\left(|f(\cdot)^p|\right)\right)(s_*)\right)^{\frac{1}{p}} \left(L\left(\left(|g(\cdot)^q|\right)\right)(s_*)\right)^{\frac{1}{q}} . \tag{11.38}$$

We also need

Theorem 11.5. *([270]) Let $N \in \mathbb{N}$ and $f, f^{(N)} \in C_B(\mathbb{R})$, $x \in \mathbb{R}$. Consider L_n a sequence of positive linear operators from $C_B(\mathbb{R})$ into itself, $n \in \mathbb{N}$, such that $L_n(1) = 1$. Assume $L_n\left(|\cdot - x|^{N+1}\right)(x) > 0$, and $f^{(i)}(x) = 0$, for $i = 1, ..., N$.*

Then

$$|L_n(f)(x) - f(x)| \leq \omega_1\left(f^{(N)}, \left(\left(L_n\left(|\cdot - x|^{N+1}\right)\right)(x)\right)^{\frac{1}{N+1}}\right)$$

$$\left[L_n\left(|\cdot - x|^N\right)(x) + \frac{\left(L_n\left(|\cdot - x|^{N+1}\right)(x)\right)^{\frac{N}{N+1}}}{(N+1)}\right] < +\infty, \quad \forall\, n \in \mathbb{N}. \tag{11.39}$$

Given that $\lim_{N\to+\infty} L_n\left(|\cdot - x|^{N+1}\right)(x) = 0$, *then* $\lim_{n\to+\infty} L_n(f)(x) = f(x)$.

If $N = 1$, *we derive*

$$|L_n(f)(x) - f(x)| \leq \omega_1\left(f', \left(\left(L_n\left((\cdot - x)^2\right)\right)(x)\right)^{\frac{1}{2}}\right)$$

$$\left[L_n(|\cdot - x|)(x) + \frac{\left(L_n\left((\cdot - x)^2\right)(x)\right)^{\frac{1}{2}}}{2}\right] < +\infty, \quad \forall\, n \in \mathbb{N}. \quad (11.40)$$

Given that $\lim_{n\to+\infty} L_n\left((\cdot - x)^2\right)(x) = 0$, *then* $\lim_{n\to+\infty} L_n(f)(x) = f(x)$.

Note 11.1. Assuming $L_n\left(|\cdot - x|^{N+1}\right)(x) > 0$, by Theorem 11.4, for $g = 1$, and L_n such that $L_n(1) = 1$, we obtain

$$L_n\left(|\cdot - x|^N\right)(x) \leq \left(L_n\left(|\cdot - x|^{N+1}\right)(x)\right)^{\frac{N}{N+1}}. \quad (11.41)$$

In case of $N = 1$, we derive

$$L_n(|\cdot - x|)(x) \leq \sqrt{\left(L_n\left((\cdot - x)^2\right)(x)\right)}. \quad (11.42)$$

We also need

Theorem 11.6. *([272]) Let* $f \in C_B(\mathbb{R})$, $x \in \mathbb{R}$. *Consider* L_n *a sequence of positive linear operators from* $C_B(\mathbb{R})$ *into itself,* $n \in \mathbb{N}$, *such that* $L_n(1) = 1$.

Assume that $L_n(|\cdot - x|)(x) > 0$. *Then*

$$|L_n(f)(x) - f(x)| \leq 2\omega_1(f, L_n(|\cdot - x|)(x)), \quad \forall\, n \in \mathbb{N}. \quad (11.43)$$

Given that $\lim_{n\to+\infty} L_n(|\cdot - x|)(x) = 0$, *and* f *is also uniformly continuous, we obtain* $\lim_{n\to+\infty} L_n(f)(x) = f(x)$.

11.4 Auxiliary Results

We have the following result.

Lemma 11.1. *Let* $m \in \mathbb{N}$. *Then*

$$0 < A_n(|\cdot - x|^m)(x) \leq$$

$$\frac{1}{n^m}\left[\frac{\tanh(\lambda)}{(m+1)}+\frac{\left(q+\frac{1}{q}\right)e^{2\lambda}m!}{(2\lambda)^m}\right](<+\infty)\to 0, \quad \textit{as } n\to+\infty, \tag{11.44}$$

$\forall\ x\in\mathbb{R}$.

And, it is

$$0<\left\|A_n\left(|\cdot-x|^m\right)(x)\right\|_\infty\leq$$

$$\frac{1}{n^m}\left[\frac{\tanh(\lambda)}{(m+1)}+\frac{\left(q+\frac{1}{q}\right)e^{2\lambda}m!}{(2\lambda)^m}\right](<+\infty)\to 0, \quad \textit{as } n\to+\infty. \tag{11.45}$$

Proof. We have that ($m\in\mathbb{N}$, $x\in\mathbb{R}$)

$$0<A_n\left(|\cdot-x|^m\right)(x)=\int_{-\infty}^{\infty}\left|\frac{u}{n}-x\right|^m\varphi(nx-u)\,du=$$

$$\frac{1}{n^m}\int_{-\infty}^{\infty}|nx-u|^m\varphi(nx-u)\,du\overset{(z:=nx-u)}{=}$$

$$\frac{1}{n^m}\int_{-\infty}^{\infty}|z|^m\varphi(z)\,dz\overset{(11.29)}{\leq} \tag{11.46}$$

$$\frac{1}{n^m}\left[\frac{\tanh(\lambda)}{(m+1)}+\frac{\left(q+\frac{1}{q}\right)e^{2\lambda}m!}{(2\lambda)^m}\right](<+\infty)\to 0, \quad \text{as } n\to+\infty.$$

□

It follows.

Lemma 11.2. *Let $m\in\mathbb{N}$. Then*

$$0<A_n^*\left(|\cdot-x|^m\right)(x)\leq$$

$$\frac{2^{m-1}}{n^m}\left[1+\left(\frac{\tanh(\lambda)}{(m+1)}+\frac{\left(q+\frac{1}{q}\right)e^{2\lambda}m!}{(2\lambda)^m}\right)\right](<+\infty)\to 0, \quad \textit{as } n\to+\infty, \tag{11.47}$$

$\forall\ x\in\mathbb{R}$.

And, it is

$$0<\left\|A_n^*\left(|\cdot-x|^m\right)(x)\right\|_\infty\leq$$

$$\frac{2^{m-1}}{n^m}\left[1+\left(\frac{\tanh(\lambda)}{(m+1)}+\frac{\left(q+\frac{1}{q}\right)e^{2\lambda}m!}{(2\lambda)^m}\right)\right](<+\infty)\to 0, \quad \textit{as } n\to+\infty. \tag{11.48}$$

Proof. We have that ($m \in \mathbb{N}$, $x \in \mathbb{R}$)

$$0 < A_n^* \left(|\cdot - x|^m\right)(x) =$$

$$n \int_{-\infty}^{\infty} \left(\int_0^{\frac{1}{n}} \left| t + \frac{u}{n} - x \right|^m dt \right) \varphi(nx - u) \, du \leq$$

$$n \int_{-\infty}^{\infty} \left(\int_0^{\frac{1}{n}} \left(|t| + \left| \frac{u}{n} - x \right| \right)^m dt \right) \varphi(nx - u) \, du \leq$$

$$\int_{-\infty}^{\infty} \left(\frac{1}{n} + \left| \frac{u}{n} - x \right| \right)^m \varphi(nx - u) \, du = \tag{11.49}$$

$$\frac{1}{n^m} \int_{-\infty}^{\infty} (1 + |nx - u|)^m \varphi(nx - u) \, du \leq$$

$$\frac{2^{m-1}}{n^m} \left[1 + \int_{-\infty}^{\infty} |nx - u|^m \varphi(nx - u) \, du \right] =$$

$$\frac{2^{m-1}}{n^m} \left[1 + \int_{-\infty}^{\infty} |z|^m \varphi(z) \, dz \right] \overset{(11.29)}{\leq}$$

$$\frac{2^{m-1}}{n^m} \left[1 + \left(\frac{\tanh(\lambda)}{(m+1)} + \frac{\left(q + \frac{1}{q}\right) e^{2\lambda} m!}{(2\lambda)^m} \right) \right] \to 0, \text{ as } n \to +\infty,$$

and it is finite. $\square$

At last we obtain

Lemma 11.3. *Let $m \in \mathbb{N}$. Then*

$$0 < \overline{A}_n \left(|\cdot - x|^m\right)(x) \leq \tag{11.50}$$

$$\frac{2^{m-1}}{n^m} \left[1 + \left(\frac{\tanh(\lambda)}{(m+1)} + \frac{\left(q + \frac{1}{q}\right) e^{2\lambda} m!}{(2\lambda)^m} \right) \right] (< +\infty) \to 0, \quad \textit{as } n \to +\infty,$$

$\forall \, x \in \mathbb{R}$.

And, it is

$$0 < \left\| \overline{A}_n \left(|\cdot - x|^m\right)(x) \right\|_\infty \leq \tag{11.51}$$

$$\frac{2^{m-1}}{n^m} \left[1 + \left(\frac{\tanh(\lambda)}{(m+1)} + \frac{\left(q + \frac{1}{q}\right) e^{2\lambda} m!}{(2\lambda)^m} \right) \right] (< +\infty) \to 0, \quad \textit{as } n \to +\infty.$$

Proof. We observe that

$$0 < \overline{A}_n \left(|\cdot - x|^m \right) (x) =$$

$$\int_{-\infty}^{\infty} \left(\sum_{i=1}^{r} w_i \left| \frac{u}{n} + \frac{i}{nr} - x \right|^m \right) \varphi (nx - u) \, du \leq$$

$$\int_{-\infty}^{\infty} \left(\sum_{i=1}^{r} w_i \left(\left| \frac{u}{n} - x \right| + \frac{i}{nr} \right)^m \right) \varphi (nx - u) \, du \leq \tag{11.52}$$

$$\int_{-\infty}^{\infty} \left(\frac{1}{n} + \left| \frac{u}{n} - x \right| \right)^m \varphi (nx - u) \, du =$$

$$\frac{1}{n^m} \int_{-\infty}^{\infty} (1 + |nx - u|)^m \varphi (nx - u) \, du.$$

The proof finishes as in the proof of Lemma 11.2. □

11.5 Main Results

We present the following results.

Theorem 11.7. *Let $N \in \mathbb{N}$ and $f, f^{(N)} \in C_B(\mathbb{R})$, $x \in \mathbb{R}$. Assume $f^{(i)}(x) = 0$, $i = 1, ..., N$. Then*

$$|\Psi_n(f)(x) - f(x)| \leq \omega_1 \left(f^{(N)}, \left(\left(\Psi_n \left(|\cdot - x|^{N+1} \right) \right) (x) \right)^{\frac{1}{N+1}} \right)$$

$$\left[\Psi_n \left(|\cdot - x|^N \right) (x) + \frac{\left(\Psi_n \left(|\cdot - x|^{N+1} \right) (x) \right)^{\frac{N}{N+1}}}{(N+1)} \right] < +\infty, \quad \forall \, n \in \mathbb{N}. \tag{11.53}$$

Hence $\lim_{n \to +\infty} \Psi_n(f)(x) = f(x)$.

If $N = 1$, we obtain

$$|\Psi_n(f)(x) - f(x)| \leq \omega_1 \left(f', \left(\left(\Psi_n \left((\cdot - x)^2 \right) \right) (x) \right)^{\frac{1}{2}} \right)$$

$$\left[\Psi_n (|\cdot - x|)(x) + \frac{\left(\Psi_n \left((\cdot - x)^2 \right) (x) \right)^{\frac{1}{2}}}{2} \right] < +\infty, \quad \forall \, n \in \mathbb{N}. \tag{11.54}$$

Again, it holds $\lim_{n \to +\infty} \Psi_n(f)(x) = f(x)$.

Proof. By Theorem 11.5 and Lemmas 11.1–11.3. □

Theorem 11.8. *Let $f \in C_B(\mathbb{R})$, $x \in \mathbb{R}$. Then*

$$|\Psi_n(f)(x) - f(x)| \leq 2\omega_1(f, \Psi_n(|\cdot - x|)(x)) < +\infty, \quad \forall\, n \in \mathbb{N}. \tag{11.55}$$

If f is also uniformly continuous, we derive $\lim_{n\to+\infty} \Psi_n(f)(x) = f(x)$.

Proof. Direct application of Theorem 11.6 and Lemmas 11.1–11.3. □

We make

Remark 11.3. By (11.41), (11.42), we derive that

$$\Psi_n\left(|\cdot - x|^N\right)(x) \leq \left(\Psi_n\left(|\cdot - x|^{N+1}\right)(x)\right)^{\frac{N}{N+1}}, \tag{11.56}$$

and

$$\Psi_n(|\cdot - x|)(x) \leq \sqrt{\left(\Psi_n\left((\cdot - x)^2\right)(x)\right)}, \quad \forall\, n \in \mathbb{N}. \tag{11.57}$$

Notation 11.1. Let $m, n \in \mathbb{N}$. Denote by

$$\rho_{1n}(m) := \frac{1}{n^m}\left[\frac{\tanh(\lambda)}{(m+1)} + \frac{\left(q + \frac{1}{q}\right)e^{2\lambda}m!}{(2\lambda)^m}\right](< +\infty) \to 0, \quad \text{as } n \to +\infty, \tag{11.58}$$

and

$$\rho_{2n}(m) := \frac{2^{m-1}}{n^m}\left[1 + \left(\frac{\tanh(\lambda)}{(m+1)} + \frac{\left(q + \frac{1}{q}\right)e^{2\lambda}m!}{(2\lambda)^m}\right)\right](< +\infty) \to 0,$$

$$\text{as } n \to +\infty. \tag{11.59}$$

We give the following explicit results.

Corollary 11.1. *Let $N \in \mathbb{N}$ and $f, f^{(N)} \in C_B(\mathbb{R})$, $x \in \mathbb{R}$. Assume $f^{(i)}(x) = 0$, $i = 1, ..., N$. Then*

$$|A_n(f)(x) - f(x)| \leq \omega_1\left(f^{(N)}, (\rho_{1n}(N+1))^{\frac{1}{N+1}}\right)$$

$$\left[\rho_{1n}(N) + \frac{(\rho_{1n}(N+1))^{\frac{N}{N+1}}}{(N+1)}\right] < +\infty, \quad \forall\, n \in \mathbb{N}. \tag{11.60}$$

Hence $\lim_{n\to+\infty} A_n(f)(x) = f(x)$.

If $N = 1$, we obtain

$$|A_n(f)(x) - f(x)| \leq \omega_1\left(f', (\rho_{1n}(2))^{\frac{1}{2}}\right)$$

$$\left[\rho_{1n}(1) + \frac{(\rho_{1n}(2))^{\frac{1}{2}}}{2}\right] < +\infty, \quad \forall\, n \in \mathbb{N}. \tag{11.61}$$

Again, it holds $\lim_{n\to+\infty} A_n(f)(x) = f(x)$.

Proof. By Theorem 11.7 and Lemma 11.1, see also (11.58). □

Corollary 11.2. *Let $N \in \mathbb{N}$ and $f, f^{(N)} \in C_B(\mathbb{R})$, $x \in \mathbb{R}$. Assume $f^{(i)}(x) = 0$, $i = 1, ..., N$. Then*

$$\left\{\begin{array}{c} |A_n^*(f)(x) - f(x)| \\ |\overline{A}_n(f)(x) - f(x)| \end{array}\right\} \leq \omega_1\left(f^{(N)}, (\rho_{2n}(N+1))^{\frac{1}{N+1}}\right)$$

$$\left[\rho_{2n}(N) + \frac{(\rho_{2n}(N+1))^{\frac{N}{N+1}}}{(N+1)}\right] < +\infty, \quad \forall\, n \in \mathbb{N}. \tag{11.62}$$

Hence $\lim_{n\to+\infty} A_n^*(f)(x) = \lim_{n\to+\infty} \overline{A}_n(f)(x) = f(x)$.

If $N = 1$, we obtain

$$\left\{\begin{array}{c} |A_n^*(f)(x) - f(x)| \\ |\overline{A}_n(f)(x) - f(x)| \end{array}\right\} \leq \omega_1\left(f', (\rho_{2n}(2))^{\frac{1}{2}}\right)$$

$$\left[\rho_{2n}(1) + \frac{(\rho_{2n}(2))^{\frac{1}{2}}}{2}\right] < +\infty, \quad \forall\, n \in \mathbb{N}. \tag{11.63}$$

Again, it holds $\lim_{n\to+\infty} A_n^*(f)(x) = \lim_{n\to+\infty} \overline{A}_n(f)(x) = f(x)$.

Proof. By Theorem 11.7 and Lemmas 11.2, 11.3, see also (11.59). □

Corollary 11.3. *Let $f \in C_B(\mathbb{R})$. Then*

$$\|A_n(f) - f\|_\infty \leq 2\omega_1(f, \rho_{1n}(1)) < +\infty, \quad \forall\, n \in \mathbb{N}. \tag{11.64}$$

If f is also uniformly continuous, we get that $\lim_{n\to+\infty} A_n(f) = f$, *pointwise and uniformly.*

Proof. By Theorem 11.8, Lemma 11.1; (11.58). □

Corollary 11.4. *Let $f \in C_B(\mathbb{R})$. Then*

$$\left\{ \begin{array}{l} \|A_n^*(f) - f\|_\infty \\ \|\overline{A}_n(f) - f\|_\infty \end{array} \right\} \leq 2\omega_1(f, \rho_{2n}(1)) < +\infty, \quad \forall\, n \in \mathbb{N}. \tag{11.65}$$

If f is also uniformly continuous, we obtain that $\lim_{n\to+\infty} A_n^*(f) = \lim_{n\to+\infty} \overline{A}_n(f) = f$, pointwise and uniformly.

Proof. By Theorem 11.8, Lemmas 11.2, 11.3; (11.59). □

We continue with simultaneous approximations.

Theorem 11.9. *Let $f^{(j)} \in C_B(\mathbb{R})$, for $j = 0, 1, ..., N \in \mathbb{N}$. Then*

$$\left\| (A_n(f))^{(j)} - f^{(j)} \right\|_\infty \leq 2\omega_1\left(f^{(j)}, \rho_{1n}(1)\right) < +\infty, \quad \forall\, n \in \mathbb{N}. \tag{11.66}$$

If $f^{(j)}$ is uniformly continuous, we get that $\lim_{n\to+\infty} (A_n(f))^{(j)} = f^{(j)}$, pointwise and uniformly.

Proof. By (11.30) and Corollary 11.3. □

Theorem 11.10. *Let $f^{(j)} \in C_B(\mathbb{R})$, for $j = 0, 1, ..., N \in \mathbb{N}$. Then*

$$\left\{ \begin{array}{l} \left\| (A_n^*(f))^{(j)} - f^{(j)} \right\|_\infty \\ \left\| \left(\overline{A}_n(f)\right)^{(j)} - f^{(j)} \right\|_\infty \end{array} \right\} \leq 2\omega_1\left(f^{(j)}, \rho_{2n}(1)\right) < +\infty, \quad \forall\, n \in \mathbb{N}. \tag{11.67}$$

If $f^{(j)}$ is uniformly continuous, we obtain that $\lim_{n\to+\infty} (A_n^*(f))^{(j)} = \lim_{n\to+\infty} \left(\overline{A}_n(f)\right)^{(j)} = f^{(j)}$, pointwise and uniformly.

Proof. By (11.30) and Corollary 11.4. □

Next we talk about iterated approximation.

Remark 11.4. Let $f \in C_B(\mathbb{R})$, $r \in \mathbb{N}$. Here Ψ_n is any of A_n, A_n^*, $\overline{A}_n$, $\forall$ $n \in \mathbb{N}$.

By Remark 11.2, see also (11.35), we have that

$$\|\Psi_n^r(f) - f\|_\infty \leq r \|\Psi_n(f) - f\|_\infty, \quad \forall\, n \in \mathbb{N}. \tag{11.68}$$

Corollary 11.5. *Let $f \in C_B(\mathbb{R})$, $r \in \mathbb{N}$. Then*

$$\|A_n^r(f) - f\|_\infty \leq r \|A_n(f) - f\|_\infty \leq 2r\omega_1(f, \rho_{1n}(1)) < +\infty, \quad \forall\, n \in \mathbb{N}. \tag{11.69}$$

If f is also uniformly continuous, we get that $\lim_{n\to+\infty} A_n^r(f) = f$, pointwise and uniformly.

And the speed of convergence of A_n^r to the unit operator is not worse than of A_n to the unit.

Proof. By (11.64) and (11.68). □

Corollary 11.6. *Let $f \in C_B(\mathbb{R})$, $r \in \mathbb{N}$. Then*

$$\left\{ \begin{array}{c} \left\| A_n^{*^r}(f) - f \right\|_\infty \\ \left\| \overline{A}_n^r(f) - f \right\|_\infty \end{array} \right\} \leq r \left\{ \begin{array}{c} \| A_n^*(f) - f \|_\infty \\ \| \overline{A}_n(f) - f \|_\infty \end{array} \right\} \tag{11.70}$$

$$\leq 2r\omega_1(f, \rho_{2n}(1)) < +\infty, \quad \forall\, n \in \mathbb{N}.$$

If f is also uniformly continuous, we get that $\lim_{n\to+\infty} A_n^{*^r}(f) = \lim_{n\to+\infty} \overline{A}_n^r(f) = f$, pointwise and uniformly.

And the speed of convergence of $A_n^{*^r}, \overline{A}_n^r$ to the unit operator is not worse than of A_n^*, $\overline{A}_n$ to the unit, respectively.

Proof. By (11.65) and (11.68). □

We finish with more general iterated approximation results.

Remark 11.5. Let $f \in C_B(\mathbb{R})$, $r \in \mathbb{N}$, and $m_1, m_2, ..., m_r \in \mathbb{N} : m_1 \leq m_2 \leq ... \leq m_r$. Here Ψ_{m_i} is any of A_{m_i}, $A_{m_i}^*$, $\overline{A}_{m_i}$.

By Remark 11.2, see also (11.37), we have that

$$\left\| \Psi_{m_r}\left(\Psi_{m_{r-1}}\left(...\Psi_{m_2}\left(\Psi_{m_1} f\right)\right)\right) - f \right\|_\infty \leq \sum_{i=1}^r \| \Psi_{m_i} f - f \|_\infty . \tag{11.71}$$

Corollary 11.7. *Let $f \in C_B(\mathbb{R})$, $r \in \mathbb{N}$, and $m_1, m_2, ..., m_r \in \mathbb{N} : m_1 \leq m_2 \leq ... \leq m_r$. Then*

$$\left\| A_{m_r}\left(A_{m_{r-1}}\left(...A_{m_2}\left(A_{m_1} f\right)\right)\right) - f \right\|_\infty \leq \sum_{i=1}^r \| A_{m_i} f - f \|_\infty \leq \tag{11.72}$$

$$2\left(\sum_{i=1}^r \omega_1\left(f, \rho_{1m_i}(1)\right) \right) \leq 2r\omega_1\left(f, \rho_{1m_1}(1)\right). \tag{11.73}$$

The speed of convergence to the unit operator of the above activated multiply iterated operators is not worse than the speed of convergence to the unit of A_{m_1}.

Proof. By (11.71), (11.64) and (11.58). □

Corollary 11.8. *Let $f \in C_B(\mathbb{R})$, $r \in \mathbb{N}$, and $m_1, m_2, ..., m_r \in \mathbb{N} : m_1 \leq m_2 \leq ... \leq m_r$. Then*

1)

$$\left\| A^*_{m_r}\left(A^*_{m_{r-1}}\left(...A^*_{m_2}\left(A^*_{m_1}f\right)\right)\right) - f \right\|_\infty \leq \sum_{i=1}^{r} \left\| A^*_{m_i}f - f \right\|_\infty \leq \tag{11.74}$$

$$2\left(\sum_{i=1}^{r} \omega_1\left(f, \rho_{2m_i}(1)\right)\right) \leq 2r\omega_1\left(f, \rho_{2m_1}(1)\right), \tag{11.75}$$

and

2)

$$\left\| \overline{A}_{m_r}\left(\overline{A}_{m_{r-1}}\left(...\overline{A}_{m_2}\left(\overline{A}_{m_1}f\right)\right)\right) - f \right\|_\infty \leq \sum_{i=1}^{r} \left\| \overline{A}_{m_i}f - f \right\|_\infty \leq \tag{11.76}$$

$$2\left(\sum_{i=1}^{r} \omega_1\left(f, \rho_{2m_i}(1)\right)\right) \leq 2r\omega_1\left(f, \rho_{2m_1}(1)\right). \tag{11.77}$$

The speed of convergence to the unit operator of the above activated multiply iterated operators is not worse than the speed of convergence to the unit of $A^*_{m_1}$, $\overline{A}_{m_1}$, respectively.

Proof. By (11.71), (11.70) and (11.59). □

Bibliography

[263] G.A. Anastassiou, *A "K-Attainable" inequality related to the convergence of Positive Linear Operators*, J. Approximation Theory, 44, 380–383 (1985).

[264] G.A. Anastassiou, *Moments in probability and approximation theory*, Pitman Research Notes in Mathematics series / Longman Scientific & Technical, Essex, New York, UK, USA, 1993.

[265] G.A. Anastassiou, *Quantitative Approximations*, Chapmen & Hall/CRC, London, New York, 2001.

[266] G.A. Anastassiou, *Intelligent Computations: Abstract Fractional Calculus, Inequalities, Approximations*, Springer, Heidelberg, New York, 2018.

[267] G.A. Anastassiou, *Parametrized, Deformed and General Neural Networks*, Springer, Heidelberg, New York, 2023.

[268] G.A. Anastassiou, *Trigonometric and Hyperbolic Generated Approximation Theory*, World Scientific, Singapore, New York, 2025.

[269] G.A. Anastassiou, *Approximation by symmetrized and perturbed hyperbolic tangent activated convolution type operators*, Mathematics, 2024, 12, 3302; `https://doi.org/10.3390/math12203302`.

[270] G.A. Anastassiou, *Neural Networks in infinite domain as positive linear operastors*, Annales Univ. Sci. Budapest, Sect. Comp., accepted, 2025.

[271] G.A. Anastassiou, S.G. Gal, *Approximation Theory*, Birkäuser, Boston, Basel, Berlin, 2000.

[272] R.A. DeVore, *The Approximation of Continuous Functions by Positive Linear Operators*, Lecture Notes in Mathematics, Vol. 293, Springer, Heidelberg, New York, 1972.

[273] Hee Sun Jung, Ryozi Sakai, *Local saturation of a positive linear convolution operator*, J. Inequal. Appl. 2014, 2014: 329, 16 p.

[274] J.J. Swetits, B. Wood, *Local L_p-saturation of positive linear convolution operators*, J. Approx. Theory 34 (1982), no. 4, 348–360.

[275] Grigor Moldovan, *Discrete convolutions in connection with functions of several variables and positive linear operators.* (Romanian) Studia Univ. Babeş-Bolyai Ser. Math.-Mech. 19 (1974), no. 1, 51–57.

[276] R. Bojanic, O. Shisha, *On the precision of uniform approximation of continuous functions by certain linear operators of convolution type.* J. Approximation Theory 8 (1973), 101–113.

[277] Grigor Moldovan, *Discrete convolutions and linear positive operators.* Ann. Univ. Sci. Budapest. Eötvös Sect. Math. 15 (1972), 31–44 (1973).

[278] A.I. Kamzolov, *The order of approximation of functions of class $Z_2(E_n)$ by positive linear convolution operators.* (Russian) Mat. Zametki 7 (1970), 723–732.

[279] G.A. Anastassiou, *Approximation by symmetrized and perturbed hyperbolic tangent activated convolutions as Positive Linear Operators*, submitted, 2025.

Chapter 12

Approximation by Symmetrized and Disturbed Generalized Logistic Energized Convolutions as Positive Linear Operators

In this work are researched further the univariate symmetrized and perturbed generalized logistic activated convolution type operators of three types. This is done with the method of positive linear operators. Their approximation properties are established by the quantitative convergence to the unit operator using the modulus of continuity. It is also presented the related simultaneous approximation, as well as the iterated approximation. It follows [296].

12.1 Introduction

The author studied extensively the quantitative approximation of positive linear operators to the unit since 1985, see for example [280]–[282], [285]. He originated from the quantitative weak convergence of finite positive measures to the unit Dirac measure, having as a method the geometric moment theory, see [281], and he produced best upper bounds, leading to attained (i.e. sharp Jackson type inequalities), e.g. see [280], [281]. These studies have been gone to all possible directions, univariate and multivariate, though in this work we stay only on the univariate approach over an infinite domain.

Our convolution operators here have as a kernel the symmetrized and perturbed generalized logistic activation function, which is used frequently in the study of neural networks, and they can be interpreted as positive linear operators.

So here our proving methods come from the theory of positive linear operators.

Thus, in Section 2, we discuss about the symmetrized and perturbed generalized logistic activation function. We also describe our activated convolution type operators and we present their properties, such as differentiation and iteration, along with positive linear operators results to be applied.

In Section 3, we derive some auxiliary results which are estimates to our operators, when applied to polynomial type functions and to be used into our main results.

In Section 4, we present our main explicit results under the lens of positive linear operators theory. We treat also the simultaneous and iterated approximation cases under the same spirit.

We are greatly inspired by our earlier works [4], [286], [288].

Furthermore, general motivation comes from the great works [290]–[295].

12.2 Essentials

Here we follow [284], pp. 395–417.

Our activation function here to be used is the q-deformed and λ-parametrized function

$$\varphi_{q,\lambda}(x) = \frac{1}{1+qA^{-\lambda x}}, \quad x \in \mathbb{R},\ q,\lambda > 0,\ A > 1. \tag{12.1}$$

This is the A-generalized logistic function.

For more read Chapter 16 of [284]: "Banach space valued ordinary and fractional neural network approximation based on q-deformed and λ-parametrized A-generalized logistic function".

This chapter motivates our current work.

The proposed "symmetrization technique" aims to use half data feed to our neural networks.

We will employ the following density function

$$G_{q,\lambda}(x) := \frac{1}{2}\left(\varphi_{q,\lambda}(x+1) - \varphi_{q,\lambda}(x-1)\right), \quad x \in \mathbb{R},\ q,\lambda > 0. \tag{12.2}$$

We have that

$$G_{q,\lambda}(-x) = G_{\frac{1}{q},\lambda}(x), \tag{12.3}$$

and

$$G_{\frac{1}{q},\lambda}(-x) = G_{q,\lambda}(x), \ \forall\, x \in \mathbb{R}. \tag{12.4}$$

Adding (12.3) and (12.4) we obtain

$$G_{q,\lambda}(-x) + G_{\frac{1}{q},\lambda}(-x) = G_{q,\lambda}(x) + G_{\frac{1}{q},\lambda}(x), \ \forall\, x \in \mathbb{R}, \tag{12.5}$$

the key to this work.

So that

$$\Phi(x) := \frac{G_{q,\lambda}(x) + G_{\frac{1}{q},\lambda}(x)}{2} > 0, \tag{12.6}$$

is an even function, symmetric with respect to the y-axis.

The global maximum of $G_{q,\lambda}$ is given by (16.18), p. 401 of [284] as

$$G_{q,\lambda}\left(\frac{\log_A q}{\lambda}\right) = \frac{A^\lambda - 1}{2(A^\lambda + 1)}. \tag{12.7}$$

And, the global max of $G_{\frac{1}{q},\lambda}$ is

$$G_{\frac{1}{q},\lambda}\left(\frac{\log_A \frac{1}{q}}{\lambda}\right) = G_{\frac{1}{q},\lambda}\left(\frac{-\log_A q}{\lambda}\right) = \frac{A^\lambda - 1}{2(A^\lambda + 1)}, \tag{12.8}$$

both sharing the same maximum at symmetric points.

By Theorem 16.1, p. 401 of [284], we have that

$$\sum_{i=-\infty}^{\infty} G_{q,\lambda}(x-i) = 1, \ \forall\, x \in \mathbb{R}, \ \lambda, q > 0, \ A > 1, \tag{12.9}$$

and

$$\sum_{i=-\infty}^{\infty} G_{\frac{1}{q},\lambda}(x-i) = 1, \ \forall\, x \in \mathbb{R}, \ \lambda, q > 0, \ A > 1. \tag{12.10}$$

Consequently, we derive that

$$\sum_{i=-\infty}^{\infty} \Phi(x-i) = 1, \ \forall\, x \in \mathbb{R}. \tag{12.11}$$

By Theorem 16.2, p. 402 of [284], we have that

$$\int_{-\infty}^{\infty} G_{q,\lambda}(x)\, dx = 1, \quad \lambda, q > 0, \ A > 1, \tag{12.12}$$

similarly it holds

$$\int_{-\infty}^{\infty} G_{\frac{1}{q},\lambda}(x)\, dx = 1, \tag{12.13}$$

so that

$$\int_{-\infty}^{\infty} \Phi(x)\, dx = 1, \tag{12.14}$$

therefore $\Phi(x)$ is a density function.

Clearly, then

$$\int_{-\infty}^{\infty} \Phi(nx-u)\,du = 1, \ \ \forall\, n \in \mathbb{N},\, x \in \mathbb{R}. \tag{12.15}$$

We give

Definition 12.1. Let $f \in C_B(\mathbb{R})$ (continuous and bounded functions on $\mathbb{R}$), $n \in \mathbb{N}$. We define the following basic activated generalized logistic perturbed convolution type operators

$$B_n(f)(x) := \int_{-\infty}^{\infty} f\left(\frac{u}{n}\right) \Phi(nx-u)\,du, \ \ \forall\, x \in \mathbb{R}. \tag{12.16}$$

In this work we examine the quantitative convergence of B_n to the unit operator.

We study similarly the activated Kantorovich type operators,

$$B_n^*(f)(x) := n \int_{-\infty}^{\infty} \left(\int_{\frac{u}{n}}^{\frac{u+1}{n}} f(t)\,dt \right) \Phi(nx-u)\,du$$

$$= n \int_{-\infty}^{\infty} \left(\int_{0}^{\frac{1}{n}} f\left(t+\frac{u}{n}\right) dt \right) \Phi(nx-u)\,du, \tag{12.17}$$

where $f \in C_B(\mathbb{R})$, $n \in \mathbb{N}$, $\forall\, x \in \mathbb{R}$,

and the activated Quadrature operators

$$\overline{B_n}(f)(x) := \int_{-\infty}^{\infty} \left(\sum_{i=1}^{r} w_i f\left(\frac{u}{n}+\frac{i}{nr}\right) \right) \Phi(nx-u)\,du, \tag{12.18}$$

where $w_i \geq 0$, $\sum\limits_{i=1}^{r} w_i = 1$; $f \in C_B(\mathbb{R})$, $n \in \mathbb{N}$, $\forall\, x \in \mathbb{R}$.

We make

Remark 12.1. By [288] we get the following important property:

$$\Phi(x) < \frac{1}{2}\left(q+\frac{1}{q}\right)(\lambda \ln A)\, A^{-\lambda(x-1)}, \ \ x \geq 1, \tag{12.19}$$

where $q, \lambda > 0$, $A > 1$.

Clearly also that (by (12.7), (12.8))

$$\Phi(x) \leq \frac{A^{\lambda}-1}{2\left(A^{\lambda}+1\right)}, \ \ \forall\, x \in \mathbb{R}. \tag{12.20}$$

We need

Theorem 12.1. *It holds* ($k \in \mathbb{N}$)

$$\int_{-\infty}^{\infty} |z|^k \Phi(z)\,dz \leq \left[\frac{(A^{\lambda}-1)}{(A^{\lambda}+1)(k+1)} + \left(q+\frac{1}{q}\right)\frac{A^{\lambda}k!}{(\lambda \ln A)^{k-1}} \right] < +\infty. \tag{12.21}$$

Proof. We can write

$$\int_{-\infty}^{\infty} |z|^k \Phi(z)\,dz = 2\int_{0}^{\infty} z^k \Phi(z)\,dz =$$

$$2\left[\int_0^1 z^k \Phi(z)\,dz + \int_1^{\infty} z^k \Phi(z)\,dz \right] \overset{\text{(by (12.19), (12.20))}}{\leq}$$

$$2\left[\frac{(A^{\lambda}-1)}{2(A^{\lambda}+1)} \int_0^1 z^k dz + \frac{1}{2}\left(q+\frac{1}{q}\right)(\lambda \ln A)\int_1^{\infty} z^k A^{-\lambda(z-1)}dz \right] \leq \tag{12.22}$$

$$2\left[\frac{(A^{\lambda}-1)}{2(A^{\lambda}+1)}\frac{1}{(k+1)} + \frac{1}{2}\left(q+\frac{1}{q}\right)(\lambda \ln A)\,A^{\lambda}\int_0^{\infty} z^k e^{-\lambda(\ln A)z}dz \right] =$$

$$2\left[\frac{(A^{\lambda}-1)}{2(A^{\lambda}+1)(k+1)} + \frac{1}{2}\left(q+\frac{1}{q}\right)\frac{(\lambda \ln A)\,A^{\lambda}}{(\lambda \ln A)^{k+1}}\int_0^{\infty} ((\lambda \ln A)\,z)^k e^{-(\lambda \ln A)z} d(\lambda \ln A)\,z \right] =$$

$$2\left[\frac{(A^{\lambda}-1)}{2(A^{\lambda}+1)(k+1)} + \frac{1}{2}\left(q+\frac{1}{q}\right)\frac{A^{\lambda}}{(\lambda \ln A)^{k-1}}\int_0^{\infty} y^k e^{-y}dy \right] = \tag{12.23}$$

$$2\left[\frac{(A^{\lambda}-1)}{2(A^{\lambda}+1)(k+1)} + \frac{1}{2}\left(q+\frac{1}{q}\right)\frac{A^{\lambda}}{(\lambda \ln A)^{k-1}}\Gamma(k+1) \right] =$$

$$\left[\frac{(A^{\lambda}-1)}{(A^{\lambda}+1)(k+1)} + \left(q+\frac{1}{q}\right)\frac{A^{\lambda}}{(\lambda \ln A)^{k-1}}k! \right] < +\infty.$$

□

The first modulus of continuity here is

$$\omega_1(f,\delta) := \sup_{\substack{x,y \in \mathbb{R}, \\ |x-y| \leq \delta}} |f(x)-f(y)|, \quad \delta > 0. \tag{12.24}$$

We make

Remark 12.2. Let $i \in \mathbb{N}$ be fixed. Assume that $f \in C^{(i)}(\mathbb{R})$, with $f^{(j)} \in C_B(\mathbb{R})$, for $j = 0, 1, ..., i$.

We have that

$$B_n(f)(x) = \int_{-\infty}^{\infty} f\left(x - \frac{z}{n}\right) \Phi(z)\, dz.$$

By applying repeatedly Leibnitz's rule we get

$$\frac{\partial^i B_n(f)(x)}{\partial x^i} = \int_{-\infty}^{\infty} f^{(i)}\left(x - \frac{z}{n}\right) \Phi(z)\, dz =$$

$$\int_{-\infty}^{\infty} f^{(i)}\left(\frac{u}{n}\right) \Phi(nx - u)\, du = B_n\left(f^{(i)}\right)(x), \quad \forall\, x \in \mathbb{R}. \tag{12.25}$$

Clearly, it is valid that

$$\frac{\partial^i B_n^*(f)(x)}{\partial x^i} = B_n^*\left(f^{(i)}\right)(x), \tag{12.26}$$

and

$$\frac{\partial^i \overline{B_n}(f)(x)}{\partial x^i} = \overline{B_n}\left(f^{(i)}\right)(x), \tag{12.27}$$

$\forall\, x \in \mathbb{R}$.

We need

Remark 12.3. About iteration:

Next, we consider the iterated analogs of B_n, B_n^*, $\overline{B_n}$, $n \in \mathbb{N}$.

We notice first that for $f \in C_B(\mathbb{R})$, we have that $B_n(f)$, $B_n^*(f)$, $\overline{B_n}(f) \in C_B(\mathbb{R})$, see [286].

Furthermore it holds

$$|B_n(f)(x)| \leq \|f\|_\infty \int_{-\infty}^{\infty} \Phi(z)\, dz = \|f\|_\infty,$$

i.e.

$$\|B_n(f)\|_\infty \leq \|f\|_\infty. \tag{12.28}$$

So B_n is a bounded positive linear operator.

Clearly it holds

$$\left\|B_n^2(f)\right\|_\infty = \|B_n(B_n(f))\|_\infty \leq \|B_n(f)\|_\infty \leq \|f\|_\infty. \tag{12.29}$$

And for $k \in \mathbb{N}$ we obtain

$$\left\|B_n^k(f)\right\|_\infty \leq \left\|B_n^{k-1}(f)\right\|_\infty \leq \left\|B_n^{k-2}(f)\right\|_\infty \leq ... \leq \|f\|_\infty, \tag{12.30}$$

so the contraction property valid and B_n^k is a bounded linear operator.

Let $r \in \mathbb{N}$. We observe that

$$B_n^r f - f = \left(B_n^r f - B_n^{r-1} f\right) + \left(B_n^{r-1} f - B_n^{r-2} f\right) + \left(B_n^{r-2} f - B_n^{r-3} f\right)$$
$$+ ... + \left(B_n^2 f - B_n f\right) + \left(B_n f - f\right). \quad (12.31)$$

Then

$$\|B_n^r f - f\|_\infty \le \left\|B_n^r f - B_n^{r-1} f\right\|_\infty + \left\|B_n^{r-1} f - B_n^{r-2} f\right\|_\infty + \left\|B_n^{r-2} f - B_n^{r-3} f\right\|_\infty$$
$$+ ... + \left\|B_n^2 f - B_n f\right\|_\infty + \|B_n f - f\|_\infty =$$
$$\left\|B_n^{r-1}(B_n f - f)\right\|_\infty + \left\|B_n^{r-2}(B_n f - f)\right\|_\infty + ... + \|B_n(B_n f - f)\|_\infty + \quad (12.32)$$
$$\|B_n f - f\|_\infty \le r \|B_n f - f\|_\infty .$$

Therefore

$$\|B_n^r f - f\|_\infty \le r \|B_n f - f\|_\infty . \quad (12.33)$$

Let now $m_1, m_2, ..., m_r \in \mathbb{N} : m_1 \le m_2 \le ... \le m_r$, and B_{m_i} as above.

$$B_{m_r}\left(B_{m_{r-1}}\left(...B_{m_2}\left(B_{m_1} f\right)\right)\right) - f = ... =$$
$$B_{m_r}\left(B_{m_{r-1}}\left(...B_{m_2}\right)\right)\left(B_{m_1} f - f\right) + B_{m_r}\left(B_{m_{r-1}}\left(...B_{m_3}\right)\right)\left(B_{m_2} f - f\right) + \quad (12.34)$$
$$B_{m_r}\left(B_{m_{r-1}}\left(...B_{m_4}\right)\right)\left(B_{m_3} f - f\right) + ... + B_{m_r}\left(B_{m_{r-1}} f - f\right) + B_{m_r} f - f.$$

Consequently it holds, as in [284], Chapter 2,

$$\left\|B_{m_r}\left(B_{m_{r-1}}\left(...B_{m_2}\left(B_{m_1} f\right)\right)\right) - f\right\|_\infty \le \sum_{i=1}^{r} \|B_{m_i} f - f\|_\infty . \quad (12.35)$$

Relations (12.28)–(12.35) are also valid for B_n^*, $\overline{B_n}$.

Indeed here B_n, B_n^*, $\overline{B_n}$ are positive linear operators form $C_B(\mathbb{R})$ into itself with the property $B_n(1) = B_n^*(1) = \overline{B_n}(1) = 1$, $\forall\, n \in \mathbb{N}$.

Call by θ_n any of B_n, B_n^*, $\overline{B_n}$, $\forall\, n \in \mathbb{N}$.

We need the following Hölder's type inequality for positive linear operators.

Theorem 12.2. *([288]) Let L be a positive linear operator from $C(\mathbb{R})$ into $C_B(\mathbb{R})$, and $f, g \in C(\mathbb{R})$, furthermore let $p, q > 1 : \frac{1}{p} + \frac{1}{q} = 1$. Assume that $L((|f(\cdot)|^p))(s_*)$, $L((|g(\cdot)|^q))(s_*) > 0$ for some $s_* \in \mathbb{R}$. Then*

$$L(|f(\cdot)g(\cdot)|)(s_*) \le \left(L((|f(\cdot)^p|))(s_*)\right)^{\frac{1}{p}} \left(L((|g(\cdot)^q|))(s_*)\right)^{\frac{1}{q}} . \quad (12.36)$$

We also need

Theorem 12.3. *([288]) Let $N \in \mathbb{N}$ and $f, f^{(N)} \in C_B(\mathbb{R})$, $x \in \mathbb{R}$. Consider L_n a sequence of positive linear operators from $C_B(\mathbb{R})$ into itself, $n \in \mathbb{N}$, such that $L_n(1) = 1$. Assume $L_n\left(|\cdot - x|^{N+1}\right)(x) > 0$, and $f^{(i)}(x) = 0$, for $i = 1, ..., N$.*

Then

$$|L_n(f)(x) - f(x)| \leq \omega_1\left(f^{(N)}, \left(\left(L_n\left(|\cdot - x|^{N+1}\right)\right)(x)\right)^{\frac{1}{N+1}}\right)$$

$$\left[L_n\left(|\cdot - x|^N\right)(x) + \frac{\left(L_n\left(|\cdot - x|^{N+1}\right)(x)\right)^{\frac{N}{N+1}}}{(N+1)}\right] < +\infty, \quad \forall\, n \in \mathbb{N}. \tag{12.37}$$

Given that $\lim_{N \to +\infty} L_n\left(|\cdot - x|^{N+1}\right)(x) = 0$, then $\lim_{n \to +\infty} L_n(f)(x) = f(x)$.

If $N = 1$, we derive

$$|L_n(f)(x) - f(x)| \leq \omega_1\left(f', \left(\left(L_n\left((\cdot - x)^2\right)\right)(x)\right)^{\frac{1}{2}}\right)$$

$$\left[L_n(|\cdot - x|)(x) + \frac{\left(L_n\left((\cdot - x)^2\right)(x)\right)^{\frac{1}{2}}}{2}\right] < +\infty, \quad \forall\, n \in \mathbb{N}. \tag{12.38}$$

Given that $\lim_{n \to +\infty} L_n\left((\cdot - x)^2\right)(x) = 0$, then $\lim_{n \to +\infty} L_n(f)(x) = f(x)$.

Note 12.1. Assuming $L_n\left(|\cdot - x|^{N+1}\right)(x) > 0$, by Theorem 12.2, for $g = 1$, and L_n such that $L_n(1) = 1$, we obtain

$$L_n\left(|\cdot - x|^N\right)(x) \leq \left(L_n\left(|\cdot - x|^{N+1}\right)(x)\right)^{\frac{N}{N+1}}. \tag{12.39}$$

In case of $N = 1$, we derive

$$L_n(|\cdot - x|)(x) \leq \sqrt{\left(L_n\left((\cdot - x)^2\right)(x)\right)}. \tag{12.40}$$

We also need

Theorem 12.4. *([289]) Let $f \in C_B(\mathbb{R})$, $x \in \mathbb{R}$. Consider L_n a sequence of positive linear operators from $C_B(\mathbb{R})$ into itself, $n \in \mathbb{N}$, such that $L_n(1) = 1$.*

Assume that $L_n(|\cdot - x|)(x) > 0$. Then

$$|L_n(f)(x) - f(x)| \leq 2\omega_1(f, L_n(|\cdot - x|)(x)), \quad \forall\, n \in \mathbb{N}. \tag{12.41}$$

Given that $\lim_{n \to +\infty} L_n(|\cdot - x|)(x) = 0$, and f is also uniformly continuous, we obtain $\lim_{n \to +\infty} L_n(f)(x) = f(x)$.

12.3 Auxiliary Results

We present the following results.

Lemma 12.1. *Let $k \in \mathbb{N}$. Then*

$$0 < B_n\left(|\cdot - x|^k\right)(x) \leq$$

$$\frac{1}{n^k}\left[\frac{(A^\lambda - 1)}{(A^\lambda + 1)(k+1)} + \left(q + \frac{1}{q}\right)\frac{A^\lambda k!}{(\lambda \ln A)^{k-1}}\right](< +\infty) \to 0, \quad \textit{as } n \to +\infty, \tag{12.42}$$

$\forall\, x \in \mathbb{R}$.

And, it is

$$0 < \left\|B_n\left(|\cdot - x|^k\right)(x)\right\|_\infty \leq$$

$$\frac{1}{n^k}\left[\frac{(A^\lambda - 1)}{(A^\lambda + 1)(k+1)} + \left(q + \frac{1}{q}\right)\frac{A^\lambda k!}{(\lambda \ln A)^{k-1}}\right](< +\infty) \to 0, \quad \textit{as } n \to +\infty. \tag{12.43}$$

Proof. We have that ($k \in \mathbb{N}$, $x \in \mathbb{R}$)

$$0 < B_n\left(|\cdot - x|^k\right)(x) = \int_{-\infty}^{\infty}\left|\frac{u}{n} - x\right|^k \Phi(nx - u)\, du =$$

$$\frac{1}{n^k}\int_{-\infty}^{\infty}|nx - u|^k \Phi(nx - u)\, du \overset{(z := nx - u)}{=}$$

$$\frac{1}{n^k}\int_{-\infty}^{\infty}|z|^k \Phi(z)\, dz \overset{(12.21)}{\leq} \tag{12.44}$$

$$\frac{1}{n^k}\left[\frac{(A^\lambda - 1)}{(A^\lambda + 1)(k+1)} + \left(q + \frac{1}{q}\right)\frac{A^\lambda k!}{(\lambda \ln A)^{k-1}}\right](< +\infty) \to 0, \quad \text{as } n \to +\infty.$$

□

It follows.

Lemma 12.2. *Let $k \in \mathbb{N}$. Then*

$$0 < B_n^*\left(|\cdot - x|^k\right)(x) \leq$$

$$\frac{2^{k-1}}{n^k}\left[1+\frac{\left(A^{\lambda}-1\right)}{\left(A^{\lambda}+1\right)(k+1)}+\left(q+\frac{1}{q}\right)\frac{A^{\lambda}k!}{\left(\lambda\ln A\right)^{k-1}}\right](<+\infty)\to 0,\ \ as\ n\to+\infty, \tag{12.45}$$

$\forall\ x\in\mathbb{R}$.

And, it is

$$0<\left\|B_n^*\left(|\cdot-x|^k\right)(x)\right\|_\infty\leq$$

$$\frac{2^{k-1}}{n^k}\left[1+\frac{\left(A^{\lambda}-1\right)}{\left(A^{\lambda}+1\right)(k+1)}+\left(q+\frac{1}{q}\right)\frac{A^{\lambda}k!}{\left(\lambda\ln A\right)^{k-1}}\right](<+\infty)\to 0,\ \ as\ n\to+\infty. \tag{12.46}$$

Proof. We have that ($k\in\mathbb{N}$, $x\in\mathbb{R}$)

$$0<B_n^*\left(|\cdot-x|^k\right)(x)=$$

$$n\int_{-\infty}^{\infty}\left(\int_0^{\frac{1}{n}}\left|t+\frac{u}{n}-x\right|^k dt\right)\Phi(nx-u)\,du\leq$$

$$n\int_{-\infty}^{\infty}\left(\int_0^{\frac{1}{n}}\left(|t|+\left|\frac{u}{n}-x\right|\right)^k dt\right)\Phi(nx-u)\,du\leq$$

$$\int_{-\infty}^{\infty}\left(\frac{1}{n}+\left|\frac{u}{n}-x\right|\right)^k\Phi(nx-u)\,du=$$

$$\frac{1}{n^k}\int_{-\infty}^{\infty}\left(1+|nx-u|\right)^k\Phi(nx-u)\,du\overset{(12.15)}{\leq} \tag{12.47}$$

$$\frac{2^{k-1}}{n^k}\left[1+\int_{-\infty}^{\infty}|nx-u|^k\,\Phi(nx-u)\,du\right]=$$

$$\frac{2^{k-1}}{n^k}\left[1+\int_{-\infty}^{\infty}|z|^k\,\Phi(z)\,dz\right]\overset{(12.21)}{\leq}$$

$$\frac{2^{k-1}}{n^k}\left[1+\frac{\left(A^{\lambda}-1\right)}{\left(A^{\lambda}+1\right)(k+1)}+\left(q+\frac{1}{q}\right)\frac{A^{\lambda}k!}{\left(\lambda\ln A\right)^{k-1}}\right](<+\infty)\to 0,\ \text{as}\ n\to+\infty.$$

$\square$

At last we obtain

Lemma 12.3. *Let $k \in \mathbb{N}$. Then*

$$0 < \overline{B_n}\left(|\cdot - x|^k\right)(x) \leq \tag{12.48}$$

$$\frac{2^{k-1}}{n^k}\left[1+\frac{\left(A^\lambda - 1\right)}{\left(A^\lambda + 1\right)(k+1)}+\left(q+\frac{1}{q}\right)\frac{A^\lambda k!}{(\lambda \ln A)^{k-1}}\right](< +\infty) \to 0, \;\; as\ n \to +\infty,$$

$\forall\ x \in \mathbb{R}$.

And, it is

$$0 < \left\|\overline{B_n}\left(|\cdot - x|^k\right)(x)\right\|_\infty \leq \tag{12.49}$$

$$\frac{2^{k-1}}{n^k}\left[1+\frac{\left(A^\lambda - 1\right)}{\left(A^\lambda + 1\right)(k+1)}+\left(q+\frac{1}{q}\right)\frac{A^\lambda k!}{(\lambda \ln A)^{k-1}}\right](< +\infty) \to 0, \;\; as\ n \to +\infty.$$

Proof. We observe that

$$0 < \overline{B_n}\left(|\cdot - x|^k\right)(x) =$$

$$\int_{-\infty}^{\infty}\left(\sum_{i=1}^{r} w_i \left|\frac{u}{n}+\frac{i}{nr}-x\right|^k\right)\Phi(nx-u)\,du \leq$$

$$\int_{-\infty}^{\infty}\left(\sum_{i=1}^{r} w_i \left(\left|\frac{u}{n}-x\right|+\frac{i}{nr}\right)^k\right)\Phi(nx-u)\,du \leq$$

$$\int_{-\infty}^{\infty}\left(\frac{1}{n}+\left|\frac{u}{n}-x\right|\right)^k\Phi(nx-u)\,du = \tag{12.50}$$

$$\frac{1}{n^k}\int_{-\infty}^{\infty}(1+|nx-u|)^k\,\Phi(nx-u)\,du.$$

The proof finishes as in the proof of Lemma 12.2. □

12.4 Main Results

We present the following results.

Theorem 12.5. *Let $N \in \mathbb{N}$ and $f, f^{(N)} \in C_B(\mathbb{R})$, $x \in \mathbb{R}$. Assume $f^{(i)}(x) = 0$, $i = 1, ..., N$. Then*

$$|\theta_n(f)(x) - f(x)| \leq \omega_1\left(f^{(N)}, \left(\left(\theta_n\left(|\cdot - x|^{N+1}\right)\right)(x)\right)^{\frac{1}{N+1}}\right)$$

$$\left[\theta_n\left(|\cdot - x|^N\right)(x) + \frac{\left(\theta_n\left(|\cdot - x|^{N+1}\right)(x)\right)^{\frac{N}{N+1}}}{(N+1)}\right] < +\infty, \quad \forall\, n \in \mathbb{N}. \tag{12.51}$$

Hence $\lim_{n\to+\infty} \theta_n(f)(x) = f(x)$.

If $N = 1$, we obtain

$$|\theta_n(f)(x) - f(x)| \leq \omega_1\left(f', \left(\left(\theta_n\left((\cdot - x)^2\right)\right)(x)\right)^{\frac{1}{2}}\right)$$

$$\left[\theta_n(|\cdot - x|)(x) + \frac{\left(\theta_n\left((\cdot - x)^2\right)(x)\right)^{\frac{1}{2}}}{2}\right] < +\infty, \quad \forall\, n \in \mathbb{N}. \tag{12.52}$$

Again, it holds $\lim_{n\to+\infty} \theta_n(f)(x) = f(x)$.

Proof. By Theorem 12.3 and Lemmas 12.1–12.3. □

Theorem 12.6. *Let $f \in C_B(\mathbb{R})$, $x \in \mathbb{R}$. Then*

$$|\theta_n(f)(x) - f(x)| \leq 2\omega_1(f, \theta_n(|\cdot - x|)(x)) < +\infty, \quad \forall\, n \in \mathbb{N}. \tag{12.53}$$

If f is also uniformly continuous, we derive $\lim_{n\to+\infty} \theta_n(f)(x) = f(x)$.

Proof. Direct application of Theorem 12.4 and Lemmas 12.1–12.3. □

We make

Remark 12.4. By (12.39), (12.40), we derive that

$$\theta_n\left(|\cdot - x|^N\right)(x) \leq \left(\theta_n\left(|\cdot - x|^{N+1}\right)(x)\right)^{\frac{N}{N+1}}, \tag{12.54}$$

and

$$\theta_n(|\cdot - x|)(x) \leq \sqrt{\left(\theta_n\left((\cdot - x)^2\right)(x)\right)}, \quad \forall\, n \in \mathbb{N}. \tag{12.55}$$

Notation 12.1. Let $k, n \in \mathbb{N}$. Denote by

$$\widetilde{\rho}_{1n}(k) := \frac{1}{n^k}\left[\frac{(A^\lambda - 1)}{(A^\lambda + 1)(k+1)} + \left(q + \frac{1}{q}\right)\frac{A^\lambda k!}{(\lambda \ln A)^{k-1}}\right](< +\infty) \to 0, \tag{12.56}$$

as $n \to +\infty$,

and

$$\widetilde{\rho}_{2n}(k) := \frac{2^{k-1}}{n^k}\left[1 + \frac{(A^\lambda - 1)}{(A^\lambda + 1)(k+1)} + \left(q + \frac{1}{q}\right)\frac{A^\lambda k!}{(\lambda \ln A)^{k-1}}\right](< +\infty) \to 0, \tag{12.57}$$

as $n \to +\infty$.

We give the following explicit results.

Corollary 12.1. *Let $N \in \mathbb{N}$ and $f, f^{(N)} \in C_B(\mathbb{R})$, $x \in \mathbb{R}$. Assume $f^{(i)}(x) = 0$, $i = 1, ..., N$. Then*

$$|B_n(f)(x) - f(x)| \leq \omega_1\left(f^{(N)}, (\widetilde{\rho}_{1n}(N+1))^{\frac{1}{N+1}}\right)$$

$$\left[\widetilde{\rho}_{1n}(N) + \frac{(\widetilde{\rho}_{1n}(N+1))^{\frac{N}{N+1}}}{(N+1)}\right] < +\infty, \quad \forall\, n \in \mathbb{N}. \tag{12.58}$$

Hence $\lim_{n \to +\infty} B_n(f)(x) = f(x)$.

If $N = 1$, we obtain

$$|B_n(f)(x) - f(x)| \leq \omega_1\left(f', (\widetilde{\rho}_{1n}(2))^{\frac{1}{2}}\right)$$

$$\left[\widetilde{\rho}_{1n}(1) + \frac{(\widetilde{\rho}_{1n}(2))^{\frac{1}{2}}}{2}\right] < +\infty, \quad \forall\, n \in \mathbb{N}. \tag{12.59}$$

Again, it holds $\lim_{n \to +\infty} B_n(f)(x) = f(x)$.

Proof. By Theorem 12.5 and Lemma 12.1, see also (12.56). ☐

Corollary 12.2. *Let $N \in \mathbb{N}$ and $f, f^{(N)} \in C_B(\mathbb{R})$, $x \in \mathbb{R}$. Assume $f^{(i)}(x) = 0$, $i = 1, ..., N$. Then*

$$\left\{\begin{array}{l} |B_n^*(f)(x) - f(x)| \\ |\overline{B}_n(f)(x) - f(x)| \end{array}\right\} \leq \omega_1\left(f^{(N)}, (\widetilde{\rho}_{2n}(N+1))^{\frac{1}{N+1}}\right)$$

$$\left[\widetilde{\rho}_{2n}(N) + \frac{(\widetilde{\rho}_{2n}(N+1))^{\frac{N}{N+1}}}{(N+1)}\right] < +\infty, \quad \forall\, n \in \mathbb{N}. \tag{12.60}$$

Hence $\lim_{n \to +\infty} B_n^*(f)(x) = \lim_{n \to +\infty} \overline{B}_n(f)(x) = f(x)$.

If $N = 1$, we obtain

$$\left\{ \begin{array}{l} |B_n^*(f)(x) - f(x)| \\ |\overline{B}_n(f)(x) - f(x)| \end{array} \right\} \leq \omega_1 \left(f', (\widetilde{\rho}_{2n}(2))^{\frac{1}{2}} \right)$$

$$\left[\widetilde{\rho}_{2n}(1) + \frac{(\widetilde{\rho}_{2n}(2))^{\frac{1}{2}}}{2} \right] < +\infty, \quad \forall\, n \in \mathbb{N}. \tag{12.61}$$

Again, it holds $\lim_{n \to +\infty} B_n^*(f)(x) = \lim_{n \to +\infty} \overline{B}_n(f)(x) = f(x)$.

Proof. By Theorem 12.5 and Lemmas 12.2, 12.3, see also (12.57). □

Corollary 12.3. *Let $f \in C_B(\mathbb{R})$. Then*

$$\|B_n(f) - f\|_\infty \leq 2\omega_1(f, \widetilde{\rho}_{1n}(1)) < +\infty, \quad \forall\, n \in \mathbb{N}. \tag{12.62}$$

If f is also uniformly continuous, we get that $\lim_{n \to +\infty} B_n(f) = f$, pointwise and uniformly.

Proof. By Theorem 12.6, Lemma 12.1; (12.56). □

Corollary 12.4. *Let $f \in C_B(\mathbb{R})$. Then*

$$\left\{ \begin{array}{l} \|B_n^*(f) - f\|_\infty \\ \|\overline{B}_n(f) - f\|_\infty \end{array} \right\} \leq 2\omega_1(f, \widetilde{\rho}_{2n}(1)) < +\infty, \quad \forall\, n \in \mathbb{N}. \tag{12.63}$$

If f is also uniformly continuous, we obtain that $\lim_{n \to +\infty} B_n^*(f) = \lim_{n \to +\infty} \overline{B}_n(f) = f$, pointwise and uniformly.

Proof. By Theorem 12.6, Lemmas 12.2, 12.3; (12.57). □

We continue with simultaneous approximations.

Theorem 12.7. *Let $f^{(j)} \in C_B(\mathbb{R})$, for $j = 0, 1, ..., N \in \mathbb{N}$. Then*

$$\left\| (B_n(f))^{(j)} - f^{(j)} \right\|_\infty \leq 2\omega_1\left(f^{(j)}, \widetilde{\rho}_{1n}(1)\right) < +\infty, \quad \forall\, n \in \mathbb{N}. \tag{12.64}$$

If $f^{(j)}$ is uniformly continuous, we get that $\lim_{n \to +\infty} (B_n(f))^{(j)} = f^{(j)}$, pointwise and uniformly.

Proof. By (12.25) and Corollary 12.3. □

Theorem 12.8. *Let $f^{(j)} \in C_B(\mathbb{R})$, for $j = 0, 1, ..., N \in \mathbb{N}$. Then*

$$\left\{ \begin{array}{l} \left\| (B_n^*(f))^{(j)} - f^{(j)} \right\|_\infty \\ \left\| (\overline{B}_n(f))^{(j)} - f^{(j)} \right\|_\infty \end{array} \right\} \leq 2\omega_1 \left(f^{(j)}, \widetilde{\rho}_{2n}(1) \right) < +\infty, \quad \forall\, n \in \mathbb{N}. \tag{12.65}$$

If $f^{(j)}$ is uniformly continuous, we obtain that $\lim_{n \to +\infty} (B_n^*(f))^{(j)} = \lim_{n \to +\infty} (\overline{B}_n(f))^{(j)} = f^{(j)}$, pointwise and uniformly.

Proof. By (12.26), (12.27) and Corollary 12.4. □

Next we talk about iterated approximation.

Remark 12.5. Let $f \in C_B(\mathbb{R})$, $r \in \mathbb{N}$. Here θ_n is any of B_n, B_n^*, $\overline{B}_n$, $\forall\, n \in \mathbb{N}$.

By Remark 12.3, see also (12.3), we have that

$$\|\theta_n^r(f) - f\|_\infty \leq r \|\theta_n(f) - f\|_\infty, \quad \forall\, n \in \mathbb{N}. \tag{12.66}$$

Corollary 12.5. *Let $f \in C_B(\mathbb{R})$, $r \in \mathbb{N}$. Then*

$$\|B_n^r(f) - f\|_\infty \leq r \|B_n(f) - f\|_\infty \leq 2r\omega_1(f, \widetilde{\rho}_{1n}(1)) < +\infty, \quad \forall\, n \in \mathbb{N}. \tag{12.67}$$

If f is also uniformly continuous, we get that $\lim_{n \to +\infty} B_n^r(f) = f$, pointwise and uniformly.

And the speed of convergence of B_n^r to the unit operator is not worse than of B_n to the unit.

Proof. By (12.62) and (12.66). □

Corollary 12.6. *Let $f \in C_B(\mathbb{R})$, $r \in \mathbb{N}$. Then*

$$\left\{ \begin{array}{l} \left\| B_n^{*^r}(f) - f \right\|_\infty \\ \left\| \overline{B}_n^r(f) - f \right\|_\infty \end{array} \right\} \leq r \left\{ \begin{array}{l} \|B_n^*(f) - f\|_\infty \\ \|\overline{B}_n(f) - f\|_\infty \end{array} \right\} \tag{12.68}$$

$$\leq 2r\omega_1(f, \widetilde{\rho}_{2n}(1)) < +\infty, \quad \forall\, n \in \mathbb{N}.$$

If f is also uniformly continuous, we get that $\lim_{n \to +\infty} B_n^{*^r}(f) = \lim_{n \to +\infty} \overline{B}_n^r(f) = f$, pointwise and uniformly.

And the speed of convergence of $B_n^{*^r}, \overline{B}_n^r$ to the unit operator is not worse than of B_n^*, $\overline{B}_n$ to the unit, respectively.

Proof. By (12.63) and (12.66). □

We finish with more general iterated approximation results.

Remark 12.6. Let $f \in C_B(\mathbb{R})$, $r \in \mathbb{N}$, and $m_1, m_2, ..., m_r \in \mathbb{N} : m_1 \leq m_2 \leq ... \leq m_r$. Here θ_{m_i} is any of B_{m_i}, $B^*_{m_i}$, $\overline{B}_{m_i}$.

By Remark 12.3, see also (12.35), we have that

$$\left\|\theta_{m_r}\left(\theta_{m_{r-1}}\left(...\theta_{m_2}\left(\theta_{m_1}f\right)\right)\right) - f\right\|_\infty \leq \sum_{i=1}^{r} \left\|\theta_{m_i}f - f\right\|_\infty . \tag{12.69}$$

Corollary 12.7. *Let $f \in C_B(\mathbb{R})$, $r \in \mathbb{N}$, and $m_1, m_2, ..., m_r \in \mathbb{N} : m_1 \leq m_2 \leq ... \leq m_r$. Then*

$$\left\|B_{m_r}\left(B_{m_{r-1}}\left(...B_{m_2}\left(B_{m_1}f\right)\right)\right) - f\right\|_\infty \leq \sum_{i=1}^{r} \left\|B_{m_i}f - f\right\|_\infty \leq \tag{12.70}$$

$$2\left(\sum_{i=1}^{r} \omega_1\left(f, \widetilde{\rho}_{1m_i}(1)\right)\right) \leq 2r\omega_1\left(f, \widetilde{\rho}_{1m_1}(1)\right). \tag{12.71}$$

The speed of convergence to the unit operator of the above activated multiply iterated operators is not worse than the speed of convergence to the unit of B_{m_1}.

Proof. By (12.69), (12.62) and (12.56). □

Corollary 12.8. *Let $f \in C_B(\mathbb{R})$, $r \in \mathbb{N}$, and $m_1, m_2, ..., m_r \in \mathbb{N} : m_1 \leq m_2 \leq ... \leq m_r$. Then*

1)

$$\left\|B^*_{m_r}\left(B^*_{m_{r-1}}\left(...B^*_{m_2}\left(B^*_{m_1}f\right)\right)\right) - f\right\|_\infty \leq \sum_{i=1}^{r} \left\|B^*_{m_i}f - f\right\|_\infty \leq \tag{12.72}$$

$$2\left(\sum_{i=1}^{r} \omega_1\left(f, \widetilde{\rho}_{2m_i}(1)\right)\right) \leq 2r\omega_1\left(f, \widetilde{\rho}_{2m_1}(1)\right), \tag{12.73}$$

and

2)

$$\left\|\overline{B}_{m_r}\left(\overline{B}_{m_{r-1}}\left(...\overline{B}_{m_2}\left(\overline{B}_{m_1}f\right)\right)\right) - f\right\|_\infty \leq \sum_{i=1}^{r} \left\|\overline{B}_{m_i}f - f\right\|_\infty \leq \tag{12.74}$$

$$2\left(\sum_{i=1}^{r} \omega_1\left(f, \widetilde{\rho}_{2m_i}(1)\right)\right) \leq 2r\omega_1\left(f, \widetilde{\rho}_{2m_1}(1)\right). \tag{12.75}$$

The speed of convergence to the unit operator of the above activated multiply iterated operators is not worse than the speed of convergence to the unit of $B^*_{m_1}$, $\overline{B}_{m_1}$, respectively.

Proof. By (12.69), (12.63) and (12.57). □

Bibliography

[280] G.A. Anastassiou, *A "K-Attainable" inequality related to the convergence of Positive Linear Operators*, J. Approximation Theory, 44, 380–383 (1985).

[281] G.A. Anastassiou, *Moments in probability and approximation theory*, Pitman Research Notes in Mathematics series / Longman Scientific & Technical, Essex, New York, UK, USA, 1993.

[282] G.A. Anastassiou, *Quantitative Approximations*, Chapmen & Hall/CRC, London, New York, 2001.

[283] G.A. Anastassiou, *Intelligent Computations: Abstract Fractional Calculus, Inequalities, Approximations*, Springer, Heidelberg, New York, 2018.

[284] G.A. Anastassiou, *Parametrized, Deformed and General Neural Networks*, Springer, Heidelberg, New York, 2023.

[285] G.A. Anastassiou, *Trigonometric and Hyperbolic Generated Approximation Theory*, World Scientific, Singapore, New York, 2025.

[286] G.A. Anastassiou, *Approximation by Parametrized logistic activated convolution type operators*, RACSAM, (2024), 118.138, `https://doi.org/10.1007/s13398-024-01635-7`.

[287] G.A. Anastassiou, *Neural Networks in infinite domain as positive linear operastors*, Annales Univ. Sci. Budapest, Sect. Comp., accepted, 2025.

[288] G.A. Anastassiou, *Generalized Logistic Neural Networks as Positive Linear Operators over infinite domain*, Neural, Parallel and Scientific Computations, accepted, 2025.

[289] R.A. DeVore, *The Approximation of Continuous Functions by Positive Linear Operators*, Lecture Notes in Mathematics, Vol. 293, Springer, Heidelberg, New York, 1972.

[290] Hee Sun Jung, Ryozi Sakai, *Local saturation of a positive linear convolution operator*, J. Inequal. Appl. 2014, 2014: 329, 16 p.

[291] J.J. Swetits, B. Wood, *Local L_p-saturation of positive linear convolution operators*, J. Approx. Theory 34 (1982), no. 4, 348–360.

[292] Grigor Moldovan, *Discrete convolutions in connection with functions of several variables and positive linear operators.* (Romanian) Studia Univ. Babeş-Bolyai Ser. Math.-Mech. 19 (1974), no. 1, 51–57.

[293] R. Bojanic, O. Shisha, *On the precision of uniform approximation of continuous functions by certain linear operators of convolution type.* J. Approximation Theory 8 (1973), 101–113.

[294] Grigor Moldovan, *Discrete convolutions and linear positive operators.* Ann. Univ. Sci. Budapest. Eötvös Sect. Math. 15 (1972), 31–44 (1973).

[295] A.I. Kamzolov, *The order of approximation of functions of class $Z_2(E_n)$ by positive linear convolution operators.* (Russian) Mat. Zametki 7 (1970), 723–732.

[296] G.A. Anastassiou, *Approximation by symmetrized and perturbed generalized logistic activated convolutions interpreted as Positive Linear Operators*, Romai Journal, accepted, 2025.

Chapter 13

Complete Approximation by Symmetrized and Disturbed Hyperbolic Tangent Energized Multivariate Convolutions as Positive Linear Operators

In this work are studied in detail the multivariate symmetrized and perturbed hyperbolic tangent activated convolution type operators of three kinds. Here this is done with the method of positive linear operators. Their alternative approximation properties are established by the quantitative convergence to the unit operator using the modulus of continuity. It also studies the related multivariate simultaneous approximation, as well as the multivariate iterated approximation. It follows [313].

13.1 Introduction

The author studied extensively the quantitative approximation of positive linear operators to the unit since 1985, see for example [297]–[299], [301]. He originated from the quantitative weak convergence of finite positive measures to the unit Dirac measure, having as a method the geometric moment theory, see [298], and he produced best upper bounds, leading to attained (i.e. sharp Jackson type inequalities), e.g. see [297], [298]. These studies have gone to all possible directions, univariate and multivariate, though in this work we stay only on the multivariate approach over an infinite domain.

Our multidimensional convolution operators here have a kernel based on the symmetrized and perturbed hyperbolic tangent activation function, which is used frequently in the study of neural networks, and they can be interpreted as positive linear operators.

So here our proving methods come from the theory of positive linear operators.

Thus in Section 2, we discuss about the symmetrized and perturbed hyperbolic tangent activation function in the multivariate setting. We also describe our activated multidimensional convolution type operators with their properties, such as differentiation and iteration.

In Section 3, we derive some auxiliary results which are estimates to our operators, when applied to basic functions and to be used into our main results.

In Section 4, we present our main explicit results under the lens of positive linear operators and we include their necessary related theory. We treat also the simultaneous and iterated approximation cases under the same spirit.

We are greatly inspired by our earlier works [302], [303].

Furthermore, general motivation comes from the great works [307]–[312].

13.2 Basics

Initially we follow [300], pp. 455–460.

Our perturbed hyperbolic tangent activation function is

$$g_{q,\lambda}(x) := \frac{e^{\lambda x} - qe^{-\lambda x}}{e^{\lambda x} + qe^{-\lambda x}}, \quad \lambda, q > 0, \ x \in \mathbb{R}. \tag{13.1}$$

Above λ is the parameter and q is the deformation coefficient.

For more read Chapter 18 of [300]: "q-deformed and λ-Parametrized Hyperbolic Tangent based Banach space Valued Ordinary and Fractional Neural Network Approximation".

The chapters 17 and 18 of [300] motivate our current work.

The proposed "symmetrization method" aims to use half data feed to our multivariate neural networks.

We will employ the following density function

$$M_{q.\lambda}(x) := \frac{1}{4}\left(g_{q,\lambda}(x+1) - g_{q,\lambda}(x-1)\right) > 0, \tag{13.2}$$

$\forall x \in \mathbb{R}$; $q, \lambda > 0$.

We have that

$$M_{q,\lambda}(-x) = M_{\frac{1}{q},\lambda}(x), \ \ \forall\, x \in \mathbb{R}; \ q, \lambda > 0, \tag{13.3}$$

and

$$M_{\frac{1}{q},\lambda}(-x) = M_{q,\lambda}(x), \ \ \forall\, x \in \mathbb{R}; \ q, \lambda > 0. \tag{13.4}$$

Adding (13.3) and (13.4) we obtain

$$M_{q,\lambda}(-x) + M_{\frac{1}{q},\lambda}(-x) = M_{q,\lambda}(x) + M_{\frac{1}{q},\lambda}(x), \tag{13.5}$$

a key to this work.

So that

$$\Phi(x) := \frac{M_{q,\lambda}(x) + M_{\frac{1}{q},\lambda}(x)}{2} > 0, \tag{13.6}$$

is an even function, symmetric with respect to the y-axis.

By (18.18) of [300], we have

$$\begin{aligned} &M_{q,\lambda}\left(\tfrac{\ln q}{2\lambda}\right) = \tfrac{\tanh(\lambda)}{2}, \\ &\text{and} \\ &M_{\frac{1}{q},\lambda}\left(-\tfrac{\ln q}{2\lambda}\right) = \tfrac{\tanh(\lambda)}{2}, \quad \lambda > 0, \end{aligned} \tag{13.7}$$

sharing the same maximum at symmetric points.

By Theorem 18.1, p. 458 of [300], we have that

$$\begin{aligned} &\sum_{i=-\infty}^{\infty} M_{q,\lambda}(x-i) = 1, \quad \forall\, x \in \mathbb{R},\ \lambda, q > 0, \\ &\text{and} \\ &\sum_{i=-\infty}^{\infty} M_{\frac{1}{q},\lambda}(x-i) = 1, \quad \forall\, x \in \mathbb{R},\ \lambda, q > 0. \end{aligned} \tag{13.8}$$

Consequently, we derive that

$$\sum_{i=-\infty}^{\infty} \Phi(x-i) = 1, \quad \forall\, x \in \mathbb{R}. \tag{13.9}$$

By Theorem 18.2, p. 459 of [300], we have that

$$\begin{aligned} &\int_{-\infty}^{\infty} M_{q,\lambda}(x)\,dx = 1, \\ &\text{and} \\ &\int_{-\infty}^{\infty} M_{\frac{1}{q},\lambda}(x)\,dx = 1, \end{aligned} \tag{13.10}$$

so that

$$\int_{-\infty}^{\infty} \Phi(x)\,dx = 1, \tag{13.11}$$

therefore Φ is a density function.

Clearly, then

$$\int_{-\infty}^{\infty} \Phi(nx-u)\,du = 1, \quad \forall\, n \in \mathbb{N},\ x \in \mathbb{R}. \tag{13.12}$$

An essential property follows:

Proposition 13.1. *([302]) It holds ($k \in \mathbb{N}$)*

$$\int_{-\infty}^{\infty} |z|^k\, \Phi(z)\,dz \leq \left[\frac{\tanh(\lambda)}{(k+1)} + \frac{\left(q + \frac{1}{q}\right) e^{2\lambda} k!}{(2\lambda)^k}\right] < \infty. \tag{13.13}$$

We mention

Definition 13.1. The modulus of continuity here is defined by

$$\omega_1(f,\delta) := \sup_{\substack{x,y\in\mathbb{R}^N:\\ \|x-y\|_\infty<\delta}} |f(x)-f(y)|, \ \ \delta>0, \tag{13.14}$$

where $f:\mathbb{R}^N\to\mathbb{R}$ bounded and continuous, denoted by $f\in C_B(\mathbb{R}^N)$, $N\in\mathbb{N}$. Similarly ω_1 is defined for $f\in C_U(\mathbb{R}^N)$ (uniformly continuous functions). We have that $f\in C_U(\mathbb{R}^N)$, iff $\omega_1(f,\delta)\to 0$ as $\delta\downarrow 0$.

Denote $\|x\|_\infty := \max\{|x_1|,...,|x_N|\}$, $x\in\mathbb{R}^N$.

We make

Remark 13.1. We introduce

$$\widetilde{Z}(x_1,...,x_N) := \widetilde{Z}(x) := \prod_{i=1}^N \Phi(x_i), \quad x=(x_1,...,x_N)\in\mathbb{R}^N, \ N\in\mathbb{N}. \tag{13.15}$$

It has the properties:

(i)

$$\widetilde{Z}(x)>0, \ \forall x\in\mathbb{R}^N; \ \widetilde{Z}(-x)=\widetilde{Z}(x),$$

(ii)

$$\int_{\mathbb{R}^N}\widetilde{Z}(x-k) = \int_{-\infty}^{\infty}...\int_{-\infty}^{\infty}\widetilde{Z}(x_1-u_1,...,x_N-u_N)\,du_1...du_N =$$

$$\prod_{i=1}^N\int_{-\infty}^{\infty}\Phi(x_i-u_i)\,du_i \overset{(13.12)}{=} 1, \ \forall x\in\mathbb{R}^N, \tag{13.16}$$

hence

(iii)

$$\int_{\mathbb{R}^N}\widetilde{Z}(nx-u)\,du = 1, \ \ \forall\, x\in\mathbb{R}^N; \ n\in\mathbb{N}, \tag{13.17}$$

and

(iv) by (13.11)

$$\int_{\mathbb{R}^N}\widetilde{Z}(x)\,dx = 1, \tag{13.18}$$

that is $\widetilde{Z}$ is a multivariate density function.

We mention a useful related result.

Theorem 13.1. *([303]) It holds* ($k \in \mathbb{N}$)

$$\int_{\mathbb{R}^N} \|x\|_\infty^k \widetilde{Z}(x)\,dx \le N^k \left[\frac{\tanh(\lambda)}{(k+1)} + \frac{\left(q+\frac{1}{q}\right) e^{2\lambda} k!}{(2\lambda)^k} \right] < \infty. \tag{13.19}$$

When $k = 0$, *(13.19) is again valid.*

We give

Definition 13.2. Let $f \in C_B(\mathbb{R}^N)$, $N \in \mathbb{N}$. We define the following activated symmetrized and perturbed hyperbolic tangent multivariate convolution type operators:

The basic one

$$\widetilde{S}_n(f)(x) := \int_{\mathbb{R}^N} f\left(\frac{u}{n}\right) \widetilde{Z}(nx-u)\,du, \quad \forall\, x \in \mathbb{R}^N, \tag{13.20}$$

the activated Kantorovich type

$$\widetilde{S}_n^*(f)(x) := n^N \int_{\mathbb{R}^N} \left(\int_{\frac{u}{n}}^{\frac{u+1}{n}} f(t)\,dt \right) \widetilde{Z}(nx-u)\,du, \quad \forall\, x \in \mathbb{R}^N. \tag{13.21}$$

Let now $\theta = (\theta_1, ..., \theta_N) \in \mathbb{N}^N$, $r = (r_1, ..., r_N) \in \mathbb{Z}_+^N$, $w_r = w_{r_1 r_2 ... r_N} \ge 0$, such that

$$\sum_{r=0}^{\theta} w_r = \sum_{r_1=0}^{\theta_1} \sum_{r_2=0}^{\theta_2} ... \sum_{r_N=0}^{\theta_N} w_{r_1 r_2 ... r_N} = 1; \quad u \in \mathbb{R}^N,$$

and

$$\delta_n(f)(u) := \delta_n(f)(u_1, ..., u_N) := \sum_{r=0}^{\theta} w_r f\left(\frac{u}{n} + \frac{r}{n\theta}\right) =$$

$$\sum_{r_1=0}^{\theta_1} \sum_{r_2=0}^{\theta_2} ... \sum_{r_N=0}^{\theta_N} w_{r_1 r_2 ... r_N} f\left(\frac{u_1}{n} + \frac{r_1}{n\theta_1}, \frac{u_2}{n} + \frac{r_2}{n\theta_2}, ..., \frac{u_N}{n} + \frac{r_N}{n\theta_N}\right), \tag{13.22}$$

where $\frac{r}{\theta} := \left(\frac{r_1}{\theta_1}, \frac{r_2}{\theta_2}, ..., \frac{r_N}{\theta_N}\right)$.

We define the activated quadrature operators

$$\widehat{S_n}(f)(x) := \widehat{S_n}(f, x_1, ..., x_N) := \int_{\mathbb{R}^N} \delta_n(f)(u)\, \widetilde{Z}(nx-u)\,du = \tag{13.23}$$

$$\int_{-\infty}^{\infty} \int_{-\infty}^{\infty} ... \int_{-\infty}^{\infty} \delta_n(f)(u_1, ..., u_N) \left(\prod_{i=1}^{N} \Phi(nx_i - u_i) \right) du_1 ... du_N, \forall x \in \mathbb{R}^N.$$

One can rewrite

$$\widetilde{S}_n(f)(x)=\int_{-\infty}^{\infty}\int_{-\infty}^{\infty}...\int_{-\infty}^{\infty}f\left(\frac{u_1}{n},\frac{u_2}{n},...,\frac{u_N}{n}\right)\left(\prod_{i=1}^{N}\Phi\left(nx_i-u_i\right)\right)du_1...du_N, \tag{13.24}$$

and

$$\widetilde{S}_n^*(f)(x)=n^N\int_{-\infty}^{\infty}\int_{-\infty}^{\infty}...\int_{-\infty}^{\infty}\left(\int_{\frac{u_1}{n}}^{\frac{u_1+1}{n}}\int_{\frac{u_2}{n}}^{\frac{u_2+1}{n}}...\int_{\frac{u_N}{n}}^{\frac{u_N+1}{n}}f(t_1,...,t_N)\,dt_1...dt_N\right)$$

$$\left(\prod_{i=1}^{N}\Phi\left(nx_i-u_i\right)\right)du_1...du_N, \tag{13.25}$$

$n\in\mathbb{N}$, $\forall\ x\in\mathbb{R}^N$.

In this work we study the approximation properties of the operators $\widetilde{S}_n$, $\widetilde{S}_n^*$ and $\widehat{S_n}$, expecially their convergence to the unit operator I, and being treated as positive linear operators.

We notice that

$$\widetilde{S}_n(1)=\widetilde{S}^*(1)=\widehat{S}(1)=1. \tag{13.26}$$

From [303] we have that $\widetilde{S}_n$, $\widetilde{S}_n^*$, $\widehat{S_n}$ map $C_B\left(\mathbb{R}^N\right)$ into itself.

Differentiation of our operators follows:

Remark 13.2. ([303]) Let $i\in\mathbb{N}$ be fixed. Assume that $f\in C^{(i)}\left(\mathbb{R}^N\right)$, $N\in\mathbb{N}$. Here f_α denotes a partial derivative of f, $\alpha:=(\alpha_1,...,\alpha_N)$, $\alpha_j\in\mathbb{Z}_+$, $j=1,...,N$, and $|\alpha|:=\sum_{j=1}^{N}\alpha_j=l$, where $l=0,1,...,i$.

We write also $f_\alpha:=\frac{\partial^\alpha f}{\partial x^\alpha}$ and we say it is of order l.

We assume that any partial $f_\alpha\in C_B\left(\mathbb{R}^N\right)$, for all $\alpha:|\alpha|=l$, $l=0,1,...,i$.

We have that

$$\left(\widetilde{S}_n(f)\right)_\alpha(x)=\int_{\mathbb{R}^N}f_\alpha\left(x-\frac{z}{n}\right)\widetilde{Z}(z)\,dz= \tag{13.27}$$

$$\int_{\mathbb{R}^N}f_\alpha\left(\frac{u}{n}\right)\widetilde{Z}(nx-u)\,du=\left(\widetilde{S}_n(f_\alpha)\right)(x),\ \ \forall\ x\in\mathbb{R}^N.$$

Similarly, we have that

$$\left(\widetilde{S}_n^*(f)\right)_\alpha(x)=\left(\widetilde{S}_n^*(f_\alpha)\right)(x), \tag{13.28}$$

and

$$\left(\widehat{S_n}(f)\right)_\alpha(x)=\left(\widehat{S_n}(f_\alpha)\right)(x),\ \ \forall\ x\in\mathbb{R}^N; \tag{13.29}$$

for all $\alpha:|\alpha|=l$, $l=0,1,...,i$.

We also need

Remark 13.3. ([303]) About iterated operators:

It holds

$$\left|\widetilde{S}_n(f)(x)\right| \leq \|f\|_\infty \int_{\mathbb{R}^N} \widetilde{Z}(z)\, dz = \|f\|_\infty ,$$

i.e.

$$\left\|\widetilde{S}_n(f)\right\|_\infty \leq \|f\|_\infty . \tag{13.30}$$

So $\widetilde{S}_n$ is a bounded positive linear operator.

Clearly it holds

$$\left\|\widetilde{S}_n^2(f)\right\|_\infty = \left\|\widetilde{S}_n\left(\widetilde{S}_n(f)\right)\right\|_\infty \leq \left\|\widetilde{S}_n(f)\right\|_\infty \leq \|f\|_\infty . \tag{13.31}$$

And for $k \in \mathbb{N}$ we obtain

$$\left\|\widetilde{S}_n^k(f)\right\|_\infty \leq \left\|\widetilde{S}_n^{k-1}(f)\right\|_\infty \leq \left\|\widetilde{S}_n^{k-2}(f)\right\|_\infty \leq ... \leq \|f\|_\infty , \tag{13.32}$$

so the contraction property valid and $\widetilde{S}_n^k$ is a bounded linear operator.

Let $r \in \mathbb{N}$, we observe that

$$\widetilde{S}_n^r f - f = \left(\widetilde{S}_n^r f - \widetilde{S}_n^{r-1} f\right) + \left(\widetilde{S}_n^{r-1} f - \widetilde{S}_n^{r-2} f\right) + \left(\widetilde{S}_n^{r-2} f - \widetilde{S}_n^{r-3} f\right)$$

$$+ ... + \left(\widetilde{S}_n^2 f - \widetilde{S}_n f\right) + \left(\widetilde{S}_n f - f\right). \tag{13.33}$$

Then

$$\left\|\widetilde{S}_n^r f - f\right\|_\infty \leq \left\|\widetilde{S}_n^r f - \widetilde{S}_n^{r-1} f\right\|_\infty + \left\|\widetilde{S}_n^{r-1} f - \widetilde{S}_n^{r-2} f\right\|_\infty + \left\|\widetilde{S}_n^{r-2} f - \widetilde{S}_n^{r-3} f\right\|_\infty$$

$$+ ... + \left\|\widetilde{S}_n^2 f - \widetilde{S}_n f\right\|_\infty + \left\|\widetilde{S}_n f - f\right\|_\infty =$$

$$\left\|\widetilde{S}_n^{r-1}\left(\widetilde{S}_n f - f\right)\right\|_\infty + \left\|\widetilde{S}_n^{r-2}\left(\widetilde{S}_n f - f\right)\right\|_\infty + ... + \left\|\widetilde{S}_n\left(\widetilde{S}_n f - f\right)\right\|_\infty +$$

$$\left\|\widetilde{S}_n f - f\right\|_\infty \leq r \left\|\widetilde{S}_n f - f\right\|_\infty .$$

Therefore

$$\left\|\widetilde{S}_n^r f - f\right\|_\infty \leq r \left\|\widetilde{S}_n f - f\right\|_\infty . \tag{13.34}$$

Let now $m_1, m_2, ..., m_r \in \mathbb{N} : m_1 \leq m_2 \leq ... \leq m_r$, and $\widetilde{S}_{m_i}$ as above.

$$\widetilde{S}_{m_r}\left(\widetilde{S}_{m_{r-1}}\left(...\widetilde{S}_{m_2}\left(\widetilde{S}_{m_1} f\right)\right)\right) - f = ... =$$

$$\widetilde{S}_{m_r}\left(\widetilde{S}_{m_{r-1}}\left(...\widetilde{S}_{m_2}\right)\right)\left(\widetilde{S}_{m_1} f - f\right) + \widetilde{S}_{m_r}\left(\widetilde{S}_{m_{r-1}}\left(...\widetilde{S}_{m_3}\right)\right)\left(\widetilde{S}_{m_2} f - f\right) +$$

$$\widetilde{S}_{m_r}\left(\widetilde{S}_{m_{r-1}}\left(...\widetilde{S}_{m_4}\right)\right)\left(\widetilde{S}_{m_3} f - f\right) + ... + \widetilde{S}_{m_r}\left(\widetilde{S}_{m_{r-1}} f - f\right) + \widetilde{S}_{m_r} f - f. \tag{13.35}$$

Consequently it holds, as in [300], Chapter 2,

$$\left\|\widetilde{S}_{m_r}\left(\widetilde{S}_{m_{r-1}}\left(...\widetilde{S}_{m_2}\left(\widetilde{S}_{m_1} f\right)\right)\right) - f\right\|_\infty \leq \sum_{i=1}^{r} \left\|\widetilde{S}_{m_i} f - f\right\|_\infty . \tag{13.36}$$

The properties (13.30)–(13.36) are also valid for the iterations of operators $\widetilde{S}_n^*$ and $\widehat{S_n}$.

Notation 13.1. We denote by K_n any of the operators $\widetilde{S}_n$, $\widetilde{S}_n^*$ and $\widehat{S_n}$, $\forall$ $n \in \mathbb{N}$.

13.3 Auxiliary Results

Here let $x_0 \in \mathbb{R}^N$, $m, n \in \mathbb{N}$. It follows:

Proposition 13.2. *It holds that*

$$\widetilde{S}_n \left(\|\cdot - x_0\|_\infty^m \right) (x_0) > 0, \tag{13.37}$$

and

$$0 < \left\| \widetilde{S}_n \left(\|\cdot - x_0\|_\infty^m \right) (x_0) \right\|_\infty \leq \tag{13.38}$$

$$\frac{N^m}{n^m} \left[\frac{\tanh(\lambda)}{(m+1)} + \frac{\left(q + \frac{1}{q}\right) e^{2\lambda} m!}{(2\lambda)^m} \right] (< +\infty) \to 0, \text{ as } n \to +\infty.$$

Proof. We have that

$$0 < \widetilde{S}_n \left(\|\cdot - x_0\|_\infty^m \right) (x_0) = \int_{\mathbb{R}^N} \left\| \frac{u}{n} - x_0 \right\|_\infty^m \widetilde{Z}(nx_0 - u)\, du =$$

$$\frac{1}{n^m} \int_{\mathbb{R}^N} \|nx_0 - u\|_\infty^m \widetilde{Z}(nx_0 - u)\, du = \tag{13.39}$$

$$\frac{1}{n^m} \int_{\mathbb{R}^N} \|x\|_\infty^m \widetilde{Z}(x)\, dx \overset{(13.19)}{\leq}$$

$$\frac{N^m}{n^m} \left[\frac{\tanh(\lambda)}{(m+1)} + \frac{\left(q + \frac{1}{q}\right) e^{2\lambda} m!}{(2\lambda)^m} \right] (< +\infty) \to 0, \text{ as } n \to +\infty.$$

□

Proposition 13.3. *It holds that*

$$\widetilde{S}_n^* \left(\|\cdot - x_0\|_\infty^m \right) (x_0) > 0, \tag{13.40}$$

and

$$0 < \left\| \widetilde{S}_n^* \left(\|\cdot - x_0\|_\infty^m \right) (x_0) \right\|_\infty \leq$$

$$\frac{2^{m-1}}{n^m} \left[1 + N^m \left(\frac{\tanh(\lambda)}{(m+1)} + \frac{\left(q + \frac{1}{q}\right) e^{2\lambda} m!}{(2\lambda)^m} \right) \right] (< +\infty) \to 0, \text{ as } n \to +\infty. \tag{13.41}$$

Proof. We have that

$$0 < \widetilde{S}_n^* \left(\|\cdot - x_0\|_\infty^m\right)(x_0) = \int_{\mathbb{R}^N} \left(n^N \int_{\left[0, \frac{1}{n}\right]^N} \left\| t + \frac{u}{n} - x_0 \right\|_\infty^m dt \right) \widetilde{Z}(nx_0 - u)\, du$$

$$\leq \int_{\mathbb{R}^N} \left(n^N \int_{\left[0, \frac{1}{n}\right]^N} \left(\left\| \frac{u}{n} - x_0 \right\|_\infty + \|t\|_\infty \right)^m dt \right) \widetilde{Z}(nx_0 - u)\, du \leq \quad (13.42)$$

$$\int_{\mathbb{R}^N} \left(n^N \int_{\left[0, \frac{1}{n}\right]^N} \left(\left\| \frac{u}{n} - x_0 \right\|_\infty + \frac{1}{n} \right)^m dt \right) \widetilde{Z}(nx_0 - u)\, du =$$

$$\int_{\mathbb{R}^N} \left(\left\| \frac{u}{n} - x_0 \right\|_\infty + \frac{1}{n} \right)^m \widetilde{Z}(nx_0 - u)\, du =$$

$$\frac{1}{n^m} \int_{\mathbb{R}^N} \left(\|nx_0 - u\|_\infty + 1 \right)^m \widetilde{Z}(nx_0 - u)\, du \leq \quad (13.43)$$

$$\frac{2^{m-1}}{n^m} \left(\int_{\mathbb{R}^N} \left(1 + \|nx_0 - u\|_\infty^m\right) \widetilde{Z}(nx_0 - u)\, du \right) \overset{(13.17)}{=}$$

$$\frac{2^{m-1}}{n^m} \left[1 + \int_{\mathbb{R}^N} \|nx_0 - u\|_\infty^m \widetilde{Z}(nx_0 - u)\, du \right] =$$

$$\frac{2^{m-1}}{n^m} \left[1 + \int_{\mathbb{R}^N} \|x\|_\infty^m \widetilde{Z}(x)\, dx \right] \overset{(13.19)}{\leq}$$

$$\frac{2^{m-1}}{n^m} \left[1 + N^m \left(\frac{\tanh(\lambda)}{(m+1)} + \frac{\left(q + \frac{1}{q}\right) e^{2\lambda} m!}{(2\lambda)^m} \right) \right] (< +\infty) \to 0, \text{ as } n \to +\infty.$$

□

Proposition 13.4. *It holds that*

$$\widehat{S_n}\left(\|\cdot - x_0\|_\infty^m\right)(x_0) > 0,$$

and

$$0 < \left\| \widehat{S_n}\left(\|\cdot - x_0\|_\infty^m\right)(x_0) \right\|_\infty \leq$$

$$\frac{2^{m-1}}{n^m} \left[1 + N^m \left(\frac{\tanh(\lambda)}{(m+1)} + \frac{\left(q + \frac{1}{q}\right) e^{2\lambda} m!}{(2\lambda)^m} \right) \right] (< +\infty) \to 0, \ \text{as } n \to +\infty. \tag{13.44}$$

Proof. We have that

$$0 < \widehat{S_n}\left(\|\cdot - x_0\|_\infty^m\right)(x_0) = \int_{\mathbb{R}^N}\left(\sum_{r=0}^{\theta} w_r \left\|\frac{u}{n} + \frac{r}{n\theta} - x_0\right\|_\infty^m\right)\widetilde{Z}(nx_0 - u)\,du \leq$$

$$\int_{\mathbb{R}^N}\left(\sum_{r=0}^{\theta} w_r \left(\left\|\frac{u}{n} - x_0\right\|_\infty + \left\|\frac{r}{n\theta}\right\|_\infty\right)^m\right)\widetilde{Z}(nx_0 - u)\,du \leq$$

$$\int_{\mathbb{R}^N}\left(\left\|\frac{u}{n} - x_0\right\|_\infty + \frac{1}{n}\right)^m \widetilde{Z}(nx_0 - u)\,du \tag{13.45}$$

(the rest as in the proof of Proposition 13.3)

$$\frac{2^{m-1}}{n^m}\left[1 + N^m\left(\frac{\tanh(\lambda)}{(m+1)} + \frac{\left(q + \frac{1}{q}\right)e^{2\lambda}m!}{(2\lambda)^m}\right)\right](< +\infty) \to 0, \text{ as } n \to +\infty.$$

□

13.4 Main Results

We need the following:

Theorem 13.2. *Let $(L_n)_{n\in\mathbb{N}}$ be a sequence of positive linear operators from $C_B(\mathbb{R}^N)$ into itself, such that $L_n(1) = 1$. Assume $0 < L_n\left(\|\cdot - x_0\|_\infty\right)(x_0) < +\infty$.*

Then

$$|L_n(f)(x_0) - f(x_0)| \leq 2\omega_1\left(f, L_n\left(\|\cdot - x_0\|_\infty\right)(x_0)\right) < +\infty. \tag{13.46}$$

Given that $\lim_{n\to+\infty} L_n\left(\|\cdot - x_0\|_\infty\right)(x_0) = 0$ and f is uniformly continuous, we get that $\lim_{n\to+\infty} L_n(f)(x_0) = f(x_0)$.

Proof. Here it is $\omega_1(f,h) < \infty$, $h > 0$.

By Lemma 7.1.1, p. 208, [298], we get that

$$|f(t) - f(x)| \leq \omega_1(f,h)\left\lceil\frac{\|t - x\|_\infty}{h}\right\rceil, \tag{13.47}$$

where $\lceil\cdot\rceil$ is the ceiling function.

But it holds

$$\left\lceil\frac{\|t - x\|_\infty}{h}\right\rceil \leq 1 + \frac{\|t - x\|_\infty}{h}. \tag{13.48}$$

Thus we get

$$\left|f(t)-f(x)\right| \leq \omega_1(f,h)\left[1+\frac{\|t-x\|_\infty}{h}\right], \quad \forall\, t \in \mathbb{R}^N. \tag{13.49}$$

The last can be written also as

$$\left|f(\cdot)-f(x)\right| \leq \omega_1(f,h)\left[1+\frac{\|\cdot-x\|_\infty}{h}\right], \quad \text{over } \mathbb{R}^N. \tag{13.50}$$

Hence we derive

$$\left|L_n(f)(x_0)-f(x_0)\right| = \left|L_n(f(\cdot)-f(x_0))(x_0)\right| \leq$$

$$L_n(\left|f(\cdot)-f(x_0)\right|)(x_0) \overset{(13.50)}{\leq} \omega_1(f,h)\left[1+\frac{L_n(\|\cdot-x_0\|_\infty)(x_0)}{h}\right] \tag{13.51}$$

(choose $h := L_n(\|\cdot-x_0\|_\infty)(x_0) > 0$)

$$= 2\omega_1(f, L_n(\|\cdot-x_0\|_\infty)(x_0)).$$

□

We derive

Theorem 13.3. *We have for $K_n = \widetilde{S}_n,\ \widetilde{S}_n^*,\ \widehat{S_n}$ that*

$$\left|K_n(f)(x_0)-f(x_0)\right| \leq 2\omega_1(f, K_n(\|\cdot-x_0\|_\infty)(x_0)) < +\infty, \tag{13.52}$$

where $x_0 \in \mathbb{R}^N$, $f \in C_B(\mathbb{R}^N)$.

Given that $\lim_{n\to+\infty} K_n(\|\cdot-x_0\|_\infty)(x_0) = 0$ and f is uniformly continuous, we get that $\lim_{n\to+\infty} K_n(f)(x_0) = f(x_0)$.

Proof. By Theorem 13.2. □

More specifically we derive:

Proposition 13.5. *It holds that ($f \in C_B(\mathbb{R}^N)$)*

$$\left\|\widetilde{S}_n(f)-f\right\|_\infty \leq 2\omega_1\left(f, \frac{N}{n}\left[\frac{\tanh(\lambda)}{2}+\frac{\left(q+\frac{1}{q}\right)e^{2\lambda}}{2\lambda}\right]\right) < +\infty. \tag{13.53}$$

If f is uniformly continuous, then $\widetilde{S}_n(f) \to f$, uniformly as $n \to +\infty$.

Proof. By Proposition 13.2, and Theorem 13.3. □

Proposition 13.6. *It holds that* $(f \in C_B(\mathbb{R}^N))$

$$\left\| \widetilde{S}_n^*(f) - f \right\|_\infty \leq 2\omega_1\left(f, \frac{1}{n}\left[1 + N\left(\frac{\tanh(\lambda)}{2} + \frac{\left(q+\frac{1}{q}\right)e^{2\lambda}}{2\lambda}\right)\right]\right) < +\infty. \tag{13.54}$$

If f *is uniformly continuous, then* $\widetilde{S}_n^*(f) \to f$, *uniformly as* $n \to +\infty$.

Proof. By Proposition 13.3, and Theorem 13.3. □

Proposition 13.7. *It holds that* $(f \in C_B(\mathbb{R}^N))$

$$\left\| \widehat{S_n}(f) - f \right\|_\infty \leq 2\omega_1\left(f, \frac{1}{n}\left[1 + N\left(\frac{\tanh(\lambda)}{2} + \frac{\left(q+\frac{1}{q}\right)e^{2\lambda}}{2\lambda}\right)\right]\right) < +\infty. \tag{13.55}$$

If f *is uniformly continuous, then* $\widehat{S_n}(f) \to f$, *uniformly as* $n \to +\infty$.

Proof. By Proposition 13.4, and Theorem 13.3. □

Next, we study our operators under convexity:

Theorem 13.4. *Let* $(L_n)_{n\in\mathbb{N}}$ *positive linear operators from* $C_B(\mathbb{R}^N)$ *into itself:* $L_n(1) = 1$. *Let* $f \in C_B(\mathbb{R}^N)$ *and the convex function* $|f(\cdot) - f(x_0)|$ *over* $\mathbb{R}^N$, $x_0 \in \mathbb{R}^N$. *Assume that* $0 < L_n(\|\cdot - x_0\|_\infty)(x_0) < +\infty$.

Then

$$|L_n(f)(x_0) - f(x_0)| \leq \omega_1(f, L_n(\|\cdot - x_0\|_\infty)(x_0)). \tag{13.56}$$

The last (13.56) is attained by $f(t) = \|t - x_0\|_\infty$, $\forall\, t \in \mathbb{R}^N$, $x_0 \in \mathbb{R}^N$, *that is sharp. Given that* f *is uniformly continuous and* $L_n(\|\cdot - x_0\|_\infty)(x_0) \to 0$, *as* $n \to \infty$, *then* $L_n(f)(x_0) \to f(x_0)$, *as* $n \to \infty$.

Proof. Based on Lemma 8.1.1, p. 243, [298], we obtain:

Let $(V, \|\cdot\|)$ be a real normed vector space and $f : V \to \mathbb{R}$ such that $|f(t) - f(x_0)|$ is convex in $t \in V$; $x_0 \in V$. Then

$$|f(t) - f(x_0)| \leq \frac{\omega_1(f, h)}{h}\|t - x_0\|, \quad \forall\, t \in V,\ h > 0. \tag{13.57}$$

That is, it holds here

$$|f(\cdot) - f(x_0)| \leq \frac{\omega_1(f, h)}{h}\|\cdot - x_0\|, \quad \text{over } \mathbb{R}^N. \tag{13.58}$$

Then, the proof of (13.56) goes as in the proof of Theorem 13.2.

Inequality (13.56) is sharp, namely it is attained by $f(t) = \|t - x_0\|_\infty$, $\forall\, t \in \mathbb{R}^N$, $x_0 \in \mathbb{R}^N$.

We have

$$|L_n(f, x_0) - f(x_0)| = L_n(\|\cdot - x_0\|_\infty)(x_0). \tag{13.59}$$

Next we observe

$$\omega_1(f, L_n(\|\cdot - x_0\|_\infty)(x_0)) = \omega_1(\|\cdot - x_0\|_\infty, L_n(\|\cdot - x_0\|_\infty)(x_0)) =$$

$$\sup_{\substack{t_1, t_2 \in \mathbb{R}^N: \\ : \|t_1 - t_2\|_\infty \le L_n(\|\cdot - x_0\|_\infty)(x_0)}} \left| \|t_1 - x_0\|_\infty - \|t_2 - x_0\|_\infty \right| \le \tag{13.60}$$

$$\sup_{\substack{t_1, t_2 \in \mathbb{R}^N: \\ : \|t_1 - t_2\|_\infty \le L_n(\|\cdot - x_0\|_\infty)(x_0)}} \|t_1 - t_2\|_\infty = L_n(\|\cdot - x_0\|_\infty)(x_0).$$

So that

$$\omega_1(f, L_n(\|\cdot - x_0\|_\infty)(x_0)) = L_n(\|\cdot - x_0\|_\infty)(x_0).$$

Thus (13.56) it is attained. ☐

Consequently we obtain:

Theorem 13.5. *Let $f \in C_B(\mathbb{R}^N)$ and the convex function $|f(\cdot) - f(x_0)|$ over $\mathbb{R}^N$, $x_0 \in \mathbb{R}^N$.*

Then

$$|K_n(f)(x_0) - f(x_0)| \le \omega_1(f, K_n(\|\cdot - x_0\|_\infty)(x_0)) < +\infty. \tag{13.61}$$

The last (13.61) is attained by $f(t) = \|t - x_0\|_\infty$, $\forall\, t \in \mathbb{R}^N$, $x_0 \in \mathbb{R}^N$, that is sharp. Given that f is uniformly continuous, we derive that $K_n(f)(x_0) \to f(x_0)$, as $n \to \infty$.

Proof. By Theorem 13.4. ☐

More specifically we derive:

Proposition 13.8. *Let $f \in C_B(\mathbb{R}^N)$ such that $|f(\cdot) - f(x_0)|$ is convex over $\mathbb{R}^N$, $x_0 \in \mathbb{R}^N$.*

Then

$$\left|\widetilde{S}_n(f)(x_0) - f(x_0)\right| \le \omega_1\left(f, \frac{N}{n}\left(\frac{\tanh(\lambda)}{2} + \frac{\left(q + \frac{1}{q}\right)e^{2\lambda}}{2\lambda}\right)\right) < +\infty. \tag{13.62}$$

If f is uniformly continuous, then $\widetilde{S}_n(f)(x_0) \to f(x_0)$, as $n \to \infty$.

Proof. By Proposition 13.2, and Theorem 13.5. □

Proposition 13.9. *Let f be as in Proposition 13.8. It holds that*

$$\left|\widetilde{S}_n^*(f)(x_0) - f(x_0)\right| \leq \omega_1\left(f, \frac{1}{n}\left[1+N\left(\frac{\tanh(\lambda)}{2} + \frac{\left(q+\frac{1}{q}\right)e^{2\lambda}}{2\lambda}\right)\right]\right) < +\infty. \tag{13.63}$$

If f is uniformly continuous, then $\widetilde{S}_n^*(f)(x_0) \to f(x_0)$, as $n \to \infty$.

Proof. By Proposition 13.3, and Theorem 13.5. □

Proposition 13.10. *Let f be as in Proposition 13.8. It holds that*

$$\left|\widehat{S_n}(f)(x_0) - f(x_0)\right| \leq \omega_1\left(f, \frac{1}{n}\left[1+N\left(\frac{\tanh(\lambda)}{2} + \frac{\left(q+\frac{1}{q}\right)e^{2\lambda}}{2\lambda}\right)\right]\right) < +\infty. \tag{13.64}$$

If f is uniformly continuous, then $\widehat{S_n}(f)(x_0) \to f(x_0)$, as $n \to \infty$.

Proof. By Proposition 13.4, and Theorem 13.5. □

We make

Remark 13.4. We consider $\left(\mathbb{R}^N, \|\cdot\|_\infty\right)$ and $\left(\mathbb{R}^N\right)^j$ denote the j-fold product space $\mathbb{R}^N \times \ldots \times \mathbb{R}^N$ endowed with the max-norm $\|x\|_{(\mathbb{R}^N)^j} := \max_{1\leq k\leq j} \|x_k\|_\infty$, where $x := (x_1, \ldots, x_j) \in \left(\mathbb{R}^N\right)^j$. Then the space $L_j := L_j\left(\left(\mathbb{R}^N\right)^j ; \mathbb{R}\right)$ of all real-valued multilinear continuous functions $g : \left(\mathbb{R}^N\right)^j \to \mathbb{R}$ is a Banach space with norm

$$\|g\| := \|g\|_{L_j} := \sup_{\left(\|x\|_{(\mathbb{R}^N)^j}=1\right)} |g(x)| = \sup \frac{|g(x)|}{\|x_1\|_\infty \ldots \|x_j\|_\infty}. \tag{13.65}$$

Let $x_0 \in \mathbb{R}^N$ be fixed, and $f : \mathbb{R}^N \to \mathbb{R}$ be a continuous and bounded function whose Fréchet derivatives

$$f^{(j)} : \mathbb{R}^N \to L_j = L_j\left(\left(\mathbb{R}^N\right)^j ; \mathbb{R}\right)$$

exist and are continuous and bounded for all $1 \leq j \leq m$, $m \in \mathbb{N}$. Call $(x - x_0)^j := (x - x_0, \ldots, x - x_0) \in \left(\mathbb{R}^N\right)^j = \underbrace{\mathbb{R}^N \times \ldots \times \mathbb{R}^N}_{-j-}$.

Then, by Taylor's formula, see [306], p. 124 and [305], we get

$$f(x)=\sum_{j=0}^{m}\frac{f^{(j)}(x_0)(x-x_0)^j}{j!}+R_m(x,x_0),\ \text{ all } x\in\mathbb{R}^N, \tag{13.66}$$

where

$$R_m(x,x_0):=\int_0^1\frac{(1-u)^{m-1}}{(m-1)!}\Big(f^{(m)}(x_0+u(x-x_0))-f^{(m)}(x_0)\Big)(x-x_0)^m\,du. \tag{13.67}$$

Here we assume that $f^{(j)}(x_0)=0$, $j=1,...,m$, then

$$f(x)-f(x_0)=\int_0^1\frac{(1-u)^{m-1}}{(m-1)!}\Big(f^{(m)}(x_0+u(x-x_0))-f^{(m)}(x_0)\Big)(x-x_0)^m\,du. \tag{13.68}$$

Consider

$$w:=\omega_1\left(f^{(m)},h\right)=\sup_{\substack{x,y\in\mathbb{R}^N\\ :\,\|x-y\|_\infty\le h}}\left\|f^{(m)}(x)-f^{(m)}(y)\right\|,\quad h>0. \tag{13.69}$$

Notice that $w<+\infty$.

We obtain that

$$\begin{gathered}\left\|\left(f^{(m)}(x_0+u(x-x_0))-f^{(m)}(x_0)\right)(x-x_0)^m\right\|\le\\ \left\|f^{(m)}(x_0+u(x-x_0))-f^{(m)}(x_0)\right\|\|x-x_0\|^m\le\\ w\|x-x_0\|^m\left\lceil\frac{u\|x-x_0\|}{h}\right\rceil,\end{gathered} \tag{13.70}$$

by Lemma 7.1.1, p. 208, [298].

Therefore for all $x\in\mathbb{R}^N$:

$$|f(x)-f(x_0)|\le w\|x-x_0\|^m\int_0^1\left\lceil\frac{u\|x-x_0\|_\infty}{h}\right\rceil\frac{(1-u)^{m-1}}{(m-1)!}du \tag{13.71}$$

$$=w\Phi_m(\|x-x_0\|_\infty),$$

by a change of variable, where

$$\Phi_m(t):=\int_0^{|t|}\left\lceil\frac{s}{h}\right\rceil\frac{(|t|-s)^{m-1}}{(m-1)!}ds=\frac{1}{m!}\left(\sum_{j=0}^{\infty}(|t|-j\cdot h)_+^m\right), \tag{13.72}$$

$t\in\mathbb{R}$; $(x)_+=\max\{x,0\}$.

For a full list of properties of the spline function Φ_m see Remark 7.1.3 of [298], p. 210.

By (7.1.18) of [298], p. 210, we have

$$\Phi_m(x) \leq \left(\frac{|x|^{m+1}}{(m+1)!h} + \frac{|x|^m}{2m!} + \frac{h|x|^{m-1}}{8(m-1)!} \right), \quad \forall\, x \in \mathbb{R}. \tag{13.73}$$

Consequently it holds ($x \in \mathbb{R}^N$)

$$|f(x) - f(x_0)| \leq \omega_1\left(f^{(m)}, h\right) \left(\frac{\|x - x_0\|_\infty^{m+1}}{(m+1)!h} + \frac{\|x - x_0\|_\infty^m}{2m!} + \frac{h\|x - x_0\|_\infty^{m-1}}{8(m-1)!} \right). \tag{13.74}$$

So that

$$|f(\cdot) - f(x_0)| \leq \omega_1\left(f^{(m)}, h\right) \left(\frac{\|\cdot - x_0\|_\infty^{m+1}}{(m+1)!h} + \frac{\|\cdot - x_0\|_\infty^m}{2m!} + \frac{h\|\cdot - x_0\|_\infty^{m-1}}{8(m-1)!} \right), \tag{13.75}$$

over $\mathbb{R}^N$.

Let $(L_n)_{n \in \mathbb{N}}$ be a sequence of positive linear operators from $C_B\left(\mathbb{R}^N\right)$ into itself such that $L_n(1) = 1$, $\forall\, n \in \mathbb{N}$.

By $|f| \leq |f|$, iff $-|f| \leq f \leq |f|$, then $-L_n(|f|) \leq L_n(f) \leq L_n(|f|)$, and $|L_n(f)| \leq L_n(|f|)$.

Hence

$$|L_n(f(\cdot) - f(x_0))(x_0)| \leq L_n(|f(\cdot) - f(x_0)|)(x_0) \overset{(13.75)}{\leq} \omega_1\left(f^{(m)}, h\right)$$

$$\left[\frac{L_n\left(\|\cdot - x_0\|_\infty^{m+1}\right)(x_0)}{(m+1)!h} + \frac{L_n\left(\|\cdot - x_0\|_\infty^m\right)(x_0)}{2m!} + \frac{hL_n\left(\|\cdot - x_0\|_\infty^{m-1}\right)(x_0)}{8(m-1)!} \right] =: (\xi). \tag{13.76}$$

We need the following Hölder's type inequality for positive linear operators.

Theorem 13.6. *([304]) Let L be a positive linear operator from $C_B\left(\mathbb{R}^N\right)$ into itself and $f, g \in C_B\left(\mathbb{R}^N\right)$, furthermore let $p, q > 1 : \frac{1}{p} + \frac{1}{q} = 1$. Assume that $L((|f(\cdot)|^p))(s_*)$, $L((|g(\cdot)|^q))(s_*) > 0$ for some $s_* \in \mathbb{R}^N$. Then*

$$L(|f(\cdot)g(\cdot)|)(s_*) \leq (L((|f(\cdot)|^p))(s_*))^{\frac{1}{p}} (L((|g(\cdot)|^q))(s_*))^{\frac{1}{q}}. \tag{13.77}$$

We continue with

Remark 13.5. Let $g = 1$ and $L_n : L_n(1) = 1$, then by (13.77) we get:

$$L_n\left(\|\cdot - x_0\|_\infty^m\right)(x_0) \le \left(L_n\left(\|\cdot - x_0\|_\infty^{m+1}\right)(x_0)\right)^{\left(\frac{m}{m+1}\right)}, \tag{13.78}$$

and

$$L_n\left(\|\cdot - x_0\|_\infty^{(m-1)}\right)(x_0) \le \left(L_n\left(\|\cdot - x_0\|_\infty^{m+1}\right)(x_0)\right)^{\left(\frac{m-1}{m+1}\right)}, \tag{13.79}$$

where $m \in \mathbb{N}$ and $L_n\left(\|\cdot - x_0\|_\infty^{m+1}\right)(x_0) > 0$.

Continuing from (13.76):

$$(\xi) \le \omega_1\left(f^{(m)}, h\right)\left[\frac{L_n\left(\|\cdot - x_0\|_\infty^{m+1}\right)(x_0)}{(m+1)!h} + \right. \tag{13.80}$$

$$\left. \frac{\left(L_n\left(\|\cdot - x_0\|_\infty^{m+1}\right)(x_0)\right)^{\frac{m}{m+1}}}{2m!} + \frac{h\left(L_n\left(\|\cdot - x_0\|_\infty^{m+1}\right)(x_0)\right)^{\frac{m-1}{m+1}}}{8(m-1)!}\right]$$

(assume $h := \left(L_n\left(\|\cdot - x_0\|_\infty^{m+1}\right)(x_0)\right)^{\frac{1}{m+1}} > 0$)

$$= \omega_1\left(f^{(m)}, \left(L_n\left(\|\cdot - x_0\|_\infty^{m+1}\right)(x_0)\right)^{\frac{1}{m+1}}\right)$$

$$\left[\frac{h^m}{(m+1)!} + \frac{h^m}{2m!} + \frac{h^m}{8(m-1)!}\right] \tag{13.81}$$

$$= \omega_1\left(f^{(m)}, \left(L_n\left(\|\cdot - x_0\|_\infty^{m+1}\right)(x_0)\right)^{\frac{1}{m+1}}\right)$$

$$\left(L_n\left(\|\cdot - x_0\|_\infty^{m+1}\right)(x_0)\right)^{\frac{m}{m+1}}\left[\frac{1}{(m+1)!} + \frac{1}{2m!} + \frac{1}{8(m-1)!}\right]. \tag{13.82}$$

We have proved that

Theorem 13.7. *Let $f \in C_B\left(\mathbb{R}^N\right)$, $x_0 \in \mathbb{R}^N$, Fréchet derivatives $f^{(j)}$ continuous and bounded with $f^{(j)}(x_0) = 0$, $j = 1, ..., m$. Then*

$$|L_n(f)(x_0) - f(x_0)| \le \omega_1\left(f^{(m)}, \left(L_n\left(\|\cdot - x_0\|_\infty^{m+1}\right)(x_0)\right)^{\frac{1}{m+1}}\right) \tag{13.83}$$

$$\left(L_n\left(\|\cdot - x_0\|_\infty^{m+1}\right)(x_0)\right)^{\frac{m}{m+1}}\left[\frac{1}{(m+1)!}+\frac{1}{2m!}+\frac{1}{8(m-1)!}\right], \quad \forall\, n \in \mathbb{N}.$$

Here $L_n(1) = 1$, L_n *positive linear operators from* $C_B\left(\mathbb{R}^N\right)$ *into itself under the assumption* $L_n\left(\|\cdot - x_0\|_\infty^{m+1}\right)(x_0) > 0$.

Let $L_n\left(\|\cdot - x_0\|_\infty^{m+1}\right)(x_0) \to 0$, *as* $n \to +\infty$, *then* $\lim_{n\to+\infty} L_n(f)(x_0) = f(x_0)$.

Next we apply Theorem 13.7 to the operators K_n.

Theorem 13.8. *Let* $f \in C_B\left(\mathbb{R}^N\right)$, $x_0 \in \mathbb{R}^N$, *Fréchet derivatives* $f^{(j)}$ *continuous and bounded with* $f^{(j)}(x_0) = 0$, $j = 1, ..., m$. *Then*

$$|K_n(f)(x_0) - f(x_0)| \leq \omega_1\left(f^{(m)}, \left(K_n\left(\|\cdot - x_0\|_\infty^{m+1}\right)(x_0)\right)^{\frac{1}{m+1}}\right) \tag{13.84}$$

$$\left(K_n\left(\|\cdot - x_0\|_\infty^{m+1}\right)(x_0)\right)^{\frac{m}{m+1}}\left[\frac{1}{(m+1)!}+\frac{1}{2m!}+\frac{1}{8(m-1)!}\right],$$

$\forall\, n \in \mathbb{N}$.

We have that $\lim_{n\to+\infty} K_n(f)(x_0) = f(x_0)$.

Proof. Here K_n any of $\widetilde{S}_n$, $\widetilde{S}_n^*$ and $\widehat{S_n}$.

By Theorem 13.7, and Propositions 13.2–13.4 we get $K_n\left(\|\cdot - x_0\|_\infty^{m+1}\right)(x_0) \to 0$, as $n \to +\infty$. □

More specifically we derive:

Proposition 13.11. *All as in Theorem 13.8,* $m \in \mathbb{N}$. *Denote by*

$$\psi_{1n}(m) := \frac{N^m}{n^m}\left[\frac{\tanh(\lambda)}{(m+1)} + \frac{\left(q+\frac{1}{q}\right)e^{2\lambda}m!}{(2\lambda)^m}\right]. \tag{13.85}$$

Then

$$\left|\widetilde{S}_n(f)(x_0) - f(x_0)\right| \leq \omega_1\left(f^{(m)}, \left(\psi_{1n}(m+1)\right)^{\frac{1}{m+1}}\right)$$

$$\left(\psi_{1n}(m+1)\right)^{\frac{m}{m+1}}\left[\frac{1}{(m+1)!}+\frac{1}{2m!}+\frac{1}{8(m-1)!}\right], \tag{13.86}$$

$\forall\, n \in \mathbb{N}$.

We have that $\lim_{n\to+\infty} \widetilde{S}_n(f)(x_0) = f(x_0)$.

Proof. By Theorem 13.8 and Proposition 13.2. □

Proposition 13.12. *All as in Theorem 13.8, $m \in \mathbb{N}$. Denote by*

$$\psi_{2n}(m) := \frac{2^{m-1}}{n^m}\left[1 + N^m\left(\frac{\tanh(\lambda)}{(m+1)} + \frac{\left(q + \frac{1}{q}\right)e^{2\lambda}m!}{(2\lambda)^m}\right)\right]. \quad (13.87)$$

Then

$$\left\{\begin{array}{l}\left|\widetilde{S}_n^*(f)(x_0) - f(x_0)\right| \\ \left|\widehat{S_n}(f)(x_0) - f(x_0)\right|\end{array}\right\} \leq \omega_1\left(f^{(m)}, (\psi_{2n}(m+1))^{\frac{1}{m+1}}\right)$$

$$(\psi_{2n}(m+1))^{\frac{m}{m+1}}\left[\frac{1}{(m+1)!} + \frac{1}{2m!} + \frac{1}{8(m-1)!}\right], \quad (13.88)$$

$\forall\, n \in \mathbb{N}$.

We have that $\lim_{n\to+\infty} \widetilde{S}_n^*(f)(x_0) = \lim_{n\to+\infty} \widehat{S_n}(f)(x_0) = f(x_0)$.

Proof. By Theorem 13.8 and Propositions 13.3, 13.4. □

Next we do simultaneous approximation.

Theorem 13.9. *Let $i \in \mathbb{N}$ be fixed and $f \in C^{(i)}(\mathbb{R}^N)$. Here f_α denotes a partial derivative of f, $\alpha := (\alpha_1, ..., \alpha_N)$, $\alpha_i \in \mathbb{Z}_+$, $j = 1, ..., N$, and $|\alpha| := \sum_{j=1}^{N} \alpha_j = l$, where $l = 0, 1, ..., i$. We assume that $f_\alpha \in C_B(\mathbb{R}^N)$, for all $\alpha : |\alpha| = l$, $l = 0, 1, ..., i$; $x_0 \in \mathbb{R}^N$. Then*

$$|(K_n(f))_\alpha(x_0) - f_\alpha(x_0)| \leq 2\omega_1(f_\alpha, K_n(\|\cdot - x_0\|_\infty)(x_0)) < +\infty. \quad (13.89)$$

Given that $\lim_{n\to+\infty} K_n(\|\cdot - x_0\|_\infty)(x_0) = 0$ *and f_α is uniformly continuous, we get that* $\lim_{n\to+\infty} (K_n(f))_\alpha(x_0) = f_\alpha(x_0)$.

Proof. By Remark 13.2 and Theorem 13.3. □

Proposition 13.13. *All as in Theorem 13.9. Then*

$$\left\|\left(\widetilde{S}_n(f)\right)_\alpha - f_\alpha\right\|_\infty \leq 2\omega_1\left(f_\alpha, \frac{N}{n}\left[\frac{\tanh(\lambda)}{2} + \frac{\left(q + \frac{1}{q}\right)e^{2\lambda}}{2\lambda}\right]\right) < +\infty. \quad (13.90)$$

If f_α is uniformly continuous, then $\left(\widetilde{S}_n(f)\right)_\alpha \to f_\alpha$, uniformly as $n \to +\infty$.

Proof. By Remark 13.2 and Proposition 13.5. □

Proposition 13.14. *All as in Theorem 13.9. Then*

$$\left\{ \begin{array}{l} \left\| \left(\widetilde{S}_n^* (f) \right)_\alpha - f_\alpha \right\|_\infty \\ \left\| \left(\widehat{S_n} (f) \right)_\alpha - f_\alpha \right\|_\infty \end{array} \right\} \leq 2\omega_1 \left(f_\alpha, \frac{1}{n} \left[1 + N \left(\frac{\tanh (\lambda)}{2} + \frac{\left(q + \frac{1}{q} \right) e^{2\lambda}}{2\lambda} \right) \right] \right) < +\infty. \tag{13.91}$$

If f_α is uniformly continuous, then $\left(\widetilde{S}_n^* (f) \right)_\alpha \to f_\alpha$, $\left(\widehat{S_n} (f) \right)_\alpha \to f_\alpha$, uniformly as $n \to +\infty$.

Proof. By Remark 13.2 and Propositions 13.6, 13.7. □

Next, we do simultaneous convexity approximation.

Theorem 13.10. *The notation is as in Theorem 13.9, with a particular $\alpha : |\alpha| = l \in \{0, 1, ..., i\}$, $f_\alpha \in C_B \left(\mathbb{R}^N \right)$ and $|f_\alpha (\cdot) - f_\alpha (x_0)|$ is convex over $\mathbb{R}^N$, $x_0 \in \mathbb{R}^N$. Then*

$$|(K_n (f))_\alpha (x_0) - f_\alpha (x_0)| \leq \omega_1 \left(f_\alpha, K_n \left(\|\cdot - x_0\|_\infty \right) (x_0) \right) < +\infty. \tag{13.92}$$

Given that f_α is uniformly continuous, we derive that $(K_n (f))_\alpha (x_0) \to f_\alpha (x_0)$, as $n \to \infty$.

Proof. By Remark 13.2 and Theorem 13.5. □

More specifically we obtain:

Proposition 13.15. *All as in Theorem 13.10. Then*

$$\left| \left(\widetilde{S}_n (f) \right)_\alpha (x_0) - f_\alpha (x_0) \right| \leq \omega_1 \left(f_\alpha, \frac{N}{n} \left[\frac{\tanh (\lambda)}{2} + \frac{\left(q + \frac{1}{q} \right) e^{2\lambda}}{2\lambda} \right] \right) < +\infty. \tag{13.93}$$

If f_α is uniformly continuous, then $\left(\widetilde{S}_n (f) \right)_\alpha (x_0) \to f_\alpha (x_0)$, as $n \to +\infty$.

Proof. By Remark 13.2, Proposition 13.8 and Theorem 13.10. □

Proposition 13.16. *All as in Theorem 13.10. Then*

$$\left\{ \begin{array}{l} \left| \left(\widetilde{S}_n^* (f) \right)_\alpha (x_0) - f_\alpha (x_0) \right| \\ \left| \left(\widehat{S_n} (f) \right)_\alpha (x_0) - f_\alpha (x_0) \right| \end{array} \right\} \leq$$

$$\omega_1 \left(f_\alpha, \frac{1}{n} \left[1 + N \left(\frac{\tanh (\lambda)}{2} + \frac{\left(q + \frac{1}{q} \right) e^{2\lambda}}{2\lambda} \right) \right] \right) < +\infty. \tag{13.94}$$

If f_α *is uniformly continuous, then* $\left(\widetilde{S}_n^*(f)\right)_\alpha(x_0) \rightarrow f_\alpha(x_0)$, $\left(\widehat{S_n}(f)\right)_\alpha(x_0) \rightarrow f_\alpha(x_0)$, *as* $n \rightarrow +\infty$.

Proof. By Remark 13.2 and Propositions 13.9, 13.10 and Theorem 13.10. □

We continue with iterated approximations:

Remark 13.6. Here $f \in C_B(\mathbb{R}^N)$; $r \in \mathbb{N}$; $m_1, ..., m_r \in \mathbb{N} : m_1 \leq m_2 \leq ... \leq m_r$; $K_n = \widetilde{S}_n, \widetilde{S}_n^*, \widehat{S_n}$.

As in (13.34) we get that

$$\|K_n^r(f) - f\|_\infty \leq r\|K_n(f) - f\|_\infty, \tag{13.95}$$

and as in (13.36) we obtain that

$$\left\|K_{m_r}\left(K_{m_{r-1}}(...K_{m_2}(K_{m_1}(f)))\right) - f\right\|_\infty \leq \sum_{i=1}^r \|K_{m_i}(f) - f\|_\infty . \tag{13.96}$$

Clearly then follows:

Proposition 13.17. *Let* $f \in C_B(\mathbb{R}^N)$; $r \in \mathbb{N}$; *and* $m_1, ..., m_r \in \mathbb{N} : m_1 \leq m_2 \leq ... \leq m_r$. *Then*

1)

$$\left\|\widetilde{S}_n^r(f) - f\right\|_\infty \leq r\left\|\widetilde{S}_n(f) - f\right\|_\infty \leq \tag{13.97}$$

$$2r\omega_1\left(f, \frac{N}{n}\left[\frac{\tanh(\lambda)}{2} + \frac{\left(q + \frac{1}{q}\right)e^{2\lambda}}{2\lambda}\right]\right) < +\infty,$$

and the speed of convergence of $\widetilde{S}_n^r$ *to the unit* I *is not worse than the speed of convergence of* $\widetilde{S}_n$ *to* I,

2)

$$\left\|\widetilde{S}_{m_r}\left(\widetilde{S}_{m_{r-1}}\left(...\widetilde{S}_{m_2}\left(\widetilde{S}_{m_1}(f)\right)\right)\right) - f\right\|_\infty \leq \sum_{i=1}^r \left\|\widetilde{S}_{m_i}(f) - f\right\|_\infty \leq \tag{13.98}$$

$$2\sum_{i=1}^r \omega_1\left(f, \frac{N}{m_i}\left[\frac{\tanh(\lambda)}{2} + \frac{\left(q + \frac{1}{q}\right)e^{2\lambda}}{2\lambda}\right]\right) \leq$$

$$2r\omega_1\left(f,\frac{N}{m_1}\left[\frac{\tanh(\lambda)}{2}+\frac{\left(q+\frac{1}{q}\right)e^{2\lambda}}{2\lambda}\right]\right)<+\infty. \tag{13.99}$$

Again, the speed of convergence of the iterated operator to I is not worse than the speed of convergence of $\widetilde{S}_{m_1}$ to I.

Proof. By (13.95), (13.96) and Proposition 13.5. □

Similarly we have:

Proposition 13.18. *Let $f \in C_B\left(\mathbb{R}^N\right)$; $r \in \mathbb{N}$; and $m_1, ..., m_r \in \mathbb{N} : m_1 \leq m_2 \leq ... \leq m_r$. Then*

1)

$$\left\{\begin{array}{l}\left\|\widetilde{S}_n^{*r}(f)-f\right\|_\infty \\ \left\|\widehat{S_n}^r(f)-f\right\|_\infty\end{array}\right\} \leq r\left\{\begin{array}{l}\left\|\widetilde{S}_n^{*}(f)-f\right\|_\infty \\ \left\|\widehat{S_n}(f)-f\right\|_\infty\end{array}\right\} \leq \tag{13.100}$$

$$2r\omega_1\left(f,\frac{1}{n}\left[1+N\left(\frac{\tanh(\lambda)}{2}+\frac{\left(q+\frac{1}{q}\right)e^{2\lambda}}{2\lambda}\right)\right]\right)<+\infty,$$

and the speed of convergence of $\widetilde{S}_n^{*r}$, $\widehat{S_n}^r$ to I is not worse than the speed of convergence of $\widetilde{S}_n^{*}$, $\widehat{S_n}$ to I,

2)

$$\left\{\begin{array}{l}\left\|\widetilde{S}_{m_r}^{*}\left(\widetilde{S}_{m_{r-1}}^{*}\left(...\widetilde{S}_{m_2}^{*}\left(\widetilde{S}_{m_1}^{*}(f)\right)\right)\right)-f\right\|_\infty \\ \left\|\widehat{S_{m_r}}\left(\widehat{S_{m_{r-1}}}\left(...\widehat{S_{m_2}}\left(\widehat{S_{m_1}}(f)\right)\right)\right)-f\right\|_\infty\end{array}\right\} \leq \sum_{i=1}^{r}\left\{\begin{array}{l}\left\|\widetilde{S}_{m_i}^{*}(f)-f\right\|_\infty \\ \left\|\widehat{S_{m_i}}(f)-f\right\|_\infty\end{array}\right\} \leq \tag{13.101}$$

$$2\sum_{i=1}^{r}\omega_1\left(f,\frac{1}{m_i}\left[1+N\left(\frac{\tanh(\lambda)}{2}+\frac{\left(q+\frac{1}{q}\right)e^{2\lambda}}{2\lambda}\right)\right]\right)\leq$$

$$2r\omega_1\left(f,\frac{1}{m_1}\left[1+N\left(\frac{\tanh(\lambda)}{2}+\frac{\left(q+\frac{1}{q}\right)e^{2\lambda}}{2\lambda}\right)\right]\right)<+\infty. \tag{13.102}$$

Again, the speed of convergence of the iterated operator to I is not worse than the speed of convergence of $\widetilde{S}_{m_1}^{*}$, $\widehat{S_{m_1}}$ to I.

Proof. By (13.95), (13.96) and Propositions 13.6, 13.7. □

We finish our work with multivariate simultaneous iterations.

Remark 13.7. Let $i \in \mathbb{N}$ be fixed. Assume that $f \in C^{(i)}\left(\mathbb{R}^N\right)$, with $f_\alpha \in C_B\left(\mathbb{R}^N\right)$, with $\alpha : |\alpha| = l$, $l = 0, 1, ..., i$; $r \in \mathbb{N}$. Then, by (13.34), we obtain

$$\left\| \widetilde{S}_n^r (f_\alpha) - f_\alpha \right\|_\infty \leq r \left\| \widetilde{S}_n (f_\alpha) - f_\alpha \right\|_\infty . \tag{13.103}$$

By (13.27) and inductively, we obtain

$$\left\| \left(\widetilde{S}_n^r (f) \right)_\alpha - f_\alpha \right\|_\infty \leq r \left\| \left(\widetilde{S}_n (f) \right)_\alpha - f_\alpha \right\|_\infty , \tag{13.104}$$

Similarly, we derive that

$$\left\| \left(\widetilde{S}_n^{*^r} (f) \right)_\alpha - f_\alpha \right\|_\infty \leq r \left\| \left(\widetilde{S}_n^* (f) \right)_\alpha - f_\alpha \right\|_\infty , \tag{13.105}$$

and

$$\left\| \left(\widehat{S_n}^r (f) \right)_\alpha - f_\alpha \right\|_\infty \leq r \left\| \left(\widehat{S_n} (f) \right)_\alpha - f_\alpha \right\|_\infty . \tag{13.106}$$

Let now $m_1, m_2, ..., m_r \in \mathbb{N} : m_1 \leq m_2 \leq ... \leq m_r$. Then, based on (13.36), we find that

$$\left\| \left(\widetilde{S}_{m_r} \left(\widetilde{S}_{m_{r-1}} \left(... \widetilde{S}_{m_2} \left(\widetilde{S}_{m_1} f \right) \right) \right) \right)_\alpha - f_\alpha \right\|_\infty \leq \sum_{i^*=1}^{r} \left\| \left(\widetilde{S}_{m_{i^*}} (f) \right)_\alpha - f_\alpha \right\|_\infty . \tag{13.107}$$

Similarly, we get that

$$\left\| \left(\widetilde{S}_{m_r}^* \left(\widetilde{S}_{m_{r-1}}^* \left(... \widetilde{S}_{m_2}^* \left(\widetilde{S}_{m_1}^* f \right) \right) \right) \right)_\alpha - f_\alpha \right\|_\infty \leq \sum_{i^*=1}^{r} \left\| \left(\widetilde{S}_{m_{i^*}}^* (f) \right)_\alpha - f_\alpha \right\|_\infty , \tag{13.108}$$

and

$$\left\| \left(\widehat{S_{m_r}} \left(\widehat{S_{m_{r-1}}} \left(... \widehat{S_{m_2}} \left(\widehat{S_{m_1}} f \right) \right) \right) \right)_\alpha - f_\alpha \right\|_\infty \leq \sum_{i^*=1}^{r} \left\| \left(\widehat{S_{m_{i^*}}} (f) \right)_\alpha - f_\alpha \right\|_\infty . \tag{13.109}$$

All the above inequalities (13.103)–(13.109) prove that our implied multivariate iterated simultaneous approximations do not have a speed worse than our basic simultaneous approximations by the activated convolution operators.

Bibliography

[297] G.A. Anastassiou, *A "K-Attainable" inequality related to the convergence of Positive Linear Operators*, J. Approximation Theory, 44, 380–383 (1985).

[298] G.A. Anastassiou, *Moments in probability and approximation theory*, Pitman Research Notes in Mathematics series / Longman Scientific & Technical, Essex, New York, UK, USA, 1993.

[299] G.A. Anastassiou, *Quantitative Approximations*, Chapmen & Hall/CRC, London, New York, 2001.

[300] G.A. Anastassiou, *Parametrized, Deformed and General Neural Networks*, Springer, Heidelberg, New York, 2023.

[301] G.A. Anastassiou, *Trigonometric and Hyperbolic Generated Approximation Theory*, World Scientific, Singapore, New York, 2025.

[302] G.A. Anastassiou, *Approximation by symmetrized and perturbed hyperbolic tangent activated convolution type operators*, Mathematics, 2024, 12, 3302; `https://doi.org/10.3390/math12203302`.

[303] G.A. Anastassiou, *Multivariate Approximation by Symmetrized and Perturbed Hyperbolic Tangent activated multidimensional convolution type operators*, Axioms, 2024, 13, 779; `https://doi.org/10.3390/axioms13110779`.

[304] G.A. Anastassiou, *Multivariate Neural Networks over infinite domain as Positive Linear Operators and a touch of measure theory*, submitted, 2025.

[305] H. Cartan, *Differential Calculus*, Herman, Paris, 1971.

[306] L.B. Rall, *Computational Solution of Nonlinear Operator Equations*, John Wiley & Sons, New York, 1969.

[307] Hee Sun Jung, Ryozi Sakai, *Local saturation of a positive linear convolution operator*, J. Inequal. Appl. 2014, 2014: 329, 16 p.

[308] J.J. Swetits, B. Wood, *Local L_p-saturation of positive linear convolution operators*, J. Approx. Theory 34 (1982), no. 4, 348–360.

[309] Grigor Moldovan, *Discrete convolutions in connection with functions of several variables and positive linear operators.* (Romanian) Studia Univ. Babeş-Bolyai Ser. Math.-Mech. 19 (1974), no. 1, 51–57.

[310] R. Bojanic, O. Shisha, *On the precision of uniform approximation of continuous functions by certain linear operators of convolution type.* J. Approximation Theory 8 (1973), 101–113.

[311] Grigor Moldovan, *Discrete convolutions and linear positive operators.* Ann. Univ. Sci. Budapest. Eötvös Sect. Math. 15 (1972), 31–44 (1973).

[312] A.I. Kamzolov, *The order of approximation of functions of class $Z_2(E_n)$ by positive linear convolution operators.* (Russian) Mat. Zametki 7 (1970), 723–732.

[313] G.A. Anastassiou, *Complete Approximation by symmetrized and perturbed hyperbolic tangent activated multidimensional convolutions as Positive Linear Operators*, Annals of Communications in Mathematics, 8(2) (2025), 234–252.

Chapter 14

Complete Approximation by Symmetrized and Disturbed Generalized Logistic Energized Multivariate Convolutions in Positive Linear Setting

In this work are studied in detail the multivariate symmetrized and perturbed generalized logistic activated convolution type operators of three types. This is done by the method of positive linear operators. Their alternative approximation properties are established by the quantitative convergence to the unit operator using the modulus of continuity. It also studies the related multivariate simultaneous approximation, as well as the multivariate iterated approximation. It follows [330].

14.1 Introduction

The author studied extensively the quantitative approximation of positive linear operators to the unit since 1985, see for example [314]–[316], [318]. He originated from the quantitative weak convergence of finite positive measures to the unit Dirac measure, having as a method the geometric moment theory, see [315], and he produced best upper bounds, leading to attained (i.e. sharp Jackson type inequalities), e.g. see [314], [315]. These studies have been gone to all possible directions, univariate and multivariate, though in this work we stay only on the multivariate approach over an infinite domain.

Our multidimensional convolution operators here have a kernel based on the symmetrized and perturbed generalized logistic activation function, which is used frequently in the study of neural networks, and they can be interpreted as positive linear operators.

So here our proving methods come from the theory of positive linear operators.

Thus in Section 2, we discuss about the symmetrized and perturbed generalized logistic activation function in the multivariate setting. We also describe our activated multidimensional convolution type operators with their properties, such as differentiation and iteration.

In Section 3, we derive some auxiliary results which are estimates to our operators, when applied to basic functions and to be used into our main results.

In Section 4, we present our main explicit results under the lens of positive linear framework and we include the necessary related theory. We treat also the simultaneous and iterated approximation cases under the same spirit.

We are greatly inspired by our earlier works [319], [320].

Furthermore, general motivation comes from the great works [324]–[329].

14.2 Basics

Here we follow [317], pp. 395–417.

Our activation function here is the q-deformed and λ-parametrized function

$$\varphi_{q,\lambda}(x) = \frac{1}{1 + qA^{-\lambda x}}, \quad x \in \mathbb{R}, \; q, \lambda > 0, \; A > 1. \tag{14.1}$$

This is the A-generalized logistic function.

For more read Chapter 16 of [317]: "Banach space valued ordinary and fractional neural network approximation based on q-deformed and λ-parametrized A-generalized logistic function".

This chapters 15, 16 of [317] motivate our current work.

The proposed "symmetrization technique" aims to use half data feed to our neural networks.

We will employ the following density function

$$G_{q,\lambda}(x) := \frac{1}{2}\left(\varphi_{q,\lambda}(x+1) - \varphi_{q,\lambda}(x-1)\right), \quad x \in \mathbb{R}, \; q, \lambda > 0. \tag{14.2}$$

We have that

$$G_{q,\lambda}(-x) = G_{\frac{1}{q},\lambda}(x), \tag{14.3}$$

and

$$G_{\frac{1}{q},\lambda}(-x) = G_{q,\lambda}(x), \; \forall \, x \in \mathbb{R}. \tag{14.4}$$

Adding (14.3) and (14.4) we obtain

$$G_{q,\lambda}(-x)+G_{\frac{1}{q},\lambda}(-x)=G_{q,\lambda}(x)+G_{\frac{1}{q},\lambda}(x),\ \ \forall\, x\in\mathbb{R}, \tag{14.5}$$

the key to this work.

So that

$$W(x):=\frac{G_{q,\lambda}(x)+G_{\frac{1}{q},\lambda}(x)}{2}>0, \tag{14.6}$$

is an even function, symmetric with respect to the y-axis.

The global maximum of $G_{q,\lambda}$ is given by (16.18), p. 401 of [317] as

$$G_{q,\lambda}\left(\frac{\log_A q}{\lambda}\right)=\frac{A^{\lambda}-1}{2(A^{\lambda}+1)}. \tag{14.7}$$

And, the global max of $G_{\frac{1}{q},\lambda}$ is

$$G_{\frac{1}{q},\lambda}\left(\frac{\log_A \frac{1}{q}}{\lambda}\right)=G_{\frac{1}{q},\lambda}\left(\frac{-\log_A q}{\lambda}\right)=\frac{A^{\lambda}-1}{2(A^{\lambda}+1)}, \tag{14.8}$$

both sharing the same maximum at symmetric points.

By Theorem 16.1, p. 401 of [317], we have that

$$\sum_{i=-\infty}^{\infty} G_{q,\lambda}(x-i)=1,\ \ \forall\, x\in\mathbb{R},\ \lambda,q>0,\ \ A>1, \tag{14.9}$$

and

$$\sum_{i=-\infty}^{\infty} G_{\frac{1}{q},\lambda}(x-i)=1,\ \ \forall\, x\in\mathbb{R},\ \lambda,q>0,\ \ A>1. \tag{14.10}$$

Consequently, we derive that

$$\sum_{i=-\infty}^{\infty} W(x-i)=1,\ \ \forall\, x\in\mathbb{R}. \tag{14.11}$$

By Theorem 16.2, p. 402 of [317], we have that

$$\int_{-\infty}^{\infty} G_{q,\lambda}(x)\,dx=1,\quad \lambda,q>0,\ \ A>1, \tag{14.12}$$

similarly it holds

$$\int_{-\infty}^{\infty} G_{\frac{1}{q},\lambda}(x)\,dx=1, \tag{14.13}$$

so that

$$\int_{-\infty}^{\infty} W(x)\,dx=1, \tag{14.14}$$

therefore $W(x)$ is a density function.

Clearly, then

$$\int_{-\infty}^{\infty} W(nx-u)\,du = 1, \quad \forall\, n \in \mathbb{N},\, x \in \mathbb{R}. \tag{14.15}$$

It holds

$$W(x) \leq \frac{A^{\lambda}-1}{2(A^{\lambda}+1)}, \quad \forall\, x \in \mathbb{R}. \tag{14.16}$$

We make

Remark 14.1. Let $x \in [1, \infty)$, and we apply the mean value theorem. We have that

$$G_{q,\lambda}(x) = \frac{1}{2}\left(\varphi_{q,\lambda}(x+1) - \varphi_{q,\lambda}(x-1)\right)$$

$$\frac{1}{2}\varphi'_{q,\lambda}(\xi)\,2 = \varphi'_{q,\lambda}(\xi) = \frac{q\lambda \ln A}{\left(1+qA^{-\lambda\xi}\right)^{2} A^{\lambda\xi}}, \tag{14.17}$$

where $0 \leq x-1 < \xi < x+1$.

We obtain that

$$G_{q,\lambda}(x) < (q\lambda \ln A)\,A^{-\lambda\xi} < (q\lambda \ln A)\,A^{-\lambda(x-1)}, \quad x \geq 1. \tag{14.18}$$

Similarly it holds

$$G_{\frac{1}{q},\lambda}(x) < \left(\frac{1}{q}\lambda \ln A\right) A^{-\lambda(x-1)}, \quad x \geq 1. \tag{14.19}$$

And, finally it holds

$$W(x) < \frac{1}{2}\left(q+\frac{1}{q}\right)(\lambda \ln A)\,A^{-\lambda(x-1)}, \quad x \geq 1, \tag{14.20}$$

where $q, \lambda > 0,\ A > 1$.

We need

Theorem 14.1. *It holds* ($k \in \mathbb{N}$)

$$\int_{-\infty}^{\infty} |z|^{k}\, W(z)\,dz \leq \left[\frac{(A^{\lambda}-1)}{(A^{\lambda}+1)(k+1)} + \left(q+\frac{1}{q}\right)\frac{A^{\lambda}k!}{(\lambda \ln A)^{k-1}}\right] < +\infty. \tag{14.21}$$

Proof. We can write

$$\int_{-\infty}^{\infty}|z|^{k}W(z)\,dz=2\int_{0}^{\infty}z^{k}W(z)\,dz=$$

$$2\left[\int_{0}^{1}z^{k}W(z)\,dz+\int_{1}^{\infty}z^{k}W(z)\,dz\right]\overset{\text{(by (14.16), (14.20))}}{\leq}$$

$$2\left[\frac{(A^{\lambda}-1)}{2(A^{\lambda}+1)}\int_{0}^{1}z^{k}dz+\frac{1}{2}\left(q+\frac{1}{q}\right)(\lambda\ln A)\int_{1}^{\infty}z^{k}A^{-\lambda(z-1)}dz\right]\leq \tag{14.22}$$

$$2\left[\frac{(A^{\lambda}-1)}{2(A^{\lambda}+1)}\frac{1}{(k+1)}+\frac{1}{2}\left(q+\frac{1}{q}\right)(\lambda\ln A)A^{\lambda}\int_{0}^{\infty}z^{k}e^{-\lambda(\ln A)z}dz\right]=$$

$$2\left[\frac{(A^{\lambda}-1)}{2(A^{\lambda}+1)(k+1)}+\frac{1}{2}\left(q+\frac{1}{q}\right)\frac{(\lambda\ln A)A^{\lambda}}{(\lambda\ln A)^{k+1}}\int_{0}^{\infty}((\lambda\ln A)z)^{k}e^{-(\lambda\ln A)z}d(\lambda\ln A)z\right]=$$

$$2\left[\frac{(A^{\lambda}-1)}{2(A^{\lambda}+1)(k+1)}+\frac{1}{2}\left(q+\frac{1}{q}\right)\frac{A^{\lambda}}{(\lambda\ln A)^{k-1}}\int_{0}^{\infty}y^{k}e^{-y}dy\right]= \tag{14.23}$$

$$2\left[\frac{(A^{\lambda}-1)}{2(A^{\lambda}+1)(k+1)}+\frac{1}{2}\left(q+\frac{1}{q}\right)\frac{A^{\lambda}}{(\lambda\ln A)^{k-1}}\Gamma(k+1)\right]=$$

$$\left[\frac{(A^{\lambda}-1)}{(A^{\lambda}+1)(k+1)}+\left(q+\frac{1}{q}\right)\frac{A^{\lambda}}{(\lambda\ln A)^{k-1}}k!\right]<+\infty.$$

□

We mention

Definition 14.1. The modulus of continuity here is defined by

$$\omega_{1}(f,\delta):=\sup_{\substack{x,y\in\mathbb{R}^{N}:\\ \|x-y\|_{\infty}<\delta}}|f(x)-f(y)|,\quad\delta>0, \tag{14.24}$$

where $f:\mathbb{R}^{N}\rightarrow\mathbb{R}$ bounded and continuous, denoted by $f\in C_{B}\left(\mathbb{R}^{N}\right)$, $N\in\mathbb{N}$. Similarly ω_{1} is defined for $f\in C_{U}\left(\mathbb{R}^{N}\right)$ (uniformly continuous functions). We have that $f\in C_{U}\left(\mathbb{R}^{N}\right)$, iff $\omega_{1}(f,\delta)\rightarrow0$ as $\delta\downarrow0$.

Denote $\|x\|_{\infty}:=\max\{|x_{1}|,...,|x_{N}|\}$, $x\in\mathbb{R}^{N}$.

We make

Remark 14.2. We introduce

$$\widetilde{Z}(x_1,...,x_N) := \widetilde{Z}(x) := \prod_{i=1}^{N} W(x_i), \quad x = (x_1,...,x_N) \in \mathbb{R}^N, \ N \in \mathbb{N}. \tag{14.25}$$

It has the properties:

(i)

$$\widetilde{Z}(x) > 0, \ \forall x \in \mathbb{R}^N; \ \widetilde{Z}(-x) = \widetilde{Z}(x),$$

(ii)

$$\int_{\mathbb{R}^N} \widetilde{Z}(x-k) = \int_{-\infty}^{\infty} ... \int_{-\infty}^{\infty} \widetilde{Z}(x_1 - u_1, ..., x_N - u_N) \, du_1 ... du_N =$$

$$\prod_{i=1}^{N} \int_{-\infty}^{\infty} W(x_i - u_i) \, du_i \overset{(14.15)}{=} 1, \ \forall x \in \mathbb{R}^N, \tag{14.26}$$

hence

(iii)

$$\int_{\mathbb{R}^N} \widetilde{Z}(nx - u) \, du = 1, \ \ \forall \ x \in \mathbb{R}^N; \ n \in \mathbb{N}, \tag{14.27}$$

and

(iv) by (14.14)

$$\int_{\mathbb{R}^N} \widetilde{Z}(x) \, dx = 1, \tag{14.28}$$

that is $\widetilde{Z}$ is a multivariate density function.

We need the following:

Theorem 14.2. *It holds* ($k \in \mathbb{N}$)

$$\int_{\mathbb{R}^N} \|x\|_{\infty}^k \widetilde{Z}(x) \, dx \leq N^k \left[\frac{(A^{\lambda} - 1)}{(A^{\lambda} + 1)(k+1)} + \left(q + \frac{1}{q} \right) \frac{A^{\lambda} k!}{(\lambda \ln A)^{(k-1)}} \right] < \infty. \tag{14.29}$$

Proof. We have that

$$\int_{\mathbb{R}^N} \|x\|_{\infty}^k \widetilde{Z}(x) \, dx \leq \int_{\mathbb{R}^N} \left(\sum_{j=1}^{N} |x_j| \right)^k \widetilde{Z}(x) \, dx \leq$$

(by a convexity argument)

$$\int_{\mathbb{R}^N} N^{k-1}\left(\sum_{j=1}^{N}|x_j|^k\right)\widetilde{Z}(x)\,dx = N^{k-1}\left(\sum_{j=1}^{N}\int_{\mathbb{R}^N}|x_j|^k\,\widetilde{Z}(x)\,dx\right) =$$

$$N^{k-1}\left(\sum_{j=1}^{N}\int_{\mathbb{R}^N}|x_j|^k\left(\prod_{i=1}^{N}W(x_i)\right)dx\right) =$$

$$N^{k-1}\left(\sum_{j=1}^{N}\left[\left(\int_{-\infty}^{\infty}|x_j|^k\,W(x_j)\,dx_j\right)\prod_{\substack{i=1\\i\neq j}}^{N}\int_{-\infty}^{\infty}W(x_i)\,dx_i\right]\right)\overset{(14.14)}{=}$$

$$N^{k-1}\left(\sum_{j=1}^{N}\int_{-\infty}^{\infty}|x_j|^k\,W(x_j)\,dx_j\right)\overset{(14.21)}{\leq}$$

$$N^{k-1}\left(\sum_{j=1}^{N}\left[\frac{(A^{\lambda}-1)}{(A^{\lambda}+1)(k+1)}+\left(q+\frac{1}{q}\right)\frac{A^{\lambda}k!}{(\lambda\ln A)^{(k-1)}}\right]\right)$$

$$= N^{k}\left[\frac{(A^{\lambda}-1)}{(A^{\lambda}+1)(k+1)}+\left(q+\frac{1}{q}\right)\frac{A^{\lambda}k!}{(\lambda\ln A)^{(k-1)}}\right] < +\infty, \qquad (14.30)$$

proving the claim. □

We give

Definition 14.2. Let $f \in C_B(\mathbb{R}^N)$, $N \in \mathbb{N}$. We define the following activated symmetrized and perturbed A-generalized logistic multivariate convolution type operators:

The basic one

$$\widetilde{S}_n(f)(x) := \int_{\mathbb{R}^N} f\left(\frac{u}{n}\right)\widetilde{Z}(nx-u)\,du, \quad \forall\, x \in \mathbb{R}^N, \qquad (14.31)$$

the activated Kantorovich type

$$\widetilde{S}_n^{*}(f)(x) := n^N\int_{\mathbb{R}^N}\left(\int_{\frac{u}{n}}^{\frac{u+1}{n}} f(t)\,dt\right)\widetilde{Z}(nx-u)\,du, \quad \forall\, x \in \mathbb{R}^N. \quad (14.32)$$

Let now $\theta = (\theta_1, ..., \theta_N) \in \mathbb{N}^N$, $r = (r_1, ..., r_N) \in \mathbb{Z}_+^N$, $w_r = w_{r_1 r_2 ... r_N} \geq 0$, such that

$$\sum_{r=0}^{\theta} w_r = \sum_{r_1=0}^{\theta_1} \sum_{r_2=0}^{\theta_2} ... \sum_{r_N=0}^{\theta_N} w_{r_1 r_2 ... r_N} = 1; \quad u \in \mathbb{R}^N,$$

and

$$\delta_n(f)(u) := \delta_n(f)(u_1, ..., u_N) := \sum_{r=0}^{\theta} w_r f\left(\frac{u}{n} + \frac{r}{n\theta}\right) =$$

$$\sum_{r_1=0}^{\theta_1} \sum_{r_2=0}^{\theta_2} ... \sum_{r_N=0}^{\theta_N} w_{r_1 r_2 ... r_N} f\left(\frac{u_1}{n} + \frac{r_1}{n\theta_1}, \frac{u_2}{n} + \frac{r_2}{n\theta_2}, ..., \frac{u_N}{n} + \frac{r_N}{n\theta_N}\right), \tag{14.33}$$

where $\frac{r}{\theta} := \left(\frac{r_1}{\theta_1}, \frac{r_2}{\theta_2}, ..., \frac{r_N}{\theta_N}\right)$.

We define the activated quadrature operators

$$\widehat{S_n}(f)(x) := \widehat{S_n}(f, x_1, ..., x_N) := \int_{\mathbb{R}^N} \delta_n(f)(u) \widetilde{Z}(nx - u)\, du = \tag{14.34}$$

$$\int_{-\infty}^{\infty} \int_{-\infty}^{\infty} ... \int_{-\infty}^{\infty} \delta_n(f)(u_1, ..., u_N) \left(\prod_{i=1}^{N} W(nx_i - u_i)\right) du_1 ... du_N, \forall x \in \mathbb{R}^N.$$

One can rewrite

$$\widetilde{S}_n(f)(x) = \int_{-\infty}^{\infty} \int_{-\infty}^{\infty} ... \int_{-\infty}^{\infty} f\left(\frac{u_1}{n}, \frac{u_2}{n}, ..., \frac{u_N}{n}\right) \left(\prod_{i=1}^{N} W(nx_i - u_i)\right) du_1 ... du_N, \tag{14.35}$$

and

$$\widetilde{S}_n^*(f)(x) = n^N \int_{-\infty}^{\infty} \int_{-\infty}^{\infty} ... \int_{-\infty}^{\infty} \left(\int_{\frac{u_1}{n}}^{\frac{u_1+1}{n}} \int_{\frac{u_2}{n}}^{\frac{u_2+1}{n}} ... \int_{\frac{u_N}{n}}^{\frac{u_N+1}{n}} f(t_1, ..., t_N) dt_1 ... dt_N\right)$$

$$\left(\prod_{i=1}^{N} W(nx_i - u_i)\right) du_1 ... du_N, \tag{14.36}$$

$n \in \mathbb{N}$, $\forall\, x \in \mathbb{R}^N$.

In this work we study the approximation properties of the operators $\widetilde{S}_n$, $\widetilde{S}_n^*$ and $\widehat{S_n}$, expecially their convergence to the unit operator I, and being treated as positive linear operators.

We notice that

$$\widetilde{S}_n(1) = \widetilde{S}^*(1) = \widehat{S}(1) = 1. \tag{14.37}$$

As in [320] we have that $\widetilde{S}_n$, $\widetilde{S}_n^*$, $\widehat{S_n}$ map $C_B(\mathbb{R}^N)$ into itself.

Differentiation of our operators follows:

Remark 14.3. As in [320] we have: Let $i \in \mathbb{N}$ be fixed. Assume that $f \in C^{(i)}(\mathbb{R}^N)$, $N \in \mathbb{N}$. Here f_α denotes a partial derivative of f, $\alpha := (\alpha_1, ..., \alpha_N)$, $\alpha_j \in \mathbb{Z}_+$, $j = 1, ..., N$, and $|\alpha| := \sum_{j=1}^{N} \alpha_j = l$, where $l = 0, 1, ..., i$.

We write also $f_\alpha := \frac{\partial^\alpha f}{\partial x^\alpha}$ and we say it is of order l.

We assume that any partial $f_\alpha \in C_B(\mathbb{R}^N)$, for all $\alpha : |\alpha| = l$, $l = 0, 1, ..., i$.

We have that

$$\left(\widetilde{S}_n(f)\right)_\alpha(x) = \int_{\mathbb{R}^N} f_\alpha\left(x - \frac{z}{n}\right) \widetilde{Z}(z)\, dz = \tag{14.38}$$

$$\int_{\mathbb{R}^N} f_\alpha\left(\frac{u}{n}\right) \widetilde{Z}(nx - u)\, du = \left(\widetilde{S}_n(f_\alpha)\right)(x), \quad \forall\, x \in \mathbb{R}^N.$$

Similarly, we have that

$$\left(\widetilde{S}_n^*(f)\right)_\alpha(x) = \left(\widetilde{S}_n^*(f_\alpha)\right)(x), \tag{14.39}$$

and

$$\left(\widehat{S_n}(f)\right)_\alpha(x) = \left(\widehat{S_n}(f_\alpha)\right)(x), \quad \forall\, x \in \mathbb{R}^N; \tag{14.40}$$

for all $\alpha : |\alpha| = l$, $l = 0, 1, ..., i$.

We also need

Remark 14.4. About iterated operators (see also [320]):

It holds

$$\left|\widetilde{S}_n(f)(x)\right| \leq \|f\|_\infty \int_{\mathbb{R}^N} \widetilde{Z}(z)\, dz = \|f\|_\infty,$$

i.e.

$$\left\|\widetilde{S}_n(f)\right\|_\infty \leq \|f\|_\infty. \tag{14.41}$$

So $\widetilde{S}_n$ is a bounded positive linear operator.

Clearly it holds

$$\left\|\widetilde{S}_n^2(f)\right\|_\infty = \left\|\widetilde{S}_n\left(\widetilde{S}_n(f)\right)\right\|_\infty \leq \left\|\widetilde{S}_n(f)\right\|_\infty \leq \|f\|_\infty. \tag{14.42}$$

And for $k \in \mathbb{N}$ we obtain

$$\left\|\widetilde{S}_n^k(f)\right\|_\infty \leq \left\|\widetilde{S}_n^{k-1}(f)\right\|_\infty \leq \left\|\widetilde{S}_n^{k-2}(f)\right\|_\infty \leq \dots \leq \|f\|_\infty, \tag{14.43}$$

so the contraction property valid and $\widetilde{S}_n^k$ is a bounded linear operator.

Let $r \in \mathbb{N}$, we observe that

$$\widetilde{S}_n^r f - f = \left(\widetilde{S}_n^r f - \widetilde{S}_n^{r-1} f\right) + \left(\widetilde{S}_n^{r-1} f - \widetilde{S}_n^{r-2} f\right) + \left(\widetilde{S}_n^{r-2} f - \widetilde{S}_n^{r-3} f\right)$$

$$+ \dots + \left(\widetilde{S}_n^2 f - \widetilde{S}_n f\right) + \left(\widetilde{S}_n f - f\right). \tag{14.44}$$

Then

$$\left\|\widetilde{S}_n^r f - f\right\|_\infty \leq \left\|\widetilde{S}_n^r f - \widetilde{S}_n^{r-1} f\right\|_\infty + \left\|\widetilde{S}_n^{r-1} f - \widetilde{S}_n^{r-2} f\right\|_\infty + \left\|\widetilde{S}_n^{r-2} f - \widetilde{S}_n^{r-3} f\right\|_\infty$$

$$+ \dots + \left\|\widetilde{S}_n^2 f - \widetilde{S}_n f\right\|_\infty + \left\|\widetilde{S}_n f - f\right\|_\infty =$$

$$\left\|\widetilde{S}_n^{r-1}\left(\widetilde{S}_n f - f\right)\right\|_\infty + \left\|\widetilde{S}_n^{r-2}\left(\widetilde{S}_n f - f\right)\right\|_\infty + \dots + \left\|\widetilde{S}_n\left(\widetilde{S}_n f - f\right)\right\|_\infty +$$

$$\left\|\widetilde{S}_n f - f\right\|_\infty \leq r \left\|\widetilde{S}_n f - f\right\|_\infty.$$

Therefore

$$\left\|\widetilde{S}_n^r f - f\right\|_\infty \leq r \left\|\widetilde{S}_n f - f\right\|_\infty. \tag{14.45}$$

Let now $m_1, m_2, \dots, m_r \in \mathbb{N} : m_1 \leq m_2 \leq \dots \leq m_r$, and $\widetilde{S}_{m_i}$ as above.

$$\widetilde{S}_{m_r}\left(\widetilde{S}_{m_{r-1}}\left(\dots\widetilde{S}_{m_2}\left(\widetilde{S}_{m_1} f\right)\right)\right) - f = \dots =$$

$$\widetilde{S}_{m_r}\left(\widetilde{S}_{m_{r-1}}\left(\dots\widetilde{S}_{m_2}\right)\right)\left(\widetilde{S}_{m_1} f - f\right) + \widetilde{S}_{m_r}\left(\widetilde{S}_{m_{r-1}}\left(\dots\widetilde{S}_{m_3}\right)\right)\left(\widetilde{S}_{m_2} f - f\right) + \tag{14.46}$$

$$\widetilde{S}_{m_r}\left(\widetilde{S}_{m_{r-1}}\left(\dots\widetilde{S}_{m_4}\right)\right)\left(\widetilde{S}_{m_3} f - f\right) + \dots + \widetilde{S}_{m_r}\left(\widetilde{S}_{m_{r-1}} f - f\right) + \widetilde{S}_{m_r} f - f.$$

Consequently it holds, as in [317], Chapter 2,

$$\left\|\widetilde{S}_{m_r}\left(\widetilde{S}_{m_{r-1}}\left(\dots\widetilde{S}_{m_2}\left(\widetilde{S}_{m_1} f\right)\right)\right) - f\right\|_\infty \leq \sum_{i=1}^{r} \left\|\widetilde{S}_{m_i} f - f\right\|_\infty. \tag{14.47}$$

The properties (14.41)–(14.47) are also valid for the iterations of operators $\widetilde{S}_n^*$ and $\widehat{S_n}$.

Notation 14.1. We denote by M_n any of the operators $\widetilde{S}_n$, $\widetilde{S}_n^*$ and $\widehat{S_n}$, $\forall$ $n \in \mathbb{N}$.

14.3 Auxiliary Results

Here let $x_0 \in \mathbb{R}^N$, $m, n \in \mathbb{N}$. It follows:

Proposition 14.1. *It holds that*

$$\widetilde{S}_n \left(\|\cdot - x_0\|_\infty^m\right)(x_0) > 0, \tag{14.48}$$

and

$$0 < \left\| \widetilde{S}_n \left(\|\cdot - x_0\|_\infty^m\right)(x_0) \right\|_\infty \leq \tag{14.49}$$

$$\frac{N^m}{n^m}\left[\frac{\left(A^\lambda - 1\right)}{\left(A^\lambda + 1\right)(m+1)} + \left(q + \frac{1}{q}\right)\frac{A^\lambda m!}{(\lambda \ln A)^{(m-1)}}\right](< +\infty) \to 0, \ \ \textit{as } n \to +\infty.$$

Proof. We have that

$$0 < \widetilde{S}_n \left(\|\cdot - x_0\|_\infty^m\right)(x_0) = \int_{\mathbb{R}^N} \left\| \frac{u}{n} - x_0 \right\|_\infty^m \widetilde{Z}(nx_0 - u)\, du =$$

$$\frac{1}{n^m} \int_{\mathbb{R}^N} \|nx_0 - u\|_\infty^m \widetilde{Z}(nx_0 - u)\, du = \tag{14.50}$$

$$\frac{1}{n^m} \int_{\mathbb{R}^N} \|x\|_\infty^m \widetilde{Z}(x)\, dx \overset{(14.29)}{\leq}$$

$$\frac{N^m}{n^m}\left[\frac{\left(A^\lambda - 1\right)}{\left(A^\lambda + 1\right)(m+1)} + \left(q + \frac{1}{q}\right)\frac{A^\lambda m!}{(\lambda \ln A)^{(m-1)}}\right](< +\infty) \to 0, \ \ \text{as } n \to +\infty.$$

□

Proposition 14.2. *It holds that*

$$\widetilde{S}_n^* \left(\|\cdot - x_0\|_\infty^m\right)(x_0) > 0, \tag{14.51}$$

and

$$0 < \left\| \widetilde{S}_n^* \left(\|\cdot - x_0\|_\infty^m\right)(x_0) \right\|_\infty \leq$$

$$\frac{2^{m-1}}{n^m}\left[1 + N^m \left(\frac{\left(A^\lambda - 1\right)}{\left(A^\lambda + 1\right)(m+1)} + \left(q + \frac{1}{q}\right)\frac{A^\lambda m!}{(\lambda \ln A)^{(m-1)}}\right)\right](< +\infty) \to 0, \tag{14.52}$$

as $n \to +\infty$.

Proof. We have that

$$0 < \widetilde{S}_n^* \left(\|\cdot - x_0\|_\infty^m\right)(x_0) = \int_{\mathbb{R}^N} \left(n^N \int_{\left[0,\frac{1}{n}\right]^N} \left\| t + \frac{u}{n} - x_0 \right\|_\infty^m dt \right) \widetilde{Z}(nx_0 - u)\, du$$

$$\leq \int_{\mathbb{R}^N} \left(n^N \int_{\left[0,\frac{1}{n}\right]^N} \left(\left\| \frac{u}{n} - x_0 \right\|_\infty + \|t\|_\infty \right)^m dt \right) \widetilde{Z}(nx_0 - u)\, du \leq \quad (14.53)$$

$$\int_{\mathbb{R}^N} \left(n^N \int_{\left[0,\frac{1}{n}\right]^N} \left(\left\| \frac{u}{n} - x_0 \right\|_\infty + \frac{1}{n} \right)^m dt \right) \widetilde{Z}(nx_0 - u)\, du =$$

$$\int_{\mathbb{R}^N} \left(\left\| \frac{u}{n} - x_0 \right\|_\infty + \frac{1}{n} \right)^m \widetilde{Z}(nx_0 - u)\, du =$$

$$\frac{1}{n^m} \int_{\mathbb{R}^N} \left(\|nx_0 - u\|_\infty + 1\right)^m \widetilde{Z}(nx_0 - u)\, du \leq \tag{14.54}$$

$$\frac{2^{m-1}}{n^m} \left(\int_{\mathbb{R}^N} \left(1 + \|nx_0 - u\|_\infty^m\right) \widetilde{Z}(nx_0 - u)\, du \right) \overset{(14.27)}{=}$$

$$\frac{2^{m-1}}{n^m} \left[1 + \int_{\mathbb{R}^N} \|nx_0 - u\|_\infty^m \widetilde{Z}(nx_0 - u)\, du \right] =$$

$$\frac{2^{m-1}}{n^m} \left[1 + \int_{\mathbb{R}^N} \|x\|_\infty^m \widetilde{Z}(x)\, dx \right] \overset{(14.29)}{\leq}$$

$$\frac{2^{m-1}}{n^m} \left[1 + N^m \left(\frac{(A^\lambda - 1)}{(A^\lambda + 1)(m+1)} + \left(q + \frac{1}{q}\right) \frac{A^\lambda m!}{(\lambda \ln A)^{(m-1)}} \right) \right] (< +\infty) \to 0,$$

as $n \to +\infty$. ☐

Proposition 14.3. *It holds that*

$$\widehat{S_n}\left(\|\cdot - x_0\|_\infty^m\right)(x_0) > 0,$$

and

$$0 < \left\| \widehat{S_n}\left(\|\cdot - x_0\|_\infty^m\right)(x_0) \right\|_\infty \leq$$

$$\frac{2^{m-1}}{n^m} \left[1 + N^m \left(\frac{(A^\lambda - 1)}{(A^\lambda + 1)(m+1)} + \left(q + \frac{1}{q}\right) \frac{A^\lambda m!}{(\lambda \ln A)^{(m-1)}} \right) \right] (< +\infty) \to 0, \tag{14.55}$$

as $n \to +\infty$.

Proof. We have that

$$0 < \widehat{S_n}\left(\|\cdot - x_0\|_\infty^m\right)(x_0) = \int_{\mathbb{R}^N}\left(\sum_{r=0}^{\theta} w_r \left\|\frac{u}{n} + \frac{r}{n\theta} - x_0\right\|_\infty^m\right)\widetilde{Z}(nx_0 - u)\,du \leq$$

$$\int_{\mathbb{R}^N}\left(\sum_{r=0}^{\theta} w_r \left(\left\|\frac{u}{n} - x_0\right\|_\infty + \left\|\frac{r}{n\theta}\right\|_\infty\right)^m\right)\widetilde{Z}(nx_0 - u)\,du \leq$$

$$\int_{\mathbb{R}^N}\left(\left\|\frac{u}{n} - x_0\right\|_\infty + \frac{1}{n}\right)^m \widetilde{Z}(nx_0 - u)\,du \qquad (14.56)$$

(the rest as in the proof of Proposition 14.3)

$$\frac{2^{m-1}}{n^m}\left[1 + N^m\left(\frac{\left(A^\lambda - 1\right)}{\left(A^\lambda + 1\right)(m+1)} + \left(q + \frac{1}{q}\right)\frac{A^\lambda m!}{(\lambda \ln A)^{(m-1)}}\right)\right](< +\infty) \to 0,$$

as $n \to +\infty$. □

14.4 Main Results

We need the following:

Theorem 14.3. *Let $\left(\overline{L_n}\right)_{n\in\mathbb{N}}$ be a sequence of positive linear operators from $C_B\left(\mathbb{R}^N\right)$ into itself, such that $\overline{L_n}(1) = 1$. Assume $0 < \overline{L_n}\left(\|\cdot - x_0\|_\infty\right)(x_0) < +\infty$.*

Then

$$\left|\overline{L_n}(f)(x_0) - f(x_0)\right| \leq 2\omega_1\left(f, \overline{L_n}\left(\|\cdot - x_0\|_\infty\right)(x_0)\right) < +\infty. \qquad (14.57)$$

Given that $\lim_{n\to+\infty} \overline{L_n}\left(\|\cdot - x_0\|_\infty\right)(x_0) = 0$ and f is uniformly continuous, we get that $\lim_{n\to+\infty} \overline{L_n}(f)(x_0) = f(x_0)$.

Proof. Here it is $\omega_1(f, h) < \infty$, $h > 0$.

By Lemma 7.1.1, p. 208, [315], we get that

$$|f(t) - f(x)| \leq \omega_1(f, h)\left\lceil \frac{\|t - x\|_\infty}{h} \right\rceil, \qquad (14.58)$$

where $\lceil\cdot\rceil$ is the ceiling function.

But it holds

$$\left\lceil \frac{\|t - x\|_\infty}{h} \right\rceil \leq 1 + \frac{\|t - x\|_\infty}{h}. \qquad (14.59)$$

Thus we get

$$|f(t)-f(x)| \leq \omega_1(f,h)\left[1+\frac{\|t-x\|_\infty}{h}\right], \quad \forall\, t \in \mathbb{R}^N. \tag{14.60}$$

The last can be written also as

$$|f(\cdot)-f(x)| \leq \omega_1(f,h)\left[1+\frac{\|\cdot-x\|_\infty}{h}\right], \quad \text{over } \mathbb{R}^N. \tag{14.61}$$

Hence we derive

$$\left|\overline{L_n}(f)(x_0)-f(x_0)\right| = \left|\overline{L_n}(f(\cdot)-f(x_0))(x_0)\right| \leq$$

$$\overline{L_n}(|f(\cdot)-f(x_0)|)(x_0) \overset{(14.61)}{\leq} \omega_1(f,h)\left[1+\frac{\overline{L_n}(\|\cdot-x_0\|_\infty)(x_0)}{h}\right] \tag{14.62}$$

(choose $h := \overline{L_n}(\|\cdot-x_0\|_\infty)(x_0) > 0$)

$$= 2\omega_1\left(f, \overline{L_n}(\|\cdot-x_0\|_\infty)(x_0)\right).$$

□

We derive

Theorem 14.4. *We have for* $M_n = \widetilde{S}_n$, $\widetilde{S}_n^*$, $\widehat{S_n}$ *that*

$$|M_n(f)(x_0)-f(x_0)| \leq 2\omega_1(f, M_n(\|\cdot-x_0\|_\infty)(x_0)) < +\infty, \tag{14.63}$$

where $x_0 \in \mathbb{R}^N$, $f \in C_B(\mathbb{R}^N)$.

Given that $\lim_{n\to+\infty} M_n(\|\cdot-x_0\|_\infty)(x_0) = 0$ *and* f *is uniformly continuous, we get that* $\lim_{n\to+\infty} M_n(f)(x_0) = f(x_0)$.

Proof. By Theorem 14.3. □

More specifically we derive:

Proposition 14.4. *It holds that* ($f \in C_B(\mathbb{R}^N)$)

$$\left\|\widetilde{S}_n(f)-f\right\|_\infty \leq 2\omega_1\left(f, \frac{N}{n}\left[\frac{(A^\lambda-1)}{(A^\lambda+1)2}+\left(q+\frac{1}{q}\right)A^\lambda\right]\right) < +\infty. \tag{14.64}$$

If f *is uniformly continuous, then* $\widetilde{S}_n(f) \to f$, *uniformly as* $n \to +\infty$.

Proof. By Proposition 14.1, and Theorem 14.4. □

Proposition 14.5. *It holds that* ($f \in C_B(\mathbb{R}^N)$)

$$\left\| \widetilde{S}_n^*(f) - f \right\|_\infty \leq 2\omega_1\left(f, \frac{1}{n}\left[1 + N\left(\frac{(A^\lambda - 1)}{(A^\lambda + 1)2} + \left(q + \frac{1}{q}\right)A^\lambda\right)\right]\right) < +\infty. \tag{14.65}$$

If f is uniformly continuous, then $\widetilde{S}_n^*(f) \to f$, uniformly as $n \to +\infty$.

Proof. By Proposition 14.2, and Theorem 14.4. □

Proposition 14.6. *It holds that* ($f \in C_B(\mathbb{R}^N)$)

$$\left\| \widehat{S_n}(f) - f \right\|_\infty \leq 2\omega_1\left(f, \frac{1}{n}\left[1 + N\left(\frac{(A^\lambda - 1)}{(A^\lambda + 1)2} + \left(q + \frac{1}{q}\right)A^\lambda\right)\right]\right) < +\infty. \tag{14.66}$$

If f is uniformly continuous, then $\widehat{S_n}(f) \to f$, uniformly as $n \to +\infty$.

Proof. By Proposition 14.3, and Theorem 14.4. □

Next, we study our operators under convexity:

Theorem 14.5. *Let $\left(\overline{L_n}\right)_{n\in\mathbb{N}}$ positive linear operators from $C_B(\mathbb{R}^N)$ into itself: $\overline{L_n}(1) = 1$. Let $f \in C_B(\mathbb{R}^N)$ and the convex function $|f(\cdot) - f(x_0)|$ over $\mathbb{R}^N$, $x_0 \in \mathbb{R}^N$. Assume that $0 < \overline{L_n}(\|\cdot - x_0\|_\infty)(x_0) < +\infty$.*

Then

$$\left|\overline{L_n}(f)(x_0) - f(x_0)\right| \leq \omega_1\left(f, \overline{L_n}(\|\cdot - x_0\|_\infty)(x_0)\right). \tag{14.67}$$

The last (14.67) is attained by $f(t) = \|t - x_0\|_\infty$, $\forall\, t \in \mathbb{R}^N$, $x_0 \in \mathbb{R}^N$, that is sharp. Given that f is uniformly continuous and $\overline{L_n}(\|\cdot - x_0\|_\infty)(x_0) \to 0$, as $n \to \infty$, then $\overline{L_n}(f)(x_0) \to f(x_0)$, as $n \to \infty$.

Proof. Based on Lemma 8.1.1, p. 243, [315], we obtain:

Let $(V, \|\cdot\|)$ be a real normed vector space and $f : V \to \mathbb{R}$ such that $|f(t) - f(x_0)|$ is convex in $t \in V$; $x_0 \in V$. Then

$$|f(t) - f(x_0)| \leq \frac{\omega_1(f, h)}{h}\|t - x_0\|, \quad \forall\, t \in V,\ h > 0. \tag{14.68}$$

That is, it holds here

$$|f(\cdot) - f(x_0)| \leq \frac{\omega_1(f, h)}{h}\|\cdot - x_0\|, \quad \text{over } \mathbb{R}^N. \tag{14.69}$$

Then, the proof of (14.67) goes as in the proof of Theorem 14.3.

Inequality (14.67) is sharp, namely it is attained by $f(t) = \|t - x_0\|_\infty$, $\forall\, t \in \mathbb{R}^N$, $x_0 \in \mathbb{R}^N$.

We have

$$\left|\overline{L_n}(f, x_0) - f(x_0)\right| = \overline{L_n}\left(\|\cdot - x_0\|_\infty\right)(x_0). \tag{14.70}$$

Next we observe

$$\omega_1\left(f, \overline{L_n}\left(\|\cdot - x_0\|_\infty\right)(x_0)\right) = \omega_1\left(\|\cdot - x_0\|_\infty, \overline{L_n}\left(\|\cdot - x_0\|_\infty\right)(x_0)\right) =$$

$$\sup_{\left\{\begin{array}{c} t_1, t_2 \in \mathbb{R}^N : \\ : \|t_1 - t_2\|_\infty \leq \overline{L_n}\left(\|\cdot - x_0\|_\infty\right)(x_0) \end{array}\right.} \left|\|t_1 - x_0\|_\infty - \|t_2 - x_0\|_\infty\right| \leq \tag{14.71}$$

$$\sup_{\left\{\begin{array}{c} t_1, t_2 \in \mathbb{R}^N : \\ : \|t_1 - t_2\|_\infty \leq \overline{L_n}\left(\|\cdot - x_0\|_\infty\right)(x_0) \end{array}\right.} \|t_1 - t_2\|_\infty = \overline{L_n}\left(\|\cdot - x_0\|_\infty\right)(x_0).$$

So that

$$\omega_1\left(f, \overline{L_n}\left(\|\cdot - x_0\|_\infty\right)(x_0)\right) = \overline{L_n}\left(\|\cdot - x_0\|_\infty\right)(x_0).$$

Thus (14.67) it is attained. □

Consequently we obtain:

Theorem 14.6. *Let $f \in C_B\left(\mathbb{R}^N\right)$ and the convex function $|f(\cdot) - f(x_0)|$ over $\mathbb{R}^N$, $x_0 \in \mathbb{R}^N$.*

Then

$$|M_n(f)(x_0) - f(x_0)| \leq \omega_1\left(f, M_n\left(\|\cdot - x_0\|_\infty\right)(x_0)\right) < +\infty. \tag{14.72}$$

The last (14.72) is attained by $f(t) = \|t - x_0\|_\infty$, $\forall\ t \in \mathbb{R}^N$, $x_0 \in \mathbb{R}^N$, that is sharp. Given that f is uniformly continuous, we derive that $M_n(f)(x_0) \to f(x_0)$, as $n \to \infty$.

Proof. By Theorem 14.5. □

More specifically we derive:

Proposition 14.7. *Let $f \in C_B\left(\mathbb{R}^N\right)$ such that $|f(\cdot) - f(x_0)|$ is convex over $\mathbb{R}^N$, $x_0 \in \mathbb{R}^N$.*

Then

$$\left|\widetilde{S}_n(f)(x_0) - f(x_0)\right| \leq \omega_1\left(f, \frac{N}{n}\left(\frac{\left(A^\lambda - 1\right)}{\left(A^\lambda + 1\right)2} + \left(q + \frac{1}{q}\right)A^\lambda\right)\right) < +\infty. \tag{14.73}$$

If f is uniformly continuous, then $\widetilde{S}_n(f)(x_0) \to f(x_0)$, as $n \to \infty$.

Proof. By Proposition 14.1, and Theorem 14.6. □

Proposition 14.8. *Let f be as in Proposition 14.7. It holds that*

$$\left|\widetilde{S}_n^*(f)(x_0) - f(x_0)\right| \leq \omega_1\left(f, \frac{1}{n}\left[1 + N\left(\frac{(A^\lambda - 1)}{(A^\lambda + 1)\,2} + \left(q + \frac{1}{q}\right)A^\lambda\right)\right]\right) < +\infty. \tag{14.74}$$

If f is uniformly continuous, then $\widetilde{S}_n^*(f)(x_0) \to f(x_0)$, as $n \to \infty$.

Proof. By Proposition 14.2, and Theorem 14.6. □

Proposition 14.9. *Let f be as in Proposition 14.7. It holds that*

$$\left|\widehat{S_n}(f)(x_0) - f(x_0)\right| \leq \omega_1\left(f, \frac{1}{n}\left[1 + N\left(\frac{(A^\lambda - 1)}{(A^\lambda + 1)\,2} + \left(q + \frac{1}{q}\right)A^\lambda\right)\right]\right) < +\infty. \tag{14.75}$$

If f is uniformly continuous, then $\widehat{S_n}(f)(x_0) \to f(x_0)$, as $n \to \infty$.

Proof. By Proposition 14.3, and Theorem 14.6. □

We make

Remark 14.5. We consider $\left(\mathbb{R}^N, \|\cdot\|_\infty\right)$ and $\left(\mathbb{R}^N\right)^j$ denote the j-fold product space $\mathbb{R}^N \times ... \times \mathbb{R}^N$ endowed with the max-norm $\|x\|_{(\mathbb{R}^N)^j} := \max_{1\leq k\leq j} \|x_k\|_\infty$, where $x := (x_1, ..., x_j) \in \left(\mathbb{R}^N\right)^j$. Then the space $L_j := L_j\left(\left(\mathbb{R}^N\right)^j; \mathbb{R}\right)$ of all real-valued multilinear continuous functions $g : \left(\mathbb{R}^N\right)^j \to \mathbb{R}$ is a Banach space with norm

$$\|g\| := \|g\|_{L_j} := \sup_{\left(\|x\|_{(\mathbb{R}^N)^j}=1\right)} |g(x)| = \sup \frac{|g(x)|}{\|x_1\|_\infty \cdots \|x_j\|_\infty}. \tag{14.76}$$

Let $x_0 \in \mathbb{R}^N$ be fixed, and $f : \mathbb{R}^N \to \mathbb{R}$ be a continuous and bounded function whose Fréchet derivatives

$$f^{(j)} : \mathbb{R}^N \to L_j = L_j\left(\left(\mathbb{R}^N\right)^j; \mathbb{R}\right)$$

exist and are continuous and bounded for all $1 \leq j \leq m$, $m \in \mathbb{N}$. Call $(x - x_0)^j := (x - x_0, ..., x - x_0) \in \left(\mathbb{R}^N\right)^j = \underbrace{\mathbb{R}^N \times ... \times \mathbb{R}^N}_{-j-}$.

Then, by Taylor's formula, see [323], p. 124 and [322], we get

$$f(x) = \sum_{j=0}^{m} \frac{f^{(j)}(x_0)(x - x_0)^j}{j!} + R_m(x, x_0), \quad \text{all } x \in \mathbb{R}^N, \tag{14.77}$$

where

$$R_m(x,x_0):=\int_0^1 \frac{(1-u)^{m-1}}{(m-1)!}\left(f^{(m)}\left(x_0+u\left(x-x_0\right)\right)-f^{(m)}\left(x_0\right)\right)\left(x-x_0\right)^m du. \tag{14.78}$$

Here we assume that $f^{(j)}(x_0)=0$, $j=1,...,m$, then

$$f(x)-f(x_0)=\int_0^1 \frac{(1-u)^{m-1}}{(m-1)!}\left(f^{(m)}\left(x_0+u\left(x-x_0\right)\right)-f^{(m)}\left(x_0\right)\right)\left(x-x_0\right)^m du. \tag{14.79}$$

Consider

$$w:=\omega_1\left(f^{(m)},h\right)=\sup_{\substack{x,y\in\mathbb{R}^N \\ :\ \|x-y\|_\infty\le h}}\left\|f^{(m)}(x)-f^{(m)}(y)\right\|,\quad h>0. \tag{14.80}$$

Notice that $w<+\infty$.

We obtain that

$$\begin{aligned}&\left\|\left(f^{(m)}\left(x_0+u\left(x-x_0\right)\right)-f^{(m)}\left(x_0\right)\right)\left(x-x_0\right)^m\right\|\le\\ &\left\|f^{(m)}\left(x_0+u\left(x-x_0\right)\right)-f^{(m)}\left(x_0\right)\right\|\left\|x-x_0\right\|^m\le \\ &w\left\|x-x_0\right\|^m\left\lceil\frac{u\left\|x-x_0\right\|}{h}\right\rceil,\end{aligned} \tag{14.81}$$

by Lemma 7.1.1, p. 208, [315].

Therefore for all $x\in\mathbb{R}^N$:

$$\begin{aligned}\left|f(x)-f(x_0)\right| &\le w\left\|x-x_0\right\|^m\int_0^1\left\lceil\frac{u\left\|x-x_0\right\|_\infty}{h}\right\rceil\frac{(1-u)^{m-1}}{(m-1)!}du \\ &= w\Phi_m\left(\left\|x-x_0\right\|_\infty\right),\end{aligned} \tag{14.82}$$

by a change of variable, where

$$\Phi_m(t):=\int_0^{|t|}\left\lceil\frac{s}{h}\right\rceil\frac{(|t|-s)^{m-1}}{(m-1)!}ds=\frac{1}{m!}\left(\sum_{j=0}^{\infty}\left(|t|-j\cdot h\right)_+^m\right), \tag{14.83}$$

$t\in\mathbb{R}$; $(x)_+=\max\{x,0\}$.

For a full list of properties of the spline function Φ_m see Remark 7.1.3 of [315], p. 210.

By (7.1.18) of [315], p. 210, we have

$$\Phi_m(x)\le\left(\frac{|x|^{m+1}}{(m+1)!h}+\frac{|x|^m}{2m!}+\frac{h|x|^{m-1}}{8(m-1)!}\right),\quad\forall\, x\in\mathbb{R}. \tag{14.84}$$

Consequently it holds ($x \in \mathbb{R}^N$)

$$\left|f(x)-f(x_0)\right| \leq \omega_1\left(f^{(m)},h\right)\left(\frac{\|x-x_0\|_\infty^{m+1}}{(m+1)!h}+\frac{\|x-x_0\|_\infty^{m}}{2m!}+\frac{h\|x-x_0\|_\infty^{m-1}}{8(m-1)!}\right).$$

So that

$$\left|f(\cdot)-f(x_0)\right| \leq \omega_1\left(f^{(m)},h\right)\left(\frac{\|\cdot-x_0\|_\infty^{m+1}}{(m+1)!h}+\frac{\|\cdot-x_0\|_\infty^{m}}{2m!}+\frac{h\|\cdot-x_0\|_\infty^{m-1}}{8(m-1)!}\right), \tag{14.85}$$

over $\mathbb{R}^N$.

Let $\left(\overline{L_n}\right)_{n\in\mathbb{N}}$ be a sequence of positive linear operators from $C_B\left(\mathbb{R}^N\right)$ into itself such that $\overline{L_n}(1)=1$, $\forall\, n\in\mathbb{N}$.

By $|f| \leq |f|$, iff $-|f| \leq f \leq |f|$, then $-\overline{L_n}(|f|) \leq \overline{L_n}(f) \leq \overline{L_n}(|f|)$, and $\left|\overline{L_n}(f)\right| \leq \overline{L_n}(|f|)$.

Hence

$$\left|\overline{L_n}(f(\cdot)-f(x_0))(x_0)\right| \leq \overline{L_n}(|f(\cdot)-f(x_0)|)(x_0) \overset{(14.85)}{\leq} \omega_1\left(f^{(m)},h\right)$$

$$\left[\frac{\overline{L_n}\left(\|\cdot-x_0\|_\infty^{m+1}\right)(x_0)}{(m+1)!h}+\frac{\overline{L_n}\left(\|\cdot-x_0\|_\infty^{m}\right)(x_0)}{2m!}+\frac{h\overline{L_n}\left(\|\cdot-x_0\|_\infty^{m-1}\right)(x_0)}{8(m-1)!}\right] =: (\xi). \tag{14.86}$$

We need the following Hölder's type inequality for positive linear operators.

Theorem 14.7. *([321]) Let $\overline{L}$ be a positive linear operator from $C_B\left(\mathbb{R}^N\right)$ into itself and $f,g \in C_B\left(\mathbb{R}^N\right)$, furthermore let $p,q>1: \frac{1}{p}+\frac{1}{q}=1$. Assume that $\overline{L}\left(\left(|f(\cdot)|^p\right)\right)(s_*)$, $\overline{L}\left(\left(|g(\cdot)|^q\right)\right)(s_*)>0$ for some $s_* \in \mathbb{R}^N$. Then*

$$\overline{L}(|f(\cdot)g(\cdot)|)(s_*) \leq \left(\overline{L}\left(\left(|f(\cdot)|^p\right)\right)(s_*)\right)^{\frac{1}{p}}\left(\overline{L}\left(\left(|g(\cdot)|^q\right)\right)(s_*)\right)^{\frac{1}{q}}. \tag{14.87}$$

We continue with

Remark 14.6. Let $g=1$ and $L_n : L_n(1)=1$, then by (14.87) we get:

$$\overline{L_n}\left(\|\cdot-x_0\|_\infty^{m}\right)(x_0) \leq \left(\overline{L_n}\left(\|\cdot-x_0\|_\infty^{m+1}\right)(x_0)\right)^{\left(\frac{m}{m+1}\right)}, \tag{14.88}$$

and

$$\overline{L_n}\left(\|\cdot-x_0\|_\infty^{(m-1)}\right)(x_0) \leq \left(\overline{L_n}\left(\|\cdot-x_0\|_\infty^{m+1}\right)(x_0)\right)^{\left(\frac{m-1}{m+1}\right)}, \tag{14.89}$$

where $m\in\mathbb{N}$ and $L_n\left(\|\cdot-x_0\|_\infty^{m+1}\right)(x_0)>0$.

Continuing from (14.86):

$$(\xi) \leq \omega_1\left(f^{(m)}, h\right)\left[\frac{\overline{L_n}\left(\|\cdot - x_0\|_\infty^{m+1}\right)(x_0)}{(m+1)!h} + \right. \tag{14.90}$$

$$\left.\frac{\left(\overline{L_n}\left(\|\cdot - x_0\|_\infty^{m+1}\right)(x_0)\right)^{\frac{m}{m+1}}}{2m!} + \frac{h\left(\overline{L_n}\left(\|\cdot - x_0\|_\infty^{m+1}\right)(x_0)\right)^{\frac{m-1}{m+1}}}{8(m-1)!}\right]$$

(assume $h := \left(\overline{L_n}\left(\|\cdot - x_0\|_\infty^{m+1}\right)(x_0)\right)^{\frac{1}{m+1}} > 0$)

$$= \omega_1\left(f^{(m)}, \left(\overline{L_n}\left(\|\cdot - x_0\|_\infty^{m+1}\right)(x_0)\right)^{\frac{1}{m+1}}\right)$$

$$\left[\frac{h^m}{(m+1)!} + \frac{h^m}{2m!} + \frac{h^m}{8(m-1)!}\right] \tag{14.91}$$

$$= \omega_1\left(f^{(m)}, \left(\overline{L_n}\left(\|\cdot - x_0\|_\infty^{m+1}\right)(x_0)\right)^{\frac{1}{m+1}}\right)$$

$$\left(\overline{L_n}\left(\|\cdot - x_0\|_\infty^{m+1}\right)(x_0)\right)^{\frac{m}{m+1}}\left[\frac{1}{(m+1)!} + \frac{1}{2m!} + \frac{1}{8(m-1)!}\right]. \tag{14.92}$$

We have proved that

Theorem 14.8. *Let $f \in C_B\left(\mathbb{R}^N\right)$, $x_0 \in \mathbb{R}^N$, Fréchet derivatives $f^{(j)}$ continuous and bounded with $f^{(j)}(x_0) = 0$, $j = 1, ..., m$. Then*

$$\left|\overline{L_n}(f)(x_0) - f(x_0)\right| \leq \omega_1\left(f^{(m)}, \left(\overline{L_n}\left(\|\cdot - x_0\|_\infty^{m+1}\right)(x_0)\right)^{\frac{1}{m+1}}\right) \tag{14.93}$$

$$\left(\overline{L_n}\left(\|\cdot - x_0\|_\infty^{m+1}\right)(x_0)\right)^{\frac{m}{m+1}}\left[\frac{1}{(m+1)!} + \frac{1}{2m!} + \frac{1}{8(m-1)!}\right], \quad \forall\, n \in \mathbb{N}.$$

Here $\overline{L_n}(1) = 1$, $\overline{L_n}$ positive linear operators from $C_B\left(\mathbb{R}^N\right)$ into itself under the assumption $\overline{L_n}\left(\|\cdot - x_0\|_\infty^{m+1}\right)(x_0) > 0$.

Let $\overline{L_n}\left(\|\cdot - x_0\|_\infty^{m+1}\right)(x_0) \to 0$, as $n \to +\infty$, then $\lim_{n\to+\infty} \overline{L_n}(f)(x_0) = f(x_0)$.

Next we apply Theorem 14.8 to the operators M_n.

Theorem 14.9. *Let $f \in C_B(\mathbb{R}^N)$, $x_0 \in \mathbb{R}^N$, Fréchet derivatives $f^{(j)}$ continuous and bounded with $f^{(j)}(x_0) = 0$, $j = 1, ..., m$. Then*

$$|M_n(f)(x_0) - f(x_0)| \leq \omega_1\left(f^{(m)}, \left(M_n\left(\|\cdot - x_0\|_\infty^{m+1}\right)(x_0)\right)^{\frac{1}{m+1}}\right) \tag{14.94}$$

$$\left(M_n\left(\|\cdot - x_0\|_\infty^{m+1}\right)(x_0)\right)^{\frac{m}{m+1}}\left[\frac{1}{(m+1)!} + \frac{1}{2m!} + \frac{1}{8(m-1)!}\right],$$

$\forall\, n \in \mathbb{N}$.

We have that $\lim_{n\to+\infty} M_n(f)(x_0) = f(x_0)$.

Proof. Here M_n any of $\widetilde{S_n}$, $\widetilde{S}_n^*$ and $\widehat{S_n}$.

By Theorem 14.8, and Propositions 14.1–14.3 we get $M_n\left(\|\cdot - x_0\|_\infty^{m+1}\right)(x_0) \to 0$, as $n \to +\infty$. □

More specifically we derive:

Proposition 14.10. *All as in Theorem 14.9, $m \in \mathbb{N}$. Denote by*

$$\widetilde{\psi_{1n}}(m) := \frac{N^m}{n^m}\left[\frac{(A^\lambda - 1)}{(A^\lambda + 1)(m+1)} + \left(q + \frac{1}{q}\right)\frac{A^\lambda m!}{(\lambda \ln A)^{(m-1)}}\right]. \tag{14.95}$$

Then

$$\left|\widetilde{S_n}(f)(x_0) - f(x_0)\right| \leq \omega_1\left(f^{(m)}, \left(\widetilde{\psi_{1n}}(m+1)\right)^{\frac{1}{m+1}}\right)$$

$$\left(\widetilde{\psi_{1n}}(m+1)\right)^{\frac{m}{m+1}}\left[\frac{1}{(m+1)!} + \frac{1}{2m!} + \frac{1}{8(m-1)!}\right], \tag{14.96}$$

$\forall\, n \in \mathbb{N}$.

We have that $\lim_{n\to+\infty} \widetilde{S_n}(f)(x_0) = f(x_0)$.

Proof. By Theorem 14.9 and Proposition 14.1. □

Proposition 14.11. *All as in Theorem 14.9, $m \in \mathbb{N}$. Denote by*

$$\widetilde{\psi_{2n}}(m) := \frac{2^{m-1}}{n^m}\left[1 + N^m\left(\frac{(A^\lambda - 1)}{(A^\lambda + 1)(m+1)} + \left(q + \frac{1}{q}\right)\frac{A^\lambda m!}{(\lambda \ln A)^{(m-1)}}\right)\right]. \tag{14.97}$$

Then

$$\left\{ \begin{array}{l} \left| \widetilde{S}_n^* (f)(x_0) - f(x_0) \right| \\ \left| \widehat{S_n}(f)(x_0) - f(x_0) \right| \end{array} \right\} \leq \omega_1 \left(f^{(m)}, \left(\widetilde{\psi_{2n}}(m+1) \right)^{\frac{1}{m+1}} \right)$$

$$\left(\widetilde{\psi_{2n}}(m+1) \right)^{\frac{m}{m+1}} \left[\frac{1}{(m+1)!} + \frac{1}{2m!} + \frac{1}{8(m-1)!} \right], \tag{14.98}$$

$\forall\ n \in \mathbb{N}$.

We have that $\lim_{n \to +\infty} \widetilde{S}_n^* (f)(x_0) = \lim_{n \to +\infty} \widehat{S_n}(f)(x_0) = f(x_0)$.

Proof. By Theorem 14.9 and Propositions 14.2, 14.3. □

Next we do simultaneous approximation.

Theorem 14.10. *Let* $i \in \mathbb{N}$ *be fixed and* $f \in C^{(i)}(\mathbb{R}^N)$. *Here* f_α *denotes a partial derivative of* f, $\alpha := (\alpha_1, ..., \alpha_N)$, $\alpha_i \in \mathbb{Z}_+$, $j = 1, ..., N$, *and* $|\alpha| := \sum_{j=1}^{N} \alpha_j = l$, *where* $l = 0, 1, ..., i$. *We assume that* $f_\alpha \in C_B(\mathbb{R}^N)$, *for all* $\alpha : |\alpha| = l$, $l = 0, 1, ..., i$; $x_0 \in \mathbb{R}^N$. *Then*

$$|(M_n(f))_\alpha (x_0) - f_\alpha (x_0)| \leq 2\omega_1 (f_\alpha, M_n (\|\cdot - x_0\|_\infty)(x_0)) < +\infty. \tag{14.99}$$

Given that $\lim_{n \to +\infty} M_n (\|\cdot - x_0\|_\infty)(x_0) = 0$ *and* f_α *is uniformly continuous, we get that* $\lim_{n \to +\infty} (M_n(f))_\alpha (x_0) = f(x_0)$.

Proof. By Remark 14.3 and Theorem 14.4. □

Proposition 14.12. *All as in Theorem 14.10. Then*

$$\left\| \left(\widetilde{S}_n(f) \right)_\alpha - f_\alpha \right\|_\infty \leq 2\omega_1 \left(f_\alpha, \frac{N}{n} \left[\frac{(A^\lambda - 1)}{(A^\lambda + 1) 2} + \left(q + \frac{1}{q} \right) A^\lambda \right] \right) < +\infty. \tag{14.100}$$

If f_α *is uniformly continuous, then* $\left(\widetilde{S}_n(f) \right)_\alpha \to f_\alpha$, *uniformly as* $n \to +\infty$.

Proof. By Remark 14.3 and Proposition 14.4. □

Proposition 14.13. *All as in Theorem 14.10. Then*

$$\left\{ \begin{array}{l} \left\| \left(\widetilde{S}_n^*(f) \right)_\alpha - f_\alpha \right\|_\infty \\ \left\| \left(\widehat{S_n}(f) \right)_\alpha - f_\alpha \right\|_\infty \end{array} \right\} \leq 2\omega_1 \left(f_\alpha, \frac{1}{n} \left[1 + N \left(\frac{(A^\lambda - 1)}{(A^\lambda + 1) 2} + \left(q + \frac{1}{q} \right) A^\lambda \right) \right] \right) < +\infty. \tag{14.101}$$

If f_α *is uniformly continuous, then* $\left(\widetilde{S}_n^*(f) \right)_\alpha \to f_\alpha$, $\left(\widehat{S_n}(f) \right)_\alpha \to f_\alpha$, *uniformly as* $n \to +\infty$.

Proof. By Remark 14.3 and Propositions 14.5, 14.6. □

Next, we do simultaneous convexity approximation.

Theorem 14.11. *The notation is as in Theorem 14.10, with a particular* $\alpha : |\alpha| = l \in \{0, 1, ..., i\}$, $f_\alpha \in C_B(\mathbb{R}^N)$ *and* $|f_\alpha(\cdot) - f_\alpha(x_0)|$ *is convex over* $\mathbb{R}^N$, $x_0 \in \mathbb{R}^N$. *Then*

$$|(M_n(f))_\alpha(x_0) - f_\alpha(x_0)| \leq \omega_1\left(f_\alpha, M_n\left(\|\cdot - x_0\|_\infty\right)(x_0)\right) < +\infty. \tag{14.102}$$

Given that f_α *is uniformly continuous, we derive that* $(M_n(f))_\alpha(x_0) \to f_\alpha(x_0)$, *as* $n \to \infty$.

Proof. By Remark 14.3 and Theorem 14.6. □

More specifically we obtain:

Proposition 14.14. *All as in Theorem 14.11. Then*

$$\left|\left(\widetilde{S}_n(f)\right)_\alpha(x_0) - f_\alpha(x_0)\right| \leq \omega_1\left(f_\alpha, \frac{N}{n}\left[\frac{(A^\lambda - 1)}{(A^\lambda + 1)2} + \left(q + \frac{1}{q}\right)A^\lambda\right]\right) < +\infty. \tag{14.103}$$

If f_α *is uniformly continuous, then* $\left(\widetilde{S}_n(f)\right)_\alpha(x_0) \to f_\alpha(x_0)$, *as* $n \to +\infty$.

Proof. By Remark 14.3, Proposition 14.7 and Theorem 14.11. □

Proposition 14.15. *All as in Theorem 14.11. Then*

$$\left\{\begin{array}{l}\left|\left(\widetilde{S}_n^*(f)\right)_\alpha(x_0) - f_\alpha(x_0)\right| \\ \left|\left(\widehat{S}_n(f)\right)_\alpha(x_0) - f_\alpha(x_0)\right|\end{array}\right\} \leq$$

$$\omega_1\left(f_\alpha, \frac{1}{n}\left[1 + N\left(\frac{(A^\lambda - 1)}{(A^\lambda + 1)2} + \left(q + \frac{1}{q}\right)A^\lambda\right)\right]\right) < +\infty. \tag{14.104}$$

If f_α *is uniformly continuous, then* $\left(\widetilde{S}_n^*(f)\right)_\alpha(x_0) \to f_\alpha(x_0)$, $\left(\widehat{S}_n(f)\right)_\alpha(x_0) \to f_\alpha(x_0)$, *as* $n \to +\infty$.

Proof. By Remark 14.3 and Propositions 14.8, 14.9 and Theorem 14.11. □

We continue with iterated approximations:

Remark 14.7. Here $f \in C_B(\mathbb{R}^N)$; $r \in \mathbb{N}$; $m_1, ..., m_r \in \mathbb{N} : m_1 \leq m_2 \leq ... \leq m_r$; $M_n = \widetilde{S}_n, \widetilde{S}_n^*, \widehat{S_n}$.

As in (14.45) we get that

$$\|M_n^r(f) - f\|_\infty \leq r\|M_n(f) - f\|_\infty, \tag{14.105}$$

and as in (14.47) we obtain that

$$\left\|M_{m_r}\left(M_{m_{r-1}}\left(...M_{m_2}\left(M_{m_1}(f)\right)\right)\right) - f\right\|_\infty \leq \sum_{i=1}^{r} \|M_{m_i}(f) - f\|_\infty. \tag{14.106}$$

Clearly then follows:

Proposition 14.16. *Let $f \in C_B(\mathbb{R}^N)$; $r \in \mathbb{N}$; and $m_1, ..., m_r \in \mathbb{N} : m_1 \leq m_2 \leq ... \leq m_r$. Then*

1)

$$\left\|\widetilde{S}_n^r(f) - f\right\|_\infty \leq r\left\|\widetilde{S}_n(f) - f\right\|_\infty \leq \tag{14.107}$$

$$2r\omega_1\left(f, \frac{N}{n}\left[\frac{(A^\lambda - 1)}{(A^\lambda + 1)2} + \left(q + \frac{1}{q}\right)A^\lambda\right]\right) < +\infty,$$

and the speed of convergence of $\widetilde{S}_n^r$ to the unit I is not worse than the speed of convergence of $\widetilde{S}_n$ to I,

2)

$$\left\|\widetilde{S}_{m_r}\left(\widetilde{S}_{m_{r-1}}\left(...\widetilde{S}_{m_2}\left(\widetilde{S}_{m_1}(f)\right)\right)\right) - f\right\|_\infty \leq \sum_{i=1}^{r}\left\|\widetilde{S}_{m_i}(f) - f\right\|_\infty \leq \tag{14.108}$$

$$2\sum_{i=1}^{r}\omega_1\left(f, \frac{N}{m_i}\left[\frac{(A^\lambda - 1)}{(A^\lambda + 1)2} + \left(q + \frac{1}{q}\right)A^\lambda\right]\right) \leq$$

$$2r\omega_1\left(f, \frac{N}{m_1}\left[\frac{(A^\lambda - 1)}{(A^\lambda + 1)2} + \left(q + \frac{1}{q}\right)A^\lambda\right]\right) < +\infty. \tag{14.109}$$

Again, the speed of convergence of the iterated operator to I is not worse than the speed of convergence of $\widetilde{S}_{m_1}$ to I.

Proof. By (14.105), (14.106) and Proposition 14.4. □

Similarly we have:

Proposition 14.17. *Let* $f \in C_B\left(\mathbb{R}^N\right)$; $r \in \mathbb{N}$; *and* $m_1, ..., m_r \in \mathbb{N} : m_1 \leq m_2 \leq ... \leq m_r$. *Then*

1)

$$\left\{ \begin{array}{l} \left\| \widetilde{S}_n^{*r}(f) - f \right\|_\infty \\ \left\| \widehat{S_n}^r(f) - f \right\|_\infty \end{array} \right\} \leq r \left\{ \begin{array}{l} \left\| \widetilde{S}_n^{*}(f) - f \right\|_\infty \\ \left\| \widehat{S_n}(f) - f \right\|_\infty \end{array} \right\} \leq \tag{14.110}$$

$$2r\omega_1\left(f, \frac{1}{n}\left[1 + N\left(\frac{\left(A^\lambda - 1\right)}{\left(A^\lambda + 1\right)2} + \left(q + \frac{1}{q}\right)A^\lambda\right)\right]\right) < +\infty,$$

and the speed of convergence of $\widetilde{S}_n^{*r}$, $\widehat{S_n}^r$ *to* I *is not worse than the speed of convergence of* $\widetilde{S}_n^{*}$, $\widehat{S_n}$ *to* I,

2)

$$\left\{ \begin{array}{l} \left\| \widetilde{S}_{m_r}^{*}\left(\widetilde{S}_{m_{r-1}}^{*}\left(...\widetilde{S}_{m_2}^{*}\left(\widetilde{S}_{m_1}^{*}(f)\right)\right)\right) - f \right\|_\infty \\ \left\| \widehat{S_{m_r}}\left(\widehat{S_{m_{r-1}}}\left(...\widehat{S_{m_2}}\left(\widehat{S_{m_1}}(f)\right)\right)\right) - f \right\|_\infty \end{array} \right\} \leq \sum_{i=1}^r \left\{ \begin{array}{l} \left\| \widetilde{S}_{m_i}^{*}(f) - f \right\|_\infty \\ \left\| \widehat{S_{m_i}}(f) - f \right\|_\infty \end{array} \right\} \leq \tag{14.111}$$

$$2\sum_{i=1}^r \omega_1\left(f, \frac{1}{m_i}\left[1 + N\left(\frac{\left(A^\lambda - 1\right)}{\left(A^\lambda + 1\right)2} + \left(q + \frac{1}{q}\right)A^\lambda\right)\right]\right) \leq$$

$$2r\omega_1\left(f, \frac{1}{m_1}\left[1 + N\left(\frac{\left(A^\lambda - 1\right)}{\left(A^\lambda + 1\right)2} + \left(q + \frac{1}{q}\right)A^\lambda\right)\right]\right) < +\infty. \tag{14.112}$$

Again, the speed of convergence of the iterated operator to I *is not worse than the speed of convergence of* $\widetilde{S}_{m_1}^{*}$, $\widehat{S_{m_1}}$ *to* I.

Proof. By (14.105), (14.106) and Propositions 14.5, 14.6. □

We finish our work with multivariate simultaneous iterations.

Remark 14.8. Let $i \in \mathbb{N}$ be fixed. Assume that $f \in C^{(i)}\left(\mathbb{R}^N\right)$, with $f_\alpha \in C_B\left(\mathbb{R}^N\right)$, with $\alpha : |\alpha| = l$, $l = 0, 1, ..., i$; $r \in \mathbb{N}$. Then, by (14.45), we obtain

$$\left\| \widetilde{S}_n^{r}(f_\alpha) - f_\alpha \right\|_\infty \leq r \left\| \widetilde{S}_n(f_\alpha) - f_\alpha \right\|_\infty. \tag{14.113}$$

By (14.38) and inductively, we can obtain

$$\left\| \left(\widetilde{S}_n^{r}(f)\right)_\alpha - f_\alpha \right\|_\infty \leq r \left\| \left(\widetilde{S}_n(f)\right)_\alpha - f_\alpha \right\|_\infty. \tag{14.114}$$

Similarly, we can derive that

$$\left\| \left(\widetilde{S}_n^{*^r} (f) \right)_\alpha - f_\alpha \right\|_\infty \leq r \left\| \left(\widetilde{S}_n^* (f) \right)_\alpha - f_\alpha \right\|_\infty, \tag{14.115}$$

and

$$\left\| \left(\widehat{S_n}^r (f) \right)_\alpha - f_\alpha \right\|_\infty \leq r \left\| \left(\widehat{S_n} (f) \right)_\alpha - f_\alpha \right\|_\infty. \tag{14.116}$$

Let now $m_1, m_2, ..., m_r \in \mathbb{N} : m_1 \leq m_2 \leq ... \leq m_r$. Then, based on (14.47), we can find that

$$\left\| \left(\widetilde{S}_{m_r} \left(\widetilde{S}_{m_{r-1}} \left(...\widetilde{S}_{m_2} \left(\widetilde{S}_{m_1} f \right) \right) \right) \right)_\alpha - f_\alpha \right\|_\infty \leq \sum_{i^*=1}^{r} \left\| \left(\widetilde{S}_{m_{i^*}} (f) \right)_\alpha - f_\alpha \right\|_\infty. \tag{14.117}$$

Similarly, we can get that

$$\left\| \left(\widetilde{S}_{m_r}^* \left(\widetilde{S}_{m_{r-1}}^* \left(...\widetilde{S}_{m_2}^* \left(\widetilde{S}_{m_1}^* f \right) \right) \right) \right)_\alpha - f_\alpha \right\|_\infty \leq \sum_{i^*=1}^{r} \left\| \left(\widetilde{S}_{m_{i^*}}^* (f) \right)_\alpha - f_\alpha \right\|_\infty, \tag{14.118}$$

and

$$\left\| \left(\widehat{S_{m_r}} \left(\widehat{S_{m_{r-1}}} \left(...\widehat{S_{m_2}} \left(\widehat{S_{m_1}} f \right) \right) \right) \right)_\alpha - f_\alpha \right\|_\infty \leq \sum_{i^*=1}^{r} \left\| \left(\widehat{S_{m_{i^*}}} (f) \right)_\alpha - f_\alpha \right\|_\infty. \tag{14.119}$$

All the above inequalities (14.113)–(14.119) prove that our implied multivariate iterated simultaneous approximations do not have a speed worse than our basic simultaneous approximations by the activated convolution operators.

Bibliography

[314] G.A. Anastassiou, *A "K-Attainable" inequality related to the convergence of Positive Linear Operators*, J. Approximation Theory, 44, 380–383 (1985).

[315] G.A. Anastassiou, *Moments in probability and approximation theory*, Pitman Research Notes in Mathematics series / Longman Scientific & Technical, Essex, New York, UK, USA, 1993.

[316] G.A. Anastassiou, *Quantitative Approximations*, Chapmen & Hall/CRC, London, New York, 2001.

[317] G.A. Anastassiou, *Parametrized, Deformed and General Neural Networks*, Springer, Heidelberg, New York, 2023.

[318] G.A. Anastassiou, *Trigonometric and Hyperbolic Generated Approximation Theory*, World Scientific, Singapore, New York, 2025.

[319] G.A. Anastassiou, *Approximation by Parametrized logistic activated convolution type operators*, RACSAM, (2024) 118.138, https:/doi.org/10.1007/s13398-024-01635-7.

[320] G.A. Anastassiou, *Multivariate approximation by parametrized logistic activated multidimensional convolution type operators*, Annals of Communications in Mathematics, 7(2) (2024), 128–159.

[321] G.A. Anastassiou, *Multivariate Neural Networks over infinite domain as Positive Linear Operators and a touch of measure theory*, submitted, 2025.

[322] H. Cartan, *Differential Calculus*, Herman, Paris, 1971.

[323] L.B. Rall, *Computational Solution of Nonlinear Operator Equations*, John Wiley & Sons, New York, 1969.

[324] Hee Sun Jung, Ryozi Sakai, *Local saturation of a positive linear convolution operator*, J. Inequal. Appl. 2014, 2014: 329, 16 p.

[325] J.J. Swetits, B. Wood, *Local L_p-saturation of positive linear convolution operators*, J. Approx. Theory 34 (1982), no. 4, 348–360.

[326] Grigor Moldovan, *Discrete convolutions in connection with functions of several variables and positive linear operators.* (Romanian) Studia Univ. Babeş-Bolyai Ser. Math.-Mech. 19 (1974), no. 1, 51–57.

[327] R. Bojanic, O. Shisha, *On the precision of uniform approximation of continuous functions by certain linear operators of convolution type.* J. Approximation Theory 8 (1973), 101–113.

[328] Grigor Moldovan, *Discrete convolutions and linear positive operators.* Ann. Univ. Sci. Budapest. Eötvös Sect. Math. 15 (1972), 31–44 (1973).

[329] A.I. Kamzolov, *The order of approximation of functions of class $Z_2(E_n)$ by positive linear convolution operators.* (Russian) Mat. Zametki 7 (1970), 723–732.

[330] G.A. Anastassiou, *Complete Approximation by symmetrized and perturbed generalized logistic activated multidimensional convolutions in Positive Linear Framework*, RACSAM, 119:96; https://doi.org/10.1007/s13398-025-01766-5.

Chapter 15

Fractional Voronovskaya Asymptotic Expansions by Quasi-Interpolation Neural Network Operators for Brownian Motion

Here we use quasi-interpolation neural network operators of one hidden layer based on sigmoidal and hyperbolic tangent activation functions. In particular we apply fractional Voronovskaya asymptotic expansions related to the error of approximation of these operators to the unit operator. These are applied to Brownian motion over the two dimensional sphere. So we produce fractional asymptotic expansions for a general expectation of Brownian motion via neural networks. We finish with several interesting specific applications. It follows [357].

15.1 Introduction

The first author in [331] and [332], see Chapters 2–5, was the first to establish neural network approximation to continuous functions with rates by very specifically defined neural network operators of Cardaliaguet-Euvrard and "Squashing" types, by employing the modulus of continuity of the engaged function or its high order derivative, and producing very tight Jackson type inequalities. He treats there both the univariate and multivariate cases. The defining these operators "bell-shaped" and "squashing" functions are assumed to be compact support. Also the first author inspired by [339], continued his studies on neural networks approximation by introducing and using the proper quasi-interpolation operators of sigmoidal and hyperbolic tangent type which resulted into [335], by treating both the univariate and multivariate cases. He did also the corresponding fractional cases [336], [337].

Here the authors based and inspired by [343], they give Voronovskaya type fractional asymptotic expansions for a general expectation of Brownian motion over the two dimensional sphere, induced by neural networks and at the end they provide many important specialized applications. We are motivated by [338]. For general knowledge about neural networks we recommend [344]–[346]. For recent studies in neural networks we refer to [347]–[356].

15.2 Basics

We need

Definition 15.1. Let $\nu > 0$, $n = \lceil \nu \rceil$ ($\lceil \cdot \rceil$ is the ceiling of the number), $f \in AC^n([a,b])$ (space of functions f with $f^{(n-1)} \in AC([a,b])$, absolutely continuous functions). We call left Caputo fractional derivative (see [340], pp. 49–52) the function

$$D_{*a}^{\nu} f(x) = \frac{1}{\Gamma(n-\nu)} \int_a^x (x-t)^{n-\nu-1} f^{(n)}(t)\, dt, \qquad (15.1)$$

$\forall\, x \in [a,b]$, where Γ is the gamma function $\Gamma(\nu) = \int_0^\infty e^{-t} t^{\nu-1} dt$, $\nu > 0$. Notice $D_{*a}^{\nu} f \in L_1([a,b])$ and $D_{*a}^{\nu} f$ exists a.e.on $[a,b]$.

We set $D_{*a}^{0} f(x) = f(x)$, $\forall\, x \in [a,b]$.

Definition 15.2. (see also [341], [342]). Let $f \in AC^m([a,b])$, $m = \lceil \alpha \rceil$, $\alpha > 0$. The right Caputo fractional derivative of order $\alpha > 0$ is given by

$$D_{b-}^{\alpha} f(x) = \frac{(-1)^m}{\Gamma(m-\alpha)} \int_x^b (\zeta - x)^{m-\alpha-1} f^{(m)}(\zeta)\, d\zeta, \qquad (15.2)$$

$\forall\, x \in [a,b]$. We set $D_{b-}^{0} f(x) = f(x)$. Notice $D_{b-}^{\alpha} f \in L_1([a,b])$ and $D_{b-}^{\alpha} f$ exists a.e.on $[a,b]$.

Convention 15.1. We assume that

$$D_{*x_0}^{\alpha} f(x) = 0, \quad \text{for } x < x_0, \qquad (15.3)$$

and

$$D_{x_0-}^{\alpha} f(x) = 0, \text{ for } x > x_0, \qquad (15.4)$$

for all $x, x_0 \in [a,b]$.

See also the related [333], [334].

We consider here the sigmoidal function of logarithmic type

$$\eta(x) = \frac{1}{1+e^{-x}}, \quad x \in \mathbb{R}.$$

It has the properties $\lim_{x \to +\infty} \eta(x) = 1$ and $\lim_{x \to -\infty} \eta(x) = 0$.

This function plays the role of an activation function in the hidden layer of neural networks.

As in [339], we consider

$$K(x) := \frac{1}{2}(\eta(x+1) - \eta(x-1)), \quad x \in \mathbb{R}. \tag{15.5}$$

We notice the following properties:

i) $K(x) > 0$, $\forall\, x \in \mathbb{R}$,
ii) $\sum_{k=-\infty}^{\infty} K(x-k) = 1$, $\forall\, x \in \mathbb{R}$,
iii) $\sum_{k=-\infty}^{\infty} K(nx-k) = 1$, $\forall\, x \in \mathbb{R}$; $n \in \mathbb{N}$,
iv) $\int_{-\infty}^{\infty} K(x)\,dx = 1$,
v) K is a density function,
vi) K is even: $K(-x) = K(x)$, $x \geq 0$.

We see that ([339])

$$K(x) = \left(\frac{e^2-1}{2e}\right)\frac{e^{-x}}{(1+e^{-x-1})(1+e^{-x+1})} = \tag{15.6}$$

$$\left(\frac{e^2-1}{2e^2}\right)\frac{1}{(1+e^{x-1})(1+e^{-x-1})}.$$

vii) By [339] K is decreasing on $\mathbb{R}_+$, and increasing on $\mathbb{R}_-$.
viii) By [336], Ch. 5, for $n \in \mathbb{N}$, $0 < \beta < 1$, we get

$$\sum_{\substack{k=-\infty \\ : |nx-k| > n^{1-\beta}}}^{\infty} K(nx-k) < \left(\frac{e^2-1}{2}\right)e^{-n^{(1-\beta)}} = 3.192e^{-n^{(1-\beta)}}. \tag{15.7}$$

Denote by $\lfloor\cdot\rfloor$ the integral part of a number. Consider $x \in [a,b] \subset \mathbb{R}$ and $n \in \mathbb{N}$ such that $\lceil na \rceil \leq \lfloor nb \rfloor$.

ix) By [336], Ch. 5, it holds

$$\frac{1}{\sum_{k=\lceil na \rceil}^{\lfloor nb \rfloor} K(nx-k)} < \frac{1}{K(1)} = 5.250312578, \ \forall\, x \in [a,b]. \tag{15.8}$$

x) It holds $\lim_{n\to\infty} \sum_{k=\lceil na \rceil}^{\lfloor nb \rfloor} K(nx-k) \neq 1$, for at least some $x \in [a,b]$. See also [336], Ch. 5.

Let $f \in C([a,b])$ and $n \in \mathbb{N}$ such that $\lceil na \rceil \leq \lfloor nb \rfloor$.

We study further (see also [336], Ch. 5) the quasi-interpolation positive linear neural network operator

$$H_n(f,x) := \frac{\sum_{k=\lceil na \rceil}^{\lfloor nb \rfloor} f\left(\frac{k}{n}\right) K(nx-k)}{\sum_{k=\lceil na \rceil}^{\lfloor nb \rfloor} K(nx-k)}, \quad x \in [a,b]. \tag{15.9}$$

For large enough n we always obtain $\lceil na \rceil \leq \lfloor nb \rfloor$. Also $a \leq \frac{k}{n} \leq b$, iff $\lceil na \rceil \leq k \leq \lfloor nb \rfloor$.

We also consider here the hyperbolic tangent function $\tanh x$, $x \in \mathbb{R}$:

$$\tanh x := \frac{e^x - e^{-x}}{e^x + e^{-x}} = \frac{e^{2x}-1}{e^{2x}+1}.$$

It has the properties $\tanh 0 = 0$, $-1 < \tanh x < 1$, $\forall\, x \in \mathbb{R}$, and $\tanh(-x) = -\tanh x$. Furthermore $\tanh x \to 1$ as $x \to \infty$, and $\tanh x \to -1$, as $x \to -\infty$, and it is strictly increasing on $\mathbb{R}$. Furthermore it holds $\frac{d}{dx}\tanh x = \frac{1}{\cosh^2 x} > 0$.

This function plays also the role of an activation function in the hidden layer of neural networks.

We further consider

$$M(x) := \frac{1}{4}\left(\tanh(x+1) - \tanh(x-1)\right) > 0, \quad \forall\, x \in \mathbb{R}. \tag{15.10}$$

We easily see that $M(-x) = M(x)$, that is M is even on $\mathbb{R}$. Obviously M is differentiable, thus continuous.

Here we follow [335]

Proposition 15.1. *$M(x)$ for $x \geq 0$ is strictly decreasing.*

Obviously $M(x)$ is strictly increasing for $x \leq 0$. Also it holds $\lim_{x \to -\infty} M(x) = 0 = \lim_{x \to \infty} M(x)$.

In fact M has the bell shape with horizontal asymptote the x-axis. So the maximum of M is at zero, $M(0) = 0.3809297$.

Theorem 15.1. *We have that $\sum_{i=-\infty}^{\infty} M(x-i) = 1$, $\forall\, x \in \mathbb{R}$.*

Thus

$$\sum_{i=-\infty}^{\infty} M(nx-i) = 1, \qquad \forall\, n \in \mathbb{N},\ \forall\, x \in \mathbb{R}.$$

Furthermore we get:

Since M is even it holds $\sum_{i=-\infty}^{\infty} M(i-x) = 1$, $\forall x \in \mathbb{R}$.

Hence $\sum_{i=-\infty}^{\infty} M(i+x) = 1$, $\forall x \in \mathbb{R}$, and $\sum_{i=-\infty}^{\infty} M(x+i) = 1$, $\forall x \in \mathbb{R}$.

Theorem 15.2. *It holds $\int_{-\infty}^{\infty} M(x)\,dx = 1$.*

So $M(x)$ is a density function on $\mathbb{R}$.

Theorem 15.3. *Let* $0 < \beta < 1$ *and* $n \in \mathbb{N}$. *It holds*

$$\sum_{\substack{k=-\infty \\ : |nx-k| \geq n^{1-\beta}}}^{\infty} M(nx-k) \leq 2e^4 \cdot e^{-2n^{(1-\beta)}}. \tag{15.11}$$

Theorem 15.4. *Let* $x \in [a,b] \subset \mathbb{R}$ *and* $n \in \mathbb{N}$ *so that* $\lceil na \rceil \leq \lfloor nb \rfloor$. *It holds*

$$\frac{1}{\sum_{k=\lceil na \rceil}^{\lfloor nb \rfloor} M(nx-k)} < 4.1488766 = \frac{1}{M(1)}. \tag{15.12}$$

Also by [335], we obtain

$$\lim_{n\to\infty} \sum_{k=\lceil na \rceil}^{\lfloor nb \rfloor} M(nx-k) \neq 1, \tag{15.13}$$

for at least some $x \in [a,b]$.

Definition 15.3. Let $f \in C([a,b])$ and $n \in \mathbb{N}$ such that $\lceil na \rceil \leq \lfloor nb \rfloor$.

We further study, as in [335], the quasi-interpolation positive linear neural network operator

$$E_n(f,x) := \frac{\sum_{k=\lceil na \rceil}^{\lfloor nb \rfloor} f\left(\frac{k}{n}\right) M(nx-k)}{\sum_{k=\lceil na \rceil}^{\lfloor nb \rfloor} M(nx-k)}, \quad x \in [a,b]. \tag{15.14}$$

We find here fractional Voronovskaya type asymptotic expansions for $H_n(f,x)$ and $E_n(f,x)$, $x \in [a,b]$.

For related work on neural networks also see [336], [337].

We need,

Theorem 15.5. *([336], Ch. 5) Let* $\alpha > 0$, $N \in \mathbb{N}$, $N = \lceil \alpha \rceil$, $f \in AC^N([a,b])$, $0 < \beta < 1$, $x \in [a,b]$, $n \in \mathbb{N}$ *large enough. Assume that* $\left\| D_{x-}^{\alpha} f \right\|_{\infty,[a,x]}$, $\left\| D_{*x}^{\alpha} f \right\|_{\infty,[x,b]} \leq \Theta$, $\Theta > 0$. *Then*

$$H_n(f,x) - f(x) = \sum_{j=1}^{N-1} \frac{f^{(j)}(x)}{j!} H_n\left((\cdot - x)^j\right)(x) + o\left(\frac{1}{n^{\beta(\alpha-\varepsilon)}}\right), \tag{15.15}$$

where $0 < \varepsilon \leq \alpha$.

If $N=1$, the sum in (15.15) collapses.
The last (15.15) implies that

$$n^{\beta(\alpha-\varepsilon)}\left[H_n(f,x)-f(x)-\sum_{j=1}^{N-1}\frac{f^{(j)}(x)}{j!}H_n\left((\cdot-x)^j\right)(x)\right]\to 0, \quad (15.16)$$

as $n\to\infty$, $0<\varepsilon\leq\alpha$.
When $N=1$, or $f^{(j)}(x)=0$, $j=1,...,N-1$, then we derive that

$$n^{\beta(\alpha-\varepsilon)}\left[H_n(f,x)-f(x)\right]\to 0$$

as $n\to\infty$, $0<\varepsilon\leq\alpha$. Of great interest is the case of $\alpha=\frac{1}{2}$.

Next we also need,

Theorem 15.6. *([336], Ch. 5) Let $\alpha>0$, $N\in\mathbb{N}$, $N=\lceil\alpha\rceil$, $f\in AC^N([a,b])$, $0<\beta<1$, $x\in[a,b]$, $n\in\mathbb{N}$ large enough. Assume that $\left\|D_{x-}^{\alpha}f\right\|_{\infty,[a,x]}$, $\left\|D_{*x}^{\alpha}f\right\|_{\infty,[x,b]}\leq\Theta$, $\Theta>0$. Then*

$$E_n(f,x)-f(x)=\sum_{j=1}^{N-1}\frac{f^{(j)}(x)}{j!}E_n\left((\cdot-x)^j\right)(x)+o\left(\frac{1}{n^{\beta(\alpha-\varepsilon)}}\right), \quad (15.17)$$

where $0<\varepsilon\leq\alpha$.
If $N=1$, the sum in (15.17) collapses.
The last (15.17) implies that

$$n^{\beta(\alpha-\varepsilon)}\left[E_n(f,x)-f(x)-\sum_{j=1}^{N-1}\frac{f^{(j)}(x)}{j!}E_n\left((\cdot-x)^j\right)(x)\right]\to 0, \quad (15.18)$$

as $n\to\infty$, $0<\varepsilon\leq\alpha$.
When $N=1$, or $f^{(j)}(x)=0$, $j=1,...,N-1$, then we derive that

$$n^{\beta(\alpha-\varepsilon)}\left[E_n(f,x)-f(x)\right]\to 0$$

as $n\to\infty$, $0<\varepsilon\leq\alpha$. Of great interest is the case of $\alpha=\frac{1}{2}$.

15.3 About Brownian Motion on 2-Dimensional Sphere

The Brownian motion ([343]) on S^n is a diffusion (Markov) process $W_t, t\geq 0$, on S^n whose transition density is a function $P(t,x,y)$ on $(0,\infty)\times S^n\times S^n$ satisfying

$$\frac{\partial P}{\partial t}=\frac{1}{2}\Delta_n P, \quad (15.19)$$

and

$$P(t,x,y) \to \delta_x(y) \quad \text{as } t \to 0^+ \tag{15.20}$$

where Δ_n is the Laplace-Beltrami operator of S^n acting on the x-variables and $\delta_x(y)$ is the delta mass at x, i.e. $P(t,x,y)$ is the **heat kernel** of S^n. The heat kernel exists, it is unique, positive, and smooth in (t,x,y).

We mention,

Proposition 15.2. *([343]) The transition density function of the Brownian motion W_t, $t \geq 0$ on S^2 with radius a it is given by the function*

$$p(t,\varphi) = \frac{1}{4\pi a^2} \sum_{n \in \mathbb{N}} (2n+1) \exp\left(-\frac{n(n+1)\sqrt{t}}{a}\right) P_n^0(\cos\varphi) \tag{15.21}$$

Theorem 15.7. *([338]) Consider function $g : \mathbb{R} \to \mathbb{R}$, which is bounded on $[0,\pi]$, i.e. there exists $M > 0$ such that $|g(\phi)| \leq M$, for every $\phi \in [0,\pi]$, and Lebesgue measurable on $\mathbb{R}$. Let also $W(t,\phi)$ be the Brownian motion on S^2. Then the expectation*

$$E(|g(W)|)(t) = \int_0^\pi |g(\phi)|\, p(t,\phi) d\phi \tag{15.22}$$

is continuous in t, and

$$E(|g(W)|)(t) \leq \pi M p(t_o,\phi_0), \tag{15.23}$$

where

$$p(t_0,\phi_0) = \max_{(t,\phi)\in[t_1,t_2]\times[0,\pi]} p(t,\phi), \ \textit{with } 0 < t_1 < t_2.$$

Here $p(t,\phi)$ is the transition density function of the Brownian motion W_t, $t \geq 0$ on S^2 given by (15.21).

Proposition 15.3. *([338]) Consider function $g : \mathbb{R} \to \mathbb{R}$, which is bounded on $[0,\pi]$ and Lebesgue measurable on $\mathbb{R}$. Let also $W(t,\phi)$ be the Brownian motion on S^2. Then the expectation*

$$E(|g(W)|)(t) = \int_0^\pi |g(\phi)|\, p(t,\phi) d\phi$$

is differentiable in t, and

$$\frac{\partial E(|g(W)|)}{\partial t} = \int_0^\pi |g(\phi)| \frac{\partial (p(t,\phi))}{\partial t} d\phi, \tag{15.24}$$

which is continuous in t.

Remark 15.1. We observe that $p(t,\varphi) \in C^\infty(\mathbb{R}_+ - \{0\})$ with respect to $t > 0$.

Acting similarly as in Proposition 15.3, we obtain that $E(|g(W)|)^{(j)} := \frac{\partial^j E(|g(W)|)}{\partial t^j}$ exist and are continuous for $t > 0$ and any $j \in \mathbb{N}$.

15.4 Main Results

Here we present the following results about Brownian Motion by neural network operators.

Proposition 15.4. *We consider $E(|g(W)|)(t)$ as in (15.22). Let $\alpha > 0$, $N \in \mathbb{N}$, $N = \lceil \alpha \rceil$, $0 < \beta < 1$, $t \in [t_1, t_2]$, where $t_1 > 0, t_1 < t_2, n \in \mathbb{N}$ large enough. Then*

$$H_n(E(|g(W)|), t) - (E(|g(W)|)(t)) = \sum_{j=1}^{N-1} \frac{E(|g(W)|)^{(j)}(t)}{j!} H_n\left((\cdot - t)^j\right)(t) + o\left(\frac{1}{n^{\beta(\alpha-\varepsilon)}}\right), \tag{15.25}$$

where $0 < \varepsilon \leq \alpha$.

If $N = 1$, the sum in (15.25) collapses.

The last (15.25) implies that

$$n^{\beta(\alpha-\varepsilon)}\left[H_n(E(|g(W)|), t) - (E(|g(W)|)(t)) - \sum_{j=1}^{N-1} \frac{E(|g(W)|)^{(j)}(t)}{j!} H_n\left((\cdot - t)^j\right)(t)\right] \to 0, \tag{15.26}$$

as $n \to \infty$, $0 < \varepsilon \leq \alpha$.

When $N = 1$, we derive that

$$n^{\beta(\alpha-\varepsilon)}\left[H_n(E(|g(W)|), t) - (E(|g(W)|)(t))\right] \to 0$$

as $n \to \infty$, $0 < \varepsilon \leq \alpha$. Of great interest is the case of $\alpha = \frac{1}{2}$.

Proof. From Theorem 15.5. □

Next we give,

Proposition 15.5. *We consider $E(|g(W)|)(t)$ as in (15.22). Let $\alpha > 0$, $N \in \mathbb{N}$, $N = \lceil \alpha \rceil$, $t \in [t_1, t_2]$, where $t_1 > 0, t_1 < t_2, n \in \mathbb{N}$ large enough. Then*

$$E_n(E(|g(W)|), t) - (E(|g(W)|)(t)) = \sum_{j=1}^{N-1} \frac{E(|g(W)|)^{(j)}(t)}{j!} E_n\left((\cdot - t)^j\right)(t) + o\left(\frac{1}{n^{\beta(\alpha-\varepsilon)}}\right), \tag{15.27}$$

where $0 < \varepsilon \leq \alpha$.

If $N = 1$, the sum in (15.27) collapses.

The last (15.27) implies that

$$n^{\beta(\alpha-\varepsilon)}\left[E_n(E(|g(W)|), t) - (E(|g(W)|)(t)) - \sum_{j=1}^{N-1} \frac{E(|g(W)|)^{(j)}(t)}{j!} E_n\left((\cdot - t)^j\right)(t)\right] \to 0, \tag{15.28}$$

as $n \to \infty$, $0 < \varepsilon \leq \alpha$.

When $N=1$, we derive that

$$n^{\beta(\alpha-\varepsilon)}\left[E_n\left(E\left(|g(W)|\right),t\right)-\left(E\left(|g(W)|\right)(t)\right)\right]\to 0$$

as $n\to\infty$, $0<\varepsilon\leq\alpha$. Of great interest is the case of $\alpha=\frac{1}{2}$.

Proof. From Theorem 15.6. □

15.5 Applications

We can apply our main results to the function $g(W)=W$.

Consider the function $g:\mathbb{R}\to\mathbb{R}$, where $g(x)=x$ for every $x\in\mathbb{R}$. Let also $W(t,\phi)$ be the Brownian motion on S^2. Then the expectation

$$E\left(|W|\right)(t)=\int_0^{\pi}\phi p(t,\phi)d\phi$$

is continuous in t.

Moreover,

Corollary 15.1. *Let $\alpha>0$, $N\in\mathbb{N}$, $N=\lceil\alpha\rceil$, $0<\beta<1$, $t\in[t_1,t_2]$, where $t_1>0,t_1<t_2,n\in\mathbb{N}$ large enough. Then*

$$H_n\left(E\left(|W|\right),t\right)-\left(E\left(|W|\right)(t)\right)=\sum_{j=1}^{N-1}\frac{E\left(|W|\right)^{(j)}(t)}{j!}H_n\left(\left(\cdot-t\right)^j\right)(t)+o\left(\frac{1}{n^{\beta(\alpha-\varepsilon)}}\right), \tag{15.29}$$

where $0<\varepsilon\leq\alpha$.

If $N=1$, the sum in (15.29) collapses.

The last (15.29) implies that

$$n^{\beta(\alpha-\varepsilon)}\left[H_n\left(E\left(|W|\right),t\right)-\left(E\left(|W|\right)(t)\right)-\sum_{j=1}^{N-1}\frac{E\left(|W|\right)^{(j)}(t)}{j!}H_n\left(\left(\cdot-t\right)^j\right)(t)\right]\to 0, \tag{15.30}$$

as $n\to\infty$, $0<\varepsilon\leq\alpha$.

When $N=1$, we derive that

$$n^{\beta(\alpha-\varepsilon)}\left[H_n\left(E\left(|W|\right),t\right)-\left(E\left(|W|\right)(t)\right)\right]\to 0$$

as $n\to\infty$, $0<\varepsilon\leq\alpha$. Of great interest is the case of $\alpha=\frac{1}{2}$.

Proof. From Proposition 15.4. □

Next we obtain,

Corollary 15.2. *We consider $N \in \mathbb{N}$, $N = \lceil \alpha \rceil$, $t \in [t_1, t_2]$, where $t_1 > 0, t_1 < t_2, n \in \mathbb{N}$ large enough. Then*

$$E_n\left(E\left(|W|\right), t\right) - \left(E\left(|W|\right)(t)\right) = \sum_{j=1}^{N-1} \frac{E\left(|W|\right)^{(j)}(t)}{j!} E_n\left(\left(\cdot - t\right)^j\right)(t) + o\left(\frac{1}{n^{\beta(\alpha-\varepsilon)}}\right), \tag{15.31}$$

where $0 < \varepsilon \leq \alpha$.

If $N = 1$, the sum in (15.31) collapses.

The last (15.31) implies that

$$n^{\beta(\alpha-\varepsilon)} \left[E_n\left(E\left(|W|\right), t\right) - \left(E\left(|W|\right)(t)\right) - \sum_{j=1}^{N-1} \frac{E\left(|W|\right)^{(j)}(t)}{j!} E_n\left(\left(\cdot - t\right)^j\right)(t)\right] \to 0, \tag{15.32}$$

as $n \to \infty$, $0 < \varepsilon \leq \alpha$.

When $N = 1$, we derive that

$$n^{\beta(\alpha-\varepsilon)} \left[E_n\left(E\left(|W|\right), t\right) - \left(E\left(|W|\right)(t)\right)\right] \to 0$$

as $n \to \infty$, $0 < \varepsilon \leq \alpha$. Of great interest is the case of $\alpha = \frac{1}{2}$.

Proof. From Proposition 15.5. □

For the next corollaries we consider the function $g : \mathbb{R} \to \mathbb{R}$, where $g(x) = cosx$ for every $x \in \mathbb{R}$. Let also $W(t, \phi)$ be the Brownian motion on S^2. Then the expectation

$$E\left(\left|\cos\left(W\right)\right|\right)(t) = \int_0^{\pi} \left|\cos\phi\right| p(t,\phi) d\phi$$

is continuous in t.

It follows,

Corollary 15.3. *Let $\alpha > 0$, $N \in \mathbb{N}$, $N = \lceil \alpha \rceil$, $0 < \beta < 1$, $t \in [t_1, t_2]$, where $t_1 > 0, t_1 < t_2, n \in \mathbb{N}$ large enough. Then*

$$H_n\left(E\left(\left|\cos\left(W\right)\right|\right), t\right) - \left(E\left(\left|\cos\left(W\right)\right|\right)(t)\right) = \sum_{j=1}^{N-1} \frac{E\left(\left|\cos\left(W\right)\right|\right)^{(j)}(t)}{j!} H_n\left(\left(\cdot - t\right)^j\right)(t) + o\left(\frac{1}{n^{\beta(\alpha-\varepsilon)}}\right), \tag{15.33}$$

where $0 < \varepsilon \leq \alpha$.

If $N = 1$, the sum in (15.33) collapses.

The last (15.33) implies that

$$n^{\beta(\alpha-\varepsilon)}\left[H_n\left(E\left(\left|\cos\left(W\right)\right|\right),t\right)-\left(E\left(\left|\cos\left(W\right)\right|\right)\left(t\right)\right)-\sum_{j=1}^{N-1}\frac{E\left(\left|\cos\left(W\right)\right|\right)^{(j)}\left(t\right)}{j!}H_n\left(\left(\cdot-t\right)^j\right)\left(t\right)\right]\to 0, \tag{15.34}$$

as $n\to\infty$, $0<\varepsilon\leq\alpha$.

When $N=1$, *we derive that*

$$n^{\beta(\alpha-\varepsilon)}\left[H_n\left(E\left(\left|\cos\left(W\right)\right|\right),t\right)-\left(E\left(\left|\cos\left(W\right)\right|\right)\left(t\right)\right)\right]\to 0$$

as $n\to\infty$, $0<\varepsilon\leq\alpha$. *Of great interest is the case of* $\alpha=\frac{1}{2}$.

Proof. From Proposition 15.4. □

Next we obtain,

Corollary 15.4. *We consider* $N\in\mathbb{N}$, $N=\lceil\alpha\rceil$, $t\in[t_1,t_2]$, *where* $t_1>0, t_1<t_2, n\in\mathbb{N}$ *large enough. Then*

$$E_n\left(E\left(\left|\cos\left(W\right)\right|\right),t\right)-\left(E\left(\left|\cos\left(W\right)\right|\right)\left(t\right)\right)=\sum_{j=1}^{N-1}\frac{E\left(\left|\cos\left(W\right)\right|\right)^{(j)}\left(t\right)}{j!}E_n\left(\left(\cdot-t\right)^j\right)\left(t\right)+o\left(\frac{1}{n^{\beta(\alpha-\varepsilon)}}\right), \tag{15.35}$$

where $0<\varepsilon\leq\alpha$.

If $N=1$, *the sum in (15.35) collapses.*

The last (15.35) implies that

$$n^{\beta(\alpha-\varepsilon)}\left[E_n\left(E\left(\left|\cos\left(W\right)\right|\right),t\right)-\left(E\left(\left|\cos\left(W\right)\right|\right)\left(t\right)\right)-\sum_{j=1}^{N-1}\frac{E\left(\left|\cos\left(W\right)\right|\right)^{(j)}\left(t\right)}{j!}E_n\left(\left(\cdot-t\right)^j\right)\left(t\right)\right]\to 0, \tag{15.36}$$

as $n\to\infty$, $0<\varepsilon\leq\alpha$.

When $N=1$, *we derive that*

$$n^{\beta(\alpha-\varepsilon)}\left[E_n\left(E\left(\left|\cos\left(W\right)\right|\right),t\right)-\left(E\left(\left|\cos\left(W\right)\right|\right)\left(t\right)\right)\right]\to 0$$

as $n\to\infty$, $0<\varepsilon\leq\alpha$. *Of great interest is the case of* $\alpha=\frac{1}{2}$.

Proof. From Proposition 15.5. □

Note 15.1. Similar results can be obtain for $g:\mathbb{R}\to\mathbb{R}$, where $g(x)=\sin x, x\in\mathbb{R}$.

Let the function $g : \mathbb{R} \to \mathbb{R}$, where $g(x) = \tanh x$ for every $x \in \mathbb{R}$. Let also $W(t, \phi)$ be the Brownian motion on S^2. Then the expectation

$$E(|\tanh(W)|)(t) = \int_0^{\pi} |\tanh(\phi)|\, p(t, \phi)\, d\phi$$

is continuous in t.

Furthermore,

Corollary 15.5. *Let $\alpha > 0$, $N \in \mathbb{N}$, $N = \lceil \alpha \rceil$, $0 < \beta < 1$, $t \in [t_1, t_2]$, where $t_1 > 0, t_1 < t_2, n \in \mathbb{N}$ large enough. Then*

$$H_n(E(|\tanh(W)|), t) - (E(|\tanh(W)|)(t)) = \sum_{j=1}^{N-1} \frac{E(|\tanh(W)|)^{(j)}(t)}{j!} H_n\left((\cdot - t)^j\right)(t) + o\left(\frac{1}{n^{\beta(\alpha-\varepsilon)}}\right), \tag{15.37}$$

where $0 < \varepsilon \leq \alpha$.

If $N = 1$, the sum in (15.37) collapses.

The last (15.37) implies that

$$n^{\beta(\alpha-\varepsilon)} \left[H_n(E(|\tanh(W)|), t) - (E(|\tanh(W)|)(t)) - \sum_{j=1}^{N-1} \frac{E(|\tanh(W)|)^{(j)}(t)}{j!} H_n\left((\cdot - t)^j\right)(t) \right] \to 0, \tag{15.38}$$

as $n \to \infty$, $0 < \varepsilon \leq \alpha$.

When $N = 1$, we derive that

$$n^{\beta(\alpha-\varepsilon)} \left[H_n(E(|\tanh(W)|), t) - (E(|\tanh(W)|)(t)) \right] \to 0$$

as $n \to \infty$, $0 < \varepsilon \leq \alpha$. Of great interest is the case of $\alpha = \frac{1}{2}$.

Proof. From Proposition 15.4. □

Next we give,

Corollary 15.6. *We consider $N \in \mathbb{N}$, $N = \lceil \alpha \rceil$, $t \in [t_1, t_2]$, where $t_1 > 0, t_1 < t_2, n \in \mathbb{N}$ large enough. Then*

$$E_n(E(|\tanh(W)|), t) - (E(|\tanh(W)|)(t)) = \sum_{j=1}^{N-1} \frac{E(|\tanh(W)|)^{(j)}(t)}{j!} E_n\left((\cdot - t)^j\right)(t) + o\left(\frac{1}{n^{\beta(\alpha-\varepsilon)}}\right), \tag{15.39}$$

where $0 < \varepsilon \leq \alpha$.

If $N = 1$, the sum in (15.39) collapses.

The last (15.39) implies that

$$n^{\beta(\alpha-\varepsilon)}\left[E_n\left(E\left(|\tanh(W)|\right),t\right)-\left(E\left(|\tanh(W)|\right)(t)\right)-\sum_{j=1}^{N-1}\frac{E\left(|\tanh(W)|\right)^{(j)}(t)}{j!}E_n\left((\cdot-t)^j\right)(t)\right]\to 0, \tag{15.40}$$

as $n\to\infty$, $0<\varepsilon\leq\alpha$.

When $N=1$, *we derive that*

$$n^{\beta(\alpha-\varepsilon)}\left[E_n\left(E\left(|\tanh(W)|\right),t\right)-\left(E\left(|\tanh(W)|\right)(t)\right)\right]\to 0$$

as $n\to\infty$, $0<\varepsilon\leq\alpha$. *Of great interest is the case of* $\alpha=\frac{1}{2}$.

Proof. From Proposition 15.5. □

For the next application we consider the generalized logistic sigmoid function $g:\mathbb{R}\to\mathbb{R}$, where $g(x)=\left(1+e^{-x}\right)^{-\delta}$, and $\delta>0$, for every $x\in\mathbb{R}$. Let also $W(t,\phi)$ be the Brownian motion on S^2. Then the expectation

$$E\left(\left(1+e^{-W}\right)^{-\delta}\right)(t)=\int_0^{\pi}\left(1+e^{-\phi}\right)^{-\delta}p(t,\phi)d\phi$$

is continuous in t.

It follows,

Corollary 15.7. *Let* $\alpha>0$, $N\in\mathbb{N}$, $N=\lceil\alpha\rceil$, $0<\beta<1$, $t\in[t_1,t_2]$, *where* $t_1>0, t_1<t_2, n\in\mathbb{N}$ *large enough. Then*

$$H_n\left(E\left(\left(1+e^{-W}\right)^{-\delta}\right),t\right)-\left(E\left(\left(1+e^{-W}\right)^{-\delta}\right)(t)\right)=$$

$$\sum_{j=1}^{N-1}\frac{E\left(\left(1+e^{-W}\right)^{-\delta}\right)^{(j)}(t)}{j!}H_n\left((\cdot-t)^j\right)(t)+o\left(\frac{1}{n^{\beta(\alpha-\varepsilon)}}\right), \tag{15.41}$$

where $0<\varepsilon\leq\alpha$.

If $N=1$, *the sum in (15.41) collapses.*

The last (15.41) implies that

$$n^{\beta(\alpha-\varepsilon)}\cdot\left[H_n\left(E\left(\left(1+e^{-W}\right)^{-\delta}\right),t\right)-\left(E\left(\left(1+e^{-W}\right)^{-\delta}\right)(t)\right)-\right.$$

$$\left.\sum_{j=1}^{N-1}\frac{E\left(\left(1+e^{-W}\right)^{-\delta}\right)^{(j)}(t)}{j!}H_n\left((\cdot-t)^j\right)(t)\right]\to 0, \tag{15.42}$$

as $n\to\infty$, $0<\varepsilon\leq\alpha$.

When $N=1$, *we derive that*

$$n^{\beta(\alpha-\varepsilon)}\left[H_n\left(E\left(\left(1+e^{-W}\right)^{-\delta}\right),t\right)-\left(E\left(\left(1+e^{-W}\right)^{-\delta}\right)(t)\right)\right]\to 0$$

as $n\to\infty$, $0<\varepsilon\leq\alpha$. *Of great interest is the case of* $\alpha=\frac{1}{2}$.

Proof. From Proposition 15.4. □

Next we derive,

Corollary 15.8. *We consider $N \in \mathbb{N}$, $N = \lceil \alpha \rceil$, $t \in [t_1, t_2]$, where $t_1 > 0, t_1 < t_2, n \in \mathbb{N}$ large enough. Then*

$$E_n\left(E\left(\left(1+e^{-W}\right)^{-\delta}\right), t\right) - \left(E\left(\left(1+e^{-W}\right)^{-\delta}\right)(t)\right) =$$

$$\sum_{j=1}^{N-1} \frac{E\left(\left(1+e^{-W}\right)^{-\delta}\right)^{(j)}(t)}{j!} E_n\left((\cdot - t)^j\right)(t) + o\left(\frac{1}{n^{\beta(\alpha-\varepsilon)}}\right), \quad (15.43)$$

where $0 < \varepsilon \leq \alpha$.

If $N = 1$, the sum in (15.43) collapses.

The last (15.43) implies that

$$n^{\beta(\alpha-\varepsilon)} \cdot \left[E_n\left(E\left(\left(1+e^{-W}\right)^{-\delta}\right), t\right) - \left(E\left(\left(1+e^{-W}\right)^{-\delta}\right)(t)\right) - \right.$$

$$\left. \sum_{j=1}^{N-1} \frac{E\left(\left(1+e^{-W}\right)^{-\delta}\right)^{(j)}(t)}{j!} E_n\left((\cdot - t)^j\right)(t)\right] \to 0, \quad (15.44)$$

as $n \to \infty$, $0 < \varepsilon \leq \alpha$.

When $N = 1$, we derive that

$$n^{\beta(\alpha-\varepsilon)} \left[E_n\left(E\left(\left(1+e^{-W}\right)^{-\delta}\right), t\right) - \left(E\left(\left(1+e^{-W}\right)^{-\delta}\right)(t)\right)\right] \to 0$$

as $n \to \infty$, $0 < \varepsilon \leq \alpha$. Of great interest is the case of $\alpha = \frac{1}{2}$.

Proof. From Proposition 15.5. □

The Gompertz function $g : \mathbb{R} \to \mathbb{R}$, with $g(x) = e^{\mu e^{-x}}, \mu < 0$ is a sigmoid function which describes growth as being slowest at the start and end of a given time period. Let $W(t, \phi)$ be the Brownian motion on S^2. Then the expectation

$$E\left(e^{\mu e^{-W}}\right)(t) = \int_0^{\pi} e^{\mu e^{-\phi}} p(t, \phi) d\phi$$

is continuous in t.

Moreover,

Corollary 15.9. *Let $\alpha > 0$, $N \in \mathbb{N}$, $N = \lceil \alpha \rceil$, $0 < \beta < 1$, $t \in [t_1, t_2]$, where $t_1 > 0, t_1 < t_2, n \in \mathbb{N}$ large enough. Then*

$$H_n\left(E\left(e^{\mu e^{-W}}\right), t\right) - \left(E\left(e^{\mu e^{-W}}\right)(t)\right) = \sum_{j=1}^{N-1} \frac{E\left(e^{\mu e^{-W}}\right)^{(j)}(t)}{j!} H_n\left((\cdot - t)^j\right)(t) + o\left(\frac{1}{n^{\beta(\alpha-\varepsilon)}}\right), \tag{15.45}$$

where $0 < \varepsilon \leq \alpha$.

If $N = 1$, the sum in (15.45) collapses.

The last (15.45) implies that

$$n^{\beta(\alpha-\varepsilon)}\left[H_n\left(E\left(e^{\mu e^{-W}}\right), t\right) - \left(E\left(e^{\mu e^{-W}}\right)(t)\right) - \sum_{j=1}^{N-1} \frac{E\left(e^{\mu e^{-W}}\right)^{(j)}(t)}{j!} H_n\left((\cdot - t)^j\right)(t)\right] \to 0, \tag{15.46}$$

as $n \to \infty$, $0 < \varepsilon \leq \alpha$.

When $N = 1$, we derive that

$$n^{\beta(\alpha-\varepsilon)}\left[H_n\left(E\left(e^{\mu e^{-W}}\right), t\right) - \left(E\left(e^{\mu e^{-W}}\right)(t)\right)\right] \to 0$$

as $n \to \infty$, $0 < \varepsilon \leq \alpha$. Of great interest is the case of $\alpha = \frac{1}{2}$.

Proof. From Proposition 15.4. □

Next we give,

Corollary 15.10. *We consider $N \in \mathbb{N}$, $N = \lceil \alpha \rceil$, $t \in [t_1, t_2]$, where $t_1 > 0, t_1 < t_2, n \in \mathbb{N}$ large enough. Then*

$$E_n\left(E\left(e^{\mu e^{-W}}\right), t\right) - \left(E\left(e^{\mu e^{-W}}\right)(t)\right) = \sum_{j=1}^{N-1} \frac{E\left(e^{\mu e^{-W}}\right)^{(j)}(t)}{j!} E_n\left((\cdot - t)^j\right)(t) + o\left(\frac{1}{n^{\beta(\alpha-\varepsilon)}}\right), \tag{15.47}$$

where $0 < \varepsilon \leq \alpha$.

If $N = 1$, the sum in (15.47) collapses.

The last (15.47) implies that

$$n^{\beta(\alpha-\varepsilon)}\left[E_n\left(E\left(e^{\mu e^{-W}}\right), t\right) - \left(E\left(e^{\mu e^{-W}}\right)(t)\right) - \sum_{j=1}^{N-1} \frac{E\left(e^{\mu e^{-W}}\right)^{(j)}(t)}{j!} E_n\left((\cdot - t)^j\right)(t)\right] \to 0, \tag{15.48}$$

as $n \to \infty$, $0 < \varepsilon \leq \alpha$.

When $N = 1$, we derive that

$$n^{\beta(\alpha-\varepsilon)}\left[E_n\left(E\left(e^{\mu e^{-W}}\right),t\right)-\left(E\left(e^{\mu e^{-W}}\right)(t)\right)\right]\to 0$$

as $n \to \infty$, $0 < \varepsilon \leq \alpha$. Of great interest is the case of $\alpha = \frac{1}{2}$.

Proof. From Proposition 15.5. □

Bibliography

[331] G.A. Anastassiou, *Rate of convergence of some neural network operators to the unit-univariate case*, J. Math. Anal. Appli. 212 (1997), 237–262.

[332] G.A. Anastassiou, *Quantitative Approximations*, Chapman & Hall/CRC, Boca Raton, New York, 2001.

[333] G.A. Anastassiou, *On Right Fractional Calculus*, Chaos, solitons and fractals, 42 (2009), 365–376.

[334] G.A. Anastassiou, *Fractional korovkin theory*, Chaos, Solitons and Fractals, Vol. 42, No. 4 (2009), 2080–2094.

[335] G.A. Anastassiou, *Inteligent Systems: Approximation by Artificial Neural Networks*, Intelligent Systems Reference Library, Vol. 19, Springer, Heidelberg, 2011.

[336] G.A. Anastassiou, *Inteligent Systems II: Complete Approximation by Neural Network Operators*, Springer, Heidelberg, New York, 2016.

[337] G.A. Anastassiou, *Nonlinearity: Ordinary and Fractional Approximations by Sublinear and Max-Product*, Springer, Heidelberg, New York, 2018.

[338] G.A. Anastassiou, D. Kouloumpou, *Brownian Motion Approximation by Neural Networks, Communications in Optimization Theory*, 38, (2023), 1–20.

[339] Z. Chen and F. Cao, *The approximation operators with sigmoidal functions*, Computers and Mathematics with Applications, 58 (2009), 758–765.

[340] K. Diethelm, *The Analysis of Fractional Differential Equations*, Lecture Notes in Mathematics 2004, Springer-Verlag, Berlin, Heidelberg, 2010.

[341] A.M.A. El-Sayed and M. Gaber, *On the finite Caputo and finite Riesz derivatives*, Electronic Journal of Theoretical Physics, Vol. 3, No. 12 (2006), 81–95.

[342] G.S. Frederico and D.F.M. Torres, *Fractional Optimal Control in the sense of Caputo and the fractional Noether's theorem*, International Mathematical Forum, Vol. 3, No. 10 (2008), 479–493.

[343] D. Kouloumpou, V.G. Papanicolaou, *Certain Calculation Regarding the Brownian Motion on the Sphere*, Journal of Concrete and Applicable Mathematics, 11(2013), (3-4), 303–316.

[344] S. Haykin, *Neural Networks: A Comprehensive Foundation* (2 ed.), Prentice Hall, New York, 1998.

[345] W. McCulloch and W. Pitts, *A logical calculus of the ideas immanent in nervous activity*, Bulletin of Mathematical Biophysics, 7 (1943), 115–133.

[346] T.M. Mitchell, *Machine Learning*, WCB-McGraw-Hill, New York, 1997.

[347] Yu, Dansheng; Cao, Feilong, *Construction and approximation rate for feed-forward neural network operators with sigmoidal functions*, J. Comput. Appl. Math. 453 (2025), paper No. 116150.

[348] Cen, Siyu; Jin, Bangti; Quan, Qimeng; Zhou, Zhi; *Hybrid neural-network FEM approximation of diffusion coeficient in elyptic and parabolic problems*, IMA J. Numer. Anal. 44 (2024), no. 5, 3059–3093.

[349] Coroianu Lucian; Costarelli, Danillo; Natale, Mariarosaria; Pantiş, Alexandra; *The approximation capabilities of Durrmeyer-type neural network operators*, J. Appl. Math. Comput. 70 (2024), no. 5, 4581–4599.

[350] Warin, Xavier; *The GroupMax neural network approximation of convex functions*, IEEE Trans. Neural Netw. Learn. Syst. 35 (2024), no. 8, 11608–11612.

[351] Fabra, Arnau; Guasch, Oriol; Baiges, Joan; Codina, Ramon; *Approximation of acoustic black holes with finite element mixed formulations and artificial neural network correction terms*, Finite Elem. Anal. Des. 241 (2024), paper No. 104236.

[352] Grohs, Philipp; Voigtlaender, Felix. *Proof of the theory-to-practice gap in deep learning via sampling complexity bounds for neural network approximation spaces*, Found. Comput. Math. 24 (2024), no. 4, 1085–1143.

[353] Basteri, Andrea; Trevisan, Dario; *Quantitative Gaussian approximation of randomly initialized deep neural networks*, Mach. Learn. 113 (2024), no. 9, 6373–6393.

[354] De Ryck, T.; Mishra, S. *Error analysis for deep neural network approximations of parametric hyperbolic conservation laws*, Math. Comp. 93 (2024), no. 350, 2643–2677.

[355] Liu, Jie; Zhang, Baoji; Lai, Yuyang; Fang, Liqiau. *Hull form optimization reserach based on multi-precision back-propagation neural network approximation model*, Internal. J. Numer. Methods Fluid 96 (2024), no. 8, 1445–1460.

[356] Yoo, Jihahm; Kim, Jaywon; Gim, Minjung; Lee, Haesung. *Error estimates of physics-informed neural networks for initial value problems.* J. Korean Soc. Ind. Appl. Math. 28 (2024), no. 1, 33–58.

[357] G.A. Anastassiou, D. Kouloumpou, *Fractional Voronovskaya Asymptotic Expansions by Quasi-Interpolation Neural Network Operators applied to Brownian Motion*, Analele Universitatii Oradea, Fasc. Matematics, accepted, 2025.

Chapter 16

Neural Network Approximation for Finite Dimension Banach Spaces

The functions under approximation here have as a domain a finite dimensional Banach space with dimension $N \in \mathbb{N}$ and are with values in $\mathbb{R}^N$. Exploiting some topological properties of the above we are able to perform neural network multivariate approximation to the above functions. The treatment is quantitative. We produce multivariate Jackson type inequalities involving the modulus of continuity of the function under approximation. The established convergences are pointwise and uniform. Perturbation and symmetrization to our operators lead to accelerated speeds of convergence. It follows [372].

16.1 Introduction

The author in [358] and [359], see Chapters 2–5 was the first to establish neural network approximation to continuous functions with rates by very specifically defined neural network operators of Cardaliaguet-Euvrard and "Squashing" types, by employing the modulus of continuity of the engaged function or its high order derivative, and producing very tight Jackson type inequalities. He treats there both the univariate and multivariate cases. The defining these operators "bell-shaped" and "squashing" functions are assumed to be of compact support.

Again the author inspired by [366], continued his studies on neural network approximation by introducing and using the proper quasi-interpolation operators of sigmoidal and hyperbolic tangent type which resulted into [360], by treating both the univariate and multivariate cases.

The author continued this trend of approximation and published the extensive monographs [361]–[365], studying in depth and as wide as possible the neural network approximations. Our activation function here is the perturbed hyperbolic tangent function.

The multivariate operators we use are: the normalized, the quasi-interpolation, the Kantorovich and the quadrature types. Our symmetrization technique accelerates immensely the speed of convergences. Our neural networks here are from a Banach space of dimension N to $\mathbb{R}^N$, as they are homeomorphic, and this is the main key to this work.

So all results here are via multivariate Jackson type inequalities, studying quantitatively the rate of convergences to the unit operator.

Here we are dealing with one hidden layer feed-forward neural networks.

A multilayer feed-forward neural network can be defined as follows (with $m \in \mathbb{N}$ hidden layers):

Let $x \in \mathbb{R}^s$; $s \in \mathbb{N}$, where $x = (x_1, ..., x_s)$; $\alpha_j, c_j \in \mathbb{R}^s$; $b_j \in \mathbb{R}$, with $0 \leq j \leq n$, $n \in \mathbb{N}$.

Here $\langle \alpha_j \cdot x \rangle$ is the inner product, thus $\sigma(\langle \alpha_j \cdot x \rangle + b_j) \in \mathbb{R}$; and $N_n(x) \in \mathbb{R}^s$, by $c_j \in \mathbb{R}^s$, as it is coming from $N_n(x) = \sum_{j=0}^{n} c_j \sigma(\langle \alpha_j \cdot x \rangle + b_j)$.

We define:

$$N_n^{(2)}(x) = \sum_{j=0}^{n} c_j \sigma(\langle \alpha_j \cdot N_n(x) \rangle + b_j) =$$

$$\sum_{j=0}^{n} c_j \sigma\left(\left\langle \alpha_j \cdot \left(\sum_{j=0}^{n} c_j \sigma(\langle \alpha_j \cdot x \rangle + b_j)\right)\right\rangle + b_j\right).$$

Furthermore, we can define

$$N_n^{(3)}(x) = \sum_{j=0}^{n} c_j \sigma\left(\left\langle \alpha_j \cdot N_n^{(2)}(x) \right\rangle + b_j\right).$$

And, in general we define:

$$N_n^{(m)}(x) = \sum_{j=0}^{n} c_j \sigma\left(\left\langle \alpha_j \cdot N_n^{(m-1)}(x) \right\rangle + b_j\right), \quad \text{for } m \in \mathbb{N}.$$

For more studies in neural networks read [367]–[371].

16.2 Basics

Initially we follow [364], pp. 455–460.

Our perturbed hyperbolic tangent activation function here to be used is

$$g_{q,\lambda}(x) := \frac{e^{\lambda x} - qe^{-\lambda x}}{e^{\lambda x} + qe^{-\lambda x}}, \quad \lambda, q > 0, \ x \in \mathbb{R}. \tag{16.1}$$

Above λ is the parameter and q is the deformation coefficient, typically it is $0 < \lambda, q \leq 1$.

For more read Chapter 18 of [364]: "q-Deformed and λ-Parametrized Hyperbolic Tangent based Banach space Valued Ordinary and Fractional Neural Network Approximation".

The chapters 17 and 18 of [364] motivate our current work.

The proposed "symmetrization method" aims to use half data feed to our multivariate neural networks.

We will employ the following density function

$$M_{q.\lambda}(x) := \frac{1}{4}\left(g_{q,\lambda}(x+1) - g_{q,\lambda}(x-1)\right) > 0, \tag{16.2}$$

$\forall x \in \mathbb{R}$; $q, \lambda > 0$.

We have that

$$M_{q,\lambda}(-x) = M_{\frac{1}{q},\lambda}(x), \ \ \forall\, x \in \mathbb{R}; \ q, \lambda > 0, \tag{16.3}$$

and

$$M_{\frac{1}{q},\lambda}(-x) = M_{q,\lambda}(x), \ \ \forall\, x \in \mathbb{R}; \ q, \lambda > 0. \tag{16.4}$$

Adding (16.3) and (16.4) we obtain

$$M_{q,\lambda}(-x) + M_{\frac{1}{q},\lambda}(-x) = M_{q,\lambda}(x) + M_{\frac{1}{q},\lambda}(x), \tag{16.5}$$

a key to this work.

So that

$$\Phi(x) := \frac{M_{q,\lambda}(x) + M_{\frac{1}{q},\lambda}(x)}{2} \tag{16.6}$$

is an even function, symmetric with respect to the y-axis.

By (18.18) of [364], we have

$$\begin{gathered} M_{q,\lambda}\left(\frac{\ln q}{2\lambda}\right) = \frac{\tanh(\lambda)}{2}, \\ \text{and} \\ M_{\frac{1}{q},\lambda}\left(-\frac{\ln q}{2\lambda}\right) = \frac{\tanh(\lambda)}{2}, \ \ \lambda > 0, \end{gathered} \tag{16.7}$$

sharing the same maximum at symmetric points.

By Theorem 18.1, p. 458 of [364], we have that

$$\begin{array}{l} \sum\limits_{i=-\infty}^{\infty} M_{q,\lambda}(x-i)=1,\ \ \forall\, x\in\mathbb{R},\ \lambda,q>0, \\ \text{and} \\ \sum\limits_{i=-\infty}^{\infty} M_{\frac{1}{q},\lambda}(x-i)=1,\ \ \forall\, x\in\mathbb{R},\ \lambda,q>0. \end{array} \tag{16.8}$$

Consequently, we derive that

$$\sum_{i=-\infty}^{\infty}\Phi(x-i)=1,\ \ \forall\, x\in\mathbb{R}. \tag{16.9}$$

By Theorem 18.2, p. 459 of [364], we have that

$$\begin{array}{l} \int_{-\infty}^{\infty} M_{q,\lambda}(x)\,dx=1, \\ \text{and} \\ \int_{-\infty}^{\infty} M_{\frac{1}{q},\lambda}(x)\,dx=1, \end{array} \tag{16.10}$$

so that

$$\int_{-\infty}^{\infty}\Phi(x)\,dx=1, \tag{16.11}$$

therefore Φ is a density function.

By Theorem 18.3, p. 459 of [364], we have:

Let $0<\alpha<1$, and $n\in\mathbb{N}$ with $n^{1-\alpha}>2$; $q,\lambda>0$. Then

$$\sum_{\begin{cases} k=-\infty \\ :|nx-k|\geq n^{1-\alpha}\end{cases}}^{\infty} M_{q,\lambda}(nx-k)<2\max\left\{q,\frac{1}{q}\right\}e^{4\lambda}e^{-2\lambda n^{(1-\alpha)}}=Te^{-2\lambda n^{(1-\alpha)}}, \tag{16.12}$$

where $T:=2\max\left\{q,\frac{1}{q}\right\}e^{4\lambda}$.

Similarly, we get that

$$\sum_{k=-\infty}^{\infty} M_{\frac{1}{q},\lambda}(nx-k)<Te^{-2\lambda n^{(1-\alpha)}}. \tag{16.13}$$

Consequently we obtain that

$$\sum_{\begin{cases} k=-\infty \\ :|nx-k|\geq n^{1-\alpha}\end{cases}}^{\infty} \Phi(nx-k)<Te^{-2\lambda n^{(1-\alpha)}}, \tag{16.14}$$

where $T:=2\max\left\{q,\frac{1}{q}\right\}e^{4\lambda}$.

Here $\lceil\cdot\rceil$ denotes the ceiling of the number, and $\lfloor\cdot\rfloor$ its integral part.

We mention

Theorem 18.4 (p. 459, [364]) Let $x \in [a,b] \subset \mathbb{R}$ and $n \in \mathbb{N}$ so that $\lceil na\rceil \leq \lfloor nb\rfloor$. For $q, \lambda > 0$, we consider $\lambda_q > z_0 > 0$, such that $M_{q,\lambda}(z_0) = M_{q,\lambda}(0)$, and $\lambda_q > 1$. Then

$$\frac{1}{\sum_{k=\lceil na\rceil}^{\lfloor nb\rfloor} M_{q,\lambda}(nx-k)} < \max\left\{\frac{1}{M_{q,\lambda}(\lambda_q)}, \frac{1}{M_{\frac{1}{q},\lambda}\left(\lambda_{\frac{1}{q}}\right)}\right\} =: \Delta(q). \tag{16.15}$$

Similarly, we consider $\lambda_{\frac{1}{q}} > z_1 > 0$, such that $M_{\frac{1}{q},\lambda}(z_1) = M_{\frac{1}{q},\lambda}(0)$, and $\lambda_{\frac{1}{q}} > 1$. Thus

$$\frac{1}{\sum_{k=\lceil na\rceil}^{\lfloor nb\rfloor} M_{\frac{1}{q},\lambda}(nx-k)} < \max\left\{\frac{1}{M_{\frac{1}{q},\lambda}\left(\lambda_{\frac{1}{q}}\right)}, \frac{1}{M_{q,\lambda}(\lambda_q)}\right\} = \Delta(q). \tag{16.16}$$

Hence

$$\sum_{k=\lceil na\rceil}^{\lfloor nb\rfloor} M_{q,\lambda}(nx-k) > \frac{1}{\Delta(q)}, \tag{16.17}$$

and

$$\sum_{k=\lceil na\rceil}^{\lfloor nb\rfloor} M_{\frac{1}{q},\lambda}(nx-k) > \frac{1}{\Delta(q)}. \tag{16.18}$$

Consequently it holds

$$\sum_{k=\lceil na\rceil}^{\lfloor nb\rfloor} \frac{\left(M_{q,\lambda}(nx-k) + M_{\frac{1}{q},\lambda}(nx-k)\right)}{2} > \frac{2}{2\Delta(q)} = \frac{1}{\Delta(q)}, \tag{16.19}$$

so that

$$\frac{1}{\sum_{k=\lceil na\rceil}^{\lfloor nb\rfloor} \frac{\left(M_{q,\lambda}(nx-k) + M_{\frac{1}{q},\lambda}(nx-k)\right)}{2}} < \Delta(q), \tag{16.20}$$

that is

$$\frac{1}{\sum_{k=\lceil na\rceil}^{\lfloor nb\rfloor} \Phi(nx-k)} < \Delta(q). \tag{16.21}$$

We have proved

Theorem 16.1. *Let* $x \in [a,b] \subset \mathbb{R}$ *and* $n \in \mathbb{N}$ *so that* $\lceil na \rceil \leq \lfloor nb \rfloor$. *For* $q, \lambda > 0$, *we consider* $\lambda_q > z_0 > 0$, *such that* $M_{q,\lambda}(z_0) = M_{q,\lambda}(0)$, *and* $\lambda_q > 1$. *Also consider* $\lambda_{\frac{1}{q}} > z_1 > 0$, *such that* $M_{\frac{1}{q},\lambda}(z_1) = M_{\frac{1}{q},\lambda}(0)$, *and* $\lambda_{\frac{1}{q}} > 1$. *Then*

$$\frac{1}{\sum_{k=\lceil na \rceil}^{\lfloor nb \rfloor} \Phi(nx-k)} < \Delta(q). \tag{16.22}$$

We make

Remark 16.1.

I) By Remark 18.5, p. 460 of [364], we have

$$\lim_{n \to \infty} \sum_{k=\lceil na \rceil}^{\lfloor nb \rfloor} M_{q,\lambda}(nx_1 - k) \neq 1, \text{ for some } x_1 \in [a,b], \tag{16.23}$$

and

$$\lim_{n \to \infty} \sum_{k=\lceil na \rceil}^{\lfloor nb \rfloor} M_{\frac{1}{q},\lambda}(nx_2 - k) \neq 1, \text{ for some } x_2 \in [a,b]. \tag{16.24}$$

Therefore it holds

$$\lim_{n \to \infty} \sum_{k=\lceil na \rceil}^{\lfloor nb \rfloor} \frac{\left(M_{q,\lambda}(nx_1 - k) + M_{\frac{1}{q},\lambda}(nx_2 - k)\right)}{2} \neq 1. \tag{16.25}$$

Hence

$$\lim_{n \to \infty} \sum_{k=\lceil na \rceil}^{\lfloor nb \rfloor} \frac{\left(M_{q,\lambda}(nx_1 - k) + M_{\frac{1}{q},\lambda}(nx_1 - k)\right)}{2} \neq 1, \tag{16.26}$$

even if

$$\lim_{n \to \infty} \sum_{k=\lceil na \rceil}^{\lfloor nb \rfloor} M_{\frac{1}{q},\lambda}(nx_1 - k) = 1, \tag{16.27}$$

because then

$$\lim_{n \to \infty} \sum_{k=\lceil na \rceil}^{\lfloor nb \rfloor} \frac{M_{q,\lambda}(nx_1 - k)}{2} + \frac{1}{2} \neq 1, \tag{16.28}$$

equivalently

$$\lim_{n\to\infty}\sum_{k=\lceil na\rceil}^{\lfloor nb\rfloor}\frac{M_{q,\lambda}\left(nx_1-k\right)}{2}\neq\frac{1}{2}, \tag{16.29}$$

true by

$$\lim_{n\to\infty}\sum_{k=\lceil na\rceil}^{\lfloor nb\rfloor}M_{q,\lambda}\left(nx_1-k\right)\neq 1. \tag{16.30}$$

II) Let $[a,b]\subset\mathbb{R}$. For large n we always have $\lceil na\rceil\leq\lfloor nb\rfloor$. Also $a\leq\frac{k}{n}\leq b$, iff $\lceil na\rceil\leq k\leq\lfloor nb\rfloor$. So in general it holds

$$\sum_{k=\lceil na\rceil}^{\lfloor nb\rfloor}\Phi\left(nx-k\right)\leq 1. \tag{16.31}$$

Next, we move on to the multivariate case, see Chapter 17 of [364], pp. 419–452, as a model of action.

We make

Remark 16.2. We introduce

$$Z_1\left(x_1,...,x_N\right):=Z_1\left(x\right):=\prod_{i=1}^{N}\Phi\left(x_i\right)=\frac{1}{2^N}\prod_{i=1}^{N}\left(M_{q,\lambda}+M_{\frac{1}{q},\lambda}\right)\left(x_i\right), \tag{16.32}$$

$x=\left(x_1,...,x_N\right)\in\mathbb{R}^N$, $\lambda,q>0$, $N\in\mathbb{N}$.

Properties:

(i)

$$Z_1\left(x\right)>0,\ \forall\ x\in\mathbb{R}^N, \tag{16.33}$$

(ii)

$$\sum_{k=-\infty}^{\infty}Z_1\left(x-k\right):=\sum_{k_1=-\infty}^{\infty}\sum_{k_2=-\infty}^{\infty}\cdots\sum_{k_N=-\infty}^{\infty}Z_1\left(x_1-k_1,...,x_N-k_N\right)=1, \tag{16.34}$$

$k:=\left(k_1,...,k_n\right)\in\mathbb{Z}^N$, $\forall\ x\in\mathbb{R}^N$,

hence

(iii)

$$\sum_{k=-\infty}^{\infty}Z_1\left(nx-k\right)=1, \tag{16.35}$$

$\forall\ x\in\mathbb{R}^N$, $n\in\mathbb{N}$,

(iv)

$$\int_{\mathbb{R}^N} Z_1(x)\,dx = 1, \tag{16.36}$$

that is Z_1 is a multivariate density function.

Here denote $\|x\|_\infty := \max\{|x_1|,...,|x_N|\}$, $x \in \mathbb{R}^N$, $\infty := (\infty,...,\infty)$, $-\infty := (-\infty,...,-\infty)$, $\lceil na \rceil := (\lceil na_1 \rceil,...,\lceil na_N \rceil)$, $\lfloor nb \rfloor := (\lfloor nb_1 \rfloor,...,\lfloor nb_N \rfloor)$, $a := (a_1,...,a_N)$, $b := (b_1,...,b_N)$.

We obviously see that

$$\sum_{k=\lceil na \rceil}^{\lfloor nb \rfloor} Z_1(nx-k) = \sum_{k=\lceil na \rceil}^{\lfloor nb \rfloor} \left(\prod_{i=1}^{N} \Phi(nx_i - k_i) \right) = \prod_{i=1}^{N} \left(\sum_{k_i=\lceil na_i \rceil}^{\lfloor nb_i \rfloor} \Phi(nx_i - k_i) \right). \tag{16.37}$$

(v) We derive that

$$\sum_{\substack{k=\lceil na \rceil \\ \left\| \frac{k}{n} - x \right\|_\infty > \frac{1}{n^\beta}}}^{\lfloor nb \rfloor} Z_1(nx-k) < Te^{-2\lambda n^{(1-\beta)}}, \quad \text{where } 0 < \beta < 1, \tag{16.38}$$

with $n \in \mathbb{N} : n^{1-\beta} > 2$, $x \in \prod_{i=1}^{N} [a_i, b_i]$.

(vi) It holds

$$0 < \frac{1}{\sum_{k=\lceil na \rceil}^{\lfloor nb \rfloor} Z_1(nx-k)} < (\Delta(q))^N, \tag{16.39}$$

$\forall\, x \in \left(\prod_{i=1}^{N} [a_i, b_i] \right)$, $n \in \mathbb{N}$.

It is clear that

(vii)

$$\sum_{\substack{k=-\infty \\ \left\| \frac{k}{n} - x \right\|_\infty > \frac{1}{n^\beta}}}^{\infty} Z_1(nx-k) < Te^{-2\lambda n^{(1-\beta)}}, \tag{16.40}$$

where $0 < \beta < 1$, $n \in \mathbb{N} : n^{1-\beta} > 2$, $x \in \mathbb{R}^N$.

Furthermore it holds

$$\lim_{n \to \infty} \sum_{k=\lceil na \rceil}^{\lfloor nb \rfloor} Z_1(nx-k) \neq 1, \tag{16.41}$$

for at least some $x \in \left(\prod_{i=1}^{N} [a_i, b_i] \right)$.

Here $\left(X, \|\cdot\|_{\gamma}\right)$ is a Banach space.

Let $f \in C\left(\prod_{i=1}^{N}[a_i, b_i], X\right)$, $x = (x_1, ..., x_N) \in \prod_{i=1}^{N}[a_i, b_i]$, $n \in \mathbb{N}$: $\lceil na_i \rceil \leq \lfloor nb_i \rfloor$, $i = 1, ..., N$.

We introduce and define the following multivariate linear normalized symmetrized neural network operator, let $x := (x_1, ..., x_N) \in \prod_{i=1}^{N}[a_i, b_i]$:

$$A_n^s(f, x_1, ..., x_N) := A_n^s(f, x) := \frac{\sum_{k=\lceil na \rceil}^{\lfloor nb \rfloor} f\left(\frac{k}{n}\right) Z_1(nx - k)}{\sum_{k=\lceil na \rceil}^{\lfloor nb \rfloor} Z_1(nx - k)} =$$

$$\frac{\sum_{k_1=\lceil na_1 \rceil}^{\lfloor nb_1 \rfloor} \sum_{k_2=\lceil na_2 \rceil}^{\lfloor nb_2 \rfloor} \cdots \sum_{k_N=\lceil na_N \rceil}^{\lfloor nb_N \rfloor} f\left(\frac{k_1}{n}, ..., \frac{k_N}{n}\right)\left(\prod_{i=1}^{N}\left(M_{q,\lambda}(nx_i - k_i) + M_{\frac{1}{q},\lambda}(nx_i - k_i)\right)\right)}{\prod_{i=1}^{N}\left(\sum_{k_i=\lceil na_i \rceil}^{\lfloor nb_i \rfloor}\left(M_{q,\lambda}(nx_i - k_i) + M_{\frac{1}{q},\lambda}(nx_i - k_i)\right)\right)}. \tag{16.42}$$

For large enough $n \in \mathbb{N}$ we always obtain $\lceil na_i \rceil \leq \lfloor nb_i \rfloor$, $i = 1, ..., N$. Also $a_i \leq \frac{k_i}{n} \leq b_i$, iff $\lceil na_i \rceil \leq k_i \leq \lfloor nb_i \rfloor$, $i = 1, ..., N$.

When $h \in C\left(\prod_{i=1}^{N}[a_i, b_i]\right)$ we define the companion operator

$$\widetilde{A}_n^s(h, x) := \frac{\sum_{k=\lceil na \rceil}^{\lfloor nb \rfloor} h\left(\frac{k}{n}\right) Z_1(nx - k)}{\sum_{k=\lceil na \rceil}^{\lfloor nb \rfloor} Z_1(nx - k)}. \tag{16.43}$$

Clearly $\widetilde{A}_n^s$ is a positive linear operator. We have that

$$\widetilde{A}_n^s(1, x) = 1, \ \ \forall\, x \in \prod_{i=1}^{N}[a_i, b_i].$$

Notice that $A_n^s(f) \in C\left(\prod_{i=1}^{N}[a_i, b_i], X\right)$ and $\widetilde{A}_n^s(h) \in C\left(\prod_{i=1}^{N}[a_i, b_i]\right)$.

Furthermore it holds

$$\|A_n^s(f, x)\|_{\gamma} \leq \widetilde{A}_n^s\left(\|f\|_{\gamma}, x\right), \tag{16.44}$$

$\forall\, x \in \prod_{i=1}^{N}[a_i, b_i]$,

and

$$A_n^s(ch, x) = c\widetilde{A}_n^s(h, x), \ \ \forall\, x \in \prod_{i=1}^{N}[a_i, b_i], \tag{16.45}$$

and

$$A_n^s(c) = c, \ \text{ any } c \in X. \tag{16.46}$$

When $f \in C_B\left(\mathbb{R}^N, X\right)$ or $f \in C_U\left(\mathbb{R}^N, X\right)$, we define

$$B_n^s(f, x) := B_n^s(f, x_1, ..., x_N) := \sum_{k=-\infty}^{\infty} f\left(\frac{k}{n}\right) Z_1(nx - k) :=$$

$$\frac{1}{2^N}\sum_{k_1=-\infty}^{\infty}\sum_{k_2=-\infty}^{\infty}\cdots\sum_{k_N=-\infty}^{\infty} f\left(\frac{k_1}{n},...,\frac{k_N}{n}\right)$$

$$\left(\prod_{i=1}^{N}\left(M_{q,\lambda}\left(nx_i-k_i\right)+M_{\frac{1}{q},\lambda}\left(nx_i-k_i\right)\right)\right), \tag{16.47}$$

$n\in\mathbb{N}$, $\forall\, x\in\mathbb{R}^N$, $N\in\mathbb{N}$, the multivariate quasi-interpolation symmetrized neural network.

Also for $f\in C_B\left(\mathbb{R}^N,X\right)$ or $f\in C_U\left(\mathbb{R}^N,X\right)$, we define the multivariate Kantorovich type symmetrized neural network operator

$$C_n^s\left(f,x\right):=C_n^s\left(f,x_1,...,x_N\right):=\sum_{k=-\infty}^{\infty}\left(n^N\int_{\frac{k}{n}}^{\frac{k+1}{n}}f\left(t\right)dt\right)Z_1\left(nx-k\right)=$$

$$\frac{1}{2^N}\sum_{k_1=-\infty}^{\infty}\sum_{k_2=-\infty}^{\infty}\cdots\sum_{k_N=-\infty}^{\infty}\left(n^N\int_{\frac{k_1}{n}}^{\frac{k_1+1}{n}}\cdots\int_{\frac{k_N}{n}}^{\frac{k_N+1}{n}}f\left(t_1,...,t_N\right)dt_1...dt_N\right)$$

$$\left(\prod_{i=1}^{N}\left(M_{q,\lambda}\left(nx_i-k_i\right)+M_{\frac{1}{q},\lambda}\left(nx_i-k_i\right)\right)\right), \tag{16.48}$$

$n\in\mathbb{N}$, $\forall\, x\in\mathbb{R}^N$.

Again for $f\in C_B\left(\mathbb{R}^N,X\right)$ or $f\in C_U\left(\mathbb{R}^N,X\right)$, we define the multivariate symmetrized neural network operator of quadrature type $D_n^s\left(f,x\right)$, $n\in\mathbb{N}$, as follows:

Let $\theta:=\left(\theta_1,...,\theta_N\right)\in\mathbb{N}^N$, $r=\left(r_1,...,r_N\right)\in\mathbb{Z}_+^n$, $w_r=w_{r_1,r_2,...,r_N}\geq 0$, such that

$$\sum_{r=0}^{\theta}w_r=\sum_{r_1=0}^{\theta_1}\sum_{r_2=0}^{\theta_2}\cdots\sum_{r_N=0}^{\theta_N}w_{r_1,r_2,...,r_N}=1;$$

$k\in\mathbb{Z}^N$, and

$$\delta_{nk}\left(f\right):=\delta_{n,k_1,...,k_N}\left(f\right):=\sum_{r=0}^{\theta}w_r f\left(\frac{k}{n}+\frac{r}{n\theta}\right)=$$

$$\sum_{r_1=0}^{\theta_1}\sum_{r_2=0}^{\theta_2}\cdots\sum_{r_N=0}^{\theta_N}w_{r_1,r_2,...,r_N}f\left(\frac{k_1}{n}+\frac{r_1}{n\theta_1},...,\frac{k_n}{n}+\frac{r_N}{n\theta_N}\right), \tag{16.49}$$

where $\frac{r}{\theta}:=\left(\frac{r_1}{\theta_1},...,\frac{r_N}{\theta_N}\right)$.

We set

$$D_n^s(f,x) := D_n^s(f,x_1,...,x_N) := \sum_{k=-\infty}^{\infty} \delta_{nk}(f) Z_1(nx-k) =$$

$$\frac{1}{2^N} \sum_{k_1=-\infty}^{\infty} \sum_{k_2=-\infty}^{\infty} \cdots \sum_{k_N=-\infty}^{\infty} \delta_{n,k_1,\ldots,k_N}(f)$$

$$\left(\prod_{i=1}^{N} \left(M_{q,\lambda}(nx_i-k_i) + M_{\frac{1}{q},\lambda}(nx_i-k_i) \right) \right), \qquad (16.50)$$

$\forall\, x \in \mathbb{R}^N$.

Definition 16.1. ([362], p. 274) Let M be a convex and compact subset of $\left(\mathbb{R}^N, \|\cdot\|_p\right)$, $p \in [1,\infty]$, and $\left(X, \|\cdot\|_\gamma\right)$ be a Banach space. Let $f \in C(M,X)$. We define the first modulus of continuity of f as

$$\omega_1(f,\delta) := \sup_{\substack{x,y \in M: \\ \|x-y\|_p \le \delta}} \|f(x)-f(y)\|_\gamma, \quad 0<\delta \le diam(M). \qquad (16.51)$$

If $\delta > diam(M)$, then

$$\omega_1(f,\delta) = \omega_1(f, diam(M)). \qquad (16.52)$$

Notice $\omega_1(f,\delta)$ is increasing in $\delta>0$. For $f \in C_B(M,X)$ (continuous and bounded functions) $\omega_1(f,\delta)$ is defined similarly.

Lemma 16.1. *([362], p. 274) We have $\omega_1(f,\delta) \to 0$ as $\delta \downarrow 0$, iff $f \in C(M,X)$, where M is a convex compact subset of $\left(\mathbb{R}^N, \|\cdot\|_p\right)$, $p \in [1,\infty]$.*

In this study we work only for the case of $p=\infty$.

Clearly we have also: $f \in C_U\left(\mathbb{R}^N, X\right)$ (uniformly continuous functions), iff $\omega_1(f,\delta) \to 0$ as $\delta \downarrow 0$, where ω_1 is defined similarly to (16.51). The space $C_B\left(\mathbb{R}^N, X\right)$ denotes the continuous and bounded functions on $\mathbb{R}^N$, $\|\cdot\|_\infty$ is the supremum norm.

In this work we treat the case of $\left(X, \|\cdot\|_\gamma\right) = \left(\mathbb{R}^N, \|\cdot\|_2\right)$, $N \in \mathbb{N}$; where $\|\cdot\|_2$ is the Euclidean norm.

Next, we describe the main ideas of this work.

Remark 16.3. Let $(E, \|\cdot\|_1)$ be a Banach space of finite dimension N, i.e. $\dim E = N$, $N \in \mathbb{N}$.

It is known that any two Banach spaces of the same finite dimension are linearly homeomorphic.

Hence $(E, \|\cdot\|_1)$ is linearly homeomorphic to $(\mathbb{R}^N, \|\cdot\|_2)$: so let $\psi : (E, \|\cdot\|_1) \to (\mathbb{R}^N, \|\cdot\|_2)$ be the linear homeomorphism.

So let $y \in E$, then $\psi(y) = x \in \mathbb{R}^N$, and back to $y = \psi^{-1}(x) \in E$, that is $\psi^{-1} : \mathbb{R}^N \to E$ is $(1-1)$ and onto map, and both ψ, ψ^{-1} are continuous maps.

Let now $f : E \to \mathbb{R}^N$ be a continuous map, then $f \circ \psi^{-1}$ is a continuous map from $\mathbb{R}^N$ into $\mathbb{R}^N$. We call $g(x) := f(\psi^{-1}(x))$, $\forall\, x \in \mathbb{R}^N$.

If $\|f(y)\|_2 \leq M$, $\forall\, y \in E$, so is $\|f(\psi^{-1}(x))\|_2 \leq M$, $\forall\, x \in \mathbb{R}^N$, where $M > 0$.

That is $\|\|f(y)\|_2\|_\infty < \infty$, implies $\|\|f \circ \psi^{-1}\|_2\|_\infty < \infty$.

In case f is a uniformly continuous map, since (see below) ψ, ψ^{-1} are also uniformly continuous, we obtain that $g = f \circ \psi^{-1}$ is a uniformly continuous map.

If $\prod_{i=1}^{N} [a_i, b_i] \subset \mathbb{R}^N$, we can consider $g|_{\prod_{i=1}^{N}[a_i, b_i]}$, which is continuous on a bounded set, that is a uniformly continuous map.

It is known that every linear function on a finite dimensional normed vector space is uniformly continuous, based on the fact that all norms there are equivalent. Thus, ψ, ψ^{-1} are uniformly continuous.

Let $f : E \to \mathbb{R}^N$ be continuous.

We define the following neural network linear operators ${}_iL_n(f) : E \to \mathbb{R}^N$, $i = 1, 2, 3, 4$; $y \in E$, as follows:

I)

$$
{}_1L_n(f)(y) = {}_1L_n(f)\left(\psi^{-1}(x)\right) := A_n^s\left(f \circ \psi^{-1}, x\right) =
$$

$$
\frac{\sum_{k=\lceil na \rceil}^{\lfloor nb \rfloor} \left(f \circ \psi^{-1}\right)\left(\frac{k}{n}\right) Z_1(nx - k)}{\sum_{k=\lceil na \rceil}^{\lfloor nb \rfloor} Z_1(nx - k)} = \tag{16.53}
$$

$$
\frac{\sum_{k_1=\lceil na_1 \rceil}^{\lfloor nb_1 \rfloor} \cdots \sum_{k_N=\lceil na_N \rceil}^{\lfloor nb_N \rfloor} \left(f \circ \psi^{-1}\right)\left(\frac{k_1}{n}, \ldots, \frac{k_N}{n}\right)\left(\prod_{i=1}^{N}\left(M_{q,\lambda}(nx_i - k_i) + M_{\frac{1}{q},\lambda}(nx_i - k_i)\right)\right)}{\prod_{i=1}^{N}\left(\sum_{k_i=\lceil na_i \rceil}^{\lfloor nb_i \rfloor}\left(M_{q,\lambda}(nx_i - k_i) + M_{\frac{1}{q},\lambda}(nx_i - k_i)\right)\right)},
$$

$\forall\, x := (x_1, \ldots, x_N) \in \prod_{i=1}^{N} [a_i, b_i] \subset \mathbb{R}^N$, $\forall\, y \in \psi^{-1}\left(\prod_{i=1}^{N} [a_i, b_i]\right) \subset E$, $n \in \mathbb{N}$.

II) When $f \in C_B\left(E, \mathbb{R}^N\right)$ or $f \in C_U\left(E, \mathbb{R}^N\right)$ we define:

$$
{}_2L_n(f)(y) = {}_2L_n(f)\left(\psi^{-1}(x)\right) := B_n^s\left(f \circ \psi^{-1}, x\right) =
$$

$$\sum_{k=-\infty}^{\infty}\left(f\circ\psi^{-1}\right)\left(\frac{k}{n}\right)Z_1\left(nx-k\right)= \tag{16.54}$$

$$\frac{1}{2^N}\sum_{k_1=-\infty}^{\infty}\sum_{k_2=-\infty}^{\infty}\cdots\sum_{k_N=-\infty}^{\infty}\left(f\circ\psi^{-1}\right)\left(\frac{k_1}{n},...,\frac{k_N}{n}\right)$$

$$\left(\prod_{i=1}^{N}\left(M_{q,\lambda}\left(nx_i-k_i\right)+M_{\frac{1}{q},\lambda}\left(nx_i-k_i\right)\right)\right),$$

$\forall\ x\in\mathbb{R}^N$, $\forall\ y\in E$, $n\in\mathbb{N}$.

III) When $f\in C_B\left(E,\mathbb{R}^N\right)$ or $f\in C_U\left(E,\mathbb{R}^N\right)$ we define:

$${}_3L_n\left(f\right)\left(y\right)=\ {}_3L_n\left(f\right)\left(\psi^{-1}\left(x\right)\right):=C_n^s\left(f\circ\psi^{-1},x\right)=$$

$$\sum_{k=-\infty}^{\infty}\left(n^N\int_{\frac{k}{n}}^{\frac{k+1}{n}}\left(f\circ\psi^{-1}\right)\left(t\right)dt\right)Z_1\left(nx-k\right)=$$

$$\frac{1}{2^N}\sum_{k_1=-\infty}^{\infty}\cdots\sum_{k_N=-\infty}^{\infty}\left(n^N\int_{\frac{k_1}{n}}^{\frac{k_1+1}{n}}\cdots\int_{\frac{k_N}{n}}^{\frac{k_N+1}{n}}\left(f\circ\psi^{-1}\right)\left(t_1,...,t_N\right)dt_1...dt_N\right)$$

$$\left(\prod_{i=1}^{N}\left(M_{q,\lambda}\left(nx_i-k_i\right)+M_{\frac{1}{q},\lambda}\left(nx_i-k_i\right)\right)\right), \tag{16.55}$$

$\forall\ x\in\mathbb{R}^N$, $\forall\ y\in E$, $n\in\mathbb{N}$.

IV) Again, when $f\in C_B\left(E,\mathbb{R}^N\right)$ or $f\in C_U\left(E,\mathbb{R}^N\right)$ we define:

$${}_4L_n\left(f\right)\left(y\right)=\ {}_4L_n\left(f\right)\left(\psi^{-1}\left(x\right)\right):=D_n^s\left(f\circ\psi^{-1},x\right)=$$

$$\sum_{k=-\infty}^{\infty}\delta_{nk}\left(f\circ\psi^{-1}\right)Z_1\left(nx-k\right)=$$

$$\frac{1}{2^N}\sum_{k_1=-\infty}^{\infty}\cdots\sum_{k_N=-\infty}^{\infty}\delta_{n,k_1,...,k_N}\left(f\circ\psi^{-1}\right)$$

$$\left(\prod_{i=1}^{N}\left(M_{q,\lambda}\left(nx_i-k_i\right)+M_{\frac{1}{q},\lambda}\left(nx_i-k_i\right)\right)\right), \tag{16.56}$$

$\forall\ x\in\mathbb{R}^N$, $\forall\ y\in E$, $n\in\mathbb{N}$.

We want to study quantitatively the multivariate approximation of ${}_iL_n\left(f\right)\left(y\right)\to f\left(y\right)$, as $n\to\infty$, for $i=1,2,3,4$; $y\in E$.

16.3 Main Results

We present our first approximation result.

Theorem 16.2. *Let* $(E, \|\cdot\|_1)$ *be a Banach space such that* $\dim E = N \in \mathbb{N}$*, and* ψ *be the corresponding homeomorphism from* E *onto* $(\mathbb{R}^N, \|\cdot\|_2)$*. Here* $f \in C(E, \mathbb{R}^N)$*,* ${}_1L_n(f)$ *is as in (16.53),* $x \in \prod_{i=1}^{N}[a_i, b_i] \subset \mathbb{R}^N$*, and* $y \in \psi^{-1}\left(\prod_{i=1}^{N}[a_i, b_i]\right) \subset E$*;* $0 < \beta < 1$*,* $n \in \mathbb{N}$ *with* $n^{1-\beta} > 2$*. Then*

(1)

$$\|{}_1L_n(f)(y) - f(y)\|_2 \leq (\Delta(q))^N$$

$$\left[\omega_1\left(f \circ \psi^{-1}, \frac{1}{n^\beta}\right) + 2\left\|\|f \circ \psi^{-1}\|_2\right\|_\infty T e^{-2\lambda n^{(1-\beta)}}\right] =: \rho_1(n), \quad (16.57)$$

and

(2)

$$\left\|\|{}_1L_n(f) - f\|_2\right\|_\infty \leq \rho_1(n). \quad (16.58)$$

We see that $\lim_{n\to\infty} {}_1L_n(f) \overset{\|\cdot\|_2}{=} f$*, pointwise and uniformly.*

Above ω_1 *is with respect to* $p = \infty$*.*

Proof. We have that

$$\|{}_1L_n(f)(y) - f(y)\|_2 =$$

$$\left\|{}_1L_n(f)\left(\psi^{-1}(x)\right) - f\left(\psi^{-1}(x)\right)\right\|_2 =$$

$$\left\|A_n^s\left(f \circ \psi^{-1}, x\right) - f\left(\psi^{-1}(x)\right)\right\|_2 = \quad (16.59)$$

$$\left\|\frac{\sum_{k=\lceil na\rceil}^{\lfloor nb\rfloor}\left(f \circ \psi^{-1}\right)\left(\frac{k}{n}\right) Z_1(nx-k)}{\sum_{k=\lceil na\rceil}^{\lfloor nb\rfloor} Z_1(nx-k)} - \frac{f\left(\psi^{-1}(x)\right)\sum_{k=\lceil na\rceil}^{\lfloor nb\rfloor} Z_1(nx-k)}{\sum_{k=\lceil na\rceil}^{\lfloor nb\rfloor} Z_1(nx-k)}\right\|_2$$

$$\overset{\text{(by (16.15), (16.39))}}{\leq} (\Delta(q))^N$$

$$\left\|\sum_{k=\lceil na\rceil}^{\lfloor nb\rfloor}\left(f \circ \psi^{-1}\right)\left(\frac{k}{n}\right) Z_1(nx-k) - f\left(\psi^{-1}(x)\right)\sum_{k=\lceil na\rceil}^{\lfloor nb\rfloor} Z_1(nx-k)\right\|_2 =$$

$$(\Delta(q))^N \left\| \sum_{k=\lceil na \rceil}^{\lfloor nb \rfloor} \left[\left(f \circ \psi^{-1}\right)\left(\frac{k}{n}\right) - \left(f \circ \psi^{-1}\right)(x) \right] Z_1(nx-k) \right\|_2 \leq$$

$$(\Delta(q))^N \left[\sum_{k=\lceil na \rceil}^{\lfloor nb \rfloor} \left\| \left(f \circ \psi^{-1}\right)\left(\frac{k}{n}\right) - \left(f \circ \psi^{-1}\right)(x) \right\|_2 Z_1(nx-k) \right] =$$

$$(\Delta(q))^N \left[\sum_{\substack{k = \lceil na \rceil \\ : \left\| \frac{k}{n} - x \right\|_\infty \leq \frac{1}{n^\beta}}}^{\lfloor nb \rfloor} \left\| \left(f \circ \psi^{-1}\right)\left(\frac{k}{n}\right) - \left(f \circ \psi^{-1}\right)(x) \right\|_2 Z_1(nx-k) \right.$$

$$\left. + \sum_{\substack{k = \lceil na \rceil \\ : \left\| \frac{k}{n} - x \right\|_\infty > \frac{1}{n^\beta}}}^{\lfloor nb \rfloor} \left\| \left(f \circ \psi^{-1}\right)\left(\frac{k}{n}\right) - \left(f \circ \psi^{-1}\right)(x) \right\|_2 Z_1(nx-k) \right] \overset{(16.35)}{\leq}$$

$$(\Delta(q))^N \left[\omega_1\left(f \circ \psi^{-1}, \frac{1}{n^\beta}\right) + 2 \left\| \left\| f \circ \psi^{-1} \right\|_2 \right\|_\infty \left(\sum_{\substack{k = \lceil na \rceil \\ : \left\| \frac{k}{n} - x \right\|_\infty > \frac{1}{n^\beta}}}^{\lfloor nb \rfloor} Z_1(nx-k) \right) \right] \overset{(16.38)}{\leq} \tag{16.60}$$

$$(\Delta(q))^N \left[\omega_1\left(f \circ \psi^{-1}, \frac{1}{n^\beta}\right) + 2 \left\| \left\| f \circ \psi^{-1} \right\|_2 \right\| T e^{-2\lambda n^{(1-\beta)}} \right].$$

□

It follows the next result all over E.

Theorem 16.3. *Let $(E, \|\cdot\|_1)$ be a Banach space with $\dim E = N \in \mathbb{N}$, and ψ be the corresponding homeomorphism from E onto $\left(\mathbb{R}^N, \|\cdot\|_2\right)$. Here $f \in C_B\left(E, \mathbb{R}^N\right)$, ${}_2L_n(f)$ is as in (16.54), $x \in \mathbb{R}^N$, and $y \in E$; $0 < \beta < 1$, $n \in \mathbb{N}$ with $n^{1-\beta} > 2$, ω_1 is for $p = \infty$. Then*

(1)

$$\left\| {}_2L_n(f)(y) - f(y) \right\|_2 \leq$$

$$\omega_1\left(f\circ\psi^{-1},\frac{1}{n^{\beta}}\right)+2\left\|\left\|f\circ\psi^{-1}\right\|_2\right\|_{\infty}Te^{-2\lambda n^{(1-\beta)}}=:\rho_2(n), \qquad (16.61)$$

and

(2)

$$\left\|\left\|{}_2L_n(f)-f\right\|_2\right\|_{\infty}\leq\rho_2(n). \qquad (16.62)$$

Given that $f\in\left(C_U\left(E,\mathbb{R}^N\right)\cap C_B\left(E,\mathbb{R}^N\right)\right)$, *we obtain* $\lim_{n\to\infty}{}_2L_n(f)=f$, *uniformly.*

Proof. We have that

$$\left\|{}_2L_n(f)(y)-f(y)\right\|_2=$$

$$\left\|{}_2L_n(f)\left(\psi^{-1}(x)\right)-f\left(\psi^{-1}(x)\right)\right\|_2=$$

$$\left\|B_n^s\left(f\circ\psi^{-1},x\right)-f\left(\psi^{-1}(x)\right)\right\|_2\overset{(\text{by }(16.35),\ (16.54))}{=}$$

$$\left\|\sum_{k=-\infty}^{\infty}\left(f\circ\psi^{-1}\right)\left(\frac{k}{n}\right)Z_1(nx-k)-f\left(\psi^{-1}(x)\right)\sum_{k=-\infty}^{\infty}Z_1(nx-k)\right\|_2=$$

$$\left\|\sum_{k=-\infty}^{\infty}\left(\left(f\circ\psi^{-1}\right)\left(\frac{k}{n}\right)-\left(f\circ\psi^{-1}\right)(x)\right)Z_1(nx-k)\right\|_2\leq$$

$$\sum_{k=-\infty}^{\infty}\left\|\left(f\circ\psi^{-1}\right)\left(\frac{k}{n}\right)-\left(f\circ\psi^{-1}\right)(x)\right\|_2Z_1(nx-k)= \qquad (16.63)$$

$$\sum_{\begin{cases}k=-\infty\\ :\left\|\frac{k}{n}-x\right\|_{\infty}\leq\frac{1}{n^{\beta}}\end{cases}}^{\infty}\left\|\left(f\circ\psi^{-1}\right)\left(\frac{k}{n}\right)-\left(f\circ\psi^{-1}\right)(x)\right\|_2Z_1(nx-k)$$

$$\left.+\sum_{\begin{cases}k=-\infty\\ :\left\|\frac{k}{n}-x\right\|_{\infty}>\frac{1}{n^{\beta}}\end{cases}}^{\infty}\left\|\left(f\circ\psi^{-1}\right)\left(\frac{k}{n}\right)-\left(f\circ\psi^{-1}\right)(x)\right\|_2Z_1(nx-k)\right]\overset{(16.35)}{\leq}$$

$$\omega_1\left(f \circ \psi^{-1}, \frac{1}{n^\beta}\right) + 2 \left\| \left\| f \circ \psi^{-1} \right\|_2 \right\|_\infty \left(\sum_{\substack{k=-\infty \\ \left\{ : \left\| \frac{k}{n} - x \right\|_\infty > \frac{1}{n^\beta} \right.}}^{\infty} Z_1 (nx - k) \right) \overset{(16.40)}{\leq} \tag{16.64}$$

$$\omega_1\left(f \circ \psi^{-1}, \frac{1}{n^\beta}\right) + 2 \left\| \left\| f \circ \psi^{-1} \right\|_2 \right\|_\infty T e^{-2\lambda n^{(1-\beta)}},$$

proving the claim. □

It follows the result for the Kantorovich type operator on E.

Theorem 16.4. *Let $(E, \|\cdot\|_1)$ be a Banach space with $\dim E = N \in \mathbb{N}$, and ψ be the corresponding homeomorphism from E onto $(\mathbb{R}^N, \|\cdot\|_2)$. Here $f \in C_B(E, \mathbb{R}^N)$, ${}_3L_n(f)$ as in (16.55), $x \in \mathbb{R}^N$ and $y \in E$; $0 < \beta < 1$, $n \in \mathbb{N}$ with $n^{1-\beta} > 2$, ω_1 is for $p = \infty$. Then*

(1)

$$\left\| {}_3L_n(f)(y) - f(y) \right\|_2 \leq$$

$$\omega_1\left(f \circ \psi^{-1}, \frac{1}{n} + \frac{1}{n^\beta}\right) + 2 \left\| \left\| f \circ \psi^{-1} \right\|_2 \right\|_\infty T e^{-2\lambda n^{(1-\beta)}} =: \rho_3(n), \tag{16.65}$$

and

(2)

$$\left\| \left\| {}_3L_n(f) - f \right\|_2 \right\|_\infty \leq \rho_3(n). \tag{16.66}$$

Given that $f \in \left(C_U(E, \mathbb{R}^N) \cap C_B(E, \mathbb{R}^N)\right)$, we obtain $\lim_{n \to \infty} {}_3L_n(f) = f$, uniformly.

Proof. We observe that

$$\int_{\frac{k}{n}}^{\frac{k+1}{n}} \left(f \circ \psi^{-1}\right)(t)\, dt =$$

$$\int_{\frac{k_1}{n}}^{\frac{k_1+1}{n}} \dots \int_{\frac{k_N}{n}}^{\frac{k_N+1}{n}} \left(f \circ \psi^{-1}\right)(t_1, \dots, t_N)\, dt_1 \dots dt_N =$$

$$\int_0^{\frac{1}{n}} \int_0^{\frac{1}{n}} \dots \int_0^{\frac{1}{n}} \left(f \circ \psi^{-1}\right)\left(t_1 + \frac{k_1}{n}, t_2 + \frac{k_2}{n}, \dots, t_N + \frac{k_N}{n}\right) dt_1 \dots dt_N =$$

$$\int_0^{\frac{1}{n}} \left(f \circ \psi^{-1}\right) \left(t + \frac{k}{n}\right) dt. \tag{16.67}$$

Therefore we can write

$$\|{}_3L_n(f)(y) - f(y)\|_2 =$$

$$\left\|{}_3L_n(f)\left(\psi^{-1}(x)\right) - f\left(\psi^{-1}(x)\right)\right\|_2 =$$

$$\left\|C_n^s\left(f \circ \psi^{-1}, x\right) - f\left(\psi^{-1}(x)\right)\right\|_2 \overset{(16.35)}{=}$$

$$\left\|\sum_{k=-\infty}^{\infty}\left(n^N \int_0^{\frac{1}{n}} \left(f \circ \psi^{-1}\right)\left(t + \frac{k}{n}\right) dt\right) Z_1(nx-k) - f\left(\psi^{-1}(x)\right)\left(\sum_{k=-\infty}^{\infty} Z_1(nx-k)\right)\right\|_2 =$$

$$\left\|\sum_{k=-\infty}^{\infty}\left(\left(n^N \int_0^{\frac{1}{n}} \left(f \circ \psi^{-1}\right)\left(t + \frac{k}{n}\right) dt\right) - \left(f \circ \psi^{-1}\right)(x)\right) Z_1(nx-k)\right\|_2 = \tag{16.68}$$

$$\left\|\sum_{k=-\infty}^{\infty}\left(n^N \int_0^{\frac{1}{n}} \left(\left(f \circ \psi^{-1}\right)\left(t + \frac{k}{n}\right) - \left(f \circ \psi^{-1}\right)(x)\right) dt\right) Z_1(nx-k)\right\|_2 \leq$$

$$\sum_{k=-\infty}^{\infty}\left(n^N \int_0^{\frac{1}{n}} \left\|\left(f \circ \psi^{-1}\right)\left(t + \frac{k}{n}\right) - \left(f \circ \psi^{-1}\right)(x)\right\|_2 dt\right) Z_1(nx-k) =$$

$$\sum_{\substack{k=-\infty \\ : \left\|\frac{k}{n} - x\right\|_\infty \leq \frac{1}{n^\beta}}}^{\infty} \left(n^N \int_0^{\frac{1}{n}} \left\|\left(f \circ \psi^{-1}\right)\left(t + \frac{k}{n}\right) - \left(f \circ \psi^{-1}\right)(x)\right\|_2 dt\right) Z_1(nx-k)$$

$$+ \sum_{\substack{k=-\infty \\ : \left\|\frac{k}{n} - x\right\|_\infty > \frac{1}{n^\beta}}}^{\infty} \left(n^N \int_0^{\frac{1}{n}} \left\|\left(f \circ \psi^{-1}\right)\left(t + \frac{k}{n}\right) - \left(f \circ \psi^{-1}\right)(x)\right\|_2 dt\right) Z_1(nx-k) \tag{16.69}$$

$$\overset{\text{(by (16.51), (16.40))}}{\leq} \omega_1\left(f \circ \psi^{-1}, \frac{1}{n} + \frac{1}{n^\beta}\right) + 2\left\|\left\|f \circ \psi^{-1}\right\|_2\right\|_\infty T e^{-2\lambda n^{(1-\beta)}},$$

proving the claim. □

The final result is for the quadrature type operator on E.

Theorem 16.5. *Let $(E, \|\cdot\|_1)$ be a Banach space with $\dim E = N \in \mathbb{N}$, and ψ be the corresponding homeomorphism from E onto $(\mathbb{R}^N, \|\cdot\|_2)$. Here $f \in C_B(E, \mathbb{R}^N)$, ${}_4L_n(f)$ is as in (16.56), $x \in \mathbb{R}^N$ and $y \in E$; $0 < \beta < 1$, $n \in \mathbb{N}$ with $n^{1-\beta} > 2$, ω_1 is for $p = \infty$. Then*

(1)

$$\|{}_4L_n(f)(y) - f(y)\|_2 \leq$$

$$\omega_1\left(f \circ \psi^{-1}, \frac{1}{n} + \frac{1}{n^\beta}\right) + 2\left\|\left\|f \circ \psi^{-1}\right\|_2\right\|_\infty Te^{-2\lambda n^{(1-\beta)}} =: \rho_4(n), \quad (16.70)$$

and

(2)

$$\left\|\|{}_4L_n(f) - f\|_2\right\|_\infty \leq \rho_4(n). \quad (16.71)$$

Given that $f \in (C_U(E, \mathbb{R}^N) \cap C_B(E, \mathbb{R}^N))$, we obtain $\lim_{n\to\infty} {}_4L_n(f) = f$, uniformly.

Proof. We have

$$\|{}_4L_n(f)(y) - f(y)\|_2 =$$

$$\left\|{}_4L_n(f)(\psi^{-1}(x)) - f(\psi^{-1}(x))\right\|_2 =$$

$$\left\|D_n^s(f \circ \psi^{-1}, x) - f(\psi^{-1}(x))\right\|_2 \overset{(16.35)}{=}$$

$$\left\|\sum_{k=-\infty}^{\infty} \delta_{nk}(f \circ \psi^{-1}) Z_1(nx - k) - f(\psi^{-1}(x))\right\|_2 =$$

$$\left\|\sum_{k=-\infty}^{\infty}\left(\sum_{r=0}^{\theta} w_r (f \circ \psi^{-1})\left(\frac{k}{n} + \frac{r}{n\theta}\right)\right) Z_1(nx - k) - \left(\sum_{k=-\infty}^{\infty} Z_1(nx - k)\right)(f \circ \psi^{-1})(x)\right\|_2 = \quad (16.72)$$

$$\left\|\sum_{k=-\infty}^{\infty}\left(\left(\sum_{r=0}^{\theta} w_r (f \circ \psi^{-1})\left(\frac{k}{n} + \frac{r}{n\theta}\right)\right) - (f \circ \psi^{-1})(x)\right) Z_1(nx - k)\right\|_2 =$$

$$\left\|\sum_{k=-\infty}^{\infty}\left(\sum_{r=0}^{\theta} w_r \left((f \circ \psi^{-1})\left(\frac{k}{n} + \frac{r}{n\theta}\right) - (f \circ \psi^{-1})(x)\right)\right) Z_1(nx - k)\right\|_2 \leq$$

$$\sum_{k=-\infty}^{\infty}\left(\sum_{r=0}^{\theta}w_r\left\|\left(f\circ\psi^{-1}\right)\left(\frac{k}{n}+\frac{r}{n\theta}\right)-\left(f\circ\psi^{-1}\right)(x)\right\|_2\right)Z_1(nx-k)=$$

$$\sum_{\substack{k=-\infty\\ \left\{:\left\|\frac{k}{n}-x\right\|_\infty\leq\frac{1}{n^\beta}\right.}}^{\infty}\left(\sum_{r=0}^{\theta}w_r\left\|\left(f\circ\psi^{-1}\right)\left(\frac{k}{n}+\frac{r}{n\theta}\right)-\left(f\circ\psi^{-1}\right)(x)\right\|_2\right)Z_1(nx-k) \tag{16.73}$$

$$+\sum_{\substack{k=-\infty\\ \left\{:\left\|\frac{k}{n}-x\right\|_\infty>\frac{1}{n^\beta}\right.}}^{\infty}\left(\sum_{r=0}^{\theta}w_r\left\|\left(f\circ\psi^{-1}\right)\left(\frac{k}{n}+\frac{r}{n\theta}\right)-\left(f\circ\psi^{-1}\right)(x)\right\|_2\right)Z_1(nx-k)$$

$$\overset{\text{(by (16.51), (16.40))}}{\leq}\omega_1\left(f\circ\psi^{-1},\frac{1}{n}+\frac{1}{n^\beta}\right)+2\left\|\left\|f\circ\psi^{-1}\right\|_2\right\|Te^{-2\lambda n^{(1-\beta)}}, \tag{16.74}$$

proving the claim. □

Bibliography

[358] G.A. Anastassiou, *Rate of convergence of some neural network operators to the unit-univariate case*, J. Math. Anal. Appl, 212 (1997), 237–262.

[359] G.A. Anastassiou, *Quantitative Approximations*, Chapman & Hall / CRC, Boca Raton, New York, 2001.

[360] G.A. Anastassiou, *Inteligent Systems: Approximation by Artificial Neural Networks*, Intelligent Systems Reference Library, Vol. 19, Springer, Heidelberg, 2011.

[361] G.A. Anastassiou, *Intelligent Systems II: Complete Approximation by Neural Network Operators*, Springer, Heidelberg, New York, 2016.

[362] G.A. Anastassiou, *Intelligent Computations: Abstract, Fractional Calculus, Inequalities, Approximations*, Springer, Heidelberg, New York, 2018.

[363] G.A. Anastassiou, *Banach Space Valued Neural Network*, Springer, Heidelberg, New York, 2023.

[364] G.A. Anastassiou, *Parametrized, deformed and general neural networks*, Springer, Heidelberg, New York, 2023.

[365] G.A. Anastassiou, *Trigonometric and Hyperbolic Generated Approximation Theory*, World Scientific, Singapore, London, New Jersey, 2025.

[366] Z. Chen and F. Cao, *The approximation operators with sigmoidal functions*, Computers and Mathematics with Applications, 58 (2009), 758–765.

[367] D. Costarelli, R. Spigler, *Approximation results for neural network operators activated by sigmoidal functions*, Neural Networks 44 (2013), 101–106.

[368] D. Costarelli, R. Spigler, *Multivariate neural network operators with sigmoidal activation functions*, Neural Networks 48 (2013), 72–77.

[369] S. Haykin, *Neural Networks: A Comprehensive Foundation* (2 ed.), Prentice Hall, New York, 1998.

[370] W. McCulloch and W. Pitts, *A logical calculus of the ideas immanent in nervous activity*, Bulletin of Mathematical Biophysics, 7 (1943), 115–133.

[371] T.M. Mitchell, *Machine Learning*, WCB-McGraw-Hill, New York, 1997.

[372] G.A. Anastassiou, *Neural Network Approximation over finite dimensional Banach spaces*, RACSAM, (2025) 119:55; https://doi.org/10.1007/s13398-025-01722-3.

Chapter 17

Generalized Logistic Neural Network Approximation in Finite Dimensional Banach Spaces

The functions under approximation here have as a domain a finite dimensional Banach space with dimension $N \in \mathbb{N}$ and are with values in $\mathbb{R}^N$. Exploiting some topological properties of the above we are able to perform Neural Network multivariate approximation to the above functions. The treatment is quantitative. We produce multivariate Jackson type inequalities involving the modulus of continuity of the function under approximation. The established convergences are pointwise and uniform. Perturbation and symmetrization to our operators lead to enhanced speeds of convergence. The activation function here is the generalized logistic. It follows [387].

17.1 Introduction

The author in [373] and [374], see Chapters 2–5 was the first to establish neural network approximation to continuous functions with rates by very specifically defined neural network operators of Cardaliaguet-Euvrard and "Squashing" types, by employing the modulus of continuity of the engaged function or its high order derivative, and producing very tight Jackson type inequalities. He treats there both the univariate and multivariate cases. The defining these operators "bell-shaped" and "squashing" functions are assumed to be of compact support.

Again the author inspired by [381], continued his studies on neural network approximation by introducing and using the proper quasi-interpolation operators of sigmoidal and hyperbolic tangent type which resulted into [375], by treating both the univariate and multivariate cases.

The author continued this trend of approximation and published the extensive monographs [376]–[380], studying in depth and as wide as possible the neural network approximations. Our activation function here is the generalized logistic function.

The multivariate operators we use are: the normalized, the quasi-interpolation, the Kantorovich and the quadrature types. Our symmetrization technique accelerates immensely the speed of convergences. Our neural networks here are from a Banach space of dimension N to $\mathbb{R}^N$, as they are homeomorphic, and this is the main key to this work.

So all results here are via multivariate Jackson type inequalities, studying quantitatively the rate of convergences to the unit operator.

Here we are dealing with one hidden layer feed-forward neural networks.

A multilayer feed-forward neural network can be defined as follows (with $m \in \mathbb{N}$ hidden layers):

Let $x \in \mathbb{R}^s$; $s \in \mathbb{N}$, where $x = (x_1, ..., x_s)$; $\alpha_j, c_j \in \mathbb{R}^s$; $b_j \in \mathbb{R}$, with $0 \le j \le n$, $n \in \mathbb{N}$.

Here $\langle \alpha_j \cdot x \rangle$ is the inner product, thus $\sigma(\langle \alpha_j \cdot x \rangle + b_j) \in \mathbb{R}$; and $N_n(x) \in \mathbb{R}^s$, by $c_j \in \mathbb{R}^s$, as it is coming from $N_n(x) = \sum_{j=0}^{n} c_j \sigma(\langle \alpha_j \cdot x \rangle + b_j)$.

We define:

$$N_n^{(2)}(x) = \sum_{j=0}^{n} c_j \sigma(\langle \alpha_j \cdot N_n(x) \rangle + b_j) =$$

$$\sum_{j=0}^{n} c_j \sigma\left(\left\langle \alpha_j \cdot \left(\sum_{j=0}^{n} c_j \sigma(\langle \alpha_j \cdot x \rangle + b_j)\right)\right\rangle + b_j\right).$$

Furthermore, we can define

$$N_n^{(3)}(x) = \sum_{j=0}^{n} c_j \sigma\left(\left\langle \alpha_j \cdot N_n^{(2)}(x)\right\rangle + b_j\right).$$

And, in general we define:

$$N_n^{(m)}(x) = \sum_{j=0}^{n} c_j \sigma\left(\left\langle \alpha_j \cdot N_n^{(m-1)}(x)\right\rangle + b_j\right), \quad \text{for } m \in \mathbb{N}.$$

For further studies in neural networks read [382]–[386].

17.2 Basics

Initially we follow [379], pp. 395–471.

Our activation function here to be used is the q-deformed and λ-parametrized function

$$\varphi_{q,\lambda}(x) = \frac{1}{1 + qA^{-\lambda x}}, \quad x \in \mathbb{R}, \ q, \lambda > 0, \ A > 1. \tag{17.1}$$

This is the A-generalized logistic function.

For more read Chapter 16 of [379]: "Banach space valued ordinary and fractional neural network approximation based on q-deformed and λ-parametrized A-generalized logistic function".

The chapters 15, 16 of [379] motivate our current work.

The proposed "symmetrization technique" aims to use half data feed to our neural networks.

We will employ the following density function

$$G_{q,\lambda}(x) := \frac{1}{2}\left(\varphi_{q,\lambda}(x+1) - \varphi_{q,\lambda}(x-1)\right), \quad x \in \mathbb{R}, \ q, \lambda > 0. \tag{17.2}$$

We have that

$$G_{q,\lambda}(-x) = G_{\frac{1}{q},\lambda}(x), \tag{17.3}$$

and

$$G_{\frac{1}{q},\lambda}(-x) = G_{q,\lambda}(x), \ \forall\, x \in \mathbb{R}. \tag{17.4}$$

Adding (17.3) and (17.4) we obtain

$$G_{q,\lambda}(-x) + G_{\frac{1}{q},\lambda}(-x) = G_{q,\lambda}(x) + G_{\frac{1}{q},\lambda}(x), \ \forall\, x \in \mathbb{R}, \tag{17.5}$$

the key to this work.

So that

$$W(x) := \frac{G_{q,\lambda}(x) + G_{\frac{1}{q},\lambda}(x)}{2} \tag{17.6}$$

is an even function, symmetric with respect to the y-axis.

The global maximum of $G_{q,\lambda}$ is given by (16.18), p. 401 of [379] as

$$G_{q,\lambda}\left(\frac{\log_A q}{\lambda}\right) = \frac{A^{\lambda} - 1}{2(A^{\lambda} + 1)}. \tag{17.7}$$

And, the global max of $G_{\frac{1}{q},\lambda}$ is

$$G_{\frac{1}{q},\lambda}\left(\frac{\log_A \frac{1}{q}}{\lambda}\right) = G_{\frac{1}{q},\lambda}\left(\frac{-\log_A q}{\lambda}\right) = \frac{A^{\lambda} - 1}{2(A^{\lambda} + 1)}, \tag{17.8}$$

both sharing the same maximum at symmetric points.

By Theorem 16.1, p. 401 of [379], we have that

$$\sum_{i=-\infty}^{\infty} G_{q,\lambda}(x-i) = 1, \ \forall\, x \in \mathbb{R},\ \lambda, q > 0,\ A > 1, \tag{17.9}$$

and

$$\sum_{i=-\infty}^{\infty} G_{\frac{1}{q},\lambda}(x-i) = 1, \ \forall\, x \in \mathbb{R},\ \lambda, q > 0,\ A > 1. \tag{17.10}$$

Consequently, we derive that

$$\sum_{i=-\infty}^{\infty} W(x-i) = 1, \ \forall\, x \in \mathbb{R}. \tag{17.11}$$

By Theorem 16.2, p. 402 of [379], we have that

$$\int_{-\infty}^{\infty} G_{q,\lambda}(x)\,dx = 1, \quad \lambda, q > 0,\ A > 1, \tag{17.12}$$

similarly it holds

$$\int_{-\infty}^{\infty} G_{\frac{1}{q},\lambda}(x)\,dx = 1, \tag{17.13}$$

so that

$$\int_{-\infty}^{\infty} W(x)\,dx = 1, \tag{17.14}$$

therefore W is a density function.

By Theorem 16.3, p. 402 of [379], we have:

Let $0 < \alpha < 1$, and $n \in \mathbb{N}$ with $n^{1-\alpha} > 2$. Then

$$\sum_{\substack{k=-\infty \\ : |nx-k| \geq n^{1-\alpha}}}^{\infty} G_{q,\lambda}(nx-k) < 2\max\left\{q, \frac{1}{q}\right\} \frac{1}{A^{\lambda\left(n^{1-\alpha}-2\right)}} = \gamma A^{-\lambda\left(n^{1-\alpha}-2\right)}, \tag{17.15}$$

where $\lambda, q > 0,\ A > 1;\ \gamma := 2\max\left\{q, \frac{1}{q}\right\}$.

Similarly, we get that

$$\sum_{\substack{k=-\infty \\ : |nx-k| \geq n^{1-\alpha}}}^{\infty} G_{\frac{1}{q},\lambda}(nx-k) < \gamma A^{-\lambda\left(n^{1-\alpha}-2\right)}. \tag{17.16}$$

Consequently we obtain that

$$\sum_{\begin{cases} k=-\infty \\ : |nx-k| \geq n^{1-\alpha} \end{cases}}^{\infty} W(nx-k) < \gamma A^{-\lambda\left(n^{1-\alpha}-2\right)}, \quad (17.17)$$

where $\gamma := 2\max\left\{q, \frac{1}{q}\right\}$.

Here $\lceil\cdot\rceil$ denotes the ceiling of the number, and $\lfloor\cdot\rfloor$ its integral part.

We mention

Theorem 16.4 (p. 402, [379]) Let $x \in [a,b] \subset \mathbb{R}$ and $n \in \mathbb{N}$ so that $\lceil na\rceil \leq \lfloor nb\rfloor$. For $q > 0$, $\lambda > 0$, $A > 1$, we consider the number $\lambda_q > z_0 > 0$ with $G_{q,\lambda}(z_0) = G_{q,\lambda}(0)$, and $\lambda_q > 1$. Then

$$\frac{1}{\sum_{k=\lceil na\rceil}^{\lfloor nb\rfloor} G_{q,\lambda}(nx-k)} < \max\left\{\frac{1}{G_{q,\lambda}(\lambda_q)}, \frac{1}{G_{\frac{1}{q},\lambda}\left(\lambda_{\frac{1}{q}}\right)}\right\} =: K(q). \quad (17.18)$$

Similarly, we consider $\lambda_{\frac{1}{q}} > z_1 > 0$, such that $G_{\frac{1}{q},\lambda}(z_1) = G_{\frac{1}{q},\lambda}(0)$, and $\lambda_{\frac{1}{q}} > 1$. Thus

$$\frac{1}{\sum_{k=\lceil na\rceil}^{\lfloor nb\rfloor} G_{\frac{1}{q},\lambda}(nx-k)} < \max\left\{\frac{1}{G_{\frac{1}{q},\lambda}\left(\lambda_{\frac{1}{q}}\right)}, \frac{1}{G_{q,\lambda}(\lambda_q)}\right\} = K(q). \quad (17.19)$$

Hence

$$\sum_{k=\lceil na\rceil}^{\lfloor nb\rfloor} G_{q,\lambda}(nx-k) > \frac{1}{K(q)}, \quad (17.20)$$

and

$$\sum_{k=\lceil na\rceil}^{\lfloor nb\rfloor} G_{\frac{1}{q},\lambda}(nx-k) > \frac{1}{K(q)}. \quad (17.21)$$

Consequently it holds

$$\sum_{k=\lceil na\rceil}^{\lfloor nb\rfloor} \frac{\left(G_{q,\lambda}(nx-k) + G_{\frac{1}{q},\lambda}(nx-k)\right)}{2} > \frac{2}{2K(q)} = \frac{1}{K(q)}, \quad (17.22)$$

so that

$$\frac{1}{\sum_{k=\lceil na\rceil}^{\lfloor nb\rfloor} \frac{\left(G_{q,\lambda}(nx-k)+G_{\frac{1}{q},\lambda}(nx-k)\right)}{2}} < K(q), \quad (17.23)$$

that is

$$\frac{1}{\sum_{k=\lceil na\rceil}^{\lfloor nb\rfloor} W(nx-k)} < K(q). \tag{17.24}$$

We have proved

Theorem 17.1. *Let $x \in [a,b] \subset \mathbb{R}$ and $n \in \mathbb{N}$ so that $\lceil na\rceil \leq \lfloor nb\rfloor$. For $q, \lambda > 0$, $A > 1$, we consider $\lambda_q > z_0 > 0$ with $G_{q,\lambda}(z_0) = G_{q,\lambda}(0)$, and $\lambda_q > 1$. Also consider $\lambda_{\frac{1}{q}} > z_1 > 0$, such that $G_{\frac{1}{q},\lambda}(z_1) = G_{\frac{1}{q},\lambda}(0)$, and $\lambda_{\frac{1}{q}} > 1$. Then*

$$\frac{1}{\sum_{k=\lceil na\rceil}^{\lfloor nb\rfloor} W(nx-k)} < K(q). \tag{17.25}$$

We make

Remark 17.1.

I) By Remark 16.5, p. 402 of [379], we have

$$\lim_{n\to\infty} \sum_{k=\lceil na\rceil}^{\lfloor nb\rfloor} G_{q,\lambda}(nx_1-k) \neq 1, \text{ for some } x_1 \in [a,b], \tag{17.26}$$

and

$$\lim_{n\to\infty} \sum_{k=\lceil na\rceil}^{\lfloor nb\rfloor} G_{\frac{1}{q},\lambda}(nx_2-k) \neq 1, \text{ for some } x_2 \in [a,b]. \tag{17.27}$$

Therefore it holds

$$\lim_{n\to\infty} \sum_{k=\lceil na\rceil}^{\lfloor nb\rfloor} \frac{\left(G_{q,\lambda}(nx_1-k) + G_{\frac{1}{q},\lambda}(nx_2-k)\right)}{2} \neq 1. \tag{17.28}$$

Hence

$$\lim_{n\to\infty} \sum_{k=\lceil na\rceil}^{\lfloor nb\rfloor} \frac{\left(G_{q,\lambda}(nx_1-k) + G_{\frac{1}{q},\lambda}(nx_1-k)\right)}{2} \neq 1, \tag{17.29}$$

even if

$$\lim_{n\to\infty} \sum_{k=\lceil na\rceil}^{\lfloor nb\rfloor} G_{\frac{1}{q},\lambda}(nx_1-k) = 1, \tag{17.30}$$

because then

$$\lim_{n\to\infty}\sum_{k=\lceil na\rceil}^{\lfloor nb\rfloor}\frac{G_{q,\lambda}\left(nx_1-k\right)}{2}+\frac{1}{2}\neq 1, \tag{17.31}$$

equivalently

$$\lim_{n\to\infty}\sum_{k=\lceil na\rceil}^{\lfloor nb\rfloor}\frac{G_{q,\lambda}\left(nx_1-k\right)}{2}\neq\frac{1}{2}, \tag{17.32}$$

true by

$$\lim_{n\to\infty}\sum_{k=\lceil na\rceil}^{\lfloor nb\rfloor}G_{q,\lambda}\left(nx_1-k\right)\neq 1. \tag{17.33}$$

II) Let $[a,b]\subset\mathbb{R}$. For large n we always have $\lceil na\rceil\leq\lfloor nb\rfloor$. Also $a\leq\frac{k}{n}\leq b$, iff $\lceil na\rceil\leq k\leq\lfloor nb\rfloor$. So in general it holds

$$\sum_{k=\lceil na\rceil}^{\lfloor nb\rfloor}W\left(nx-k\right)\leq 1. \tag{17.34}$$

Next, we move on to the multivariate case, see Chapter 15 of [379], pp. 365–394, as a model of action.

We make

Remark 17.2. We introduce

$$Z_q\left(x_1,...,x_N\right):=Z_q\left(x\right):=\prod_{i=1}^{N}W\left(x_i\right)=\frac{1}{2^N}\prod_{i=1}^{N}\left(G_{q,\lambda}+G_{\frac{1}{q},\lambda}\right)\left(x_i\right), \tag{17.35}$$

$x=\left(x_1,...,x_N\right)\in\mathbb{R}^N$, $\lambda,q>0$, $N\in\mathbb{N}$.

Properties:

(i)

$$Z_q\left(x\right)>0,\ \forall\ x\in\mathbb{R}^N, \tag{17.36}$$

(ii)

$$\sum_{k=-\infty}^{\infty}Z_q\left(x-k\right):=\sum_{k_1=-\infty}^{\infty}\sum_{k_2=-\infty}^{\infty}...\sum_{k_N=-\infty}^{\infty}Z_q\left(x_1-k_1,...,x_N-k_N\right)=1, \tag{17.37}$$

$k:=\left(k_1,...,k_n\right)\in\mathbb{Z}^N$, $\forall\ x\in\mathbb{R}^N$,

hence

(iii)

$$\sum_{k=-\infty}^{\infty} Z_q(nx-k) = 1, \tag{17.38}$$

$\forall\, x \in \mathbb{R}^N$, $n \in \mathbb{N}$,

(iv)

$$\int_{\mathbb{R}^N} Z_q(x)\, dx = 1, \tag{17.39}$$

that is Z_q is a multivariate density function.

Here denote $\|x\|_\infty := \max\{|x_1|, ..., |x_N|\}$, $x \in \mathbb{R}^N$, $\infty := (\infty, ..., \infty)$, $-\infty := (-\infty, ..., -\infty)$, $\lceil na \rceil := (\lceil na_1 \rceil, ..., \lceil na_N \rceil)$, $\lfloor nb \rfloor := (\lfloor nb_1 \rfloor, ..., \lfloor nb_N \rfloor)$, $a := (a_1, ..., a_N)$, $b := (b_1, ..., b_N)$.

We obviously see that

$$\sum_{k=\lceil na \rceil}^{\lfloor nb \rfloor} Z_q(nx-k) = \sum_{k=\lceil na \rceil}^{\lfloor nb \rfloor} \left(\prod_{i=1}^{N} W(nx_i - k_i) \right) = \prod_{i=1}^{N} \left(\sum_{k_i=\lceil na_i \rceil}^{\lfloor nb_i \rfloor} W(nx_i - k_i) \right). \tag{17.40}$$

(v) We derive that

$$\sum_{\substack{k = \lceil na \rceil \\ \left\| \frac{k}{n} - x \right\|_\infty > \frac{1}{n^\beta}}}^{\lfloor nb \rfloor} Z_q(nx-k) < \gamma A^{-\lambda\left(n^{1-\beta}-2\right)}, \text{ where } 0 < \beta < 1, \tag{17.41}$$

$0 < \beta < 1$, $n \in \mathbb{N} : n^{1-\beta} > 2$, $x \in \prod_{i=1}^{N} [a_i, b_i]$.

(vi) It holds

$$0 < \frac{1}{\sum_{k=\lceil na \rceil}^{\lfloor nb \rfloor} Z_q(nx-k)} < (K(q))^N, \tag{17.42}$$

$\forall\, x \in \left(\prod_{i=1}^{N} [a_i, b_i] \right)$, $n \in \mathbb{N}$.

It is clear that

(vii)

$$\sum_{\substack{k = -\infty \\ \left\| \frac{k}{n} - x \right\|_\infty > \frac{1}{n^\beta}}}^{\infty} Z_q(nx-k) < \gamma A^{-\lambda\left(n^{1-\beta}-2\right)}, \tag{17.43}$$

$0 < \beta < 1$, $n \in \mathbb{N} : n^{1-\beta} > 2$, $x \in \prod_{i=1}^{N} [a_i, b_i]$.

Furthermore it holds

$$\lim_{n\to\infty}\sum_{k=\lceil na\rceil}^{\lfloor nb\rfloor} Z_q(nx-k)\neq 1, \tag{17.44}$$

for at least some $x\in\left(\prod_{i=1}^{N}[a_i,b_i]\right)$.

Here $\left(X,\|\cdot\|_{\gamma}\right)$ is a Banach space.

Let $f\in C\left(\prod_{i=1}^{N}[a_i,b_i],X\right)$, $x=(x_1,...,x_N)\in\prod_{i=1}^{N}[a_i,b_i]$, $n\in\mathbb{N}$: $\lceil na_i\rceil\leq\lfloor nb_i\rfloor$, $i=1,...,N$.

We introduce and define the following multivariate linear normalized symmetrized neural network operator, let $x:=(x_1,...,x_N)\in\prod_{i=1}^{N}[a_i,b_i]$:

$$A_n^s(f,x_1,...,x_N):=A_n^s(f,x):=\frac{\sum_{k=\lceil na\rceil}^{\lfloor nb\rfloor} f\left(\frac{k}{n}\right)Z_q(nx-k)}{\sum_{k=\lceil na\rceil}^{\lfloor nb\rfloor} Z_q(nx-k)}=$$

$$\frac{\sum_{k_1=\lceil na_1\rceil}^{\lfloor nb_1\rfloor}\sum_{k_2=\lceil na_2\rceil}^{\lfloor nb_2\rfloor}\cdots\sum_{k_N=\lceil na_N\rceil}^{\lfloor nb_N\rfloor} f\left(\frac{k_1}{n},...,\frac{k_N}{n}\right)\left(\prod_{i=1}^{N}\left(G_{q,\lambda}(nx_i-k_i)+G_{\frac{1}{q},\lambda}(nx_i-k_i)\right)\right)}{\prod_{i=1}^{N}\left(\sum_{k_i=\lceil na_i\rceil}^{\lfloor nb_i\rfloor}\left(G_{q,\lambda}(nx_i-k_i)+G_{\frac{1}{q},\lambda}(nx_i-k_i)\right)\right)}. \tag{17.45}$$

For large enough $n\in\mathbb{N}$ we always obtain $\lceil na_i\rceil\leq\lfloor nb_i\rfloor$, $i=1,...,N$. Also $a_i\leq\frac{k_i}{n}\leq b_i$, iff $\lceil na_i\rceil\leq k_i\leq\lfloor nb_i\rfloor$, $i=1,...,N$.

When $f\in C_B\left(\mathbb{R}^N,X\right)$ or $f\in C_U\left(\mathbb{R}^N,X\right)$, we define

$$B_n^s(f,x):=B_n^s(f,x_1,...,x_N):=\sum_{k=-\infty}^{\infty} f\left(\frac{k}{n}\right)Z_q(nx-k):=$$

$$\frac{1}{2^N}\sum_{k_1=-\infty}^{\infty}\sum_{k_2=-\infty}^{\infty}\cdots\sum_{k_N=-\infty}^{\infty} f\left(\frac{k_1}{n},...,\frac{k_N}{n}\right)$$

$$\left(\prod_{i=1}^{N}\left(G_{q,\lambda}(nx_i-k_i)+G_{\frac{1}{q},\lambda}(nx_i-k_i)\right)\right), \tag{17.46}$$

$n\in\mathbb{N}$, $\forall\, x\in\mathbb{R}^N$, $N\in\mathbb{N}$, the multivariate quasi-interpolation symmetrized neural network.

Also for $f\in C_B\left(\mathbb{R}^N,X\right)$ or $f\in C_U\left(\mathbb{R}^N,X\right)$, we define the multivariate Kantorovich type symmetrized neural network operator

$$C_n^s(f,x):=C_n^s(f,x_1,...,x_N):=\sum_{k=-\infty}^{\infty}\left(n^N\int_{\frac{k}{n}}^{\frac{k+1}{n}} f(t)\,dt\right)Z_q(nx-k)=$$

$$\frac{1}{2^N}\sum_{k_1=-\infty}^{\infty}\sum_{k_2=-\infty}^{\infty}...\sum_{k_N=-\infty}^{\infty}\left(n^N\int_{\frac{k_1}{n}}^{\frac{k_1+1}{n}}...\int_{\frac{k_N}{n}}^{\frac{k_N+1}{n}}f(t_1,...,t_N)\,dt_1...dt_N\right)$$

$$\left(\prod_{i=1}^{N}\left(G_{q,\lambda}(nx_i-k_i)+G_{\frac{1}{q},\lambda}(nx_i-k_i)\right)\right), \tag{17.47}$$

$n\in\mathbb{N}$, $\forall\ x\in\mathbb{R}^N$.

Again for $f\in C_B\left(\mathbb{R}^N,X\right)$ or $f\in C_U\left(\mathbb{R}^N,X\right)$, we define the multivariate symmetrized neural network operator of quadrature type $D_n^s(f,x)$, $n\in\mathbb{N}$, as follows:

Let $\theta:=(\theta_1,...,\theta_N)\in\mathbb{N}^N$, $r=(r_1,...,r_N)\in\mathbb{Z}_+^n$, $w_r=w_{r_1,r_2,...,r_N}\geq 0$, such that

$$\sum_{r=0}^{\theta}w_r=\sum_{r_1=0}^{\theta_1}\sum_{r_2=0}^{\theta_2}...\sum_{r_N=0}^{\theta_N}w_{r_1,r_2,...,r_N}=1;$$

$k\in\mathbb{Z}^N$, and

$$\delta_{nk}(f):=\delta_{n,k_1,...,k_N}(f):=\sum_{r=0}^{\theta}w_rf\left(\frac{k}{n}+\frac{r}{n\theta}\right)=$$

$$\sum_{r_1=0}^{\theta_1}\sum_{r_2=0}^{\theta_2}...\sum_{r_N=0}^{\theta_N}w_{r_1,r_2,...,r_N}f\left(\frac{k_1}{n}+\frac{r_1}{n\theta_1},...,\frac{k_n}{n}+\frac{r_N}{n\theta_N}\right), \tag{17.48}$$

where $\frac{r}{\theta}:=\left(\frac{r_1}{\theta_1},...,\frac{r_N}{\theta_N}\right)$.

We set

$$D_n^s(f,x):=D_n^s(f,x_1,...,x_N):=\sum_{k=-\infty}^{\infty}\delta_{nk}(f)\,Z_q(nx-k)=$$

$$\frac{1}{2^N}\sum_{k_1=-\infty}^{\infty}\sum_{k_2=-\infty}^{\infty}...\sum_{k_N=-\infty}^{\infty}\delta_{n,k_1,...,k_N}(f)$$

$$\left(\prod_{i=1}^{N}\left(G_{q,\lambda}(nx_i-k_i)+G_{\frac{1}{q},\lambda}(nx_i-k_i)\right)\right), \tag{17.49}$$

$\forall\ x\in\mathbb{R}^N$.

Definition 17.1. ([377], p. 274) Let M be a convex and compact subset of $\left(\mathbb{R}^N, \|\cdot\|_p\right)$, $p \in [1, \infty]$, and $\left(X, \|\cdot\|_\gamma\right)$ be a Banach space. Let $f \in C(M, X)$. We define the first modulus of continuity of f as

$$\omega_1(f, \delta) := \sup_{\substack{x, y \in M: \\ \|x-y\|_p \leq \delta}} \|f(x) - f(y)\|_\gamma, \quad 0 < \delta \leq diam(M). \tag{17.50}$$

If $\delta > diam(M)$, then

$$\omega_1(f, \delta) = \omega_1(f, diam(M)). \tag{17.51}$$

Notice $\omega_1(f, \delta)$ is increasing in $\delta > 0$. For $f \in C_B(M, X)$ (continuous and bounded functions) $\omega_1(f, \delta)$ is defined similarly.

Lemma 17.1. *([377], p. 274) We have $\omega_1(f, \delta) \to 0$ as $\delta \downarrow 0$, iff $f \in C(M, X)$, where M is a convex compact subset of $\left(\mathbb{R}^N, \|\cdot\|_p\right)$, $p \in [1, \infty]$.*

In this study we work only for the case of $p = \infty$.

Clearly we have also: $f \in C_U\left(\mathbb{R}^N, X\right)$ (uniformly continuous functions), iff $\omega_1(f, \delta) \to 0$ as $\delta \downarrow 0$, where ω_1 is defined similarly to (17.50). The space $C_B\left(\mathbb{R}^N, X\right)$ denotes the continuous and bounded functions on $\mathbb{R}^N$, $\|\cdot\|_\infty$ is the supremum norm.

In this work we treat the case of $\left(X, \|\cdot\|_\gamma\right) = \left(\mathbb{R}^N, \|\cdot\|_2\right)$, $N \in \mathbb{N}$; where $\|\cdot\|_2$ is the Euclidean norm.

Next, we describe the main ideas of this work.

Remark 17.3. Let $(E, \|\cdot\|_1)$ be a Banach space of finite dimension N, i.e. $\dim E = N$, $N \in \mathbb{N}$.

It is known that any two Banach spaces of the same finite dimension are linearly homeomorphic.

Hence $(E, \|\cdot\|_1)$ is linearly homeomorphic to $\left(\mathbb{R}^N, \|\cdot\|_2\right)$: so let $\phi : (E, \|\cdot\|_1) \to \left(\mathbb{R}^N, \|\cdot\|_2\right)$ be the linear homeomorphism.

So let $y \in E$, then $\phi(y) = x \in \mathbb{R}^N$, and back to $y = \phi^{-1}(x) \in E$, that is $\phi^{-1} : \mathbb{R}^N \to E$ is $(1-1)$ and onto map, and both ϕ, ϕ^{-1} are continuous maps.

Let now $f : E \to \mathbb{R}^N$ be a continuous map, then $f \circ \phi^{-1}$ is a continuous map from $\mathbb{R}^N$ into $\mathbb{R}^N$. We call $g(x) := f\left(\phi^{-1}(x)\right)$, $\forall\, x \in \mathbb{R}^N$.

If $\|f(y)\|_2 \leq M$, $\forall\, y \in E$, so is $\left\|f\left(\phi^{-1}(x)\right)\right\|_2 \leq M$, $\forall\, x \in \mathbb{R}^N$, where $M > 0$.

That is $\|\|f(y)\|_2\|_\infty < \infty$, implies $\left\|\left\|f \circ \phi^{-1}\right\|_2\right\|_\infty < \infty$.

In case f is a uniformly continuous map, since (see below) ϕ, ϕ^{-1} are also uniformly continuous, we obtain that $g = f \circ \phi^{-1}$ is a uniformly continuous map.

If $\prod_{i=1}^{N} [a_i, b_i] \subset \mathbb{R}^N$, we can consider $g|_{\prod_{i=1}^{N} [a_i, b_i]}$, which is continuous on a bounded set, that is a uniformly continuous map.

It is known that every linear function on a finite dimensional normed vector space is uniformly continuous, based on the fact that all norms there are equivalent. Thus, ϕ, ϕ^{-1} are uniformly continuous.

Let $f : E \to \mathbb{R}^N$ be continuous.

We define the following neural network linear operators ${}_iL_n(f) : E \to \mathbb{R}^N$, $i = 1, 2, 3, 4$; $y \in E$, as follows:

I)

$$ {}_1L_n(f)(y) = {}_1L_n(f)\left(\phi^{-1}(x)\right) := A_n^s\left(f \circ \phi^{-1}, x\right) = $$

$$ \frac{\sum_{k=\lceil na \rceil}^{\lfloor nb \rfloor} \left(f \circ \psi^{-1}\right)\left(\frac{k}{n}\right) Z_q(nx - k)}{\sum_{k=\lceil na \rceil}^{\lfloor nb \rfloor} Z_q(nx - k)} = \tag{17.52} $$

$$ \frac{\sum_{k_1=\lceil na_1 \rceil}^{\lfloor nb_1 \rfloor} \cdots \sum_{k_N=\lceil na_N \rceil}^{\lfloor nb_N \rfloor} \left(f \circ \phi^{-1}\right)\left(\frac{k_1}{n}, ..., \frac{k_N}{n}\right)\left(\prod_{i=1}^{N}\left(G_{q,\lambda}(nx_i - k_i) + G_{\frac{1}{q},\lambda}(nx_i - k_i)\right)\right)}{\prod_{i=1}^{N}\left(\sum_{k_i=\lceil na_i \rceil}^{\lfloor nb_i \rfloor}\left(G_{q,\lambda}(nx_i - k_i) + G_{\frac{1}{q},\lambda}(nx_i - k_i)\right)\right)}, $$

$\forall\, x := (x_1, ..., x_N) \in \prod_{i=1}^{N} [a_i, b_i] \subset \mathbb{R}^N$, $\forall\, y \in \phi^{-1}\left(\prod_{i=1}^{N} [a_i, b_i]\right) \subset E$, $n \in \mathbb{N}$.

II) When $f \in C_B\left(E, \mathbb{R}^N\right)$ or $f \in C_U\left(E, \mathbb{R}^N\right)$ we define:

$$ {}_2L_n(f)(y) = {}_2L_n(f)\left(\phi^{-1}(x)\right) := B_n^s\left(f \circ \phi^{-1}, x\right) = $$

$$ \sum_{k=-\infty}^{\infty} \left(f \circ \phi^{-1}\right)\left(\frac{k}{n}\right) Z_q(nx - k) = \tag{17.53} $$

$$ \frac{1}{2^N} \sum_{k_1=-\infty}^{\infty} \sum_{k_2=-\infty}^{\infty} \cdots \sum_{k_N=-\infty}^{\infty} \left(f \circ \phi^{-1}\right)\left(\frac{k_1}{n}, ..., \frac{k_N}{n}\right) $$

$$ \left(\prod_{i=1}^{N}\left(G_{q,\lambda}(nx_i - k_i) + G_{\frac{1}{q},\lambda}(nx_i - k_i)\right)\right), $$

$\forall\, x \in \mathbb{R}^N$, $\forall\, y \in E$, $n \in \mathbb{N}$.

III) When $f \in C_B(E, \mathbb{R}^N)$ or $f \in C_U(E, \mathbb{R}^N)$ we define:

$$
{}_3L_n(f)(y) = {}_3L_n(f)\left(\phi^{-1}(x)\right) := C_n^s\left(f \circ \phi^{-1}, x\right) =
$$

$$
\sum_{k=-\infty}^{\infty} \left(n^N \int_{\frac{k}{n}}^{\frac{k+1}{n}} \left(f \circ \phi^{-1}\right)(t)\, dt \right) Z_q(nx - k) =
$$

$$
\frac{1}{2^N} \sum_{k_1=-\infty}^{\infty} \cdots \sum_{k_N=-\infty}^{\infty} \left(n^N \int_{\frac{k_1}{n}}^{\frac{k_1+1}{n}} \cdots \int_{\frac{k_N}{n}}^{\frac{k_N+1}{n}} \left(f \circ \phi^{-1}\right)(t_1, \ldots, t_N)\, dt_1 \ldots dt_N \right)
$$

$$
\left(\prod_{i=1}^{N} \left(G_{q,\lambda}(nx_i - k_i) + G_{\frac{1}{q},\lambda}(nx_i - k_i) \right) \right), \tag{17.54}
$$

$\forall\, x \in \mathbb{R}^N$, $\forall\, y \in E$, $n \in \mathbb{N}$.

IV) Again, when $f \in C_B(E, \mathbb{R}^N)$ or $f \in C_U(E, \mathbb{R}^N)$ we define:

$$
{}_4L_n(f)(y) = {}_4L_n(f)\left(\phi^{-1}(x)\right) := D_n^s\left(f \circ \phi^{-1}, x\right) =
$$

$$
\sum_{k=-\infty}^{\infty} \delta_{nk}\left(f \circ \phi^{-1}\right) Z_q(nx - k) =
$$

$$
\frac{1}{2^N} \sum_{k_1=-\infty}^{\infty} \cdots \sum_{k_N=-\infty}^{\infty} \delta_{n,k_1,\ldots,k_N}\left(f \circ \phi^{-1}\right)
$$

$$
\left(\prod_{i=1}^{N} \left(G_{q,\lambda}(nx_i - k_i) + G_{\frac{1}{q},\lambda}(nx_i - k_i) \right) \right), \tag{17.55}
$$

$\forall\, x \in \mathbb{R}^N$, $\forall\, y \in E$, $n \in \mathbb{N}$.

We want to study quantitatively the multivariate approximation of ${}_iL_n(f)(y) \to f(y)$, as $n \to \infty$, for $i = 1, 2, 3, 4$; $y \in E$.

17.3 Main Results

We present our first approximation result.

Theorem 17.2. *Let* $(E, \|\cdot\|_1)$ *be a Banach space such that* $\dim E = N \in \mathbb{N}$*, and* ϕ *be the corresponding homeomorphism from* E *onto* $(\mathbb{R}^N, \|\cdot\|_2)$.

Here $f \in C\left(E, \mathbb{R}^N\right)$*,* ${}_1L_n(f)$ *is as in (17.52),* $x \in \prod_{i=1}^{N}[a_i, b_i] \subset \mathbb{R}^N$*, and* $y \in \phi^{-1}\left(\prod_{i=1}^{N}[a_i, b_i]\right) \subset E$*;* $0 < \beta < 1$*,* $n \in \mathbb{N}$ *with* $n^{1-\beta} > 2$*. Then*

(1)

$$\left\|{}_1L_n(f)(y) - f(y)\right\|_2 \leq (K(q))^N$$

$$\left[\omega_1\left(f \circ \phi^{-1}, \frac{1}{n^\beta}\right) + 2\left\|\left\|f \circ \phi^{-1}\right\|_2\right\|_\infty \gamma A^{-\lambda\left(n^{1-\beta}-2\right)}\right] =: \lambda_1(n), \quad (17.56)$$

and

(2)

$$\left\|\left\|{}_1L_n(f) - f\right\|_2\right\|_\infty \leq \lambda_1(n). \quad (17.57)$$

We see that $\lim_{n\to\infty} {}_1L_n(f) \overset{\|\cdot\|_2}{=} f$*, pointwise and uniformly.*

Above ω_1 *is with respect to* $p = \infty$*.*

Proof. We have that

$$\left\|{}_1L_n(f)(y) - f(y)\right\|_2 =$$

$$\left\|{}_1L_n(f)\left(\phi^{-1}(x)\right) - f\left(\phi^{-1}(x)\right)\right\|_2 =$$

$$\left\|A_n^s\left(f \circ \phi^{-1}, x\right) - f\left(\phi^{-1}(x)\right)\right\|_2 = \quad (17.58)$$

$$\left\|\frac{\sum_{k=\lceil na\rceil}^{\lfloor nb\rfloor}\left(f \circ \psi^{-1}\right)\left(\frac{k}{n}\right) Z_q(nx-k)}{\sum_{k=\lceil na\rceil}^{\lfloor nb\rfloor} Z_q(nx-k)} - \frac{f\left(\psi^{-1}(x)\right)\sum_{k=\lceil na\rceil}^{\lfloor nb\rfloor} Z_q(nx-k)}{\sum_{k=\lceil na\rceil}^{\lfloor nb\rfloor} Z_q(nx-k)}\right\|_2$$

$$\overset{\text{(by (17.42))}}{\leq} (K(q))^N$$

$$\left\|\sum_{k=\lceil na\rceil}^{\lfloor nb\rfloor}\left(f \circ \phi^{-1}\right)\left(\frac{k}{n}\right) Z_q(nx-k) - f\left(\phi^{-1}(x)\right)\sum_{k=\lceil na\rceil}^{\lfloor nb\rfloor} Z_q(nx-k)\right\|_2 =$$

$$(K(q))^N\left\|\sum_{k=\lceil na\rceil}^{\lfloor nb\rfloor}\left[\left(f \circ \phi^{-1}\right)\left(\frac{k}{n}\right) - \left(f \circ \phi^{-1}\right)(x)\right] Z_q(nx-k)\right\|_2 \leq$$

$$(K(q))^N\left[\sum_{k=\lceil na\rceil}^{\lfloor nb\rfloor}\left\|\left(f \circ \phi^{-1}\right)\left(\frac{k}{n}\right) - \left(f \circ \phi^{-1}\right)(x)\right\|_2 Z_q(nx-k)\right] =$$

$$(K(q))^N \left[\sum_{\substack{k=\lceil na \rceil \\ : \left\| \frac{k}{n} - x \right\|_\infty \le \frac{1}{n^\beta}}}^{\lfloor nb \rfloor} \left\| \left(f \circ \phi^{-1}\right)\left(\frac{k}{n}\right) - \left(f \circ \phi^{-1}\right)(x) \right\|_2 Z_q(nx-k) \right.$$

$$\left. + \sum_{\substack{k=\lceil na \rceil \\ : \left\| \frac{k}{n} - x \right\|_\infty > \frac{1}{n^\beta}}}^{\lfloor nb \rfloor} \left\| \left(f \circ \phi^{-1}\right)\left(\frac{k}{n}\right) - \left(f \circ \phi^{-1}\right)(x) \right\|_2 Z_q(nx-k) \right] \overset{\text{by (17.38)}}{\le}$$

$$(K(q))^N \left[\omega_1\left(f \circ \phi^{-1}, \frac{1}{n^\beta}\right) + 2 \left\| \left\| f \circ \phi^{-1} \right\|_2 \right\|_\infty \left(\sum_{\substack{k=\lceil na \rceil \\ : \left\| \frac{k}{n} - x \right\|_\infty > \frac{1}{n^\beta}}}^{\lfloor nb \rfloor} Z_q(nx-k) \right) \right] \overset{\text{by (17.41)}}{\le} \tag{17.59}$$

$$(K(q))^N \left[\omega_1\left(f \circ \phi^{-1}, \frac{1}{n^\beta}\right) + 2 \left\| \left\| f \circ \phi^{-1} \right\|_2 \right\| \gamma A^{-\lambda\left(n^{1-\beta}-2\right)} \right].$$

□

It follows the next result all over E.

Theorem 17.3. *Let $(E, \|\cdot\|_1)$ be a Banach space with $\dim E = N \in \mathbb{N}$, and ϕ be the corresponding homeomorphism from E onto $(\mathbb{R}^N, \|\cdot\|_2)$. Here $f \in C_B(E, \mathbb{R}^N)$, ${}_2L_n(f)$ is as in (17.53), $x \in \mathbb{R}^N$, and $y \in E$; $0 < \beta < 1$, $n \in \mathbb{N}$ with $n^{1-\beta} > 2$, ω_1 is for $p = \infty$. Then*

(1)

$$\left\| {}_2L_n(f)(y) - f(y) \right\|_2 \le$$

$$\omega_1\left(f \circ \phi^{-1}, \frac{1}{n^\beta}\right) + 2 \left\| \left\| f \circ \phi^{-1} \right\|_2 \right\|_\infty \gamma A^{-\lambda\left(n^{1-\beta}-2\right)} =: \lambda_2(n), \tag{17.60}$$

and

(2)

$$\left\| \left\| {}_2L_n(f) - f \right\|_2 \right\|_\infty \le \lambda_2(n). \tag{17.61}$$

Given that $f \in \left(C_U(E, \mathbb{R}^N) \cap C_B(E, \mathbb{R}^N)\right)$, we obtain $\lim_{n \to \infty} {}_2L_n(f) = f$, uniformly.

Proof. We have that

$$\left\|{}_2L_n(f)(y) - f(y)\right\|_2 =$$

$$\left\|{}_2L_n(f)\left(\phi^{-1}(x)\right) - f\left(\phi^{-1}(x)\right)\right\|_2 =$$

$$\left\|B_n^s\left(f\circ\phi^{-1}, x\right) - f\left(\phi^{-1}(x)\right)\right\|_2 \overset{(\text{by } (17.38),\ (17.53))}{=}$$

$$\left\|\sum_{k=-\infty}^{\infty}\left(f\circ\phi^{-1}\right)\left(\frac{k}{n}\right)Z_q(nx-k) - f\left(\phi^{-1}(x)\right)\sum_{k=-\infty}^{\infty}Z_q(nx-k)\right\|_2 =$$

$$\left\|\sum_{k=-\infty}^{\infty}\left(\left(f\circ\phi^{-1}\right)\left(\frac{k}{n}\right) - \left(f\circ\phi^{-1}\right)(x)\right)Z_q(nx-k)\right\|_2 \leq$$

$$\sum_{k=-\infty}^{\infty}\left\|\left(f\circ\phi^{-1}\right)\left(\frac{k}{n}\right) - \left(f\circ\phi^{-1}\right)(x)\right\|_2 Z_q(nx-k) = \tag{17.62}$$

$$\sum_{\substack{k=-\infty \\ :\left\|\frac{k}{n}-x\right\|_\infty \leq \frac{1}{n^\beta}}}^{\infty}\left\|\left(f\circ\phi^{-1}\right)\left(\frac{k}{n}\right) - \left(f\circ\phi^{-1}\right)(x)\right\|_2 Z_q(nx-k)$$

$$\left.+\sum_{\substack{k=-\infty \\ :\left\|\frac{k}{n}-x\right\|_\infty > \frac{1}{n^\beta}}}^{\infty}\left\|\left(f\circ\phi^{-1}\right)\left(\frac{k}{n}\right) - \left(f\circ\phi^{-1}\right)(x)\right\|_2 Z_q(nx-k)\right] \overset{(17.38)}{\leq}$$

$$\omega_1\left(f\circ\phi^{-1}, \frac{1}{n^\beta}\right) + 2\left\|\left\|f\circ\phi^{-1}\right\|_2\right\|_\infty\left(\sum_{\substack{k=-\infty \\ :\left\|\frac{k}{n}-x\right\|_\infty > \frac{1}{n^\beta}}}^{\infty} Z_q(nx-k)\right) \overset{(17.43)}{\leq} \tag{17.63}$$

$$\omega_1\left(f\circ\phi^{-1}, \frac{1}{n^\beta}\right) + 2\left\|\left\|f\circ\phi^{-1}\right\|_2\right\|_\infty \gamma A^{-\lambda\left(n^{1-\beta}-2\right)},$$

proving the claim. □

It follows the result for the Kantorovich type operator on E.

Theorem 17.4. *Let $(E, \|\cdot\|_1)$ be a Banach space with $\dim E = N \in \mathbb{N}$, and ϕ be the corresponding homeomorphism from E onto $(\mathbb{R}^N, \|\cdot\|_2)$. Here $f \in C_B(E, \mathbb{R}^N)$, ${}_3L_n(f)$ as in (17.54), $x \in \mathbb{R}^N$ and $y \in E$; $0 < \beta < 1$, $n \in \mathbb{N}$ with $n^{1-\beta} > 2$, ω_1 is for $p = \infty$. Then*

(1)

$$\|{}_3L_n(f)(y) - f(y)\|_2 \leq$$

$$\omega_1\left(f \circ \phi^{-1}, \frac{1}{n} + \frac{1}{n^\beta}\right) + 2\left\|\left\|f \circ \phi^{-1}\right\|_2\right\|_\infty \gamma A^{-\lambda\left(n^{1-\beta}-2\right)} =: \lambda_3(n), \tag{17.64}$$

and

(2)

$$\left\|\|{}_3L_n(f) - f\|_2\right\|_\infty \leq \lambda_3(n). \tag{17.65}$$

Given that $f \in \left(C_U(E, \mathbb{R}^N) \cap C_B(E, \mathbb{R}^N)\right)$, we obtain $\lim_{n\to\infty} {}_3L_n(f) = f$, uniformly.

Proof. We observe that

$$\int_{\frac{k}{n}}^{\frac{k+1}{n}} \left(f \circ \phi^{-1}\right)(t)\, dt =$$

$$\int_{\frac{k_1}{n}}^{\frac{k_1+1}{n}} \ldots \int_{\frac{k_N}{n}}^{\frac{k_N+1}{n}} \left(f \circ \phi^{-1}\right)(t_1, \ldots, t_N)\, dt_1 \ldots dt_N =$$

$$\int_0^{\frac{1}{n}} \int_0^{\frac{1}{n}} \ldots \int_0^{\frac{1}{n}} \left(f \circ \phi^{-1}\right)\left(t_1 + \frac{k_1}{n}, t_2 + \frac{k_2}{n}, \ldots, t_N + \frac{k_N}{n}\right) dt_1 \ldots dt_N =$$

$$\int_0^{\frac{1}{n}} \left(f \circ \phi^{-1}\right)\left(t + \frac{k}{n}\right) dt. \tag{17.66}$$

Therefore we can write

$$\|{}_3L_n(f)(y) - f(y)\|_2 =$$

$$\left\|{}_3L_n(f)\left(\phi^{-1}(x)\right) - f\left(\phi^{-1}(x)\right)\right\|_2 =$$

$$\left\|C_n^s\left(f \circ \phi^{-1}, x\right) - f\left(\phi^{-1}(x)\right)\right\|_2 \overset{(17.38)}{=}$$

$$\left\| \sum_{k=-\infty}^{\infty} \left(n^N \int_0^{\frac{1}{n}} \left(f \circ \phi^{-1}\right)\left(t + \frac{k}{n}\right) dt \right) Z_q(nx - k) - f\left(\phi^{-1}(x)\right) \left(\sum_{k=-\infty}^{\infty} Z_q(nx - k) \right) \right\|_2 =$$

$$\left\| \sum_{k=-\infty}^{\infty} \left(\left(n^N \int_0^{\frac{1}{n}} \left(f \circ \phi^{-1}\right)\left(t + \frac{k}{n}\right) dt \right) - \left(f \circ \phi^{-1}\right)(x) \right) Z_q(nx - k) \right\|_2 =$$

$$(17.67)$$

$$\left\| \sum_{k=-\infty}^{\infty} \left(n^N \int_0^{\frac{1}{n}} \left(\left(f \circ \phi^{-1}\right)\left(t + \frac{k}{n}\right) - \left(f \circ \phi^{-1}\right)(x) \right) dt \right) Z_q(nx - k) \right\|_2 \le$$

$$\sum_{k=-\infty}^{\infty} \left(n^N \int_0^{\frac{1}{n}} \left\| \left(f \circ \phi^{-1}\right)\left(t + \frac{k}{n}\right) - \left(f \circ \phi^{-1}\right)(x) \right\|_2 dt \right) Z_q(nx - k) =$$

$$\sum_{\substack{k = -\infty \\ : \left\|\frac{k}{n} - x\right\|_\infty \le \frac{1}{n^\beta}}}^{\infty} \left(n^N \int_0^{\frac{1}{n}} \left\| \left(f \circ \phi^{-1}\right)\left(t + \frac{k}{n}\right) - \left(f \circ \phi^{-1}\right)(x) \right\|_2 dt \right) Z_q(nx - k)$$

$$+ \sum_{\substack{k = -\infty \\ : \left\|\frac{k}{n} - x\right\|_\infty > \frac{1}{n^\beta}}}^{\infty} \left(n^N \int_0^{\frac{1}{n}} \left\| \left(f \circ \phi^{-1}\right)\left(t + \frac{k}{n}\right) - \left(f \circ \phi^{-1}\right)(x) \right\|_2 dt \right) Z_q(nx - k)$$

$$(17.68)$$

$$\overset{\text{(by (17.50), (17.43))}}{\le} \omega_1\left(f \circ \phi^{-1}, \frac{1}{n} + \frac{1}{n^\beta}\right) + 2\left\|\left\|f \circ \phi^{-1}\right\|_2\right\|_\infty \gamma A^{-\lambda\left(n^{1-\beta}-2\right)},$$

proving the claim. □

The final result is for the quadrature type operator on E.

Theorem 17.5. *Let $(E, \|\cdot\|_1)$ be a Banach space with $\dim E = N \in \mathbb{N}$, and ϕ be the corresponding homeomorphism from E onto $\left(\mathbb{R}^N, \|\cdot\|_2\right)$. Here $f \in C_B\left(E, \mathbb{R}^N\right)$, ${}_4L_n(f)$ is as in (17.55), $x \in \mathbb{R}^N$ and $y \in E$; $0 < \beta < 1$, $n \in \mathbb{N}$ with $n^{1-\beta} > 2$, ω_1 is for $p = \infty$. Then*

(1)

$$\|{}_4L_n(f)(y) - f(y)\|_2 \le$$

$$\omega_1\left(f \circ \phi^{-1}, \frac{1}{n} + \frac{1}{n^\beta}\right) + 2\left\|\left\|f \circ \phi^{-1}\right\|_2\right\|_\infty \gamma A^{-\lambda\left(n^{1-\beta}-2\right)} =: \lambda_4(n),$$

$$(17.69)$$

and

(2)

$$\left\| \left\| {}_4L_n(f) - f \right\|_2 \right\|_\infty \leq \lambda_4(n). \tag{17.70}$$

Given that $f \in \left(C_U\left(E, \mathbb{R}^N\right) \cap C_B\left(E, \mathbb{R}^N\right)\right)$, *we obtain* $\lim_{n\to\infty} {}_4L_n(f) = f$, *uniformly.*

Proof. We have

$$\left\| {}_4L_n(f)(y) - f(y) \right\|_2 =$$

$$\left\| {}_4L_n(f)\left(\phi^{-1}(x)\right) - f\left(\phi^{-1}(x)\right) \right\|_2 =$$

$$\left\| D_n^s\left(f \circ \phi^{-1}, x\right) - f\left(\phi^{-1}(x)\right) \right\|_2 \overset{(17.55)}{=}$$

$$\left\| \sum_{k=-\infty}^{\infty} \delta_{nk}\left(f \circ \phi^{-1}\right) Z_q(nx - k) - f\left(\phi^{-1}(x)\right) \right\|_2 \overset{(17.38)}{=}$$

$$\left\| \sum_{k=-\infty}^{\infty} \left(\sum_{r=0}^{\theta} w_r \left(f \circ \phi^{-1}\right)\left(\frac{k}{n} + \frac{r}{n\theta}\right)\right) Z_q(nx - k) - \left(\sum_{k=-\infty}^{\infty} Z_q(nx - k)\right)\left(f \circ \phi^{-1}\right)(x) \right\|_2 = \tag{17.71}$$

$$\left\| \sum_{k=-\infty}^{\infty} \left(\left(\sum_{r=0}^{\theta} w_r \left(f \circ \phi^{-1}\right)\left(\frac{k}{n} + \frac{r}{n\theta}\right)\right) - \left(f \circ \phi^{-1}\right)(x) \right) Z_q(nx - k) \right\|_2 =$$

$$\left\| \sum_{k=-\infty}^{\infty} \left(\sum_{r=0}^{\theta} w_r \left(\left(f \circ \phi^{-1}\right)\left(\frac{k}{n} + \frac{r}{n\theta}\right) - \left(f \circ \phi^{-1}\right)(x) \right) \right) Z_q(nx - k) \right\|_2 \leq$$

$$\sum_{k=-\infty}^{\infty} \left(\sum_{r=0}^{\theta} w_r \left\| \left(f \circ \phi^{-1}\right)\left(\frac{k}{n} + \frac{r}{n\theta}\right) - \left(f \circ \phi^{-1}\right)(x) \right\|_2 \right) Z_q(nx - k) =$$

$$\sum_{\begin{cases} k = -\infty \\ : \left\| \frac{k}{n} - x \right\|_\infty \leq \frac{1}{n^\beta} \end{cases}}^{\infty} \left(\sum_{r=0}^{\theta} w_r \left\| \left(f \circ \phi^{-1}\right)\left(\frac{k}{n} + \frac{r}{n\theta}\right) - \left(f \circ \phi^{-1}\right)(x) \right\|_2 \right) Z_q(nx - k) \tag{17.72}$$

$$+ \sum_{\begin{cases} k = -\infty \\ : \left\| \frac{k}{n} - x \right\|_\infty > \frac{1}{n^\beta} \end{cases}}^{\infty} \left(\sum_{r=0}^{\theta} w_r \left\| \left(f \circ \phi^{-1}\right)\left(\frac{k}{n} + \frac{r}{n\theta}\right) - \left(f \circ \phi^{-1}\right)(x) \right\|_2 \right) Z_q(nx - k)$$

$$\overset{\text{(by (17.50), (17.43))}}{\leq} \omega_1\left(f \circ \phi^{-1}, \frac{1}{n} + \frac{1}{n^\beta}\right) + 2 \left\| \left\| f \circ \phi^{-1} \right\|_2 \right\| \gamma A^{-\lambda\left(n^{1-\beta} - 2\right)}, \tag{17.73}$$

proving the claim. □

Bibliography

[373] G.A. Anastassiou, *Rate of convergence of some neural network operators to the unit-univariate case*, J. Math. Anal. Appl, 212 (1997), 237–262.

[374] G.A. Anastassiou, *Quantitative Approximations*, Chapman & Hall / CRC, Boca Raton, New York, 2001.

[375] G.A. Anastassiou, *Inteligent Systems: Approximation by Artificial Neural Networks*, Intelligent Systems Reference Library, Vol. 19, Springer, Heidelberg, 2011.

[376] G.A. Anastassiou, *Intelligent Systems II: Complete Approximation by Neural Network Operators*, Springer, Heidelberg, New York, 2016.

[377] G.A. Anastassiou, *Intelligent Computations: Abstract, Fractional Calculus, Inequalities, Approximations*, Springer, Heidelberg, New York, 2018.

[378] G.A. Anastassiou, *Banach Space Valued Neural Network*, Springer, Heidelberg, New York, 2023.

[379] G.A. Anastassiou, *Parametrized, deformed and general neural networks*, Springer, Heidelberg, New York, 2023.

[380] G.A. Anastassiou, *Trigonometric and Hyperbolic Generated Approximation Theory*, World Scientific, Singapore, London, New Jersey, 2025.

[381] Z. Chen and F. Cao, *The approximation operators with sigmoidal functions*, Computers and Mathematics with Applications, 58 (2009), 758–765.

[382] D. Costarelli, R. Spigler, *Approximation results for neural network operators activated by sigmoidal functions*, Neural Networks 44 (2013), 101–106.

[383] D. Costarelli, R. Spigler, *Multivariate neural network operators with sigmoidal activation functions*, Neural Networks 48 (2013), 72–77.

[384] S. Haykin, *Neural Networks: A Comprehensive Foundation* (2 ed.), Prentice Hall, New York, 1998.

[385] W. McCulloch and W. Pitts, *A logical calculus of the ideas immanent in nervous activity*, Bulletin of Mathematical Biophysics, 7 (1943), 115–133.

[386] T.M. Mitchell, *Machine Learning*, WCB-McGraw-Hill, New York, 1997.

[387] G.A. Anastassiou, *Generalized logistic Neural Network Approximation over finite dimension Banach spaces*, Issues of Analysis, 14(32)(2), 2025, 3–24.

Conclusion

In this monograph summation Neural Network Operators and activated convolution operators are treated under the general theory of Positive Linear Operators regarding their approximation to the unit operator. The produced quantitative results are to all possible directions and are very interesting and different than those of treating directly and individually the examined operators. This general unified approach is presented for the first time.

www.ingramcontent.com/pod-product-compliance
Lightning Source LLC
LaVergne TN
LVHW010341220926
848484LV00004BB/1009